Multiple Imputation of Missing Data in Practice

Multiple Imputation of Missing Data in Practice

Basic Theory and Analysis Strategies

Yulei He
Guangyu Zhang
Chiu-Hsieh Hsu

CRC Press
Taylor & Francis Group
Boca Raton London New York

CRC Press is an imprint of the
Taylor & Francis Group, an **informa** business
A CHAPMAN & HALL BOOK

First edition published 2022
by CRC Press
6000 Broken Sound Parkway NW, Suite 300, Boca Raton, FL 33487-2742

and by CRC Press
2 Park Square, Milton Park, Abingdon, Oxon, OX14 4RN

CRC Press is an imprint of Taylor & Francis Group, LLC

The findings and conclusions in this book are those of the authors and do not necessarily represent the official position of the National Center for Health Statistics, U.S. Centers for Disease Control and Prevention, and University of Arizona.

ISBN: 9781498722063 (hbk)
ISBN: 9781032136899 (pbk)
ISBN: 9780429156397 (ebk)

DOI: 10.1201/9780429156397

Publisher's Note: This book has been prepared from camera-ready copy provided by the author.

Yulei He: to my wife, Shujun, and my children, Michelle and Elizabeth, for ensuring that I have a life outside of my work and writing, and to my parents, Guanghui and Xiaohui, for their long-standing encouragement and support for me.

Guangyu Zhang: to Huanyuan and Rod.

Chiu-Hsieh Hsu: to my family and friends.

Contents

Foreword

We are delighted to see this new book on multiple imputation by three accomplished statisticians, He, Zhang, and Hsu (HZH). The authors have substantial experience conducting research on multiple imputation and applying it in a variety of settings, and they are thus eminently qualified to write a book in this area. The book covers a large number of important topics relevant for any applied researcher. Many books in this area have been published, yet there are still unmet needs. This is a sign of growth in interest and continued advances through research and applications. We certainly welcome this book, which discusses many recent developments.

When we started working on multiple imputation in the early 1980s, there was a lot of skepticism and even hostility concerning the approach, which was seen as impractical and problematic. This was understandable given the computational limitations of those days as well as the fact that multiple imputation deviates from the traditional and still somewhat prevalent notion of obtaining the single "best" value to substitute for each missing datum. Over the years, however, the idea of simulating multiple values from the posterior predictive distribution of the missing data given the observed data, to create multiple plausible complete data sets, has gained wide acceptance. The main goal of multiple imputation is to use all available information in the best possible manner, while fully reflecting the inherent uncertainty in doing so. Fortunately, computational technology supporting the approach has also improved greatly over the years, as is evidenced by the many software packages for creating and analyzing multiply imputed data sets.

The topics covered in this book are bread and butter for any applied researcher. HZH have made life easy by including necessary statistical background in Chapter 2, and this should be read first, as a brush-up, by anybody using the book. Chapter 3 provides a very nice basic heuristic and technical background on multiple imputation. Many of the remaining chapters discuss multiple imputation in a wide variety of scenarios, including complicated situations such as longitudinal data, measurement error, sample surveys, combining information from multiple data sets, and nonignorable missing data. Of special general importance is the chapter on diagnostics for multiple imputation, for which there have been many recent developments. Just as with any statistical method, using diagnostic methods is very important for ensuring appropriateness. We have seen in applications of imputation that diagnostics are sometimes given short shrift, so we are very happy to see this chapter.

Throughout the book, HZH have included important pointers for applied researchers, alternatives to consider, checks for pitfalls, and potential remedies. The book is loaded with real-life examples and simulation studies to help reinforce the points being made. Although no particular software is emphasized in the book, many important hints are provided using, for example, WinBUGS. Those hints will be useful regardless of which software is employed by the reader. HZH have also included over four hundred references to work in the field, including many publications from the past decade. It is great to have this new contribution to the theory and practice of multiple imputation.

Finally, our delight at the publication of this book is also personal. As mentioned in the preface, the authors consider us to be among their mentors. One of the greatest satisfactions for mentors occurs when their mentees write a book on a particular topic to which the mentors have devoted a large amount of work, not to mention when the mentors are asked to write the foreword! We are simply beaming with pride to see the publication of this book as an important contribution by HZH and as an indication of many more contributions to come!!

Trivellore Raghunathan and Nathaniel Schenker

Preface

Missing data problems are common and difficult to handle in data analysis. Ad hoc methods such as simply removing cases with missing values can lead to invalid analysis results. This book focuses on the multiple imputation analysis approach that is a principled missing data strategy. Under this approach, missing values in a dataset are imputed (i.e., filled in) by multiple sets of plausible values. These imputed values are random numbers generated from predictive distributions of missing values conditional on observed data and relevant information, which can be formulated by statistical models. This process results in multiple completed datasets, each of which includes a set of imputed values and original observed values. Each completed dataset is then analyzed separately using established complete-data analysis procedures to generate statistics of interest (e.g., regression coefficients and standard errors, test statistics, p-values, etc). Finally, the multiple sets of statistics are combined to form a single statistical inference based on simple rules.

Multiple imputation started in the 1970s as a proposal to handle nonresponse problems in large surveys produced by organizations so that data users could analyze completed datasets released by these organizations. In the past 40 years or so, there exists vibrant research that advances both the theory and application of multiple imputation. Nowadays multiple imputation is arguably the most popular and versatile statistical approach to missing data problems and is widely used across different disciplines. Although there exists a large volume of literature on multiple imputation, the demand of learning about this strategy and using it appropriately and effectively for a wide variety of missing data problems, both routine and new ones, continues to be strong.

The purpose of this book is to describe general structures of missing data problems, present the motivation of multiple imputation analysis and its methodological reasoning, and elaborate upon its applications to a wide class of problems via examples. Illustrative datasets and sample programming are either included in the book or available at a github site (https://github.com/he-zhang-hsu/multiple_imputation_book). In the following we summarize some of the main features included in the book.

1. Multiple imputation sits at the intersection of two fundamental statistical frameworks, frequentist and Bayesian statistics. To make the book more self-contained, we include necessary material (e.g., Chapter 2) on basic statistical concepts and analysis methods that can be helpful for understanding technical components of multiple imputation.

2. We believe that it is important that researchers and practitioners not only know "how" to use certain imputation methods, but also know "why" these methods work. In addition, provided that there often exist multiple modeling options for an imputation analysis, multiple imputation users need to know which method to choose. This book includes examples on assessing and comparing the performance of alternative methods using simulation studies, thus providing some insights for understanding these issues and making appropriate decisions. Some simulation studies are also used to demonstrate the basic theory of multiple imputation.

3. Many multiple imputation methods are implemented in popular statistical software packages (e.g., SAS and R). We do not focus on one particular package in the book, as programs in different packages can be used to implement the same method in many cases. In some examples, we demonstrate the use of Bayesian analysis software for carrying out the joint modeling imputation (e.g., Chapter 7). The software package and sample code used in these examples are from WinBUGS. Applying this strategy can sometimes save users from deriving and coding complicated posterior predictive distributions needed in imputation. WinBUGS (and its more recent variants such as OpenBUGS and R NIMBLE) has been a popular programming language for learning and researching in applied Bayesian statistics. In other examples, we mainly mention the name of the software package used. As as supplement of this book, sample code for most of the examples, written in SAS or R, can be found in (https://github.com/he-zhang-hsu/multiple_imputation_book).

4. There exists a wide range of modeling and analysis techniques used in multiple imputation. This book includes material on some common parametric methods (Chapter 4) and robust imputation techniques (Chapter 5) for typical univariate missing data problems. It also includes imputation strategies targeted to multivariate missing data problems and those embedded within special types of data and study designs (e.g., Chapters 6-11).

5. Both survival data analysis and longitudinal data analysis are important fields in biostatistics. This book includes some methods in the application of multiple imputation for survival data (Chapter 8) and longitudinal data (Chapter 9).

6. This book also contains material on applying multiple imputation to survey data (Chapter 10). Topics discussed include not only modeling and analysis techniques, but also issues related to data editing, processing, and release from the organizations' perspective. The latter is especially relevant if the goal of the project is to produce completed datasets for public use, which goes back to the original motivation of multiple imputation when it was first proposed.

7. Many statistical problems can be framed and approached from the perspective of missing data. In this book we choose to present the topic on handling measurement error problems using multiple imputation (Chapter 11). This topic can also be connected with the general idea of combining information from multiple data sources.

8. Besides modeling techniques, we believe that it is crucial and beneficial to include all the necessary information (i.e., variables) from observed data and sometimes those from extra data sources for imputation. That is, it is preferable to have a general imputation model. This principle is the so called "inclusive imputation strategy" and is emphasized throughout the book.

9. "All models are wrong, but some are useful" (George E. P. Box). Some techniques targeted to imputation diagnostics and checking are discussed in Chapter 12. The demand of conducting imputation diagnostics is increasing given the popularity of multiple imputation nowadays. We believe the purpose of imputation diagnostics is not to identify the perfect method per se, but to improve the imputation from some baseline models and produce reasonably good analysis results.

10. The field of multiple imputation analysis is evolving fast in the era of big data and data science, which pose new and challenging missing data problems as well as analysis needs. In Chapter 14 we briefly touch upon several advanced research topics. We hope this book can stimulate more related research!

Readers are recommended to have entry-level graduate statistical or biostatistical training including regression analysis, mathematical statistics, applied Bayesian statistics, and basic statistical programming. The primary audience consists of researchers and practitioners who have to deal with missing data problems in data analysis. We hope they will enjoy handling missing data using multiple imputation after reading this book! We also hope some of the material can motivate readers to develop new multiple imputation ideas and methods. As a companion to existing literature, this book can be used as a reference for both research and teaching.

Our interests in multiple imputation stemmed from the missing data analysis course (taught by Rod Little and Trivellore Raghunathan) that we took as graduate students in biostatistics at the University of Michigan (Go Blue!). In the past 20 years or so, we have researched and applied multiple imputation analysis in a wide range of projects. We will certainly keep doing this in the future! We have also in the past taught the topics of missing data and multiple imputation, as well as related topics such as longitudinal and survival data analysis.

We thank our employers, the National Center for Health Statistics in the U.S. Centers for Disease Control and Prevention (for Yulei He and Guangyu

Zhang) and the University of Arizona (for Chiu-Hsieh Hsu), and our colleagues for supporting the writing of the book. Many of the ideas and examples in the book are stimulated from our working projects. The findings and conclusions in this book are those of the authors and do not necessarily represent the official position of our employers. The book inevitably contains errors and shortcomings, for which we take full responsibility.

We extend our thanks to several mentors in teaching us about missing data analysis and multiple imputation: Trivellore Raghunathan, Alan Zaslavsky, Nat Schenker, Rod Little, and Jeremy Taylor. We also thank collaborators in our research and investigations of multiple imputation: Juan Albertorio, Alan Dorfman, Joseph Kang, Benmei Liu, Qiyuan Pan, Yi Pan, Jennifer Parker, Van Parsons, Lihong Qi, Jenny Rammon, Iris Shimizu, Juned Siddique, Ying-Fang Wang, Rong Wei, Mandi Yu, Ying Yuan, Recai Yucel, and Xiao Zhang. Thanks are also due to colleagues who read and commented to the draft of the book: Don Malec, John Pleis, Amy Branum, Alex Strashny, Anjel Vahratian, Isabelle Horton, Jonaki Bose, and Naman Ahluwalia, as well as four anonymous reviewers. We hope that they are not the only readers, and readers like you will find the book useful! In addition, we thank John Kimmel, Lee Baldwin, Shashi Kumar, Lara Spieker, and Robin Lloyd-Starkes from CRC Press/Taylor & Francis group for their advice and help.

We thank our family members for their warm and ongoing support, and for allowing us to devote spare time, often evenings and weekends, to work on this book. Last but not least, Yulei He thanks God for His patience, forgiveness, and blessings during the writing.

1

Introduction

1.1 A Motivating Example

Missing (or incomplete) data problems often occur in studies that involve collecting and analyzing data. How to handle missing data appropriately in statistical analysis is a major challenge for researchers and practitioners. For example, income (personal or family) is an important socioeconomic variable and frequently used in economic, social, and health research. Information about income is collected in many studies, yet income data are frequently subject to missing values or nonresponses. A real-world example is given below.

Example 1.1. *Survey nonresponse of family income from the National Health Interview Survey*

The National Health Interview Survey (NHIS) is an annual multi-purpose survey that is the principal source of information on the health of the civilian, noninstitutionalized household population of the United States (U.S.). NHIS has monitored the health of the U.S. since 1957. It is administered by the National Center for Health Statistics (NCHS) at the U.S. Centers for Disease Control and Prevention (CDC). NCHS contracts with the U.S. Bureau of the Census to conduct the survey fieldwork. NHIS has gone through periodic changes of the questionnaire and sampling designs over time. In this example and throughout the book, we focus on survey questionnaire and sampling design used from 1997 to 2018. During that period, the base sample of the survey included about 35000 households with 87500 persons each year, with some additional samples being added in different years. Detailed information about NHIS can be found in (https://www.cdc.gov/nchs/nhis/index.htm).

Besides many other objectives, NHIS provides a rich source of data for studying relationships between income and health and for monitoring health and health care for persons at different income levels. The survey collects a wide array of variables of survey participants' economic and wealth status. We look at one of the key income items, total combined family income for all family members including children. For illustration, we use the family component of NHIS 2016 public-use data (https://www.cdc.gov/nchs/nhis/nhis_2016_data_release.htm) which contains information collected from the survey at the family level. The question in the survey is, "What is your best estimate of {your total income/the total income of all family members} from all

DOI: 10.1201/9780429156397-1

sources, before taxes, in {last calendar year}?" Survey participants are expected to answer it in dollars, and this variable is labeled as FAMINCI2 in the public-use data. For participants who choose "don't know" or "refused", their FAMINCI2 values are treated as missing.

In NHIS 2016, the nonresponse rate of FAMINCI2 is 20.6%. Fig. 1.1, which includes the histogram and normal quantile-quantile (QQ) plot, shows the distribution of the observed income values, which is highly skewed to the right. The spike at the right end of the distribution is due to the top-coding of the public-use data.

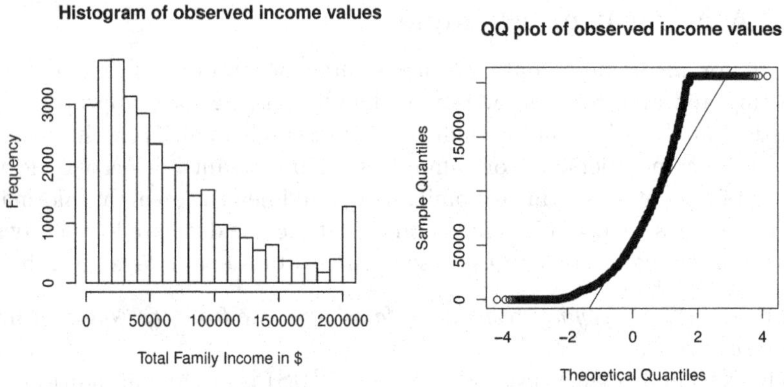

FIGURE 1.1
Example 1.1. Left: the histogram of FAMINCI2 in NHIS 2016; right: the QQ plot.

How do we deal with the missing data problem in this variable? "Obviously the best way to treat missing data is not to have them" (Orchard and Woodbury 1972, Page 697). Some attempts can be made in terms of data collection to obtain the ranges of the income. In NHIS 2016, for example, if the participant did not provide the amount, then a series of income bracketing questions were asked starting with, "Was your total family income from all sources less than {250% of poverty threshold} or {250% of poverty threshold} or more?" Additional follow up questions of income ranges were asked based on the respondent's answer, with the goal of placing the income into several other detailed income categories. The poverty threshold used in these questions are based on the size of their family, the number of children, and the presence of a person aged 66 years or over. The poverty threshold dollar amounts are adjusted each year. Variables for these bracketed income questions are useful in recovering some of the information lost in the nonresponses of FAMINCI2 because they provide possible ranges of income values for nonrespondents. However, an exact dollar amount, not the range of income, is still preferred in many statistical analyses. And bracketed income variables are still subject to nonresponse in the survey.

In terms of analyzing the incomplete income variable, one quick method is to remove cases with missing FAMINCI2 values and proceed with the analysis. This strategy, often referred to as the complete-case analysis or case-wise/list-wise deletion, can be problematic. For example, there is some evidence suggesting that survey participants from different groups might have different nonresponse rates (Example 1.10). Suppose participants with lower actual income are more likely to have missing income data; then the mean estimate by simply removing these missing cases would overestimate the true mean of the income for the target population, resulting in a positive nonresponse bias.

An appealing analytic strategy is to impute, that is, to fill in, all the missing values, so that FAMINCI2 is completed. Routine complete-data analysis procedures can then be easily applied to the completed income variable. This is how the issue is approached within the NHIS. However and very importantly, the imputation is done multiple times using the multiple imputation analysis approach (Rubin 1987).

Example 1.2. *Multiply imputed family income from the National Health Interview Survey*

Continuing with Example 1.1, NCHS multiply imputes the family total income variable of NHIS starting from 1997 and releases five (sets of) completed income variables annually to external data users for carrying out analyses. Table 1.1 shows a small subset of multiply imputed FAMINCI2 for illustration. Across the five imputations, the observed values are identical, and the imputed values (italic) are varying. For NHIS 2016, more relevant information on the multiple imputation can be found in (https://www.cdc.gov/nchs/data/nhis/tecdoc16.pdf). Schenker et al. (2006) provided technical details of the methodology. See also Example 10.4.

Why is the imputation done multiple times? How are the imputed values generated? How should we analyze these five (or multiple) completed datasets? How do we know whether these imputed values are good enough? Is multiple imputation a good missing data analysis method in general? Can we apply this strategy to other types of missing data problems, and if so, how do we do it? Researchers and practitioners might have many relevant questions. In this book we aim to provide an overview of the multiple imputation analysis approach and offer answers and discussions for these questions.

Before we dive into more details (starting from Chapter 2), we first go through some general background about missing data problems. Section 1.2 defines missing data. Section 1.3 classifies missing data patterns. Section 1.4 introduces the missing data mechanism, one of the most important concepts in missing data analysis. Section 1.5 summarizes the structure of the book.

TABLE 1.1
Example 1.2. Multiply imputed FAMINCI2 from an illustrative subset of NHIS 2016

Imputation 1	Imputation 2	Imputation 3	Imputation 4	Imputation 5
...	...	...	...	...
14000	14000	14000	14000	14000
50000	50000	50000	50000	50000
30000	30000	30000	30000	30000
35058	*37258*	*33829*	*33308*	*38651*
47000	47000	47000	47000	47000
140000	140000	140000	140000	140000
4000	4000	4000	4000	4000
8000	8000	8000	8000	8000
5000	5000	5000	5000	5000
3000	3000	3000	3000	3000
206000	206000	206000	206000	206000
59729	*62608*	*66525*	*52792*	*73603*
50000	50000	50000	50000	50000
77000	77000	77000	77000	77000
100000	100000	100000	100000	100000
45000	45000	45000	45000	45000
89908	*87911*	*80668*	*82118*	*97226*
42339	*14292*	*40298*	*20121*	*5211*
21730	*20105*	*55292*	*9536*	*13236*
...	...	...	...	...

Note: Imputed values are in italics.

1.2 Definition of Missing Data

We follow the definition of missing data provided by Little and Rubin (2020, Page 4): "Missing data are unobserved values that would be meaningful for analysis if observed; in other words, a missing value hides a meaningful value." This general definition is connected with both the scientific question posed and analyses applied for the data, and may have different implications within different contexts. The survey nonresponse problem for the family income question in NHIS (Example 1.1) is a classic example of missing data: every eligible survey participant should have an exact family income value. Typical analyses for the income variable (e.g., to obtain the mean or median of the family income) would need to involve the actual income value of each participant. In the following we provide a few other examples of missing data.

Example 1.3. *Partial missingness: grouped or censored data*

Continuing with Example 1.1, some income nonrespondents provide their income categories for the bracketed income questions in NHIS. This situation is an example of grouped data. For these cases, although their specific income values are unknown, we have some ideas about their ranges or income groups defined by the bracketed income questions. These ranges provide partial information about the income distribution. For example, if we know that the family income from an income nonrespondent is greater than \$75000, then a reasonable imputation for the missing FAMINCI2 value should be greater than \$75000.

A somewhat similar scenario is the problem of censored data often encountered in survival data analysis. The outcome of a survival analysis is typically a time-to-event type of variable. For example in many medical studies, a participant is followed until an event (e.g., death) occurs or he/she exits the study. The latter situation is termed as "right censoring", and the individual is still free of event (e.g., being alive) at the censoring time when he/she leaves the study. For a subject with right censoring, his/her time-to-event is unknown (i.e., missing) but if it is imputed, the value shall be greater than the censoring time. Such survival data are often referred to as right-censored data. Further discussion about the multiple imputation for survival data analysis can be found in Chapter 8.

Example 1.4. *Variables subject to measurement or reporting error*

Many health studies involve variables recorded with measurement or reporting error. A useful analysis strategy is to treat true values of variables subject to measurement error as missing data (Carroll et al. 2006, Page 30), especially when some validation data are available. For instance, in surveys or administrative databases, variables on the receipt of health services can be subject to misreporting or incompleteness. We consider a specific example here. The Cancer Care Outcomes Research and Surveillance (CanCORS) Consortium is a multi-site and multi-wave study of the quality and patterns of care delivered to population-based cohorts of newly diagnosed patients with lung and colorectal cancer (Ayanian et al. 2004). For stage II/III colorectal cancer patients in the CanCORS study, their receipts of adjuvant cancer treatments were collected from cancer registries. However, there existed evidence that these variables might be subject to underreporting at the time of study. That is, for a patient who had received the treatment, it is possible that the registry system did not capture it yet when data were collected for the study. Therefore, the true treatment status of patients can be treated as missing data. More information about the multiple imputation approach to handling measurement error problems can be found in Chapter 11.

Example 1.5. *Not applicable*

Studies for human subjects frequently include "not applicable". Values of "not applicable" could result from skip patterns in survey data. For instance, survey questions about tobacco use of participants often start by asking, "Have you smoked at least 100 cigarettes in your entire life?" (the gateway question). For participants who answer "No" to this question (i.e., never smokers), they would be skipped from a series of follow-up questions such as, "How often do you now smoke cigarettes? Every day, some days, or not at all?" For an analysis that is only targeted to current and former smokers, excluding these skips (never smokers) generally make sense. However, missing values in the gateway question variable impose some uncertainty: If the true answer for the participant had been "No", then his/her answers to the follow-up questions would need to be skipped; if "Yes", then his/her answers to the follow-up questions would not be skipped in principle and would need to be included in the analysis. More discussions about handling skip patterns in multiple imputation can be found in Chapters 7 and 10.

In certain cases the definition of missing values is not clear-cut. Sometimes the missingness, the status of being missing, might be a legitimate analysis category, as can be seen from the following example.

Example 1.6. *"Don't know/refused" for questions soliciting opinions towards end-of-life cancer care*

For the CanCORS study, a patient survey was used to collect information on the cancer care received by the patients. Some questions were aimed to solicit patients' opinions and preferences towards cancer treatment options including end-of-life care. An example is, "Do you prefer receiving treatment that would improve your symptoms but not necessarily prolong your life?" Possible answers include "Yes", "No", and "Don't know/refused". For patients choosing the "Don't know/refused" category, can we simply treat them as missing values in a similar way to that for the income nonresponse problem? Not necessarily. Some participants might have a clear preference, yes or no, but do not want reveal it, and this scenario is similar to the income nonresponse problem. For others, choosing this category reflects their realistic attitude (i.e., not sure or undecided) towards this difficult question. In the latter situation, "Don't know/refused" might constitute a separate group besides "Yes" and "No." In general, nonresponses in questions soliciting opinions have complicated implications. See also Little and Rubin (2020, Example 1.3) for a discussion.

As a useful summary, Raghunathan (2016, Section 1.2) suggested that a practical way to define the (imputable) missing values for a variable in a specific analysis is to consider whether or not one should replace these missing items with observed values. In this book we focus on imputable missing data problems such as those in Examples 1.1, 1.3, and 1.4. In addition, many statistical problems that do not appear to have missing data can be approached from the incomplete data perspective using the multiple imputation strategy. Examples include the counterfactual model of causal inference and data synthesis for minimizing data disclosure risks. These topics are not covered in this book. See Gelman and Meng (2004) for a related discussion.

1.3 Missing Data Patterns

For a single variable, "missing data" generally means multiple missing values. Missing data often occur for multiple variables. Although the missingness may occur in some random fashion, the pattern of missing data, which describes the locations of the missing values relative to observed ones, can provide some useful insights into the process leading to missing data and therefore help us better formulate missing data problems. This section describes several common missing data patterns based on established mathematical notations.

We follow the scheme established by Little and Rubin (2020, Section 1.2). A typical population study includes subjects (units) for whom some variables (measurements or items) are taken. Thus a dataset usually consists of a sample collection of subjects, each of which provides information on a collection of variables. In a cross-sectional questionnaire survey (e.g., NHIS), the subjects are individuals participating in the survey, and the variables are their answers to the survey questions. Sometimes more complex data structure might arise, for example, in studies with multilevel nested structure (Chapter 9). Yet for simplicity, it is often straightforward to describe the structure of a dataset as a rectangular matrix, in which the rows correspond to the subjects, and the columns correspond to the variables.

To mathematically describe the pattern of missing data, let $Y = \{y_{ij}\}$, $i = 1, \ldots, n$, $j = 1, \ldots p$ denote a $n \times p$ rectangular dataset without missing values (i.e., complete or fully observed dataset). Its ith row $y_i = (y_{i1}, y_{i2}, \ldots, y_{ip})$ where y_{ij} is the value of variable Y_j $(j = 1, \ldots, p)$ for subject i. When missing data occur, some of the y_{ij}'s are not observed. To describe this, we can define the response indicator matrix $R = \{r_{ij}\}$, where $r_{ij} = 1$ if y_{ij} is observed and $r_{ij} = 0$ if y_{ij} is missing. Note that the observed information from the dataset includes both the observed elements of the Y-matrix and the fully known R-matrix.

Among possible missing data patterns, a fundamental one is the so-called "univariate missingness". In univariate missing data problems, one variable is incomplete and all other variables are fully observed in the dataset. For simplicity, we can assume that Y_1 has some missing values and Y_j $(j = 2, \ldots p)$ are fully observed, as illustrated in Table 1.2. This pattern is well suited for scenarios where we have a main variable of analytical interests yet subject to missing values. In this book, many basic ideas and methods of multiple imputation are illustrated in the setup of univariate missing data problems (e.g., Chapters 4 and 5). In addition, methods for univariate missing data problems are often the starting point for developing methods applicable to more complicated, multivariate missing data problems (e.g., Chapters 6 and 7).

In typical univariate missing data problems, some cases have observed Y_1-values. However, Y_1 can be fully unobserved in some measurement error problems (Table 1.3). For instance, in Example 1.4, Y_1 can be set as the true treatment status which is unknown and subject to reporting error. Its reported version might be recorded as one of the fully observed Y_j's (say Y_2).

A multivariate missing data problem occurs if at least two variables have missing values. In surveys, unit nonresponses occur when a subset of sampled subjects does not complete the survey questionnaire because of non-contact, refusal, or some other reasons. This would result in a pattern in which survey design variables (e.g., list or frame information) are available for both respondents and nonrespondents (e.g., Y_1 and Y_2 in Table 1.4), and all or most of the variables in the survey questionnaire are missing (e.g., Y_3, Y_4, etc.) for

TABLE 1.2
A schematic table of a univariate missingness pattern

Y_1	Y_2	Y_3	Y_4	...
...	...	...	...	...
O	O	O	O	...
O	O	O	O	...
O	O	O	O	...
M	O	O	O	...
M	O	O	O	...
M	O	O	O	...
...	...	...	...	...

Note: "O" symbolizes observed values; "M" symbolizes missing values.

TABLE 1.3
A schematic table of a univariate missingness pattern where the missing variable is fully unobserved

Y_1	Y_2	Y_3	Y_4	...
...	...	...	...	...
M	O	O	O	...
M	O	O	O	...
M	O	O	O	...
M	O	O	O	...
M	O	O	O	...
M	O	O	O	...
...	...	...	...	...

Note: "O" symbolizes observed values; "M" symbolizes missing values; Y_1 is the variable of true values; Y_2 is the reported version of Y_1 subject to measurement error.

unit nonrespondents. Survey unit nonresponse problems are often handled by weighting (Chapter 10).

TABLE 1.4
A schematic table of survey unit nonresponse

Y_1	Y_2	Y_3	Y_4	...
...	...	...	...	...
O	O	O	O	...
O	O	O	O	...
O	O	O	O	...
O	O	M	M	M
O	O	M	M	M
O	O	M	M	M
...	...	...	...	...

Note: "O" symbolizes observed values; "M" symbolizes missing values; Y_1 and Y_2 are variables of survey sampling frame information; Y_3, Y_4, etc. are survey questionnaire variables, and they are all missing for unit nonrespondents.

Another type of survey nonresponse is item nonresponse. For item nonresponse, a sampled subject fills out the survey questionnaire yet does not provide information for all the survey variables. (e.g., the income nonresponse problem in Example 1.1)

One of the special multivariate missingness patterns is called the monotone pattern (Table 1.5). That is, $Y_{j+1}, \ldots, Y_p$ are all missing for cases where Y_j is missing, $j = 1, \ldots p-1$. The monotone missingness pattern occurs a lot in longitudinal studies, in which Y_j denotes the subject's measurements at wave/time j. Subjects might have fully observed data at the baseline (e.g., Y_1), and yet some of them might gradually drop out from the study at later waves and therefore produce missing data (e.g., for Y_2-Y_4). These attritional subjects usually do not return once they exit the study. Additional discussion about the multiple imputation for longitudinal missing data can be found in Chapter 9.

In some cases, a single construct is measured by two different instruments in two datasets. This can be viewed as a measurement error problem as in Table 1.3. However, the analytic interest might not be to recover the unknown true values. Instead, the goal might be to enhance the comparability of the two datasets by converting one instrument to the other and vice versa. In Table 1.6, Y_1 and Y_2 can be viewed as the two different instruments measuring the same construct. In dataset 1 ($S = 1$), only Y_1 is used (observed), and the corresponding Y_2 is missing; in dataset 2 ($S = 2$), only Y_2 is observed and the corresponding Y_1 is missing. Can we compare the same construct between the two datasets despite that this construct is measured by different instruments? Studies like that often have a bridge datasest ($S = B$) in which both Y_1 and Y_2

are observed. In other cases, there might exist no bridge study, which would result in a pattern called statistical file matching (Table 1.7). Applications of multiple imputation to these problems can be found in Chapter 11.

It is often easier to develop imputation strategies for missing data problems with somewhat structured patterns such as the monotone missing data. In practice, however, most of multivariate missing data problems do not have structured missing data patterns, and we usually refer to them as the general missing data pattern, as illustrated in Table 1.8.

Example 1.7. *The nonresponse pattern for four variables in NHIS 2016*

TABLE 1.5
A schematic table of monotone missingness pattern

Y_1	Y_2	Y_3	Y_4	...
...	...	...	...	...
O	O	O	O	...
O	O	O	O	...
O	O	O	O	M
O	O	O	M	M
O	O	M	M	M
O	M	M	M	M
...	...	...	...	...

Note: "O" symbolizes observed values; "M" symbolizes missing values.

TABLE 1.6
A schematic table of bridge study used to link two instruments for the same construct

Y_1	Y_2	Y_3	Y_4	S
...	...	...	...	...
O	M	O	...	1
O	M	O	...	1
O	O	O	...	B
O	O	O	...	B
M	O	O	...	2
M	O	O	...	2
...	...	...	...	...

Note: "O" symbolizes observed values; "M" symbolizes missing values; $S = 1$ and $S = 2$ symbolize two datasets; Y_1 is the instrument used in $S = 1$; Y_2 is the instrument used in $S = 2$; $S = B$ symbolizes a bridge dataset where both Y_1 and Y_2 are included.

TABLE 1.7
A schematic table of statistical file matching

Y_1	Y_2	Y_3	...	S
...	...	...	...	...
O	M	O	...	1
O	M	O	...	1
O	M	O	...	1
M	O	O	...	2
M	O	O	...	2
M	O	O	...	2
...	...	...	...	...

Note: "O" symbolizes observed values; "M" symbolizes missing values; $S = 1$ and $S = 2$ symbolize two datasets; Y_1 is the instrument used in $S = 1$; Y_2 is the instrument used in $S = 2$; there exists no bridge data where both Y_1 and Y_2 are observed.

TABLE 1.8
A schematic table of general missingness pattern

Y_1	Y_2	Y_3	Y_4	...
...	...	...	...	...
O	M	O	O	...
O	M	O	M	...
M	M	O	O	...
M	O	M	O	...
O	O	M	O	...
M	O	O	O	...
...	...	...	...	...

Note: "O" symbolizes observed values; "M" symbolizes missing values.

Besides the family income variable, other variables in NHIS also have nonresponses. For illustration, Table 1.9 lists the missingness pattern for four variables in the family file of NHIS 2016. In addition to FAMINCI2, the other three variables include, "Does the family have working phone inside home?" (CURWRKN), "Any family member receiving dividends from stocks, etc.?" (FDIVDYN), and "Education of adult with highest education in family" (FM_EDU1), where the variable labels for the public-use data are inside the parentheses. The whole dataset can be divided into 14 groups according to the distribution of response indicators of the four variables. The missingness pattern of the four selected variables is a general pattern. Inclusion of other variables will also lead to a general missingness pattern.

TABLE 1.9
Example 1.7. The missingess pattern of four variables from the family file of NHIS 2016

Group	FAMINCI2	CURWRKN	FDIVDYN	FM_EDU1	Sample size
1	O	O	O	O	31957
2	O	O	O	M	14
3	O	O	M	O	112
4	O	M	O	O	357
5	O	M	O	M	1
6	O	M	M	O	3
7	M	O	O	O	7215
8	M	O	O	M	74
9	M	O	M	O	829
10	M	O	M	M	81
11	M	M	O	O	186
12	M	M	O	M	6
13	M	M	M	O	33
14	M	M	M	M	7

Note: "O" symbolizes observed values; "M" symbolizes missing values.

1.4 Missing Data Mechanisms

Missing data mechanisms concern why missing data occur. In the case of income nonresponse (Example 1.1), it is possible that some nonrespondents might not be able to recollect the exact income amount and thus just answer "Don't know" even if they would like to respond. In addition, the income status is always a sensitive issue so that some nonrespondents might not want to disclose their information and thus choose to answer "Refused." There might exist other reasons that prompt nonrespondents not to provide their income values.

In general, there could be many reasons leading to missing data. Some of them might occur in a random fashion. In longitudinal studies, for example, subjects may have missed a visit for a practical reason, or data may not have been collected on a particular time because of equipment failure. Missing data can also arise from study designs (i.e. planned missingness), which purposefully do not collect data at certain scenarios. For example, matrix sampling applies different versions of the same survey instruments to different subgroups (e.g., Raghunathan and Grizzle 1995; Graham 2012).

Although in many cases we are not able to pinpoint exact reasons behind missing data, the notion of missing data mechanisms can be used to formulate the possible probability process that leads to missing data. As pointed out by Rubin (1976), every data point has some probability to be missing. Therefore the missingness of a variable is perceived as a random variable. The missingness mechanism can be formulated as a statistical model for the response (or missingness) indicator matrix, $R = \{r_{ij}\}$, given data. In this book, the terms "missing data mechanism", "missingness mechanism", "missingness model", "response/nonresponse mechanism", and "response/nonresponse model" have the same meaning and are used interchangeably.

In practical terms, the missingness model states the relationship between the missingness indicator R and data Y. Let $Y = (Y_{obs}, Y_{mis})$, where Y_{obs} and Y_{mis} are the observed and missing parts of Y. Let ϕ denote the parameter for the missingness model; then a general expression of the missing data mechanism is $f(R|Y_{obs}, Y_{mis}, \phi)$, where $f(\cdot|\cdot)$ denotes a probability distribution (or density) function.

To illustrate the idea, we consider a univariate missing data problem (Table 1.2). With a little abuse of notations, here Y_{mis} are confined to a single incomplete variable (say Y_1), and R is the response indicator (vector) of Y_1; Y_{obs} includes the fully observed variables in the data matrix (say Y_2 to Y_p). Also note that modeling $f(R = 0)$ is no different from modeling $f(R = 1)$ because $f(R = 1|Y_{obs}, Y_{mis}, \phi) = 1 - f(R = 0|Y_{obs}, Y_{mis}, \phi)$. In addition, we drop the subject-level index i in describing the mechanisms assuming independent and identical distributions across subjects.

First, the missing data are said to be missing completely at random (MCAR) if $f(R = 0|Y_{obs}, Y_{mis}, \phi) = f(R = 0|\phi)$. That is, the probability of being missing is not related to the data and is only dependent on some parameter ϕ. Suppose Y_1 is complete in the first place; a simple example of MCAR would be drawing a simple random sample from Y_1 and then deleting the nonsampled cases. While simple to understand, MCAR is often unrealistic for actual data.

Second, the missing data are said to be missing at random (MAR) if $f(R = 0|Y_{obs}, Y_{mis}, \phi) = f(R = 0|Y_{obs}, \phi)$. That is, the missingness probability is only related to the observed variables in the data. A generally exchangeable term for MAR is ignorable missingness. MAR also implies that the missingness probability is not related to the missing variable in the data after conditioning on the observed variables. Note that MCAR is a special case of MAR. On the other hand, if Y_{mis} is MCAR within groups defined Y_{obs}, then Y_{mis} is MAR conditional on Y_{obs}. That is, in certain cases, MAR can be viewed as a collection of MCARs among groups defined by the observed data.

Example 1.8. *An algebraic illustration of the connection between MCAR and MAR*

Consider a dataset with two variables: an incomplete Y_1 and a fully observed, binary Y_2. Suppose $f(R=0|Y_2=1)=\phi_1$ and $f(R=0|Y_2=0)=\phi_0$, where ϕ_0 and ϕ_1 are proportions $\in (0,1)$. This indicates that the missingness mechanism of Y_1 is MCAR within each group of Y_2. The missingness mechanism can also be expressed as $f(R=0|Y_2)=\phi_0+(\phi_1-\phi_0)\times I(Y_2=1)$, where $I(\cdot)$ is the identity function. It is an MAR mechanism because the missingness is related to Y_2 in general. When $\phi_0 \neq \phi_1$, the missingness mechanism is MAR but not MCAR. When $\phi_0=\phi_1$, this MAR is reduced to MCAR since the missingness of Y_1 is then independent of Y_2.

Lastly, the missing data are said to be missing not at random (MNAR) if $f(R|Y_{obs}, Y_{mis}, \phi)$ cannot be simplified as either MCAR or MAR. MNAR means that the missingness probability is related to missing values of the incomplete variable, which are unknown to us, even after conditioning on observed information. "Missing not at random" is also often termed as "not missing at random" (NMAR) or nonignorable missingness in literature.

Example 1.9. *A simulated example to illustrate MCAR, MAR, and MNAR*

We use a simulated example to contrast between different missingness mechanisms. We consider two variables X and Y. Let $X \sim N(1,1)$, and $Y=-2+X+\epsilon$, where $\epsilon \sim N(0,1)$. The joint distribution for X and Y follows a bivariate normal model. We generate 1000 pairs of random numbers of X and Y. We then let X be fully observed and randomly delete Y by generating R from random Bernoulli distributions as follows:

(a) MCAR: $f(R=0|X,Y)=0.4$, that is, randomly set Y as missing with a 40% chance;

(b) MAR: $f(R=0|X,Y)=\frac{exp(-1.4+X)}{1+exp(-1.4+X)}$: a larger X has a higher probability of having its corresponding Y to be missing. The missingness model can also be expressed as $logit(f(R=0|X,Y))=-1.4+X$, where $logit(a)=log(a/(1-a))$. A logit function is also referred to as logistic function or log-odds; the inverse of a logit function is $logit^{-1}(a)=exp(a)/(1+exp(a))$.

(c) MNAR: $logit(f(R=0|X,Y))=0.2+0.2X+Y$: a larger X or Y has a higher probability of missing Y.

To make the scenarios comparable, the coefficients in the logit function in (b) and (c) are chosen to generate around 40% missing cases. Fig. 1.2 presents

Box plots of Y before-deletion (BD) (i.e., complete Y) and Y_{obs} generated under different missingness mechanisms. Under MCAR, the distribution of observed Y-values is similar to that of the complete data, with similar mean and variance. However, this is not the case for MAR and MNAR. In both scenarios, since X and Y are positively correlated, missing cases are more likely to have larger unobserved Y-values, leading to a lower sample average from only observed Y-values. To assess the relationship between X and Y, Fig. 1.3 shows the scatter plots of two variables before and after deletion. A fitted line between Y and X is also imposed as a summary of the relationship. In both MCAR and MAR, the fitted lines are similar to that of complete data. However in MNAR, the fitted line is somewhat different, as the slope is attenuated (i.e., moving close to 0). These patterns are confirmed with actual estimates of means and regressions shown in Table 1.10.

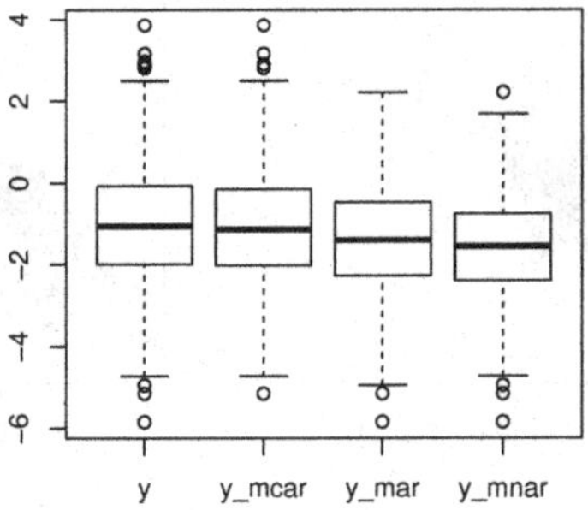

FIGURE 1.2
Example 1.9. Box plots of Y before and after deletion. y: complete data; *y_mcar*: Y_{obs} under MCAR; *y_mar*: Y_{obs} under MAR; *y_mnar*: Y_{obs} under MNAR.

The fact that the relationship between Y and X is well retained for observed cases in (a) and (b) implies a very important feature of missing data under MAR (and MCAR): the relationship between Y and X can be characterized as a statistical model using observed cases, and this model can then be used to impute the missing Y-values. However, this is not the case under MNAR. This idea will be further discussed in the book.

If Y is the incomplete variable and X contains fully observed covariates, MAR for the missingness of Y can also be expressed as $Y \perp R|X$ in some literature. This states that the missing variable, Y, and its missingness indicator, R, are conditionally independent given observed covariates, X. A review of this notation system can be found in Dawid (2006).

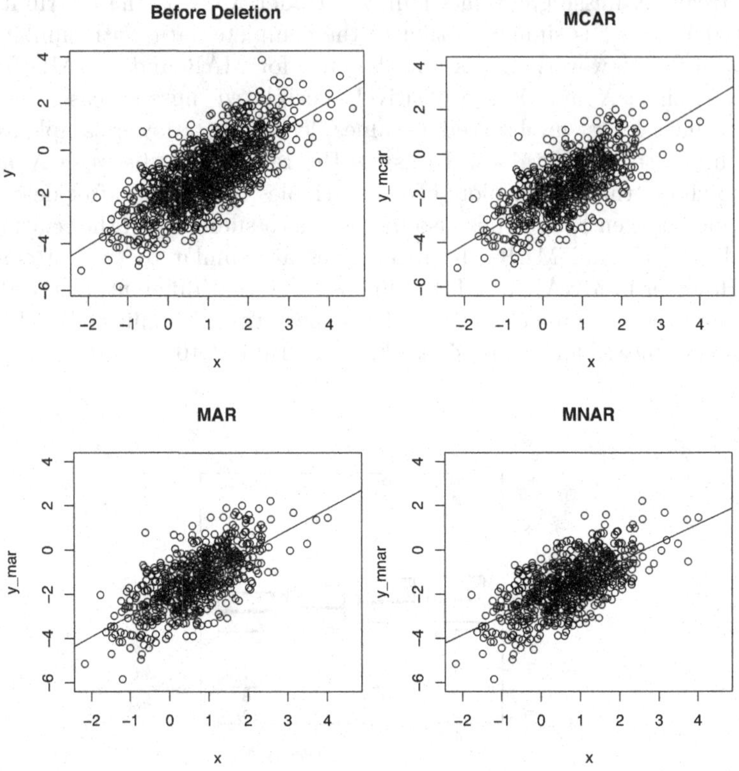

FIGURE 1.3
Example 1.9. Scatter plots of X and Y and their fitted lines. Top left: before deletion; top right: MCAR; bottom left: MAR; bottom right: MNAR.

For multivariate missing data problems with a general missingness pattern, it can be difficult to state and express the missingness mechanism in simple functional forms as in univariate missing data problems such as in Examples 1.8 and 1.9. The concept of MAR can be understood as: $f(R|Y_{obs}, Y_{mis}) = g(R, Y_{obs})$, a conditional distribution that is not a function of Y_{mis}. Little and Rubin (2020, Chapter 1) provided a more technical discussion and some examples.

Having a plausible MAR assumption is important because principled missing data methods, including multiple imputation, often start with the MAR assumption. For the purpose of making statistical inference on the distribution of Y, there is no need to consider the missingness model, $f(R = 0|Y_{obs}, Y_{mis}, \phi)$, under MAR. This would presumably make the inferential task simpler. However under MNAR, we usually need to account for the missingness model in the estimation. As expected, doing this is often very difficult and requires lots of speculation because we do not observe missing values in the first place! More detailed discussions relating missingness mechanisms to statistical inference can be found in Section 2.4.

It might be of practical interest to test these assumptions and determine the possible missingness mechanism for data at hand. One simple strategy to assess whether the missingness is MCAR is to compare nonresponse rates of the missing variable across groups formed by observed variables, as implied in Example 1.8. We can also regress the missingness indicator R on observed covariates, typically through a binary logistic regression. If coefficients for some covariates are significantly different from 0, then MCAR is unlikely to hold. For simulated data in Example 1.9, Table 1.10 shows the logistic regression coefficient on X. For (a), the coefficient is not significant at .05 level, yet for (b) and (c), the coefficients are highly significant (p-values not shown), suggesting that MCAR is unlikely to hold in both cases.

However, results from such tests can only help assess whether the missingness is MCAR. Unlike in Example 1.9, where we know the complete data before the missingness is applied using simulation, in practice we are not able to distinguish between MAR and MNAR based only on the information from observed data. Imagine in Fig. 1.3 or Table 1.10, how do we know whether (b) or (c) is MAR without knowing results from the before-deletion data ? Therefore, arguments that missing data are likely MNAR need evidence from external information, prior knowledge, or subject-matter input, all of which can be speculative. For instance, some literature (e.g., Gelman and Hill 2007, Section 25.1) suggested an MNAR assumption for income nonresponse problems: people with higher income are less likely to reveal their income in surveys. It would be interesting to see whether this is the case for some real data.

TABLE 1.10
Example 1.9. Mean and regression estimates

Scenario	$\overline{Y}$	$\hat{\beta_0}$	$\hat{\beta_1}$	$\hat{\alpha_1}$
BD	−1.02 (0.05)	−2.04 (0.04)	1.02 (0.03)	
MCAR	−1.05 (0.06)	−2.03 (0.05)	1.04 (0.04)	0.12 (0.06)
MAR	−1.38 (0.05)	−2.01 (0.05)	0.97 (0.04)	1.02 (0.08)
MNAR	−1.60 (0.05)	−2.18 (0.04)	0.82 (0.04)	0.96 (0.08)

Note: A linear regression of $E(Y) = \beta_0 + \beta_1 X$ is fitted; the estimates and standard errors (in parentheses) for β_0 and β_1 are shown. A logistic regression of $logit(f(R = 0)) = \alpha_0 + \alpha_1 X$ is fitted; the estimates and standard errors (in parentheses) for α_1 are shown.

Example 1.10. *Illustrative analyses of nonresponse mechanisms of family income in NHIS 2016*

In the NHIS family file, responses from bracketed-income variables might help us gauge the relation between the missingness of income and its actual values. We focus on one income category variable, labeled as INCGRP5 in the NHIS 2016 public-use data. Table 1.11 lists nonresponse rates of total family income in different categories of INCGRP5. This bracketed income variable is also incomplete, resulting in approximately 12% of participants who failed to respond to the value of FAMINCI2 and INCGRP5 at the same time. Despite that, the results appear to show that the nonresponse rate of family income is not evenly distributed across the income groups: the lower-income group appears to have a considerably higher nonresponse rate. For example, nonresponse rate is 13.7% for participants whose family income is less than \$35000, and decreases to around 9% for the rest of the income groups. These results appear to suggest that the nonresponse of FAMINCI2 is not MCAR and a likely MNAR: participants with lower income are less likely to respond to the family income question.

TABLE 1.11
Example 1.10. Nonresponse rates of family income across income brackets in NHIS 2016

INCGRP5	Percentage (%)	Nonresponse of FAMINCI2 (%)
\$0 - \$34999	31.7	13.7
\$35000 - \$74999	25.9	9.3
\$75000 - \$99999	10.0	9.5
$\geq$ \$100000	20.6	9.2
Missing	11.8	100

What will happen if we consider other variables in the nonresponse mechanism? Recall that in Example 1.9 the missingness mechanism for MNAR is related to both missing Y and other covariates X. We first consider a variable measuring the highest family education level (i.e., educational attainment), which is labeled as FM_EDUC1 and almost fully observed in NHIS 2016. For simplicity, we create a new education variable (FM_EDUC1_RECODE) by collapsing some of the original categories. We examine its association with the nonresponse rate of FAMINCI2 after dropping around 0.5% of cases with the missing education variable. Table 1.12 lists the missingness rates of the family income on different levels of education. The results show that the missingness rate of income is also related to the educational level. For participants whose education level is high school or less, the corresponding nonresponse rate for income is higher than the rest of participants (24.7% vs. 19.7%).

We then stratify the sample by the education level and calculate the nonresponse rate of family income across the bracketed income groups. For simplicity, we exclude cases with missing bracketed income. Table 1.13 shows the results. The dependency of the income nonresponse on bracketed income still appears to exist on the two education groups (high school or less and some college). However, such dependency is considerably reduced for the group with college or graduate school education, ranging between around 9% and 10%. That is, within this group, the nonresponse of family income is closer to being MCAR. The reduction of this dependency can also be reflected by running a logistic regression of the nonresponse indicator using both the bracketed income groups and education level as predictors. Table 1.14 shows that, after adding the education variable, the odds ratio of the nonresponse probability in the lowest income group is reduced from 1.566 to 1.394, although still significant at the 0.05 level (p-values not shown). The odds ratios for the other two groups are little changed and remain nonsignificant.

Exploratory analysis results so far seem to show that lower-income and lower-education survey participants were less likely to respond to the income question in NHIS 2016. It is consistent with the results from the multiple imputation: the mean estimate of imputed family income is lower than that of observed cases after accounting for the information from these nonrespondents (Example 3.3).

More importantly, by including (or adjusting for) more observed variables, the dependency of the income nonresponse on the actual income values is

TABLE 1.12
Example 1.10. Nonresponse rates of family income across education level in NHIS 2016

FM_EDUC1_RECODE	Percentage (%)	Nonresponse of FAMINCI2 (%)
High school or less	27.1	24.7
Some college	32.2	19.7
College or graduate school	40.2	19.7

TABLE 1.13
Example 1.10. Nonresponse rates (%) of family income across income groups and education level in NHIS 2016

	High school or less	Some college	College or graduate school
\$0 - \$34999	15.7	12.1	10.3
\$35000 - \$74999	11.6	8.1	8.9
\$75000 - \$99999	10.4	9.2	9.4
$\geq$ \$100000	8.0	7.3	9.7

likely to get further reduced as parts of the dependency might be explained through additional variables. That is, the nonresponse probabilities would become more homogeneous in groups defined by more observed variables, allowing the MAR assumption of income to be more plausible. To see this pattern, we add another variable with little missing data, FSALYN (does any family member receive income from wages or salary: Yes/No). After including this variable to the logistic regression predicting the nonresponse, the odds ratio in the lowest income group gets further reduced (from 1.394 to 1.237). The odds ratios in the other two income groups are still not significantly different from 1. Results are included in Table 1.14.

TABLE 1.14
Example 1.10. Nonresponse logistic regression analysis for family income using selected variables in NHIS 2016

Income group	No extra covariate	Plus education	Plus education and wage
\$0-\$34999	1.566*	1.394*	1.237*
\$35000-\$74999	1.015	0.967	0.933
\$75000-\$99999	1.035	1.022	0.989

Note: The table shows odds ratio estimates associated with income groups from running a logistic regression for the nonresponse indicator of family income; the reference group is $\geq$ \$100000. "No extra covariate" only includes income groups as the predictor; "Plus education" includes both income groups and education level as predictors; "Plus education and wage" includes income groups, education level, and receipt of wage/salary as predictors. $*$ significant at the 0.05 level.

All analysis results in Example 1.10 account for the survey design of NHIS. This example shows that by including more observed variables, the missingness of family income is likely to move closer to being MAR, which is desired for the purpose of multiple imputation and other principled missing data methods. This is because some, if not all, of these variables might also be related to the income nonresponse and therefore by adjusting for them, the association between the nonresponse probability and missing values would be attenuated. In addition, including more observed variables, some of which can be also predictive of the missing income value, has the potential to improve imputations for better statistical inference. In Example 1.10, both education and receipt of wage/salary are associated with the family income: participants with higher education or receiving wage/salary are more likely to have higher income. In the actual NHIS income imputation project, many more variables are used to impute the missing income variable. This general idea of including many variables in the imputation is termed as "inclusive strategy" (Collins et al. 2001), which will be discussed further in later chapters.

1.5 Structure of the Book

Missing data problems are common for data analysis across different types of research and studies. They can also take a variety of forms and patterns, depending on the scientific interests and analytic goals. Missingness mechanisms are assumptions describing the relation between the nonresponse probability and variables in the data.

We use the income nonresponse problem in NHIS 2016 as a motivating example to start the book. More generally, a main goal of this book is to present multiple imputation as both a principled and applied missing data strategy for researchers and practitioners. We aim to explain fundamental ideas behind multiple imputation and discuss how to use it in practice. Illustrations will be based on both actual and simulated data, as well as examples from past literature. Some of the discussed topics might also stimulate more research ideas to further expand the utility of multiple imputation.

There exists abundant literature for missing data analysis and multiple imputation. To name a few here, a classic book by Little and Rubin (1987; 2002; 2020) is the first to provide a systematic treatment of analysis of missing data. Molenberghs et al. (2015) includes a wide variety of advanced research topics on missing data methodology. Raghunathan (2016) provides an excellent summary of missing data methods for practitioners. Raghunathan et al. (2018) can be viewed as a companion of Raghunathan (2016), using IVEware (https://www.src.isr.umich.edu/software/) to illustate the multiple imputation approach. Rubin (1987) lays the foundation of the multiple imputation approach. Van Buuren (2012; 2018) provides a vivid description and review of various multiple imputation methods and illustrates them using R mice, which is a library from R software (https://www.r-project.org/). Carpenter and Kenward (2013) illustrated multiple imputation using numerous examples from health and clinical research. A very recent book on multiple imputation for researchers in social and behavioral sciences can be found in Kleinke et al. (2020). We hope that our book can add to the growing literature to advance the methods and practice of multiple imputation analysis.

The rest of the book is organized as follows. Chapter 2 provides some statistical background and briefly reviews alternative missing data methods. Chapter 3 reviews the basic theory and procedure of multiple imputation analysis. Chapters 4 and 5 introduce some commonly used imputation methods for univariate missing data problems: Chapter 4 focuses on parametric models, and Chapter 5 focuses on robust techniques. Chapters 6 and 7 tackle multivariate missing data problems: Chapter 6 focuses on the joint modeling approach, and Chapter 7 focuses on the fully conditional specification approach. Chapters 8-11 discuss imputation methods applied in a variety of settings: Chapter 8 is for survival analysis, Chapter 9 is for longitudinal data, Chapter 10 is for sample survey data, and Chapter 11 is for data subject to measurement error.

Chapter 12 presents ideas for imputation diagnostics. Chapter 13 discusses strategies for dealing with nonignorable missing data. Chapter 14 concludes this book by listing several advanced research topics.

2

Statistical Background

2.1 Introduction

Chapter 2 presents some statistical background for multiple imputation analysis. We briefly discuss two main frameworks for drawing statistical inference: frequentist (Section 2.2) and Bayesian (Section 2.3). Multiple imputation sits at the intersection of the frequentist and Bayesian frameworks. Section 2.4 introduces the likelihood-based approach to missing data problems under different assumptions of missingness mechanisms. Section 2.5 recommends generally avoiding ad hoc approaches to missing data problems. Section 2.6 describes some basic principles of conducting simulation studies. Section 2.7 provides a summary.

2.2 Frequentist Theory

2.2.1 Sampling Experiment

We start the discussion by considering a concrete scientific question. For example, suppose we are interested in the health insurance coverage of an (infinite) target population during a specified period and assume that everyone has the identical probability of being covered. We can imagine a theoretical experiment that consists of randomly selecting n independent subjects from the population and asking every participant whether he/she has been covered by any health insurance. The data would consist of ones (Yes) and zeros (No). For simplicity, we assume that all sampled subjects respond (i.e., no missing data). Suppose the dataset has Y (Y is the sum of all ones) subjects having insurance and $n - Y$ subjects not having it. Here Y is a random variable: its possible values are $\{0, 1, 2, \ldots, n\}$, which is called the sample space, namely a collection of all possible results from the experiment.

In general, the frequentist method is an approach to statistical inference that draws conclusions from sample data by emphasizing the frequency or proportion of the data. Frequentist inference has been associated with the frequentist interpretation of probability, specifically that any given experiment

DOI: 10.1201/9780429156397-2

can be considered as one of an infinite sequence of possible repetitions of the same experiment, each capable of producing statistically independent results.

In the context of our example, we can imagine that this experiment is to be conducted many times, each time sampling n individuals from the population and collecting their responses to the question on health insurance coverage. It is expected that the corresponding Y-value can vary from experiment to experiment and this variability constitutes the randomness of Y.

2.2.2 Model, Parameter, and Estimation

Once the data are available, the next step in statistical inference is to parameterize the estimand of interest and posit a statistical model for the data. An estimand is a quantity of scientific interest that can be calculated if the entire population is known. For example, we use θ, the parameter of interest, to denote the proportion of population having health insurance coverage. Basically, a statistical model describes the probability of observing the experimental data based on certain distributional assumptions. With a little abuse of notation, we use $f(\cdot|\cdot)$ to denote the distribution function for a random variable throughout this book: for a continuous/discrete random variable, $f(\cdot|\cdot)$ is the density/probability distribution function. In addition, we do not distinguish between the random variable Y and its observed value y in these expressions unless we note it otherwise.

In the example, the event of observing Y subjects with health insurance in the sample of size n can be described by a binomial distribution $f(Y|\theta)$:

$$f(Y|\theta) = \frac{n!}{Y!(n-Y)!}\theta^Y(1-\theta)^{n-Y}. \tag{2.1}$$

Obviously the range of θ is between 0 and 1 (i.e., $\theta \in [0,1]$), which is referred to as the parameter space. Under the frequentist framework, the true value θ is unknown yet assumed to be a fixed quantity; that is, θ should not change across the conducted experiments. Based on the sample data, what we can infer about the value of θ is a point estimator or estimate of θ, denoted as $\hat{\theta}$. Generally speaking, the term "estimator" emphasizes the mathematical functional form, and the term "estimate" emphasizes the actual values calculated. From our introductory statistical course, a natural point estimator $\hat{\theta}$ based on Model (2.1) is $\hat{\theta} = Y/n$, the proportion of having health insurance in the sample.

Note that, in general, the value of $\hat{\theta}$ estimate can vary across repeated experiments due to the random variability associated with Y. The collection of $\hat{\theta}$'s across all of the possible experiments constitutes a distribution of $\hat{\theta}$. This is termed as the sampling distribution of $\hat{\theta}$ and can be denoted as $f(\hat{\theta}|\theta)$. Statistical properties of the estimator, $\hat{\theta}$, can be displayed from its sampling distributions, $f(\hat{\theta}|\theta)$. In general, a good estimator $\hat{\theta}$ would possess several desirable properties. One is unbiasedness. We can define $Bias(\hat{\theta}) = E(\hat{\theta}|\theta) - \theta$, where $E(\hat{\theta}|\theta)$ denotes the expectation of $\hat{\theta}$ over its sampling distribution. The estimator is called unbiased if the average of its sampling distribution is equal to θ. This can be stated as

$$E(\hat{\theta}|\theta) = \theta, \tag{2.2}$$

or $Bias(\hat{\theta}) = 0$.

The second property is the minimum variance and can be described as follows. Suppose that $\tilde{\theta}$ is any other unbiased estimator for estimating θ. The variance of $\tilde{\theta}$ across repeated experiments (or its sampling distribution) is no less than that of $\hat{\theta}$. This can be stated as

$$Var(\hat{\theta}|\theta) <= Var(\tilde{\theta}|\theta). \tag{2.3}$$

In the example of the binomial random experiment, it can be verified that $\hat{\theta} = Y/n$ is both unbiased and with the minimum variance.

However, except in rare situations, the variance of an $\hat{\theta}$ is unknown and has to be estimated as well because it often involves the unknown θ. The sampling variance estimate (or simply the sampling variance) of $\hat{\theta}$ is defined as an estimator of $Var(\hat{\theta}|\theta)$, here denoted by $\hat{Var}(\hat{\theta}|\theta)$. The standard error estimate is the square root of the sampling variance estimate, $\sqrt{\hat{Var}(\hat{\theta}|\theta)}$. In the binomial random experiment, the variance of $\hat{\theta} = Y/n$ is $\theta(1-\theta)/n$ based on the property of the binomial distribution. However, since θ is unknown, we can replace it by the estimate $\hat{\theta} = Y/n$. Therefore a variance estimator for $\hat{\theta}$ is $\hat{Var}(\hat{\theta}|\theta) = \hat{\theta}(1-\hat{\theta})/n = \frac{Y/n(1-Y/n)}{n}$, which can be calculated from the data.

A variance estimator also has its sampling distribution. If a variance estimator is unbiased, it means

$$E(\hat{Var}(\hat{\theta}|\theta)) = Var(\hat{\theta}|\theta). \tag{2.4}$$

In many cases we might not get strictly unbiased estimators for either point or variance estimation. However for some estimators, when the sample size increases, the corresponding bias might decrease and approach zero when the sample size approaches infinity. Loosely speaking, we call these estimators consistent (or unbiased in the asymptotic sense), which can be expressed as

$$E(\hat{\theta}|\theta) \rightarrow \theta, \tag{2.5}$$
$$E(\hat{Var}(\hat{\theta}|\theta)) \rightarrow Var(\hat{\theta}|\theta), \tag{2.6}$$

as $n \rightarrow \infty$.

For simplicity from now on we use $E(\hat{\theta})$ and $Var(\hat{\theta})$ to denote the mean and variance of the estimator for θ with respect to its sampling distribution, respectively.

Variance estimation further allows us to construct an interval estimate that is deemed to cover the true value θ. This is termed as the confidence interval in the frequentist framework. The confidence interval can be formulated as a range defined by $(\hat{\theta}_L, \hat{\theta}_U)$ so that $\hat{\theta}_L < \hat{\theta} < \hat{\theta}_U$, where L denotes the lower bound and U denotes the upper bound. The use of the sampling distribution of $\hat{\theta}$ is the most common way to construct a confidence interval. In many real scenarios, we would use $\hat{\theta}_L = \hat{\theta} - t_{n-1,1-\alpha/2}\sqrt{\hat{Var}(\hat{\theta})}$, and $\hat{\theta}_U = \hat{\theta} + t_{n-1,1-\alpha/2}\sqrt{\hat{Var}(\hat{\theta})}$, where $0 < \alpha < 1$ and $t_{n-1,1-\alpha/2}$ denotes the $100(1-\alpha/2)$ percentile of the t distribution with $n-1$ degrees of freedom.

When the random experiment is carried over and over, the frequency that θ is covered by the confidence intervals constructed from data of experiments is probability $1-\alpha$. In an equation it can be expressed as

$$\sum_{Y=0}^{n} I(\theta \in (\hat{\theta}_L, \hat{\theta}_U))f(Y|\theta) = 1 - \alpha, \tag{2.7}$$

where $I(\cdot)$ is the identity function.

For example, if $\alpha = 0.05$, then the interval is called a 95% confidence interval, and the coverage is called a nominal coverage. In general, it is also a nice property for $\hat{\theta}$ to have at least a coverage probability (or coverage rate) of $1-\alpha$ with a prespecified α, that is, $Pr(\hat{\theta}_L \leq \theta \leq \hat{\theta}_U) \geq 1-\alpha$.

In summary, for estimating θ, we aim to establish a point, variance, and interval estimation procedure with good frequentist properties that are unbiased or consistent with desirable coverage probabilities.

Finally, we briefly introduce a very useful statistical technique, often referred to as the delta method, to calculate the variance of a smooth function of a random variable Y, say $f(Y)$. Suppose $E(Y) = \mu$ and applying Taylor linear approximation to $f(Y)$ around μ, we have

$$f(Y) \approx f(\mu) + (Y-\mu)f'(\mu), \tag{2.8}$$

where $f'(\mu)$ is the first-order derivative of $f(\cdot)$ evaluated at μ. Based on Eq. (2.8), we have

$$Var(f(Y)) \approx Var(Y)[f'(\mu)]^2. \tag{2.9}$$

Therefore we can conveniently calculate the variance of $f(Y)$ using Eq. (2.9) once we know the mean and variance of Y. The delta method is very useful when we apply transformation to variables in statistical analysis. In addition, since estimator $\hat{\theta}$ is also a random variable, the delta method can be used to calculate the variance for functions of parameter estimates.

2.2.3 Hypothesis Testing

In addition to parameter estimation, another major inferential task in the frequentist framework is hypothesis testing. That is, we test a preconceived hypothesis about the possible values of the target quantity of interest. Following the health insurance coverage example, we might formulate a null hypothesis, H_0: $\theta < \theta_0$ against the alternative hypothesis, H_A: $\theta \geq \theta_0$, where θ_0 is a preset number such as 0.90. Through hypothesis testing, we aim to make a decision:

either reject or fail to reject the null hypothesis. The common procedure in hypothesis testing is to calculate the test statistics and obtain the p-value, which can be understood as the probability of obtaining a result as extreme as or more extreme than the testing statistics obtained given that the null hypothesis is true. A smaller p-value typically suggests that the evidence is stronger for rejecting the null hypothesis.

However, there always exists a probability (i.e., error) that our conclusion from the testing procedure deviates from the underlying truth. This is because the test statistics we use in hypothesis testing are subject to random variation. To quantify the error, $Pr(\text{Reject } H_0 | H_0 \text{ is true})$ is defined as a type-I error, and $Pr(\text{Fail to Reject } H_0 | H_A \text{ is true})$ is defined as a type-II error. In many contexts the question of hypothesis testing is equivalent to checking whether the preconceived value (e.g., θ_0) is included in the interval estimate by the duality of a $100(1-\alpha)\%$ confidence interval and a significance test at level (i.e., type-I error) α.

2.2.4 Resampling Methods: The Bootstrap Approach

In many cases an estimator, $\hat{\theta}$, can be nonlinear forms of statistics, and its variance estimator might not have closed algebraic forms. Resampling methods can be applied to compute the variance estimate numerically. One of the commonly used resampling approaches is the bootstrap method. To describe the basic idea, let $\hat{\theta}$ be a consistent estimate of a parameter θ based on a sample data S which consists of n independent observations. Let $S^{(b)}$ be a sample size n obtained from the original sample S by simple random sampling with replacement, and let $\hat{\theta}^{(b)}$ be the estimate of θ obtained by applying the original estimation method to $S^{(b)}$, where b indexes the b-th bootstrap sample of S. Let $(\hat{\theta}^{(1)}, \ldots, \hat{\theta}^{(B)})$ be the set of estimates obtained by repeating this procedure B times. A simple nonparametric bootstrap estimate of θ is then the average of the bootstrap estimates:

$$\hat{\theta}_{BOOT} = \frac{1}{B} \sum_{b=1}^{B} \hat{\theta}^{(b)}.$$

Large-sample variance estimates can be obtained from the bootstrap distribution of $\hat{\theta}^{(b)}$'s. In particular, the bootstrap estimate of the variance of $\hat{\theta}_{BOOT}$ is

$$\hat{V}_{BOOT} = \frac{1}{B-1} \sum_{b=1}^{B} (\hat{\theta}^{(b)} - \hat{\theta}_{BOOT})^2.$$

Under quite general conditions, both $\hat{\theta}_{BOOT}$ and $\hat{V}_{BOOT}$ are consistent estimates for θ and $Var(\hat{\theta}_{BOOT})$ as n and B increase to infinity. Moreover, if the bootstrap distribution is approximately normal, a $100(1-\alpha)\%$ bootstrap confidence interval for θ can be computed as $\hat{\theta} +/- z_{1-\alpha/2}\sqrt{\hat{V}_{BOOT}}$, where $z_{1-\alpha/2}$

is the $100(1-\alpha/2)$ percentile of the standard normal distribution. Alternatively if the bootstrap distribution is nonnormal, then a $100(1-\alpha)\%$ bootstrap confidence interval for θ can be computed as $(\hat{\theta}^L_{BOOT}, \hat{\theta}^U_{BOOT})$, where $\hat{\theta}^L_{BOOT}$ and $\hat{\theta}^U_{BOOT}$ are the empirical $100\alpha/2$ and $100(1-\alpha/2)$ percentile of the bootstrap distribution of $\hat{\theta}^{(b)}$'s.

The bootstrap method can be used as a brute forth, computationally intensive approach to variance estimation, avoiding sometimes complex or intractable formulas. The required number of bootstrap replicates, B, is often on the scale of hundreds or thousands. Given the affordable computational resources nowadays, this strategy is widely applied in practice. There exists a large body of literature for the bootstrap method. More information can be found, for example, in Efron and Tibshirani (1993) and Davison and Hinkley (1997).

In the context of this book, the main usage of the bootstrap method is to approximate the posterior distribution of model parameters in Bayesian estimation, as will be shown in Example 2.2.

Example 2.1. *Estimating the proportion of health insurance coverage from a web survey*

Traditional survey data collection methods include in-person interviews, telephone interviews, and mail surveys. For example, the first two methods have been the primary data collection methods for NHIS. Nowadays, the increasing access to the internet in the U.S. households has led to new opportunities to conduct online or web surveys. Some survey organizations manage probability-based subject panels that can be used to conduct web surveys as well as telephone surveys. Through the Research and Development Survey (RANDS), which consists of a series of studies using survey data being collected from these panels, NCHS has been investigating the utility of using commercial probability panel-based web surveys to measure and estimate health outcomes. Additional information about RANDS can be found in (https://www.cdc.gov/nchs/rands).

RANDS I (e.g., He et al. 2020) was conducted in the third quarter of 2015 and composed of exclusively web survey data from 2304 participants. In RANDS I there was a question, "Are you covered by any kind of health insurance or some kind of health plans?" The actual survey design of RANDS I is complicated and includes sampling with unequal probabilities (estimation for complex probability survey data will be deferred to Chapter 10). Survey nonresponses also occur. The survey weighted estimate for the insurance coverage variable is 0.934. However for illustration, here we treat the RANDS I data as a random sample from an infinite population to match with the theoretical binomial experiment. In this setup, $n = 2304$ (the denominator) and $Y = 2304 \times 0.934 \approx 2152$ (the numerator). Using the aforementioned formulas, we obtain:

Mean estimate: $\hat{\theta} = Y/n = 2152/2304 = 0.934$

Variance (standard error) estimate: $\hat{Var}(\hat{\theta}) = \hat{\theta}(1-\hat{\theta})/n = \frac{Y/n(1-Y/n)}{n} = 2.676 \times 10^{-5}$. The standard error estimate is $\sqrt{2.676 \times 10^{-5}} = 0.00517$.

95% confidence interval estimate: we use the normal approximation of the binomial distribution so that $\hat{\theta}_L = 0.934 - Z_{.975} * 0.00517 = 0.924$ and $\hat{\theta}_U = 0.934 + Z_{.975} * 0.00517 = 0.944$.

We also try the bootstrap method with $B = 10000$ bootstrap replications. The mean, standard error, and confidence interval estimates based on bootstrap samples are almost identical to the above estimates, with no difference for three digits after the decimal point. Results are omitted here.

To illustrate hypothesis testing, we set the null and alternative hypothesis as H_0: $\theta < 0.90$ vs. H_A: $\theta \geq 0.90$. Note that this is a one-side test. The p-value is calculated as $p = 1 - \Phi(\frac{0.934-0.90}{0.00517}) < 0.0001$, which is significant at the 0.05 level. Therefore we reject H_0.

2.3 Bayesian Analysis

2.3.1 Rudiments

An alternative inferential framework is the Bayesian analysis. Like the frequentist approach, the Bayesian analysis is also based on the idea of probability and accepts that there is a true, unknown value θ. However, from the Bayesian perspective, since θ is unknown, there is *a priori* uncertainty about its value. This uncertainty should be expressed in a form of a prior distribution for θ, with density $\pi(\theta)$ defined on the parameter space. This viewpoint also implies that θ can be treated as a random variable rather than a fixed quantity as in the frequentist analysis.

It is often straightforward to illustrate the Bayesian analysis based on statistical models. Still using the binomial experiment example, the distribution of the random variable Y given θ is expressed as $f(Y|\theta)$; then the joint distribution of Y and θ can be expressed as $f(Y, \theta) = f(Y|\theta)\pi(\theta)$, where $\pi(\theta)$ is the prior distribution for θ. From the joint distribution, we are able to obtain two more distributions. One is the marginal distribution of Y, $f(Y) = \int_0^1 f(Y, \theta)d\theta$. The other is the conditional distribution of θ given Y, $f(\theta|Y)$. By the famous Bayes rule,

$$f(\theta|Y) = \frac{f(Y, \theta)}{f(Y)} = \frac{f(Y|\theta)\pi(\theta)}{f(Y)} \propto f(Y|\theta)\pi(\theta). \tag{2.10}$$

In Eq. (2.10), $f(\theta|Y)$ is called the posterior distribution of θ given data Y, which is proportional to the product of $\pi(\theta)$ and $f(Y|\theta)$. In Bayesian analysis, the posterior distribution essentially contains all the information that can be inferred on θ. For estimating θ, we can use the posterior mean,

$$E(\theta|Y) = \int_0^1 \theta f(\theta|Y)d\theta, \tag{2.11}$$

which is the average value of θ over its posterior distribution. The mode of the posterior distribution can also be used to estimate θ.

For the variance estimation, we can use the posterior variance,

$$Var(\theta|Y) = \int_0^1 (\theta - E(\theta|Y))^2 f(\theta|Y)d\theta, \tag{2.12}$$

to express the uncertainty of the posterior mean estimate.

There are several approaches to constructing the Bayesian interval estimate. For example, suppose that (a, b) is such that $Pr(a \leq \theta \leq b) = \int_a^b f(\theta|Y)d\theta = 1 - \alpha$, and for any value of θ in the interval (a, b) and θ^* outside the interval (a, b), $f(\theta|Y) > f(\theta^*|Y)$. That is, the values of θ inside the interval have higher values of the posterior distribution than those outside the interval. The interval (a, b) is called the $100(1-\alpha)\%$ highest posterior credible interval. Its interpretation is that the probability of having θ being between a and b is $1 - \alpha$.

2.3.2 Prior Distribution

In general, prior distributions in Bayesian analysis can be constructed from pilot data, data from similar studies, or some subject-matter knowledge. Continuing with the binomial experiment example, it is natural to assume all values of θ are equally possible before the experiment. In an algebraic form, this assumption can be translated to a uniform prior distribution: $\pi(\theta) = 1$ if $\theta \in [0, 1]$ and $\pi(\theta) = 0$ otherwise. Under the uniform prior, the posterior distribution of θ can be shown as

$$f(\theta|Y) = \frac{\Gamma(Y + 1 + n - Y + 1)}{\Gamma(Y + 1)\Gamma(n - Y + 1)} \theta^{Y+1-1}(1 - \theta)^{n-Y+1-1}, \tag{2.13}$$

where $\Gamma(\cdot)$ is the Gamma function ($\Gamma(z) = \int_0^\infty x^{z-1}e^{-x}dx$). Eq. (2.13) shows that the posterior distribution of θ is a Beta distribution with parameter $Y + 1$ and $n - Y + 1$. After some algebra, we can show that the posterior mean, $E(\theta|Y) = \tilde{\theta}(Y) = \frac{Y+1}{n+2} \approx \frac{Y}{n}$, and the posterior variance, $Var(\theta|Y) = \frac{\tilde{\theta}(Y)(1-\tilde{\theta}(Y))}{n+3} \approx \frac{Y/n(1-Y/n)}{n}$. Note that here both the posterior mean and variance estimates are very close to their counterparts in the frequentist analysis (Section 2.2.2), especially when the sample size n is large.

The uniform prior distribution is a so-called noninformative prior distribution because $\pi(\theta) \propto 1$, essentially contributing no information to the posterior distribution. In many cases, suppose there exists little or vague prior knowledge about the behavior of model parameters; then it is often recommended to assign noninformative or diffuse prior distributions. As can be seen in Eq. (2.10), having a diffuse prior distribution is almost equivalent to working directly with $f(Y|\theta)$.

2.3.3 Bayesian Computation

Although the idea of Bayesian analysis is straightforward by focusing on the posterior distribution, $f(\theta|Y)$, the actual implementation might not always be straightforward. In many complicated problems, $f(\theta|Y)$ does not have tractable algebraic forms, and the use of Bayesian methods could have been constrained. This is especially the case when θ contains multiple elements. For example, suppose $\theta = (\theta_1, \theta_2, \ldots, \theta_p)$; then the posterior distribution of θ_1 is $f(\theta_1|Y) = \int f(\theta|Y)d\theta_2, \ldots, d\theta_p$, which would involve a multivariate integration.

These problems now have been readily addressed by stochastic simulation methods that aim at taking numerical draws/samples (i.e., random numbers) from $f(\theta|Y)$, rather than at trying to compute it analytically. These posterior samples of θ can be used to estimate the characteristics of the posterior distribution in a numerical way. In many cases, numerically drawing θ from $f(\theta|Y)$ can be achieved by using a technique known as Markov Chain Monte Carlo (MCMC) (e.g., Gilks et al. 1996). By using MCMC, we can set up an iterative sampling algorithm to draw a sequence (chain) of parameter values $\theta^{(0)}$, $\theta^{(1)}$, ..., $\theta^{(t)}$, ..., whose stationary distribution is the posterior distribution, $f(\theta|Y)$. Thus, running this algorithm from initial values, after discarding early values of the chain, which is known as the "burn in" period, we end up with a posterior sample of the target distribution. However, successive draws of θ often suffer from autocorrelations. A practical procedure to handle the within-sequence correlation is to thin the sequence by keeping every kth simulation draw and discarding the rest after approximate convergence of the chain is reached, where k is a large number (e.g., on the scale of hundreds or thousands). For the ease of notation, after thinning we denote the posterior sample as $(\theta^{(t)}, t = 1, \ldots T)$. Therefore, the mean and variance of the posterior distribution of scalar θ can be estimated as the sample mean and variance of the T draws. In addition, the 95% posterior credible interval for θ can be estimated as the 2.5th and 97.5th percentiles of the empirical distribution of $\theta^{(t)}$'s.

There are many established MCMC posterior sampling algorithms. One of the commonly used algorithms in applied Bayesian statistics is the Gibbs sampler (Gelfand et al. 1990). Suppose $\theta = (\theta_1, \ldots, \theta_p)$, a vector of p components. The Gibbs sampler is rather effective if directly drawing θ from $f(\theta|Y)$ is hard to implement, but draws from each univariate conditional distribution, $f(\theta_j|\theta_1, \ldots, \theta_{j-1}, \theta_{j+1}, \ldots, \theta_j, Y)$, $j = 1, \ldots, p$, are relatively easy to compute. First, initial values for these parameters $\theta_1^{(0)}$, ..., $\theta_p^{(0)}$ are chosen following some practical guidelines. Then given values $\theta_1^{(t)}$, ..., $\theta_p^{(t)}$ at iteration t, new draws of θ can be obtained by drawing from the following sequence of p conditional distributions:

$$\begin{aligned}
\theta_1^{(t+1)} &\sim f(\theta_1|\theta_2^{(t)}, \theta_3^{(t)}, \ldots, \theta_p^{(t)}, Y), \\
\theta_2^{(t+1)} &\sim f(\theta_2|\theta_1^{(t+1)}, \theta_3^{(t)}, \ldots, \theta_p^{(t)}, Y), \\
&\vdots \\
\theta_j^{(t+1)} &\sim f(\theta_j|\theta_1^{(t+1)}, \theta_2^{(t+1)}, \ldots, \theta_{j-1}^{(t+1)}, \theta_{j+1}^{(t)}, \ldots, \theta_p^{(t)}, Y), \\
&\vdots \\
\theta_p^{(t+1)} &\sim f(\theta_p|\theta_1^{(t+1)}, \theta_2^{(t+1)}, \ldots, \theta_{p-1}^{(t+1)}, Y). \qquad (2.14)
\end{aligned}$$

It can be shown that, under quite general conditions, the sequence of iterates $\theta^{(t)} = (\theta_1^{(t)}, \ldots, \theta_p^{(t)})$ converges to a draw from the joint distribution of $f(\theta|Y)$.

Note that in the Gibbs sampler, an individual component, θ_j, can also be a multi-parameter vector, not necessarily a scalar quantity. In Eqs. (2.14), if one of the univariate conditional distributions is not simple to draw, then we can use other Bayesian sampling algorithms such as the Metropolis-Hastings algorithm (e.g., Hastings 1970; Chib and Greenberg 1995).

From a practical point of view, nowadays applied statisticians often do not have to code these Bayesian sampling steps from scratch, which can be rather sophisticated when the model is complicated. Instead they can rely on well developed Bayesian software packages such as WinBUGS (Lunn et al. 2000), which provide user-friendly coding environments to execute complicated MCMC sampling steps under many established Bayesian analysis models. In this book we will show that multiple imputation can be conducted using some of these packages.

Finally, the general idea behind the Gibbs sampler, that is, to divide a p-dimensional problem into p one-dimensional problems and approach them sequentially, can also be applied in the multiple imputation for multivariate missing data. See Chapter 7 for more details.

2.3.4 Asymptotic Equivalence between Frequentist and Bayesian Estimates

Aside from the philosophical distinctions between the frequentist and Bayesian analyses, their connections are also obvious as both rely on the likelihood function of the observed data. In general, it is well accepted that the likelihood function is the best "summary" of information in the data about the parameters (Section 2.4). As a matter of fact, with a large sample size, there is little difference between the two types of estimates when we use diffuse prior distributions for the Bayesian analysis. More specifically, let $\hat{\theta}$ denote a maximum likelihood estimate of θ based on data Y, or the posterior mode in a corresponding Bayesian analysis, and suppose that the model is correctly specified.

The most important practical property of $\hat{\theta}$ is that, in many cases, especially with large samples and under some regularity conditions as listed in Gelman et al. (2013, Chapter 4), the following approximation can be applied:

$$(\theta - \hat{\theta}) \text{ or } (\hat{\theta} - \theta) \sim N(0, C), \tag{2.15}$$

where C is the covariance matrix for $(\theta - \hat{\theta})$ (or $(\hat{\theta} - \theta)$).

Eq. (2.15) has both a frequentist and Bayesian interpretation. The Bayesian interpretation treats θ as a random variable and $\hat{\theta}$ as the mode of the posterior distribution, fixed by the observed data. Likewise, $C = I^{-1}(\hat{\theta}|Y)$ is the inverse of the observed information evaluated at $\hat{\theta}$, also statistics fixed at their observed values. The interpretation is that, conditional on $f(\cdot|\cdot)$ and data, the posterior distribution of θ is normal with mean $\hat{\theta}$ and covariance matrix C, where $\hat{\theta}$ and C are both statistics fixed at their observed values.

On the other hand, the frequentist interpretation of Eq. (2.15) is that, under $f(\cdot|\cdot)$ in repeated samples with fixed θ, $\hat{\theta}$ will be approximately normally distributed with mean equal to the true value of θ and covariance matrix C, which has lower order variability than $\hat{\theta}$. Here the lower order variability can be understood as that the covariance matrix estimate is assumed to be roughly equal to the true value.

Thus, the asymptotic equivalence of the two types of estimates implies that the Bayesian approach generally yields methods that have desirable frequentist properties in a broad definition of "good" procedures, as argued by Rubin (1984). This fact provides a key justification for multiple imputation analysis (Chapter 3). Additional discussions can be found in Little and Rubin (2020, Section 6.1.3).

Example 2.2. *Bayesian estimation of the proportion of health insurance coverage from a web survey*

Continuing with Example 2.1, we first estimate the posterior mean and variance using Eq. (2.13). By plugging in $Y = 2152$ and $n = 2304$, we obtain the posterior mean: $\tilde{\theta} = \frac{2152+1}{2304+2} = 0.934$, and posterior variance: $\hat{Var}(\tilde{\theta}) = 0.934 \times (1 - 0.934)/(2304 + 2) = 2.686 \times 10^{-5}$. We also simulate 10000 samples from the posterior distribution of θ (a Beta distribution with parameters $a = 2153$ and $b = 153$). Based on these posterior draws, the posterior mean and variance estimates are 0.934 and 2.749×10^{-5}, and the 95% posterior credible interval for θ is (0.923, 0.944).

For Bayesian estimation, numerical estimates using actual posterior samples are very close to those calculated using formulas. In fact, the former can be made as close as possible to the latter by taking sufficiently large posterior samples. In addition, these Bayesian estimates are nearly identical to their frequentist counterparts in Example 2.1, illustrating the asymptotic equivalence between the two types of estimates.

Fig. 2.1 shows the histogram and QQ plots of the 10000 posterior samples of $f(\theta|Y)$. It also includes corresponding plots for 10000 bootstrap samples of

$\hat{\theta}$ from Example 2.1. The former is very close to a normal distribution, and the latter seems to be slightly less smooth. However, the two distributions are similar. This is not a coincidence. In many cases the bootstrap distribution represents an (approximate) nonparametric, noninformative posterior distribution of model parameters (Hastie et al. 2008, Section 8.4). By perturbing the data, the bootstrap procedure approximates the Bayesian effect of perturbing the parameters. Yet the bootstrap resampling can be sometimes simpler to carry out if the posterior distribution of the parameter is rather complicated. This nice property of the bootstrap resampling can be useful for devising less complex multiple imputation algorithms (Section 3.4.3).

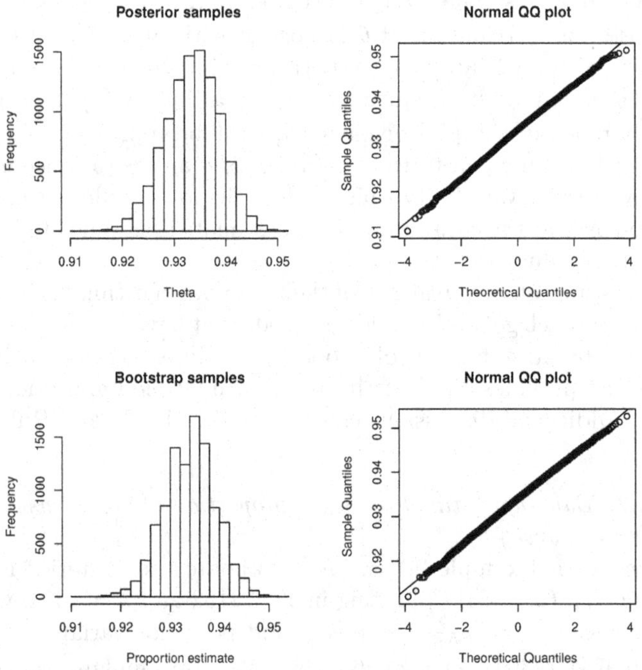

FIGURE 2.1
Example 2.2. Top left: the histogram of 10000 posterior samples of θ; top right: the QQ plot of the posterior samples; bottom left: the histogram of 10000 bootstrap samples of $\hat{\theta}$; bottom right: the QQ plot of the bootstrap samples.

2.4 Likelihood-based Approaches to Missing Data Analysis

In general, for complete data Y and parameter θ, we can define a likelihood function $L(\theta|Y)$, which is any function of θ in the parameter space that is proportional to $f(Y|\theta)$, the distribution function under a posited statistical model. Note that $L(\theta|Y)$ is a function of the parameter θ for fixed Y, whereas $f(Y|\theta)$ is a function of Y for fixed θ. Little and Rubin (2020, Chapter 6) summarized the likelihood-based strategy as a principled estimation approach to missing data analysis. From the frequentist perspective, the estimation can be conducted using the maximum likelihood method, which identifies the $\hat{\theta}$ that maximizes $L(\theta|Y)$. From the Bayesian perspective, the estimation can be conducted via the posterior distribution, $f(\theta|Y) \propto L(\theta|Y)\pi(\theta)$, after incorporating the information from a prior distribution of θ.

In Chapter 1 we introduced different missingness mechanisms including MCAR, MAR, and MNAR. Here we examine their implications to the likelihood-based approach based on some heuristic arguments. As before, let $Y = (Y_{obs}, Y_{mis})$, where Y_{obs} and Y_{mis} are the observed and missing components of Y, respectively. Let R be the response indicator. Let θ be the parameter of the complete-data model, $f(Y|\theta)$, which characterizes the distribution of the data in the absence of missing values. Let ϕ be the parameter governing the nonresponse model, $f(R|Y,\phi)$. Since the observed information from the data includes both Y_{obs} and R, the full likelihood function of observed data can be expressed as

$$\begin{aligned} L(\theta,\phi|Y_{obs},R) &\propto f(Y_{obs},R|\theta,\phi) \\ &= \int f(Y,R|\theta,\phi)dY_{mis} \\ &= \int f(Y|\theta)f(R|Y,\phi)dY_{mis} \\ &= \int f(Y_{obs},Y_{mis}|\theta)f(R|Y_{obs},Y_{mis},\phi)dY_{mis}, \end{aligned} \tag{2.16}$$

which essentially integrates out Y_{mis} from the joint distribution of Y and R, $f(Y,R|\theta,\phi)$.

For most missing data analyses, the inferential interest centers on θ. From Eq. (2.16), if $f(R|Y_{obs},Y_{mis},\phi) = f(R|Y_{obs},\phi)$ for all Y_{mis} (i.e., MAR), and the parameters θ and ϕ are distinct, then it can be shown that

$$\begin{aligned} f(Y_{obs},R|\theta,\phi) &= f(R|Y_{obs},\phi)\int f(Y_{obs},Y_{mis}|\theta)dY_{mis} \\ &= f(R|Y_{obs},\phi)f(Y_{obs}|\theta). \end{aligned} \tag{2.17}$$

A missingness mechanism is called ignorable if it is MAR and the parameters

θ and ϕ are distinct. Eq. (2.17) shows that under an ignorable missingness mechanism, we only need to use $f(Y_{obs}|\theta)$ instead of $f(Y_{obs}, R|\theta, \phi)$ for drawing inference about θ. That is, it is sufficient to work on the observed-data likelihood function, $L(\theta|Y_{obs}) \propto f(Y_{obs}|\theta)$, ignoring the missingness mechanism.

Similarly, the idea of ignorability applies to the Bayesian inference in the presence of missing data. The joint posterior distribution of θ and ϕ given Y_{obs} and R is

$$f(\theta, \phi|Y_{obs}, R) \propto \pi(\theta, \phi)L(\theta, \phi|Y_{obs}, R), \tag{2.18}$$

where $\pi(\theta, \phi)$ denotes the joint prior distribution for (θ, ϕ). Under MAR and also assuming that θ and ϕ are distinct as well as *a priori* independent, that is, $\pi(\theta, \phi) = \pi(\theta)\pi(\phi)$, then it can be shown that

$$\begin{aligned} f(\theta, \phi|Y_{obs}, R) &\propto [\pi(\theta)L(\theta|Y_{obs})][\pi(\phi)L(\phi|Y_{obs}, R)] \\ &\propto f(\theta|Y_{obs})f(\phi|Y_{obs}, R), \end{aligned} \tag{2.19}$$

that is, θ and ϕ are *a posteriori* independent. Bayesian inference about θ can then be based on the posterior distribution, $f(\theta|Y_{obs})$, ignoring the missing data mechanism.

In brief summary, the ignorable missingness assumption allows us to work on the complete-data model, $f(Y|\theta)$ (note that $f(Y_{obs}|\theta) = \int f(Y|\theta)dY_{mis}$), and ignore the nonresponse model to draw inference about θ. The most important condition that leads to the ignorability is MAR, as other conditions (e.g., θ and ϕ are distinct in the likelihood framework) often hold in general. Therefore, we often do not distinguish between MAR and ignorable missingness. On the other hand, when MAR does not hold, we cannot rely only on the complete-data model and have to consider the nonresponse model, $f(R|Y_{obs}, Y_{mis}, \phi)$, to infer about θ. However, handling missing data under nonignorability assumptions is much more difficult because Y_{mis} are not known. See Chapter 13 for more details.

From a practical perspective, the term "ignorable" means that we can ignore the process (and its related parameters) that generates missing data and only focus on modeling the distribution of the hypothetical complete data. However, being "ignorable" does not mean that we can be entirely careless about missing values, such as removing all incomplete cases before doing the analysis (i.e., complete-case analysis).

In Example 1.9 we simulated numbers from a bivariate normal model/distribution and remove some cases for one variable to illustrate different missingness mechanisms. We continue using this setup to illustrate some features of the likelihood-based approach. Before diving into details we first review some of the properties of a bivariate normal distribution.

Example 2.3. *The factorization of a bivariate normal distribution*

Suppose X and Y are from a bivariate normal distribution, that is, $(X, Y) \sim N_2(\mu_{X,Y}, \Sigma_{X,Y})$, where $\mu_{X,Y} = (\mu_X, \mu_Y)$ and $\Sigma = \begin{pmatrix} \sigma_X^2, \sigma_{XY} \\ \sigma_{XY}, \sigma_Y^2 \end{pmatrix}$.

The joint distribution of X and Y, $f(X, Y|\mu_{X,Y}, \Sigma_{X,Y})$, can be factorized as

$$f(X, Y|\mu_{X,Y}, \Sigma_{X,Y}) = f_X(X|\mu_X, \sigma_X^2) f_{Y|X}(Y|X, \beta_0, \beta_1, \sigma_{Y|X}^2),$$

where $f_X(X|\mu_X, \sigma_X^2)$ is the marginal distribution of X, which is a normal distribution, and $f_{Y|X}(Y|X, \beta_0, \beta_1, \sigma_{Y|X}^2)$ is the conditional distribution of Y given X, which is also a normal distribution.

More specifically,

$$f_X(X|\mu_X, \sigma_X^2) = \frac{1}{\sqrt{2\pi}\sigma_X} e^{-\frac{(X-\mu_X)^2}{2\sigma_X^2}},$$

and

$$f_{Y|X}(Y|X, \beta_0, \beta_1, \sigma_{Y|X}^2) = \frac{1}{\sqrt{2\pi}\sigma_{Y|X}} e^{-\frac{(Y-\beta_0-\beta_1 X)^2}{2\sigma_{Y|X}^2}}.$$

In this factorization, $\beta_1 = \sigma_{XY}/\sigma_X^2$, $\beta_0 = \mu_Y - \beta_1\mu_X$, and $\sigma_{Y|X}^2 = \sigma_Y^2 - \sigma_{XY}^2/\sigma_X^2$.

The bivariate normal distribution can also be factorized the other way around. That is,

$$f(X, Y|\mu_{X,Y}, \Sigma_{X,Y}) = f_Y(Y|\mu_Y, \sigma_Y^2) f_{X|Y}(X|Y, \alpha_0, \alpha_1, \sigma_{X|Y}^2),$$

where $f_Y(Y|\mu_Y, \sigma_Y^2)$ is the marginal normal distribution of Y, and $f_{X|Y}(X|Y, \alpha_0, \alpha_1, \sigma_{X|Y}^2)$ is the conditional normal distribution of X given Y. In this factorization, $\alpha_1 = \sigma_{XY}/\sigma_Y^2$, $\alpha_0 = \mu_X - \alpha_1\mu_Y$, and $\sigma_{X|Y}^2 = \sigma_X^2 - \sigma_{XY}^2/\sigma_Y^2$.

In both factorizations, the conditional distribution implies a normal linear regression model for one variable using the other variable as the regressor. For example, $f_{Y|X}(Y|X, \beta_0, \beta_1, \sigma_{Y|X}^2)$ implies a linear regression model for Y on X as $Y = \beta_0 + \beta_1 X + \epsilon$, where $\epsilon \sim N(0, \sigma_{Y|X}^2)$.

Example 2.4. *A bivariate normal sample with one variable subject to missing data*

We consider a sample of n independent cases for (x_i, y_i), where $i = 1, \ldots n$. Suppose that x_i's are fully observed and missing values occur for some y_i's. For simplicity, assume that the first n_1 cases have observed y-values and the remaining cases have missing y-values. That is, $r_i = 1$ for $i = 1, \ldots, n_1$ and $r_i = 0$ for $i = n_1 + 1, \ldots, n$. Suppose that the nonresponse model can be characterized as $logit(f(r_i = 1)) = \eta_0 + \eta_1 x_i + \eta_2 y_i$.

Take the factorization,

$$f(X, Y|\mu_{X,Y}, \Sigma_{X,Y}) = f_X(X|\mu_X, \sigma_X^2) f_{Y|X}(Y|X, \beta_0, \beta_1, \sigma_{Y|X}^2),$$

where X denotes the collection of x_i's and Y denotes the collection of y_i's. Let $\theta = (\mu_X, \sigma_X^2, \beta_0, \beta_1, \sigma_{Y|X}^2)$ and $\phi = (\eta_0, \eta_1, \eta_2)$. As shown in Example 2.3, the joint distribution of the complete-data can be expressed as

$f(X,Y|\theta)=\prod_{i=1}^{n}\frac{1}{\sqrt{2\pi}\sigma_X}e^{-\frac{(x_i-\mu_X)^2}{2\sigma_X^2}}\frac{1}{\sqrt{2\pi}\sigma_{Y|X}}e^{-\frac{(y_i-\beta_0-\beta_1x_i)^2}{2\sigma_{Y|X}^2}}$. The distribution for the response indicators is $f(R|X,Y,\phi)=\prod_{i=1}^{n}p_i^{r_i}(1-p_i)^{1-r_i}$, where R denotes the collection of r_i's and $p_i=f(r_i=1)=\frac{exp(\eta_0+\eta_1x_i+\eta_2y_i)}{1+exp(\eta_0+\eta_1x_i+\eta_2y_i)}$.

The full likelihood function therefore is

$$
\begin{aligned}
L(\theta,\phi|Y_{obs},X,R) &\propto \int f(X,Y|\theta)f(R|X,Y,\phi)dY_{mis} \\
&= \int\prod_{i=1}^{n}\{\frac{1}{\sqrt{2\pi}\sigma_X}e^{-\frac{(x_i-\mu_X)^2}{2\sigma_X^2}}\frac{1}{\sqrt{2\pi}\sigma_{Y|X}}e^{-\frac{(y_i-\beta_0-\beta_1x_i)^2}{2\sigma_{Y|X}^2}} \qquad (2.20) \\
&(\frac{exp(\eta_0+\eta_1x_i+\eta_2y_i)}{1+exp(\eta_0+\eta_1x_i+\eta_2y_i)})^{r_i}(\frac{1}{1+exp(\eta_0+\eta_1x_i+\eta_2y_i)})^{1-r_i}\}dy_{n_1+1}\ldots dy_n.
\end{aligned}
$$

When $\eta_2=0$, that is, MAR for missing y's, Eq. (2.20) can be simplified as

$$L(\theta,\phi|Y_{obs},X,R)=f_1(Y_{obs}|X,\beta_0,\beta_1,\sigma_{Y|X}^2)f_2(X;\mu_X,\sigma_X^2,\eta_0,\eta_1), \qquad (2.21)$$

where $f_1(Y_{obs}|X,\beta_0,\beta_1,\sigma_{Y|X}^2)=\prod_{i=1}^{n_1}\frac{1}{\sqrt{2\pi}\sigma_{Y|X}}e^{-\frac{(y_i-\beta_0-\beta_1x_i)^2}{2\sigma_{Y|X}^2}}$ indicating a conditional normal distribution, and $f_2(X;\mu_X,\sigma_X^2,\eta_0,\eta_1)$ is a function of X that does not involve parameters β_0, β_1, and $\sigma_{Y|X}^2$.

Suppose the inferential interest is on β_0 and β_1, the regression relationship between Y on X, Eq. (2.21) suggests that it is sufficient to make the inferences based on the distribution function $f_1(Y_{obs}|X,\beta_0,\beta_1,\sigma_{Y|X}^2)$, ignoring $f_2(X;\mu_X,\sigma_X^2,\eta_0,\eta_1)$. That is, we can run a linear regression of Y on X using only the first n_1 (complete) cases to obtain valid estimates of the intercept and slope. Intuitively it makes sense because the statistical information on the regression relation between Y and X is only contained in the n_1 cases with both variables observed. Since Y is the response variable of the regression, the cases with missing Y-values do not contribute any extra information for inferring about this regression and can therefore be discarded. This is demonstrated by the numerical results in Example 1.9, where the intercept and slope estimates under either MCAR and MAR based on observed cases are little different from those before deletion. A more systematic simulation study will be presented in Example 2.5.

However, if $\eta_2\neq 0$, that is, MNAR (or nonignorability) for missing y's, Eq. (2.20) in general cannot be simplified as something similar to Eq. (2.21). Special techniques are needed to estimate the parameters (Chapter 13).

Practitioners might be interested in multiple estimands from the same data. In Example 2.4, although under MAR, valid estimates for regressing Y on X can be obtained using complete cases, it is not straightforward to infer about other estimands of interest, such as the mean and other distribution functions (e.g., percentiles) of Y, as well as estimates for regressing X on Y, based on the likelihood function in Eq. (2.21). Some special techniques can be applied (e.g., transformation of parameters under multivariate normal models by Little and Rubin (2020, Chapter 7)), yet they might not be easily accessible to practitioners. This would become more of an issue when both missing data patterns and underlying models get more complex.

In general, although the likelihood-based estimation is a principled approach to missing data problems, it arguably lacks the flexibility that allows practitioners apply different analyses to the same data. In many cases the practical implementation can also be challenging because the estimation often involves nontrivial likelihood functions and algorithms (e.g., the EM algorithm by Dempster et al. 1977) tailored to specific problems. On the other hand, as the focus of this book, multiply imputing missing values and making the dataset completed is an appealing option, allowing different complete-data analyses to be applied.

2.5 Ad Hoc Missing Data Methods

Missing data analysis is not simple. Our brief discussion on the likelihood-based approach shows that appropriate assumptions about the missingness mechanism and complete-data model are necessary to make valid inferences. In practice, however, many simple and convenient methods (i.e., ad hoc methods) have been frequently used to handle missing data. Some of them are listed below:

1. Complete-case analysis by removing units with any missing data, which is the default approach to missing data in most of statistical software packages.

2. Simple single imputation methods (i.e., impute the missing data once), which can include:

   2.1. Replacing missing values using means, medians, or some specific observed values such as the last observation carry forward imputation in longitudinal data settings.

   2.2. Replacing missing values by predictions from some regression models.

3. Missingness indicator method that codes the missing value as a distinct category when the missing variable is used as a covariate in a regression analysis.

There exist some situations where certain ad hoc methods might yield valid or acceptable results. In Example 2.4, estimates for regressing Y on X are valid using only complete cases under the MAR assumption. More generally, Carpenter and Kenward (2013, Table 1.9) documented the pattern of biases of complete-case analysis in both linear and logistic regressions with two predictors. In some scenarios, depending on the underlying missingness mechanism, coefficient estimates of the predictors can be unbiased. In addition, ad hoc methods might be used to explore the data to gain some general insights about the direction and magnitude of the estimates. For example, estimates from ad hoc methods can be used as initial values for some estimation algorithms based on more principled methods.

However, numerous literature (e.g., Little and Rubin 2020, Chapter 3; Van Buuren 2018, Section 1.3) have documented the suboptimality of ad hoc methods compared with more principled methods such as multiple imputation. It is recommended to avoid using ad hoc missing data methods in general, especially for the main statistical analyses that need to meet high quality standards.

2.6 Monte Carlo Simulation Study

In Section 2.2, we mentioned a few desirable properties of a statistical estimator under the frequentist framework: both the point and variance estimate are unbiased or consistent, and the 95% confidence interval has a nominal coverage under repeated sampling. In certain cases, such properties might be demonstrated by mathematical derivations. However in many other scenarios, especially if estimators are complicated, a more feasible approach to evaluating their performances is through Monte Carlo simulation studies, based on which we can numerically generate sampling distributions of the estimator of interest.

Designing and conducting high-quality simulation studies for evaluation purposes is a rather involved topic. In Example 1.9 we conduct a very simple simulation study. Here we provide some general ideas in the context of missing data analysis. The basic to-do steps can include:

1. Define the estimands of interest, which are usually the population quantities estimated from statistical analyses (e.g., means and regression coefficients).

2. Generate a complete dataset under a complete-data model (or data-generating model) for the target population.

3. Posit a plausible missingness mechanism and assign missing values to the complete data generated in Step 2.

4. Apply the missing data method to the incomplete dataset generated in Step 3 to obtain both the point and variance estimate, as well as the 95% confidence interval. In addition, the before-deletion method, which

analyzes the complete data before missing values are generated, is also applied.

5. Repeat Steps 2 to 4 independently for a large number of replicates/simulations.

6. Evaluate the performance of the missing data method by calculating key accuracy and precision quantities for the collection of estimates in Step 5. Estimates from the before-deletion method can be used as the benchmark in the evaluation.

Suppose θ is the true value of the estimand. In a simulation study, θ is known and preset. Let B be the number of replicates conducted in the simulation study. For a missing data method, let $\hat{\theta}_i$ be the estimate of θ and $SE(\hat{\theta}_i)$ be the standard error estimate within each of the $i = 1, \ldots, B$ replicates. The mean estimate is defined as $\overline{\hat{\theta}} = \frac{\sum_{i=1}^{B} \hat{\theta}_i}{B}$, the empirical standard deviation is defined as $SD(\hat{\theta}) = \sqrt{\frac{\sum_{i=1}^{B}(\hat{\theta}_i - \overline{\hat{\theta}})^2}{B-1}}$, and the average standard error is defined as $\overline{SE}(\hat{\theta}) = \frac{\sum_{i=1}^{B} SE(\hat{\theta}_i)}{B}$.

There exist multiple evaluation criteria for assessing the performance of the method. In this book, we focus on the several of them as follows:

1. Bias= $\overline{\hat{\theta}} - \theta$; the bias is the deviation in an estimate from the true quantity.

2. Mean-squared-error (MSE)= $\frac{\sum_{i=1}^{B}(\hat{\theta}_i - \theta)^2}{B} = \text{Bias}^2 + SD(\hat{\theta})^2$. MSE provides a global measure of the estimation error, as it consists of measures of both bias and variability.

3. Coverage rate (COV): proportion of times the $100(1-\alpha)\%$ confidence interval of $\hat{\theta}_i$ (e.g., $\hat{\theta}_i + / - Z_{1-\alpha/2} SE(\hat{\theta}_i)$ for large samples) includes θ for $i = 1, \ldots B$. Typically α is set as 0.05.

4. Other evaluation criteria derived from the above three measures, such as the relative bias (RBIAS)=Bias/$\theta \times 100\%$, standardized bias= Bias/$SD(\hat{\theta})$, root-mean-squared-error= $\sqrt{\text{MSE}}$, and average length of the confidence interval (CI) $= 2Z_{1-\alpha/2}\overline{SE}(\hat{\theta})$.

Ideally, a good missing data method is expected to have little bias for $\overline{\hat{\theta}}$, little bias for the variance estimate (i.e., $\overline{SE}(\hat{\theta}) \approx SD(\hat{\theta})$), a smaller mean-squared-error and a narrower confidence interval compared with alternative methods, and a coverage rate close to the nominal level. Among different criteria, it is generally believed that the issue of bias is more crucial than that of the variance estimate. In fact, it can be argued that providing a valid variance estimate is worse than providing no estimate at all if the point estimator has a large bias that dominates the mean-squared-error. (Little and Rubin 2020, Chapter 5). In addition, methods having a coverage rate higher than the nominal level are deemed conservative. On the other hand, methods having a lower-than-nominal coverage rate are often more problematic because they can lead to an inflated type I error. A low coverage rate is often due to a sizable bias and/or underestimation of the variance. Choices of criteria in simulation studies are not fixed, as they depend on specific goals in assessment.

There exist multiple subtle issues in designing a good simulation study such as choosing the appropriate sample size and number of replicates, as well as more general topics including using, interpreting, and presenting simulation results. Detailed discussion can be found, for example, in Burton et al. (2006) and Morris et al. (2019).

Example 2.5. *A simulation study for illustrating the performance of some ad hoc missing data methods*

The suboptimal performance of some ad hoc missing data methods can be demonstrated using a simulation study. We use the setup in Example 1.9 and consider three estimands: the mean of Y, the slope coefficient from regressing Y on X, and the slope coefficient from regressing X on Y. Since X and Y are from a bivariate normal model, both regression models are normal linear regressions.

The complete-data model is $X \sim N(1,1)$, and $Y = -2 + X + \epsilon$, where $\epsilon \sim N(0,1)$. We let X be fully observed and generate missing cases of Y assuming (I) MCAR : randomly set Y as missing with a 40% chance; and (II) MAR : $logit(f(R = 0|X,Y)) = -1.4 + X$. The complete-data sample size is set as $n = 1000$, and the simulation includes 1000 replicates.

For estimating the mean of Y and slope coefficient of Y on X, we apply two missing data methods: complete-case analysis (CC) and a regression prediction method (RP). In RP, we first run a linear regression for Y on X based on complete cases and then use the predicted Y-value for the missing cases as the imputation. That is, let $\hat{\beta}_0$ and $\hat{\beta}_1$ be the intercept and slope estimate of the regression based on complete cases, the imputed value is $Y_i^* = \hat{\beta}_0 + \hat{\beta}_1 X_i$, where i indexes cases with missing Y-values. For estimating the slope coefficient of X on Y, we apply a missingness indicator method (INDI) in addition to CC and RP. In INDI, the response indicator, R, is coded as a distinct category; we fit a linear regression model $X = \gamma_0 + \gamma_1 R + \gamma_2 R \times Y + \epsilon$, where $R \times Y = Y$ if $R = 1$ and $R \times Y = 0$ if $R = 0$, and use $\hat{\gamma}_2$ as the estimate. Analysis results from before-deletion (BD) are used as a yardstick in the assessment.

Table 2.1 summarizes the simulation results under MCAR (Scenario I). For all estimands, CC yields little bias and a coverage rate close to the nominal level. For the mean of Y, RP has little bias. However, its coverage rate (87.6%) is considerably lower than the nominal level. Its variance is also underestimated, which can be seen from the fact that $\overline{\text{SE}}$ is lower than SD ($\overline{\text{SE}}$=0.040 < SD=0.051). Its confidence interval is even shorter than that of BD (0.157 vs. 0.175). The pattern is similar for the slope coefficient of regressing Y on X. For the slope coefficient of regressing X on Y, RP has a large bias (0.126) and performs badly. INDI has little bias for estimating the slope coefficient of regressing X on Y. Its coverage rate is slightly higher than the nominal level, and its variance is overestimated ($\overline{\text{SE}}$=0.024 > SD=0.020).

TABLE 2.1
Example 2.5. Simulation results under MCAR

Method	Bias	SD	$\overline{\text{SE}}$	MSE	CI	COV (%)
			Mean of Y			
BD	0	0.043	0.045	0.00188	0.175	96.1
CC	−0.001	0.057	0.058	0.00322	0.226	94.7
RP	−0.001	0.051	0.040	0.00259	0.157	87.6
			Slope coefficient of regressing Y on X			
BD	0	0.032	0.032	0.000995	0.124	95.0
CC	−0.001	0.039	0.041	0.00152	0.160	96.0
RP	−0.001	0.039	0.024	0.00152	0.096	78.4
			Slope coefficient of regressing X on Y			
BD	0	0.016	0.016	0.000251	0.062	95.0
CC	0	0.020	0.020	0.000406	0.080	95.2
INDI	0	0.020	0.024	0.000406	0.095	98.1
RP	0.126	0.019	0.015	0.0162	0.060	0

Note: SD: empirical standard deviation; $\overline{\text{SE}}$: the average of standard errors; MSE: mean-squared-error; CI: the average length of the 95% confidence intervals; COV: the coverage rate of the 95% confidence intervals.

Table 2.2 summarizes the simulation results under MAR (Scenario II). For the mean of Y, CC is badly biased (Bias= −0.350). For the slope coefficient of regressing Y on X, CC yields little bias and performs well, and this is consistent with the theoretical argument in Example 2.3. Interestingly, for the slope coefficient of regressing X on Y, CC has a sizable bias (Bias= −0.045) which results in a bad coverage rate (41.5%). The performance of RP for all estimands is similar to that under MCAR. For the slope coefficient of regressing X on Y, INDI has an identical bias to that of CC, and its variance is again overestimated. Some theoretical explanations for the behavior of these methods will be given in Example 2.6. In summary, none of the ad hoc missing data methods performs well for all estimands in both scenarios.

The problem of RP used in Example 2.5 can be further demonstrated by plotting the observed, missing, and predicted values. The top row of Fig. 2.2 shows histograms of deleted values of Y and the corresponding predicted values from a randomly chosen subset (sample size 250) of one replicate under MAR. Compared with the missing values, the predicted values preserve their mean well, yet the latter have a more concentrated distribution. The bottom row shows the scatter plots between X and Y. The predicted values (filled triangles) all fall on the fitted regression line of Y on X, preserving the mean of the regression relation. However they miss the noise part of the conditional distribution of Y given X, when contrasted with the removed values (filled circles). Therefore, regression predictions of missing values might preserve the mean structure of the data but cannot fully recover the original joint distribution of X and Y. In addition, analyzing predicted values as if they had not been missing also ignores the uncertainty from the missing values. This is the problem for all single imputation methods. Both facts can result in an underestimated variance and a low coverage rate. That is, the estimates can have inflated precision and type-I error.

Example 2.6. *The missing covariate problem in a bivariate normal sample*
In Example 2.5, why cannot we regress X on Y just using complete cases? We still use the setup of Example 2.5, assuming that X is fully observed and the 1st n_1 cases have Y-values observed. The analysis interest is to regress X on Y so here Y is a missing covariate in the regression where X is the response variable. The observed-data likelihood function can be written as

$$L(\theta|Y_{obs}, X) = L_1 L_2. \tag{2.22}$$

In Eq. (2.22), $L_1 = \prod_{i=1}^{n_1} \frac{1}{\sqrt{2\pi}\sigma_Y} e^{-\frac{(y_i-\mu_Y)^2}{2\sigma_Y^2}} \frac{1}{\sqrt{2\pi}\sigma_{X|Y}} e^{-\frac{(x_i-\alpha_0-\alpha_1 y_i)^2}{2\sigma_{X|Y}^2}}$, where μ_Y and σ_Y^2 are the marginal mean and variance of Y, respectively, α_0 and α_1 are the regression coefficients of X on Y, and $\sigma_{X|Y}^2$ is the residual variance of the regression; $L_2 = \prod_{i=n_1+1}^{n} \frac{1}{\sqrt{2\pi}\sigma_X} e^{-\frac{(X_i-\mu_X)^2}{2\sigma_X^2}}$, where $\mu_X = \alpha_0+\alpha_1\mu_Y$ and $\sigma_X^2 = \alpha_1^2\sigma_Y^2 + \sigma_{X|Y}^2$ are the marginal mean and variance of X, respectively. Here the

TABLE 2.2
Example 2.5. Simulation results under MAR

Method	Bias	SD	$\overline{\text{SE}}$	MSE	CI	COV (%)
			Mean of Y			
BD	0	0.043	0.045	0.00188	0.175	96.1
CC	−0.350	0.056	0.056	0.1255	0.220	0
RP	−0.001	0.056	0.040	0.00308	0.156	83.0
			Slope coefficient of regressing Y on X			
BD	0	0.032	0.032	0.000995	0.124	95.0
CC	−0.001	0.045	0.045	0.00204	0.178	95.7
RP	−0.001	0.045	0.024	0.00204	0.095	70.4
			Slope coefficient of regressing X on Y			
BD	0	0.016	0.016	0.000251	0.062	950
CC	−0.045	0.021	0.021	0.00246	0.081	41.5
INDI	−0.045	0.021	0.024	0.00246	0.094	53.1
RP	0.132	0.020	0.015	0.0179	0.060	0

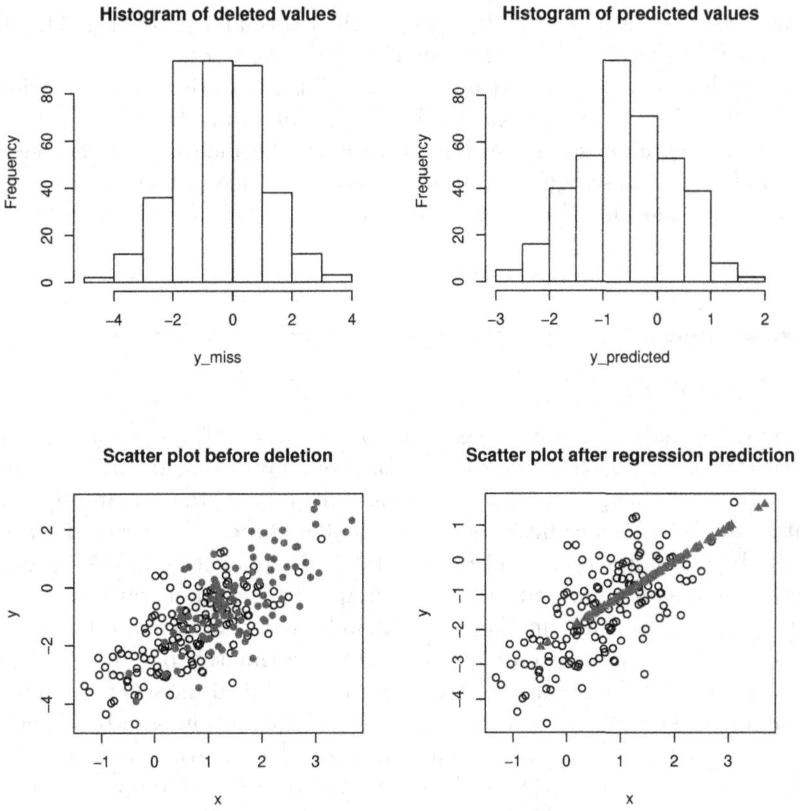

FIGURE 2.2
Example 2.5. Plots of X and Y from a random subset of one simulation under MAR. Top left: histogram of deleted values; top right: histogram of regression predictions; bottom left: scatter plot before deletion; bottom right: scatter plot after regression prediction. Circle: observed values; filled circle: deleted values; filled triangle: imputed values by regression prediction.

cases with missing Y-values ($i = n_1+1, \ldots, n$) still contain information about parameters α_0 and α_1, based on the likelihood function L_2. Therefore, unlike the problem of regressing incomplete Y on X, regressing X on incomplete Y using the first n_1 cases (i.e., by only using L_1) does not necessarily lead to valid inferences in general: under MCAR the estimate is unbiased yet less efficient; under MAR the estimate is biased.

In addition, the missingness indicator method in Example 2.5 is essentially fitting the data using two models. For the cases with $R = 1$ (i.e., the first n_1 cases), the regression is $X = \gamma_0 + \gamma_1 + \gamma_2 Y + \epsilon$; for the cases with $R = 0$

(i.e., the last $n - n_1$ cases), the regression is $X = \gamma_0 + \epsilon$, where ϵ from the two regressions is assumed to follow an identical normal distribution. The first regression is identical to the regression of X on Y using complete cases, as can be seen by letting $\alpha_0 = \gamma_0 + \gamma_1$ and $\alpha_1 = \gamma_2$. Therefore the slope coefficients from INDI and CC are identical in the simulation study from Example 2.5. The variance estimates can be different due to the inclusion of the second regression in the missingness indicator method, which constrains the error variance to be identical for two regressions.

2.7 Summary

Missing data analysis is a complex area no matter whether the statistical inference is made using the frequentist or Bayesian approach. We briefly review some statistical background that can be useful to understanding multiple imputation analysis. A nice introduction to the foundation of statistical estimation can be found in Casella and Berger (1992). Gelman et al. (2013) provided detailed discussion about various topics in applied Bayesian statistics.

In contrast to ad hoc methods, likelihood-based methods are the principled approach to missing data problems. The estimated parameters nicely summarize the available information under the assumed models for complete data and missing data mechanisms. The model assumptions can be explicitly stated and evaluated, and in many cases it is possible to estimate the standard error of the estimates. Little and Rubin (2020) provides a comprehensive review and discussion of the likelihood-based approach to missing data problems. For example, as generalizations of Examples 2.3 and 2.4, the likelihood-based estimation is provided for missing data problems under multivariate normal models (Little and Rubin 2020). As will be shown in Chapter 3, the multiple imputation approach is tightly connected with the likelihood-based approach yet can provide more flexibility for practical analysis.

Finally, it is helpful to distinguish among frequently used terms including "complete cases", "observed data", "complete data", and "completed data". For a rectangular dataset with missing values, observed data contain all the observed data elements, including those from units for which not all the variables are observed. And observed data more generally also include the response indicators for each data element (observed or missing). Complete cases are part of the observed data, only including units for which every variable is observed. Complete data usually imply data elements before missing values occur and are mostly hypothetical in the context of missing data analysis. In simulation studies complete data are known. Completed data usually refer to the datasets with missing values imputed.

3

Multiple Imputation Analysis: Basics

3.1 Introduction

Imputation is an appealing approach to missing data problems because it fills in missing values and thus allows regular complete-data procedures to be applied fairly easily. However, imputation has to be conducted in a correct manner. The single imputation method based on regression prediction (Example 2.5) fails to fully capture the distributional relation between the missing and observed variables. In addition, using direct variance estimates from the completed data created by single imputation methods can underestimate the true variability of observed values because it also fails to account for the uncertainty due to missing values.

Multiple imputation (Rubin 1987) is a principled imputation and missing data analysis strategy. Briefly speaking, it consists of three stages. In the imputation stage, missing values are replaced by draws from their posterior predictive distribution, $f(Y_{mis}|Y_{obs})$, or its approximations. This process is independently repeated multiple (say M) times, resulting in M completed datasets. In the analysis stage, each completed dataset is analyzed by conventional complete-data statistical procedures, resulting in M sets of results including point estimates, standard errors, and test statistics. Finally in the combining stage, these M sets of results are then pooled together to form a single set of inference using the so-called "Rubin's combining rules". Fig. 3.1 shows a schematic plot of a multiple imputation process with $M = 5$ imputations.

The idea of multiple imputation was originally developed as a solution to survey nonresponse problems faced by organizations which produce public-use data (Rubin 1978). The NHIS income multiple imputation project (Examples 1.1 and 1.2) is one of the successful applications following the original idea. Over time, multiple imputation has gained steady attention from other scientific fields and has been applied to many different types of missing data problems (e.g., clinical trials, observational studies, administrative data).

Chapter 3 presents some fundamentals of multiple imputation. Section 3.2 justifies the idea from a Bayesian analysis perspective. Section 3.3 presents Rubin's combining rules. Section 3.4 discusses some general imputation algorithms. Section 3.5 touches on the current status of practical implementations. Section 3.6 provides a summary.

DOI: 10.1201/9780429156397-3

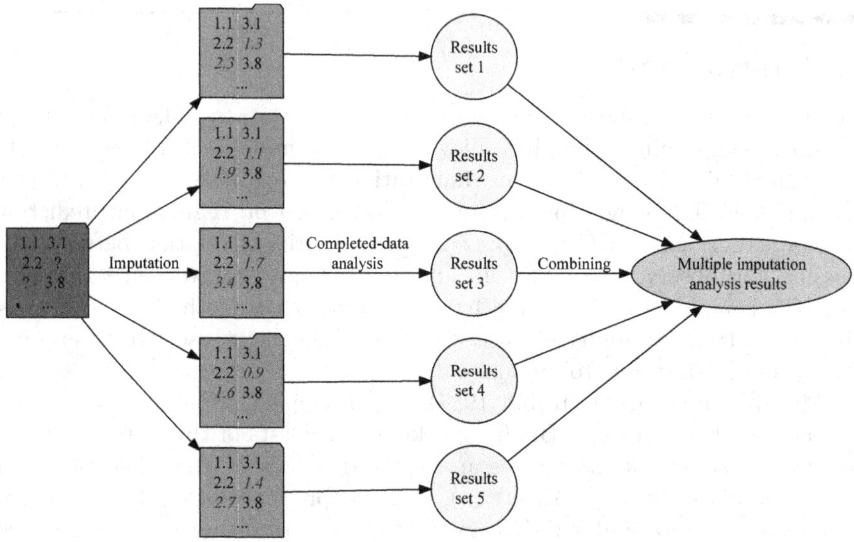

FIGURE 3.1
A schematic plot of a multiple imputation analysis with $M = 5$. "?" are missing values that are multiply imputed and analyzed. Imputed values are in italics.

3.2 Basic Idea

3.2.1 Bayesian Motivation

To lay out the basic idea, denote the target estimand of analysis as Q. Here we can also view Q as the parameter, θ, governing the complete-data model. Assuming that the missingness is ignorable, a Bayesian analysis of Q requires the calculation of the posterior distribution, $f(Q|Y_{obs})$ (Section 2.4). In many problems it is complicated and does not have closed forms. Even if it is solvable, in practice solving it may still require tailored programming to perform the necessary computations and thus can often be burdensome. However, using the idea of "integrating out missing data", we have

$$f(Q|Y_{obs}) = \int f(Q, Y_{mis}|Y_{obs})dY_{mis} \tag{3.1}$$

$$= \int f(Q|Y_{obs}, Y_{mis})f(Y_{mis}|Y_{obs})dY_{mis} \tag{3.2}$$

$$= \frac{1}{M}\sum_{m=1}^{M\to\infty} f(Q|Y_{obs}, Y_{mis}^{(m)}) \tag{3.3}$$

$$\approx \frac{1}{M}\sum_{m=1}^{M} f(Q|Y_{obs}, Y_{mis}^{(m)}), \tag{3.4}$$

where $M > 1$ and $Y_{mis}^{(m)}$ is drawn from $f(Y_{mis}|Y_{obs})$ for $m = 1, \ldots M$. In Eqs (3.1)-(3.4), $f(Q|Y_{obs}, Y_{mis})$ is the posterior distribution of Q given completed data, and $f(Y_{mis}|Y_{obs})$ is the posterior predictive distribution of missing values.

Eqs. (3.1)-(3.4) state that, in the imputation stage, multiple imputation uses an arithmetic average to approximate the integral. If we can generate or impute (a finite) M sets of values $Y_{mis}^{(m)}$ from $f(Y_{mis}|Y_{obs})$, then the posterior distribution, $f(Q|Y_{obs})$, may be approximated using information from M sets of complete-data, which are denoted by $\{Y_{obs}, Y_{mis}^{(m)}\}_{m=1}^{M}$. Interestingly, this reasoning also suggests that we cannot simply impute missing values just once (i.e., M has to be greater than one). It also implies that multiple imputations are not aimed to recover the actual missing values or to make up data. When data are indeed missing, the originally complete data cannot be reconstructed from the observed incomplete data, and the imputations are not the actual values from the nonrespondents. However, multiple imputations are aimed to create plausible versions of the dataset sampled from the population distribution, accounting for the missing data uncertainty.

How do we proceed from Eq. (3.4)? First we need to link $f(Q|Y_{obs}, Y_{mis}^{(m)})$ with estimates obtained from conventional complete-data analysis procedures. Note that practical analyses often concern the first two moments (i.e., mean and variance) of the parameter estimates. Suppose Q is a scalar quantity; let $\hat{Q}_m$ ($m = 1, \ldots M$) be the point estimate from the m-th completed dataset and U_m be the variance estimate of $\hat{Q}_m$, both of which are obtained by applying the complete-data procedure. In many cases, $f(Q|Y_{obs}, Y_{mis}^{(m)}) \approx N(\hat{Q}_m, U_m)$, a normal distribution defined by the point and variance estimates of Q from the completed data. This is justified by the asymptotic equivalence between frequentist and Bayesian estimates and their asymptotic normality (Section 2.3.4).

Next, how do we approximate $f(Q|Y_{obs})$ by $f(Q|\{\hat{Q}_m, U_m\}_{m=1}^M)$? Let $\{\hat{Q}_m, U_m\}_{m=1}^M$ denote the M sets of completed-data point estimate and its variance for Q. They can be viewed as sufficient statistics from multiply imputed datasets for drawing inference on Q in the presence of missing data. Again, a basic assumption behind the approximation is that once the sample size is sufficiently large, $f(Q|Y_{obs})$ can be well approximated by a normal distribution (Section 2.3.4), which is completely specified by its first and second moments. The solution to this approximation problem is the multiple imputation combining procedure (Rubin 1987), which will be elaborated more in the next section.

3.2.2 Basic Combining Rules and Their Justifications

We first provide the basic rules that pool results from M sets of completed datasets into a single set of inference. For multiple imputation analysis, we can use a single estimate $\overline{Q}_M$ and its variance estimate T_M, calculated as follows, to make the statistical inference:

$$\overline{Q}_M = \frac{1}{M}\sum_{m=1}^{M} \hat{Q}_m, \tag{3.5}$$

$$T_M = \overline{U}_M + (1 + \frac{1}{M})B_M, \tag{3.6}$$

where $\overline{U}_M = \frac{1}{M}\sum_{m=1}^{M} U_m$ is the within-imputation variance, $B_M = \frac{1}{M-1}\sum_{m=1}^{M}(\hat{Q}_m - \overline{Q}_M)^2$ is the between-imputation variance, and T_M is often referred to as the total variance.

We now provide some reasoning behind the combining procedure. To derive $f(Q|\{\hat{Q}_m, U_m\}_{m=1}^M)$, first let $M \to \infty$ (i.e., an infinite number of imputations) so that approximation in Eq. (3.4) approaches the equality. Then we can define $\overline{Q}_\infty$, $\overline{U}_\infty$, and B_∞ by setting them as $\overline{Q}_M$, $\overline{U}_M$, and B_M when $M \to \infty$, respectively. By the rules of iterated expectations:

$$E(Q|Y_{obs}) = E(E(Q|Y_{obs}, Y_{mis})|Y_{obs}), \tag{3.7}$$

$$Var(Q|Y_{obs}) = Var(E(Q|Y_{obs}, Y_{mis})|Y_{obs}) + E(Var(Q|Y_{obs}, Y_{mis})|Y_{obs}), \tag{3.8}$$

where $Y_{mis} \sim f(Y_{mis}|Y_{obs})$.

By the asymptotic equivalence between Bayesian and frequentist estimates (for the first two moments), Eq. (3.7) suggests that $\overline{Q}_\infty = E(Q|Y_{obs})$ and Eq. (3.8) suggests that $\overline{U}_\infty = E(Var(Q|Y_{obs}, Y_{mis})|Y_{obs})$ and $B_\infty = Var(E(Q|Y_{obs}, Y_{mis})|Y_{obs})$. Then $T_\infty = \overline{U}_\infty + B_\infty$ is the estimate of $Var(Q|Y_{obs})$, or the variance estimate for $\overline{Q}_\infty$. As we reasoned before, $f(Q|Y_{obs}) = f(Q|\{\hat{Q}_m, U_m\}_{m=1}^{M\to\infty})$, and the latter can be approximated by $f(Q|\overline{Q}_\infty, \overline{U}_\infty, B_\infty) = N(\overline{Q}_\infty, T_\infty = \overline{U}_\infty + B_\infty)$.

In practice the number of imputations cannot be infinity, and a finite M has to be used. So the next question is to identify $f(Q|\{\hat{Q}_m, U_m\}_{m=1}^{M}) = f(Q|\overline{Q}_M, \overline{U}_M, B_M)$, where $\overline{Q}_M$, $\overline{U}_M$, B_M can be calculated by the completed-data sufficient statistics, $\{\hat{Q}_m, U_m\}_{m=1}^{M}$. Simply replacing $\overline{Q}_M$, $\overline{U}_M$, and B_M with their counterparts as $M \to \infty$ would result in $f(Q|\overline{Q}_M, \overline{U}_M, B_M) \approx N(\overline{Q}_M, \overline{U}_M + B_M)$. This approximation is tempting but can be inappropriate because of the uncertainty of the completed-data statistics with a finite M. Some additional adjustments need to be made.

Assuming that for a finite M, the following three conditions hold approximately (Rubin 1987, Chapter 3):

$$\frac{\overline{Q}_M - \overline{Q}_\infty}{\sqrt{B_\infty/M}} \sim N(0,1), \tag{3.9}$$

$$\frac{(M-1)B_M}{B_\infty} \sim \chi^2_{M-1}, \tag{3.10}$$

$$\overline{U}_M \approx \overline{U}_\infty, \tag{3.11}$$

where χ^2_{M-1} denotes a chi-square distribution with $M-1$ degrees of freedom.

Eq. (3.9) states the fact that $\overline{Q}_M$ itself is an estimate of $\overline{Q}_\infty$ with a finite M. It suggests that $E(\overline{Q}_\infty|\overline{Q}_M) = \overline{Q}_M$, and $Var(\overline{Q}_\infty|\overline{Q}_M) = B_\infty/M$. Eq. (3.10) uses the common chi-square distribution to account for the variability of the between-imputation variance estimate. It suggests that $f(B_\infty|B_M) = \frac{(M-1)B_M}{\chi^2_{M-1}}$. Eq. (3.11) basically states no variability for the within-imputation variance estimate for a finite M, that is $\overline{U}_\infty = \overline{U}_M$. Based on these assumptions,

$$\begin{aligned} & f(Q|\overline{Q}_M, \overline{U}_M, B_M) \\ & = \int f(Q|\overline{Q}_\infty, \overline{U}_\infty, B_\infty) f(\overline{Q}_\infty, \overline{U}_\infty, B_\infty|\overline{Q}_M, \overline{U}_M, B_M) \\ & d\overline{Q}_\infty d\overline{U}_\infty dB_\infty \\ & = \int N(\overline{Q}_M, \overline{U}_M + (1+\frac{1}{M})B_\infty) f(B_\infty|\overline{Q}_M, \overline{U}_M, B_M) dB_\infty \\ & \approx t_\nu(\overline{Q}_M, T_M). \end{aligned} \tag{3.12}$$

Eq. (3.12) states that $f(Q|\overline{Q}_M, \overline{U}_M, B_M)$ can be approximated by a t-distribution with mean $\overline{Q}_M$ and variance T_M. Using a t-distribution for the approximation follows established customs in statistical inference: the variance of $\overline{Q}_M$ is unknown and has to be estimated by T_M. Its degrees of freedom $\nu = \frac{M-1}{[\frac{(1+1/M)B_M}{T_M}]^2}$, which can be derived by matching $\frac{\nu[\overline{U}_M+(1+\frac{1}{M})B_M]}{\overline{U}_M+(1+\frac{1}{M})B_\infty}$ with a chi-square distribution with ν degrees of freedom. Note that as $M \to \infty$, $t_\nu(\overline{Q}_M, T_M) \to N(\overline{Q}_\infty, T_\infty)$. Now we arrive at the solution to the approximation of $f(Q|Y_{obs})$ in the multiple imputation analysis framework.

In practical terms, Eq. (3.6) suggests that the variance estimate T_M for the combining estimate $\overline{Q}_M$ consists of three sources:

1. $\overline{U}_M$: the average of the variance estimated from a single completed-dataset, and this reflects the sampling variability from taking a sample (including both observed and missing cases) from the population.

2. B_M: the extra variance induced by the fact that there are missing data in the sample, and this reflects the sampling variability from having only observed values from the original sample. It also reflects the variation in the predictive distribution of the missing data and thus gauges the uncertainty due to imputation.

3. B_M/M: the extra simulation variance caused by the fact that $\overline{Q}_M$ itself is an estimate for a finite M. Apparently this component approaches 0 if $M \to \infty$. However, it cannot be simply ignored for a small or moderate M or a large amount of missing data.

Example 3.1. *Multiple imputation for an univariate incomplete sample from a normal distribution*

We consider a univariate incomplete sample from a normal distribution $N(\mu, \sigma^2)$, assuming that both μ and σ^2 are unknown. Suppose that there are n independent observations of $y_i \sim N(\mu, \sigma^2)$, $i = 1, \ldots, n$. Among them, only n_{obs} observations are observed and assume that the missingness is MCAR. We assume that the 1st n_{obs} cases are observed, that is, $Y_{obs} = (y_1, \ldots, y_{n_{obs}})^t$ and $Y_{mis} = (y_{n_{obs}+1}, \ldots, y_n)^t$. Let $n_{mis} = n - n_{obs}$ and $p_{obs} = n_{obs}/n \in (0, 1)$, the fraction of observed cases.

We impose a noninformative prior distribution for μ and σ^2 as $\pi(\mu, \sigma^2) \propto \sigma^{-2}$. Define $\overline{Y}_{obs} = \sum_{i=1}^{n_{obs}} y_i/n_{obs}$ (sample mean) and $s_{obs}^2 = \sum_{i=1}^{n_{obs}} (y_i - \overline{Y}_{obs})^2/(n_{obs} - 1)$ (sample variance). From a Bayesian perspective, it can be shown that $f(\mu|Y_{obs}) = t_{n_{obs}-1}(\overline{Y}_{obs}, \frac{s_{obs}^2}{n_{obs}})$, a t-distribution. The multiple imputation algorithm for the missing y's is as follows:

1. Draw μ and σ^2 from their posterior distributions: $f(\sigma^{*2(m)}|Y_{obs}) \sim (n_{obs} - 1)s_{obs}^2/\chi_{n_{obs}-1}^2$, and $f(\mu^{*(m)}|Y_{obs}, \sigma^{*2(m)}) \sim N(\overline{Y}_{obs}, \sigma^{*2(m)}/n_{obs})$.

2. Impute the missing values: $f(y_i^{*(m)}|Y_{obs}, \mu^{*(m)}, \sigma^{*2(m)}) \sim N(\mu^{*(m)}, \sigma^{*2(m)})$ independently for $i = n_{obs} + 1, \ldots n$.

3. Repeat Steps 1 and 2 for $m = 1, \ldots M$ independently (i.e. with different random seeds) to obtain M sets of imputations.

Suppose that the estimand of interest is the population mean μ. If the data are MCAR, then using the mean of observed cases, $\overline{Y}_{obs}$ would be an unbiased and most efficient estimator as $E(\overline{Y}_{obs}) = \mu$ and $Var(\overline{Y}_{obs}) = \sigma^2/n_{obs}$. On the other hand, the multiple imputation estimator is $\hat{\mu}_{MI} = \sum_{m=1}^{M} \hat{\mu}_{SI}^{(m)}$, where $\hat{\mu}_{SI}^{(m)} = \frac{1}{n}(n_{obs}\overline{Y}_{obs} + \sum_{i=n_{obs}+1}^{n} y_i^{*(m)})$ and "SI" here symbolizes estimates from a single set of imputed data. It can be shown that $E(\hat{\mu}_{MI}) = E(\hat{\mu}_{SI}^{(m)}) = \mu$ and thus the multiple imputation estimator is unbiased.

The within-imputation variance from the m-th imputed dataset is $U_m = \frac{\sum_{i=1}^{n_{obs}}(y_i - \hat{\mu}_{SI}^{(m)})^2 + \sum_{i=n_{obs}+1}^{n}(y_i^{*(m)} - \hat{\mu}_{SI}^{(m)})^2}{(n-1)n}$, and the between-imputation variance $B_M = \sum_{m=1}^{M}(\hat{\mu}_{SI}^{(m)} - \hat{\mu}_{MI})^2/(M-1)$. Some involved algebra is needed to understand the variance property of $\hat{\mu}_{MI}$ and the corresponding variance estimate using the combining rules. Details can be found in Rubin and Schenker (1986), Carpenter and Kenward (2013, Section 2.7) and Kim and Shao (2014, Example 4.5). Here we only provide a summary as follows:

$$Var(\hat{\mu}_{MI}) = \frac{\sigma^2}{n_{obs}} + \frac{n_{obs}-1}{n_{obs}-3}\frac{1-p_{obs}}{n_{obs}M}\sigma^2, \tag{3.13}$$

$$E(U_m) = E(\overline{U}_M) = \frac{\sigma^2}{n}\frac{n-3}{n-1}\frac{n_{obs}-1}{n_{obs}-3}, \tag{3.14}$$

$$E(B_M) = \frac{1-p_{obs}}{n_{obs}}\frac{n_{obs}-1}{n_{obs}-3}\sigma^2. \tag{3.15}$$

All the expectations in Eqs. (3.13)-(3.15) are over the sampling distribution of Y. Several remarks can be drawn from these results. First we note that $Var(\hat{\mu}_{MI}) \rightarrow \frac{\sigma^2}{n_{obs}}$ as $M \rightarrow \infty$ for fixed n_{obs} and n. This indicates that asymptotically the multiple imputation estimator is equivalent to using the complete-case average, $\overline{Y}_{obs}$. We also note that $\frac{E(\overline{U}_M)}{\sigma^2/n} \rightarrow 1$ as $n \rightarrow \infty$. This is consistent with the assumption that the within-imputation variance estimate would be asymptotically unbiased if the data had been complete. This also shows that if we only analyze the singly imputed data, the variance tend to be underestimated as $\sigma^2/n < \sigma^2/n_{obs}$, that is, the estimates would be too precise.

From Eqs (3.14) and (3.15) and using the combining rules, we get

$$E(T_M) = \frac{\sigma^2}{n_{obs}}\frac{n_{obs}-1}{n_{obs}-3}(1 - \frac{2n_{obs}}{n(n-1)} + \frac{1-p_{obs}}{M}). \tag{3.16}$$

The bias in T_M (as the variance estimate of $\hat{\mu}_{MI}$) is therefore

$$E(T_M) - Var(\hat{\mu}_{MI}) = \frac{2\sigma^2}{n_{obs}(n_{obs}-3)}(1 - \frac{n_{obs}(n_{obs}-1)}{n(n-1)}). \tag{3.17}$$

Clearly the bias $\rightarrow 0$ when n (and n_{obs}) $\rightarrow \infty$. This suggests that using Rubin's combining rules can yield an asymptotically unbiased variance estimator for $\hat{\mu}_{MI}$.

Example 3.2. *A simulation evaluation for Example 3.1*

We simulate 1000 observations, $y_i \sim N(1,1), i = 1, \ldots, 1000$. We randomly set 50% of the observations to be missing. Based on observed cases, $\overline{Y}_{obs} = 0.963$ and $\hat{Var}(\overline{Y}_{obs}) = 2.23 \times 10^{-3}$. We run the multiple imputation algorithm in Example 3.1 for different choices of M. Table 3.1 shows the corresponding results for $\overline{Q}_M$, $\overline{U}_M$, B_M, and T_M. As M increases, $\overline{Q}_M \to \overline{Y}_{obs}$ and $T_M \to \hat{Var}(\overline{Y}_{obs})$, as predicted by the theoretical results from Example 3.1. The mean estimate $\overline{Q}_M$ is not very different from $\overline{Y}_{obs}$ even when M is as small as two. On the other hand, for a small M (e.g., $M = 2$), the corresponding T_M can be large due to a large B_M. When M increases, the corresponding T_M is moving close to $\hat{Var}(\overline{Y}_{obs})$. Compared with $\overline{Q}_M$ and $\overline{U}_M$, B_M is more sensitive to M.

TABLE 3.1
Example 3.2. Simulation results

M	$\overline{Q}_M$	$\overline{U}_M$	B_M	T_M
2	0.945	0.00116	0.00477	0.00831
5	0.946	0.00114	0.00127	0.00266
20	0.955	0.00112	0.00083	0.00200
50	0.962	0.00113	0.00091	0.00205
200	0.967	0.00113	0.00119	0.00231
1000	0.965	0.00112	0.00115	0.00227
10000	0.963	0.00112	0.00112	0.00223

Note: $\overline{Y}_{obs} = 0.963$ and $\hat{Var}(\overline{Y}_{obs}) = 0.00223$.

3.2.3 Why Does Multiple Imputation Work?

Rubin (1978) termed multiple imputation analysis as a "phenomenological Bayesian" approach to survey nonresponse problems. In other words, multiple imputation analysis can be viewed as a combination of both frequentist and Bayesian methods. The imputation stage usually follows the Bayesian paradigm to generate/simulate imputed values that are random numbers from some predictive distributions. However, in the analysis and combining stages, the inference paradigm is mainly frequentist since most of the complete-data procedures fall into this category. Some discussion of using Bayesian procedures to analyze completed datasets can be found in Section 14.3.2. Again, the motivation of creating multiple imputations is neither to recover the original missing values nor to make up data.

In Section 2.3.4, we briefly mentioned the asymptotic equivalence between Bayesian and frequentist estimates and their asymptotic normality. In the context of multiple imputation analysis, this fact further implies that if we analyze multiply imputed data, which are generated by correctly formulated Bayesian models, use common complete-data statistical procedures for completed data and apply the combining procedures, the resultant estimates are expected to have good frequentist properties as if we directly estimate the parameters using a correctly formulated likelihood function of the observed data. Obviously, this nice feature of multiple imputation analysis is desirable for data analysts who are used to regular complete-data statistical procedures. In addition, much research on multiple imputation has also demonstrated that analyzing a few completed datasets is a rather good approximation of the authentic Bayesian analysis that is based on the infinite or a large number of posterior draws.

3.3 Statistical Inference on Multiply Imputed Data

3.3.1 Scalar Inference

This section provides more details on making statistical inference using multiply imputed data. We first consider the case where Q is a scalar quantity. Eqs. (3.5) and (3.6) present the point and variance estimate from a multiple imputation analysis. Table 3.2 lists several key statistics that are functions of B_M, $\overline{U}_M$, and M and are often encountered in the combining process.

TABLE 3.2
Some key statistics for multiple imputation inference of scalar Q

Statistic	Remark
$\lambda = \frac{(1+1/M)B_M}{T_M}$	The ratio of the between-imputation variance to the total variance; it measures the proportion of the variation attributable to missing data
$r = \frac{(1+1/M)B_M}{\overline{U}_M} = \lambda/(1-\lambda)$	The relative increase of the variance due to nonresponse; it measures the ratio of the between-imputation variance to the within-imputation variance
$\nu = (M-1)/\lambda^2$	Degrees of freedom * of the t-distribution for $\overline{Q}_M$
$\gamma = \frac{r+2/(\nu+3)}{1+r}$	The fraction of missing information; it measures the ratio of the between-imputation variance to the total variance

Note: * An adjustment needs to be made by replacing ν with ν_{adj} if the sample size is small.

With a finite M, the multiple imputation estimator $\overline{Q}_M$ follows a t-distribution as an approximation to $f(Q|Y_{obs})$. Univariate tests are based on the approximation:

$$\frac{\overline{Q}_M - Q}{\sqrt{T_M}} \sim t_\nu, \tag{3.18}$$

where t_ν is a t-distribution with ν degrees of freedom (Table 3.2). The $100(1-\alpha)\%$ confidence interval for $\overline{Q}$ is calculated as

$$\overline{Q}_M \pm t_{\nu,1-\alpha/2}\sqrt{T_M}, \tag{3.19}$$

where $t_{\nu,1-\alpha/2}$ is the percentile corresponding to the probability of $1-\alpha/2$ of t_ν.

Suppose we test the null hypothesis $Q = Q_0$ for some prespecified value Q_0. We can find the p-value of the test as the probability:

$$P_s = Pr(F_{1,\nu} > \frac{(Q_0 - \overline{Q}_M)^2}{T_M}), \tag{3.20}$$

where $F_{1,\nu}$ is an F-distribution with 1 and ν degrees of freedom.

With a small complete-data sample size n, some adjustments need to be applied to ν. This is because $M-1 < \nu < \infty$, and as $M \to \infty$, $\nu \to \infty$, which can be considerably greater than n. Barnard and Rubin (1999) made some revision of ν to adapt to small sample size. To summarize their findings and recommendations, let ν_{com} be the degrees of freedom of $\hat{Q}$ in the hypothetically complete data. In models that fit k parameters for data with a sample size of n, we may set $\nu_{com} = n - k$. The corresponding observed-data degrees of freedom would be $\nu_{obs} = \frac{\nu_{com}+1}{\nu_{com}+3}\nu_{com}(1-\lambda)$. The adjusted degrees of freedom to be used for the small sample size would then be $\nu_{adj} = \frac{\nu_{obs}\nu_{com}}{\nu_{obs}+\nu_{com}}$, which is always less than or equal to ν_{com}. If $\nu_{com} \to \infty$ (for a moderate or large sample size n), then $\nu_{adj} \to \nu$. However in general, to use ν_{adj}, we need to specify ν_{com} for the completed-data analysis. For additional references, see Lipsitz et al. (2002) and Wagstaff and Harel (2011).

Example 3.3. *Estimating the mean of family total income from NHIS 2016 public-use data*

We illustrate the aforementioned combining rules using the NHIS income imputation public-use files (Example 1.2). Here $M = 5$ and we let Q be the population mean of the family total income. The complete-data sample size $n = 40875$. The mean estimates from five completed datasets are $\{66044, 66107, 65999, 66091, 66044\}$ (in dollars). The corresponding standard error estimates are $\{690.58, 701.29, 684.69, 690.83, 696.10\}$. Survey designs of NHIS are accounted for when calculating these estimates. From these numbers we get $\overline{Q}_5 = 66057$, $\overline{U}_5 = 479862$, $B_5 = 1839.5$, and $T_5 = 482069.4$ (the standard error is $\sqrt{482069.4} = 694.31$). The 95% confidence interval for the mean estimate is (64696, 67418). In addition, $r = 0.0046$, $\lambda = 0.0046$, $\nu = 33533$, and $\gamma = 0.0046$. Moreover, the average of imputed income values is 64419, lower than that of the observed sample (66613). This is consistent with the exploratory nonresponse analysis results showing that nonrespondents are more likely to have lower income (Example 1.10).

Example 3.4. *A simulation study for the mean estimate of an incomplete univariate normal sample*

We conduct a simulation study to demonstrate the performance of the combining rules. Based on the setup of Example 3.2, we generate $Y \sim N(1,1)$ for 1000 observations. A MCAR mechanism is applied to generate around 50% of missing cases on Y. We report the mean estimates from the before-deletion (BD), complete-case (CC), and multiple imputation (MI). We vary M in the simulation: $M = \{2, 5, 20, 50, 200\}$. We also analyze the imputed data from single imputation (SI) (i.e., $M = 1$). The simulation is replicated 1000 times.

Table 3.3 shows the simulation results. Since the missingness is MCAR, CC yields an unbiased mean estimate with a nominal coverage rate. SI has a small bias for the mean estimate (Bias= 0.030), and its variance is underestimated ($\overline{\text{SE}} = 0.032 <$ SD=0.045), which leads to a low coverage rate (73.8%). SI is problematic because it underestimates the uncertainty of missing values. For MI with $M = 2$, the bias of the mean estimate is reduced compared with SI. On the other hand, the variance is overestimated ($\overline{\text{SE}} = 0.089 >$ SD=0.045), which leads to a perfect coverage rate (100%) and a rather wide confidence interval (CI=1.308). When M increases to 5, the bias of the variance estimate is reduced ($\overline{\text{SE}} = 0.050$ vs SD=0.045), and therefore the coverage rate (97.8%) moves closer to the nominal level. As M further increases, the bias of the mean estimate quickly diminishes, the variance estimate becomes less biased, the coverage rate moves further to the nominal level, and the performance of MI approaches that of CC. In this example, the results from MI seem to achieve some stability as M increases to 20.

TABLE 3.3
Example 3.4. Simulation results

Method	Bias	SD	$\overline{\text{SE}}$	MSE	CI	COV (%)
BD	0	0.032	0.032	0.00103	0.124	94.3
CC	0	0.045	0.045	0.00203	0.175	95.5
SI	0.030	0.045	0.032	0.00296	0.125	73.8
MI ($M = 2$)	−0.017	0.045	0.089	0.00234	1.308	100
MI ($M = 5$)	−0.016	0.045	0.050	0.00227	0.219	97.8
MI ($M = 20$)	−0.007	0.045	0.042	0.00208	0.168	94.1
MI ($M = 50$)	0	0.045	0.043	0.00203	0.169	94.9
MI ($M = 200$)	0.004	0.045	0.046	0.00205	0.179	95.9

A frequently used statistic in multiple imputation analysis is the fraction of missing information (FMI) (γ in Table 3.2), which is also closely related to the ratio of the between-imputation variance to the total variance (r in Table 3.2). Note that the seemingly awkward form in γ is due to the density form of the t distribution (Rubin 1987). Rubin (1987) also defined the population fraction of missing information as

$$\gamma_\infty = \frac{B_\infty}{T_\infty} = \frac{B_\infty}{B_\infty + U_\infty}, \tag{3.21}$$

which is the limit of γ as $M \rightarrow \infty$.

FMI can be used as a measure to gauge the efficiency gain of using multiple imputation analysis compared with complete-case analysis.

Example 3.5. *The fraction of missing information in a univariate missing data problem assuming MCAR*

Following from Example 3.1, it can be verified that as $M \rightarrow \infty$ and $n \rightarrow \infty$, $E(B_M)/E(T_M) \rightarrow 1 - p_{obs}$. That is, the population fraction of missing information is asymptotically equivalent to the fraction of missing cases. This fact confirms that using multiple imputation analysis does not gain any efficiency compared with using complete cases for estimating the mean in that setup, as is numerically shown in Example 3.4.

However, the information contained in FMI can be more complicated in general situations. It depends on the estimand of interest and the imputation model, especially when the model includes strong predictors of the missing variable. More related discussion can be found in Section 12.5.

3.3.2 Multi-parameter Inference

Suppose the estimand of interest Q is a $K \times 1$ vector ($K > 1$), say $Q = (Q_1, \ldots, Q_K)$. A direct strategy of combining estimates from multiply imputed data is to apply the combining rules for the scalar estimand to each element of the Q vector. To our knowledge, this approach is perhaps mostly used in practice. However, sometimes there is a need to conduct one statistical test that involves all the K elements of Q at once, following the hypothesis testing framework. There are a number of methods developed for combining test statistics involving multiple parameters. This section briefly reviews some of them.

Li, Raghunathan, et al. (1991) proposed a procedure to combine the Wald test statistic from each completed dataset, requiring the M sets of completed-data estimates, $\hat{Q}_m$ (the $K \times 1$ vector of estimates) and U_m (the $K \times K$ variance-covariance matrix of $\hat{Q}_m$) ($m = 1, \ldots M$) to be available. However, if the dimension K becomes large, the covariance matrices might not be readily available from the statistical software. In these cases, if the likelihood ratio statistics can be calculated from the software, Meng and Rubin (1992) proposed to combine the likelihood ratio tests, which only require the point estimates to be available. These two procedures are asymptotically equivalent. A even simpler procedure is to combine the completed-data chi-square test (χ^2)

statistics and their associated p-values, proposed by Li, Meng, et al. (1991). This procedure only requires M-sets of χ^2 statistics and their corresponding p-values. Although simple to use, since this procedure does not use information from the point estimates and their covariance matrices, its performance is expected to be less reliable than the other two procedures.

Suppose the null hypothesis is $H_0 : Q = Q_0$, Table 3.4 summarizes the established procedures of combining test statistics for multivariate estimands. In these calculations, $\overline{Q}_M = \sum_{m=1}^{M} \hat{Q}_m/M$ is the vector of the average of M completed-data estimates, $\overline{U}_M = \sum_{m=1}^{M} U_m/M$ is the within-imputation variance-covariance matrix, and $B_M = \sum_{m=1}^{M}(\hat{Q}_m - \overline{Q}_M)(\hat{Q}_m - \overline{Q}_M)^t/(M-1)$ is the between-imputation variance-covariance matrix.

For the procedure combining the likelihood ratio test statistics, $\overline{S}_L$ is the average of the likelihood ratio test statistics across M completed datasets, $\tilde{S}_L$ is the average of the M likelihood ratio statistics, $S_{L,1}, \ldots S_{L,M}$, evaluated using the average multiple imputation parameter estimates and the average of the estimates from a model fitted subject to the null hypothesis.

For the procedure combining the χ^2 statistics, $S_{C,1}, \ldots S_{C,M}$ are χ^2 statistics associated with testing the null hypothesis on each completed dataset, $\overline{S}_C$ is the average of M sets of χ^2 statistics, and $\overline{\sqrt{S_C}}$ is the average of the square root of these statistics.

These procedures have complicated formulae, yet they are implemented in multiple imputation software packages to enable automatic calculations. Comparisons and decisions can be made by applying all procedures if the required statistics are available. In addition, there is ongoing research on improving the performance of the combining procedures. For example, Reiter (2007) proposed a small-sample adjustment for ν_1 used in the Wald test (Table 3.4). Licht (2010) considered several new ideas including applying a Z-transformation to multiple p-values and then combining transformed Z-values. Reviews targeted for practitioners can be found in Marshall et al. (2009) and Van Buuren (2018, Section 5.3.4).

Example 3.6. *Combining chi-square test results from NHIS 2016 public-use data*

For illustration, we consider the association between the family total income and highest family education level using the NHIS income imputation public-use files (Example 1.2). Table 3.5 shows the distribution of education level by different income groups from five completed datasets. There exists a clear pattern that subjects with higher income tend to have a higher education level. Across five datasets, the χ^2-test statistics and the associated p-values all confirm such significant association, although the exact numbers are slightly different due to the imputation uncertainty. Calculation for both the frequency and χ^2 statistics account for the survey design information. To combine these test results, we use the procedure outlined in Table 3.4: for the F-test, $M = 5$, $K = (4-1) \times (3-1) = 6$, $\nu_3 = 4.49$, the combined F-statistic=675.28, and $p < 0.0001$.

TABLE 3.4
Summary of combining test statistics from multiply imputed data

Statistic	Wald test	Likelihood ratio	χ^2 test
T	$T_1 = \frac{(Q_0 - \overline{Q}_M)\overline{U}_M^{-1}(Q_0 - \overline{Q}_M)^t}{K(1+r_1)}$	$T_2 = \frac{\tilde{S}_L}{K(1+r_2)}$	$T_3 = \frac{\frac{\overline{S}_C}{K} - \frac{M+1}{M-1} r_3}{1+r_3}$
df	$\nu_1 = \begin{cases} 4 + (l-4)(1 + \frac{1-\frac{2}{l}}{r_1})^2 & \text{if } l > 4 \\ \frac{1}{2} l (1 + \frac{1}{K})(1 + \frac{l}{r_1})^2 & \text{otherwise} \end{cases}$ where $l = K(M-1)$	$\nu_2 = \begin{cases} 4 + (l-4)(1 + \frac{1-\frac{2}{l}}{r_2})^2 & \text{if } l > 4 \\ \frac{1}{2} l (1 + \frac{1}{K})(1 + \frac{l}{r_2})^2 & \text{otherwise} \end{cases}$ where $l = K(M-1)$	$\nu_3 = K^{\frac{-3}{M}}(M-1)(1+r_3)^2$
r	$r_1 = \frac{Tr(B_M \overline{U}_M^{-1})}{(1+M^{-1})K}$	$r_2 = \frac{(M+1)(\overline{S}_L - \tilde{S}_L)}{K(M-1)}$	$r_3 = \frac{(M+1)\sum_{m=1}^{M}(\sqrt{S_{C,m}} - \overline{\sqrt{S_C}})^2}{M(M-1)}$
F	F_{K,ν_1}	F_{K,ν_2}	F_{K,ν_3}

Note: T: test statistics; df: the degrees of freedom in the F-test; r: the relative increase of variance; F: the critical value from the F-distribution, for which the F-test statistic is compared to.

TABLE 3.5
Example 3.6. Frequency (%) of highest family education by completed family income

Imputation number	Completed family income (group)	Education			χ^2 statistics and p-values
		High school or less	Some college	College or graduate school	
1	\$0-\$34999	47.45	37.06	15.49	8837.57
	\$35000-\$74999	25.67	37.15	37.18	$p < 0.0001$
	\$75000-\$99999	13.22	32.91	53.87	
	$\geq$ \$100000	5.55	18.53	75.92	
2	\$0-\$34999	47.52	36.99	15.49	9192.08
	\$35000-\$74999	25.62	37.17	37.21	$p < 0.0001$
	\$75000-\$99999	13.28	32.80	53.91	
	$\geq$ \$100000	5.49	18.73	75.78	
3	\$0-\$34999	47.40	37.07	15.53	9103.73
	\$35000-\$74999	25.60	37.00	37.40	$p < 0.0001$
	\$75000-\$99999	13.41	33.42	53.17	
	$\geq$ \$100000	5.55	18.44	76.02	
4	\$0-\$34999	47.34	36.90	15.76	9211.00
	\$35000-\$74999	25.67	37.22	37.11	$p < 0.0001$
	\$75000-\$99999	13.22	32.82	53.96	
	$\geq$ \$100000	5.49	18.73	75.78	
5	\$0-\$34999	47.27	36.88	15.85	8815.56
	\$35000-\$74999	25.70	37.24	37.06	$p < 0.0001$
	\$75000-\$99999	13.12	33.05	53.83	
	$\geq$ \$100000	5.70	18.62	75.68	

Note: For the F-test combining χ^2-tests: $M = 5$, $K = (4 - 1) \times (3 - 1) = 6$, $\nu_3 = 4.49$, the combined F-statistic=675.28, and $p < 0.0001$.

3.3.3 How to Choose the Number of Imputations

In Examples 3.2 and 3.4 we see that results of multiple imputation analysis can vary somewhat depending on the choice of M especially for the variance estimates. Multiple imputation is essentially a simulation technique, and hence the estimate $\overline{Q}_M$ and its variance estimate T_M are subject to simulation error caused by using a finite number for M. Setting $M = \infty$ would cause all simulation error to disappear yet is impractical. Therefore a frequently asked question in practice is, "How many imputations should we use?"

First, from the Bayesian perspective and for many applied problems, $f(Q|Y_{obs})$ can be well approximated by normal distributions. This is especially true when the sample size is large, guaranteed by the Bayesian asymptotics (Section 2.3.4). Thus we need only sufficient, but not necessarily many, draws from the posterior predictive distribution of missing data to reliably estimate the posterior mean and variance of Q conditional on the completed data. Second, with a moderate or large sample size, conventional complete-data procedures would yield asymptotically equivalent results to those from Bayesian methods. As a result, the number of draws (M) of averaging the posterior distributions does not have to be large in many cases to achieve reasonably good results.

So what is the main factor that decides the choice of M? Intuitively, if the dataset for an analysis only has a small proportion of missing values, then the information from observed cases would dominate the completed-data inference. In addition, we expect that this part would be similar across the M imputations since the relative contribution of the information from imputed missing cases would be small. Thus the needed M is unlikely to be large because we would not expect to see much between-imputation variation as compared with the within-imputation variation. If this is the case, increasing M would not effectively bring substantial changes to the final inference.

Rubin (1987) quantified this intuition by using FMI to guide the choice of M. The idea is to use $M = \infty$ as the yardstick (i.e., the ideal scenario) and assess how much estimation efficiency would be lost by using a finite M. The ratio of T_M over T_∞ (Rubin 1987, Eq. (4.5.20)), can be written as

$$T_M = T_\infty(1 + \frac{\gamma_\infty}{M}), \tag{3.22}$$

where γ_∞ is the population fraction of missing information given by Eq. (3.21). On the scale of standard deviation, the relative efficiency (RE) of using a finite number compared with an infinite number for M is therefore

$$RE = \frac{T_M}{T_\infty} = (1 + \frac{\gamma_\infty}{M})^{-1/2} \tag{3.23}$$

One early criterion for choosing M is based on a value M that $T_M \approx T_\infty$. Table 3.6 shows the trend of RE (rounded to two decimal places) when M and γ_∞ change. Apparently for a fixed M, a higher γ_∞ leads to a lower RE. This suggests that for a large γ_∞, a larger number for M is needed to reduce the simulation error. In addition, for a fixed γ_∞, increasing M would make RE approach 1. Based on Table 3.6, however, a notable fact is that even for $\gamma_\infty = 0.5$ and choosing $M = 5$, the corresponding RE can still be as high as 95%.

Typically, for multiple imputations generated under a reasonable model, the FMI would be no greater than the rates of missingness (Section 12.5). Therefore, Rubin (1987) concluded that $M = 5$ would be appropriate for most analyses with a moderate amount (e.g., less than 50%) of missing data. And this guideline had been followed by many researchers and practitioners for a while. From a historical perspective, the guideline of setting $M = 5$ turned out to be a welcome choice at the time (from late 1970s to late 1990s) when computing and file storage were still major limiting factors in applied data analysis. Nowadays when it is not inconvenient to set M as high as 100, it can be seen that RE can be sufficiently high even for the rare case where $\gamma_\infty = 0.9$.

TABLE 3.6
Large-sample relative efficiency (RE) (in %) when using a finite number of imputations, M, intestad of an infinite number, as function of the population fraction of missing information, γ_∞.

	γ_∞								
M	0.1	0.2	0.3	0.4	0.5	0.6	0.7	0.8	0.9
1	95	91	88	85	82	79	77	75	73
2	98	95	93	91	89	88	86	85	83
3	98	97	95	94	93	91	90	89	88
5	99	98	97	96	95	94	94	93	92
10	100	99	99	98	98	97	97	96	96
20	100	100	99	99	99	99	98	98	98
50	100	100	100	100	100	99	99	99	99
100	100	100	100	100	100	100	100	100	100

Interestingly, although Eq. (3.23) shows the importance of γ_∞ in determining the appropriate M, the true value of γ_∞ is actually unknown and has to be estimated by γ with some finite M. Therefore, using RE statistics from Table 3.6 in practice is still subject to some uncertainty. A relevant discussion can be found in Section 12.5.

Several recent papers (e.g., Hershberger and Fisher 2003; Royston 2004; Graham et al. 2007; Bodner 2008; Pan et al. 2014) have conducted more detailed investigations on the choice of M. They assessed different criteria from RE, targeting for a wide variety of analytic interests. A general conclusion is that the typical $M = 5$ rule might not satisfy in certain cases, such as having reproducible statistical power estimates or p-values even with a moderate amount of missing data. Thus, increasing M from 5 to a number above 20 or even on the scale of hundreds can be more assuring. At the same time, computing need and file storage are generally not issues nowadays. Therefore, researchers and practitioners do not have to be constrained by the old custom of setting $M = 5$.

In most of our own applications, we typically use the rule of thumb, "The number of imputations M should be similar to the percentages of cases that are incomplete." (Von Hippel 2009; White et al. 2011) Therefore, for a dataset with average 20% (or 50%) missing cases, we would choose $M = 20$ (or $M = 50$). We also suggest running a sensitivity analysis with different M, with additional specifics as follows:

1. Use $M = 5$ during the exploratory and model building stages of the analysis.

2. When the imputation model and post-imputation analyses are finalized, start with M close to the percentage of incomplete cases for the main analysis.

3. Perform a sensitivity analysis to assess the robustness of the results against different random seeds under the same M as well as under different M and make a decision.

For example, if the percentage of incomplete cases in a dataset is around 25%, we would first try $M = 25$ and compare analysis results under different random seeds for generating imputations. If the analysis results are similar, we then increase M, such as setting $M = 50$ and $M = 100$. If there is little change, we would finally report results from $M = 25$ and document that such results are little changed with a larger M. If the results are sensitive, we then need to use a larger M.

If the primary goal of a multiple imputation analysis is to obtain results of some specific analyses, then the storage of multiply imputed datasets for a large M is of less concern because only the M sets of estimates and their variances are needed. That is, it is not necessary to keep and store the M sets of imputed datasets. To make the results reproducible, however, the random seeds for generating imputations should be retained. Yet when the primary goal of a multiple imputation project is to release the completed datasets for data users, data processing and managing might become a major component in the project. This could impact the choice of M. More relevant discussion can be found in Section 10.5.

3.4 How to Create Multiple Imputations

3.4.1 Bayesian Imputation Algorithm

Assuming the missingness is ignorable, Eqs. (3.1)-(3.4) state that multiple imputations for missing values need to be drawn from their posterior predictive distribution as

$$Y_{mis}^{(m)} \sim f(Y_{mis}|Y_{obs}), \tag{3.24}$$

where $m = 1, \ldots M$. Eq. (3.24) simply suggests that we should impute missing values as random draws from certain distributions derived under the imputation model. Therefore, the idea of multiple imputation is not to estimate or predict missing values from some equations such as the regression prediction method assessed in Example 2.5. Little and Rubin (2020, Chapter 4) provided more details about distinctions between two imputation strategies (i.e., using draws vs. using predictions).

Let θ be the parameter of the complete-data model $f(Y|\theta)$. From a Bayesian perspective, a prior distribution for θ, $\pi(\theta)$, is also needed to complete the model specification. Directly formulating $f(Y_{mis}|Y_{obs})$ is difficult because θ is not included in the posterior predictive distribution. However, the imputation algorithm has to involve θ in certain ways. In Example 3.1, we let $\theta = (\mu, \sigma^2)$, impose $\pi(\theta) \propto \sigma^{-2}$, and provide an imputation algorithm for an incomplete univariate normal sample. This idea can be generalized. Note that

$$f(Y_{mis}|Y_{obs}) = \int f(Y_{mis}, \theta|Y_{obs})d\theta = \int f(Y_{mis}|Y_{obs}, \theta)f(\theta|Y_{obs})d\theta. \quad (3.25)$$

Eq. (3.25) suggests that conditional on a draw of θ, Y_{mis} can be drawn from $f(Y_{mis}|Y_{obs}, \theta)$. This strategy can be sketched in an algorithmic manner. In this book it is referred to as the direct Bayesian (DB) imputation algorithm:

1. Derive the observed-data posterior distribution, $f(\theta|Y_{obs})$, under a prior $\pi(\theta)$.
2. Draw a value of the parameter $\theta^{(m)}$ from $f(\theta|Y_{obs})$.
3. Draw values of missing data from the conditional distribution: $Y_{mis}^{(m)} \sim f(Y_{mis}|Y_{obs}, \theta^{(m)})$.
4. Repeat Steps 2 and 3 independently M times to create multiple completed datasets $(Y_{mis}^{(m)}, Y_{obs})$ $(m = 1, \ldots, M)$.

In many cases, however, deriving $f(\theta|Y_{obs})$ and drawing parameters from it can be difficult if the missing data pattern is not simple or the complete-data model is complicated. A more general Bayesian imputation algorithm is the data augmentation (DA) strategy. The DA algorithm considers the joint posterior distribution of θ and Y_{mis} conditional on Y_{obs}, $f(\theta, Y_{mis}|Y_{obs})$, and obtains draws in an iterative manner (Tanner and Wong 1987). To understand the idea, first note that

$$\begin{aligned} f(\theta, Y_{mis}|Y_{obs}) &= f(\theta|Y_{mis}, Y_{obs})f(Y_{mis}|Y_{obs}) \\ &= f(Y_{mis}|\theta, Y_{obs})f(\theta|Y_{obs}). \end{aligned} \quad (3.26)$$

Second, recall the idea of Gibbs sampling (Section 2.3.3), if we view Y_{mis} also as a "parameter", then we can draw $f(\theta, Y_{mis}|Y_{obs})$ by iterating between drawing from $f(\theta|Y_{mis}, Y_{obs})$ and $f(Y_{mis}|Y_{obs}, \theta)$. These draws then constitute iterations of Monte Carlo Markov chains (MCMC). When draws from both $f(\theta|Y_{mis}, Y_{obs})$ and $f(Y_{mis}|Y_{obs}, \theta)$ converge, we also obtain draws of Y_{mis} that converge to the posterior predictive distribution, $f(Y_{mis}|Y_{obs})$, therefore creating multiple imputations.

The DA strategy can be sketched in an algorithmic manner as follows:

1. Derive the complete-data posterior distribution, $f(\theta|Y) = f(\theta|Y_{obs}, Y_{mis})$, under a prior $\pi(\theta)$.

2. Begin with an estimate or guess of the parameter θ, say $\theta^{*(t)}$ (where $t = 0$ at the 1st iteration).

3. Draw values of missing data from the conditional distribution: $Y_{mis}^{*(t)} \sim f(Y_{mis}|Y_{obs}, \theta^{*(t)})$ (the I-step, for "Imputation").

4. Draw a new value of the parameter from its complete-data posterior distribution, "plugging in" the newly drawn value of Y_{mis}: $\theta^{*(t+1)} \sim f(\theta|Y_{obs}, Y_{mis}^{*(t)})$ (the P-step, for "Posterior").

5. Repeat Steps 3 and 4 (the I-step and P-step) until the convergence for $(\theta^{*(t)}, Y_{mis}^{*(t)})$ is satisfied, say at $t = T$. The draws of $Y_{mis}^{*(T)}$ constitute the m-th set of imputations, $Y_{mis}^{(m)}$.

6. Repeat Steps 2 to 5 independently M times to create multiple sets of imputations.

An advantage of the DA algorithm over the DB algorithm is that working on $f(\theta|Y_{mis}, Y_{obs})$ is often easier than $f(\theta|Y_{obs})$ because the former is conditional on completed data and not constrained by the missing data pattern. Note that unlike the DB algorithm, the DA algorithm requires iterations between the I-step and P-step. When θ contains multiple components, the P-step of the DA algorithm can consist of multiple steps drawing each parameter conditional on other parameters and missing values as in the Gibbs sampler (Section 2.3.3).

Example 3.7. *The data augmentation algorithm for Example 3.1*

A DB imputation algorithm is given in Example 3.3. Although a DA imputation algorithm is not necessary, here we provide a sketch of it for illustration. Let $\overline{Y}_{obs} = \sum_{i=1}^{n_{obs}} y_i/n_{obs}$ and $s_{obs}^2 = \sum_{i=1}^{n_{obs}} (y_i - \overline{Y}_{obs})^2/(n_{obs} - 1)$.

1. Set some initial values: $\mu^{*(0)} = \overline{Y}_{obs} + \epsilon_1$ and $\sigma^{2*(0)} = s_{obs}^2 + \epsilon_2$ where ϵ_1 and ϵ_2 are small random numbers.

2. Impute the missing values as $y_i^{*(t)} \sim N(\mu^{*(t)}, \sigma^{*2(t)})$ independently for $i = n_{obs} + 1, \ldots n$. Calculate the completed-data average $\overline{Y}_{com}^{(t)} = (\sum_{i=1}^{n_{obs}} y_i + \sum_{i=n_{obs}+1}^{n} y_i^{*(t)})/n$ and sum of squares $\sum_{i=1}^{n} (y_i - \overline{Y}_{com}^{(t)})^2$.

3. Draw $\mu^{*(t)}$ and $\sigma^{2*(t)}$ from their respective completed-data posterior distributions: $\sigma^{*2(t)} \sim (n-1)s_{com}^{2(t)}/\chi_{n-1}^2$, where $s_{com}^{2(t)} = \frac{\sum_{i=1}^{n}(y_i - \overline{Y}_{com}^{(t)})^2}{n-1}$, and $\mu^{*(t)} \sim N(\overline{Y}_{com}^{(t)}, \sigma^{*2(t)}/n)$.

4. Repeat Steps 2 and 3 until the convergence of $\mu^{*(t)}$ and $\sigma^{*2(t)}$. Collect the imputed missing values at the last iteration T as $Y_{mis}^{(m)}$.

5. Repeat Steps 1 to 4 independently M times.

It is natural to run the DA algorithm independently M times and collect the imputed values at the last iteration from each of the MCMC chains to create multiple imputations. Or we can run the DA algorithm only one time to collect M sets imputations at different iterations. In the past, it was advised that there should exist enough iterations among the M sets of imputations within a single chain to minimize their autocorrelations (Schafer 1997). However, recent research (Hu et al. 2013) suggests that consecutive draws of Y_{mis} from a single chain might be used as multiply imputed data with a sufficiently large M (e.g., on the scale of hundreds).

Besides creating multiple imputations, another product of the DB or DA algorithm is the posterior distribution of θ given observed data, that is, $f(\theta|Y_{obs})$. From the perspective of a Bayesian analysis with missing data (Section 2.4), the posterior inference of θ can be obtained directly from $f(\theta|Y_{obs})$, and there is no need to use imputed values for Y_{mis} from the algorithm. However, as we discussed before, one desirable feature of multiple imputation analysis is to generate completed datasets that can be used by ordinary data users for implementing conventional complete-data analysis procedures. These data users do not have to be familiar with Bayesian analysis. In addition, multiply imputed datasets allow users to apply a wide variety of analyses that are not limited to those directly implied by the complete-data model. That is, the estimand of interest Q in multiple imputation analysis can be more general than θ used in formulating the complete-data model. For instance, after we impute the missing data in the univariate incomplete sample as in Example 3.1, we can make inferences not only on the population mean and variance, but on other quantities of the distribution function (e.g., percentiles). We will provide more discussion about this feature later in the book.

3.4.2 Proper Multiple Imputation

From a Bayesian perspective, creating multiple imputations using the DB or DA algorithm is a theoretically preferable approach because it accommodates the uncertainty of the parameter θ under the assumed complete-data model using random draws. Rubin (1987) defined such a strategy of drawing multiple imputations as "proper". In general, creating proper multiple imputations would guarantee the validity of the multiple imputation estimates: the point estimate $\overline{Q}_M$ is unbiased or consistent; the variance estimate T_M is unbiased or consistent, or it overestimates the true variance, which might result in conservative statistical inferences (Rubin 1996). Additional discussion of some of the subtle issues can be found in Van Buuren (2018, Section 2.3.3) and references therein.

On the contrary, "improper" multiple imputation methods generally refer to those that do not appropriately account for the uncertainty of the parameter. In many cases, this can be understood as a repeated imputation procedure by fixing the parameter at its estimate. Specifically, an improper imputation algorithm would be, instead of drawing Y_{mis} from $f(Y_{mis}|Y_{obs})$ using the DB or DA algorithm, drawing Y_{mis} from $f(Y_{mis}|Y_{obs}, \theta = \hat{\theta})$, where θ or some of its components is fixed at $\hat{\theta}$ (e.g., the maximum likelihood estimate of θ using observed data or complete cases). Using the combining rules, the variance from improper multiple imputations can be underestimated and thus lead to a lower coverage rate than the nominal level for the estimate. In principle, improper multiple imputation methods or algorithms need to be avoided.

Example 3.8. *Improper multiple imputation methods for the univariate missing data problem*

To illustrate the idea of improper multiple imputation and its consequences, we use the same simulation setting in Example 3.4. The DB imputation algorithm is proper because it accounts for the uncertainty of both μ and σ^2. We consider three improper multiple imputation methods sketched as follows:

1. Fixing both μ and σ^2: impute the missing values as $y_i^{*(m)} \sim N(\mu^{*(m)} = \overline{Y}_{obs}, \sigma^{*2(m)} = s_{obs}^2)$.

2. Fixing μ only: draw σ^2 from its posterior distribution and impute the missing values as $y_i^{*(m)} \sim N(\mu = \overline{Y}_{obs}, \sigma^{*2(m)})$.

3. Fixing σ^2 only: draw μ from its posterior distribution and impute the missing values as $y_i^{*(m)} \sim N(\mu^{*(m)}, \sigma^{*2(m)} = s_{obs}^2)$.

Table 3.7 shows the simulation results for both the proper and improper imputation methods based on $M = 50$. The biases for both the proper and improper imputations are tiny. When both μ and σ^2 are fixed at their sample estimates in the imputation, the variance is considerably underestimated ($\overline{\text{SE}} = 0.037 <$ SD=0.045). This leads to a low coverage rate (89.9%) and a seemingly narrower confidence interval than that of the proper imputation (0.144 vs. 0.169). When the improper imputation algorithm only fixes μ, the variance is still underestimated ($\overline{\text{SE}} = 0.039 <$ SD=0.045) and the coverage rate (91.8%) is still somewhat below the nominal level. Lastly, if the improper imputation algorithm only fixes σ^2, the results seem to be similar to those from the proper imputation and are satisfactory. In this example, the uncertainty of σ^2 in the model has a relatively small effect on the variance of the mean estimate.

3.4.3 Alternative Strategies

Both DB and DA algorithms are fully Bayesian. These imputation algorithms for many statistical models are developed, which will be discussed later in the book. In some scenarios, however, it might be challenging to devise and apply fully Bayesian algorithms. For example, in situations for multivariate data with nonlinear relations and complicated model structure, building one reasonable Bayesian model for the joint distributions of the variables, deriving the posterior distributions, programming the DB or DA algorithm, and assessing convergence of the draws may be difficult and time-consuming for researchers and practitioners. This section briefly lists several alternative strategies that are not fully Bayesian.

Once the complete-data model is defined, the most challenging component in the imputation is deriving the complete-data posterior distribution, $f(\theta|Y)$, and generating random draws from it. That is, the P-step in the DA algorithm. The I-step is usually more straightforward to implement. In Section 2.3.4, we show that Bayesian estimates are asymptotically equivalent to their frequentist counterparts in many cases. In addition, there exists the similarity between the posterior distribution and the bootstrap distribution of the estimate. These facts suggest that we might approximate $f(\theta|Y)$ using two strategies.

The first strategy is to use the asymptotic distribution of the maximum likelihood (ML) estimate of θ, $N(\hat{\theta}_{ML}, \hat{V}(\hat{\theta}_{ML}))$, to approximate $f(\theta|Y)$. A practical advantage is that in many statistical software packages, both the ML estimate and the covariance matrix estimate for established statistical models are readily available. Therefore if necessary, Step 4 of the DA algorithm can be replaced by:

4*. Estimate θ and its covariance matrix from the t-th version of completed data using ML methods, and denote them as $\hat{\theta}_{ML}^{(t)}$ and $\hat{V}(\hat{\theta}_{ML}^{(t)})$, respectively. Draw a new value of the parameter from an approximation of its complete-data posterior distribution, $\theta^{*(t+1)} \sim N(\hat{\theta}_{ML}^{(t)}, \hat{V}(\hat{\theta}_{ML}^{(t)}))$.

The second strategy is to use the bootstrap distribution of $\hat{\theta}_{ML}$, that is, the distribution of the ML estimate over bootstrap samples of Y, to approximate $f(\theta|Y)$. Unlike the first strategy, here the covariance matrix of the ML estimate is not needed. If necessary, Step 4 of the DA algorithm can be replaced by

4**. Draw a bootstrap sample of $(Y_{mis}^{(t)}, Y_{obs})$ as $Y_{com}^{(b)}$. Apply the ML method to $Y_{com}^{(b)}$ to obtain a new value of θ as $\theta^{*(t+1)}$.

Although not fully Bayesian, both strategies aim to approximate $f(\theta|Y)$ by propagating the uncertainty of the estimate of θ, and thus produce proper

TABLE 3.7
Example 3.8. Simulation results

MI Method	Bias	SD	$\overline{\text{SE}}$	MSE	CI	COV (%)
Proper	0	0.045	0.043	0.00203	0.169	94.9
Improper fixing μ and σ^2	0.003	0.045	0.037	0.00204	0.144	89.9
Improper fixing μ	−0.001	0.045	0.039	0.00203	0.152	91.8
Improper fixing σ^2	−0.001	0.045	0.043	0.00203	0.171	94.9

multiple imputations. Similar ideas can be applied to the DB algorithm in some cases. More related discussion can be found in Little and Rubin (2020, Section 10.2.3).

3.5 Practical Implementation

In the early days of researching and applying multiple imputation, researchers and practitioners had to write their own computer code to implement the imputation algorithms. Nowadays, common statistical software packages, such as SAS (https://www.sas.com/en_us/home.html), R (https://www.r-project.org/), and STATA (https://www.stata.com/), have incorporated automatic multiple imputation routines or libraries. Many of the established imputation models and algorithms are included, and therefore users do not have to program the detailed imputation steps by themselves. These imputation programs also include automatic steps for combining analysis results from multiply imputed datasets (e.g., SAS PROC MIANALYZE). The availability of multiple imputation software programs greatly promotes the use of multiple imputation analysis in practice.

Several books on multiple imputation also focus on specific imputation programs. Examples include R mice in Van Buuren (2018), IVEware in Raghunathan et al. (2018), and SAS PROC MI/PROC MIANALYZE in Berglund and Heeringa (2014). Imputation software programs are often accompanied by manuals and documents that explain both the general concepts and specific syntaxes. On the other hand, new ideas and methods for multiple imputation keep coming and are being incorporated in these programs. Users are recommended to do a literature search to obtain the most updated information. This book contains sample program code for some examples. For other examples, sample code can be found in (https://github.com/he-zhang-hsu/multiple_imputation_book).

In many cases, we believe that the same imputation method can be implemented by alternative software packages. We also note that imputation programs have multiple options of models or methods for the same type of missing data problems. Therefore users often need to make a reasonable choice based on correct understanding of the statistical concepts and ideas behind multiple imputation analysis. Correspondingly, a major focus of this book is to provide explanations and illustrations of commonly used imputation models and their targeted missing data problems. It is our hope that practitioners can have a solid understanding of these methods and related issues before implementing them, either using existing software programs or writing their own code.

Another topic concerns the presentation or publication of multiple imputation analysis. If the research work centers on the theory and methodology, the paper can surely include enough technical details. On the other hand, the report of the multiple imputation analysis for subject-matter papers needs to be both precise and concise. In many cases, the methods used might be based on existing literature so it is unnecessary to provide all the details. However, citations of the major reference (e.g., Rubin 1987) or the software procedures (e.g., SAS PROC MI) are very important. Van Buuren (2018, Section 12.2) provided some guidelines and templates for reporting multiple imputation analysis for subject-matter papers. A review of current practices in the medical field can be found in Rezvan et al. (2015).

3.6 Summary

An excellent description of the early history of multiple imputation can be found in Van Buuren (2018, Sections 2.1). Multiple imputation can be well justified as an approximation to a Bayesian analysis with missing data. However, the separation between the imputation and analysis/combining stages is a unique feature that brings in many advantages to analysts. These features will be shown in later chapters of the book. In principle, multiple imputations are random draws from the posterior predictive distribution (or its approximations) of missing values, and they are not the model-based predictions. The DA algorithm is a general-purpose strategy of creating imputations and making them proper. The combining rules are needed in multiple imputation analysis. For regular analyses, the combining procedure is rather straightforward to apply: the only quantities needed are the point estimates and their variance estimates from multiple completed datasets. The combined statistical inferences tend to have desirable statistical properties. Some new ideas of combining analysis results will be discussed in Section 14.3. Automatic procedures for many imputation models and analyses are enabled in multiple statistical software packages.

In this chapter we mainly use examples from a univariate missing data problem without any covariate. Such problem rarely happens in actual settings, yet many of the implied basic ideas and properties of multiple imputation analysis also apply for more complex and realistic missing data problems. Starting from Chapter 4, we will provide detailed illustration and discussion for missing data problems that are more practically relevant.

4

Multiple Imputation for Univariate Missing Data: Parametric Methods

4.1 Overview

In Chapters 4 and 5 we focus on missing data problems with one incomplete variable and other variables fully observed, as depicted in Table 1.2. Chapter 4 presents some multiple imputation methods based on parametric models. Section 4.2 introduces the imputation method for a continuous variable using normal linear models. Section 4.3 presents imputation methods for a non-continuous variable using generalized linear models. Section 4.4 focuses on the problem of having a missing covariate in a targeted analysis. Section 4.5 provides a summary.

4.2 Imputation for Continuous Data Based on Normal Linear Models

Suppose Y is an incomplete variable and there exist p ($p > 1$) fully observed other variables (i.e., covariates) in the data. Suppose the complete-data sample size is n; denote the $n \times (p+1)$ covariate matrix as $X = (\underline{1}, X_1, \ldots X_p)$, where $\underline{1}$ is the vector of 1's (for the intercept). We also assume that the missingness of Y is ignorable, that is, $f(R = 1|Y, X) = f(R = 1|X)$, where R is the response indicator of Y. If Y is continuous, then a commonly used imputation model is a normal linear regression model for Y conditional on X: for subject i ($i = 1, \ldots, n$),

$$y_i = x_i\beta + \epsilon_i, \tag{4.1}$$

where x_i is the i-th row of the X-covariate matrix, β is a $(p+1) \times 1$ vector of regression coefficients, and $\epsilon_i \sim N(0, \sigma^2)$. From a Baysian perspective, we can impose a noninformative prior distribution, $\pi(\beta, \sigma^2) \propto \sigma^{-2}$, to complete the model specification, although other prior distributions can also be considered.

Because of the univariate missingness pattern, the imputation algorithm can follow the direct Bayesian (DB) strategy (Section 3.4.1). It is sufficient

DOI: 10.1201/9780429156397-4

to use the information from cases with both Y and X observed to infer about β and σ^2, as shown in Example 2.4. We first fit Model (4.1) using the data from complete cases and obtain the least-squares estimate for β: $\hat{\beta}_{obs} = (X_{obs}^T X_{obs})^{-1} X_{obs}^T Y_{obs}$, where Y_{obs} is the vector of observed y's and X_{obs} is the corresponding subset of the X-covariate matrix. The residual sum of squares, or the sum of squared error denoted as SSE_{obs}, is calculated over n_{obs} observed cases as $SSE_{obs} = \sum_{i,obs}(y_i - x_{i,obs}\hat{\beta}_{obs})^2$. The DB imputation algorithm (Rubin 1987, Section 5.3) consists of the following steps:

1. Draw β and σ^2 from their posterior distributions: $\sigma^{2(m)} \sim SSE_{obs}/\chi^2_{n_{obs}-p-1}$ and $\beta^{(m)} \sim N(\hat{\beta}_{obs}, (X_{obs}^T X_{obs})^{-1}\sigma^{2(m)})$.

2. Impute the missing values as $y_{i,mis}^{(m)} \sim N(x_{i,mis}\beta^{(m)}, \sigma^{2(m)})$ independently for missing y_i's.

3. Repeat Steps 1 and 2 independently M times from $m = 1, \ldots M$.

Step 1 of the imputation algorithm ensures that the multiple imputations are proper (Section 3.4.2). This is because both the parameters, β and σ^2, and imputations for missing values are random draws from their posterior (predictive) distributions. Detailed theoretical properties of the multiple imputation estimator based on normal linear models can be found in Schenker and Welsh (1988) and Kim (2004). In practice, both Model (4.1) and the imputation algorithm have been included in a variety of imputation software programs.

Example 4.1. *A simulation study for the normal linear regression imputation*

In Example 2.5, we used a simulation study to show that none of the ad hoc missing data methods (i.e., CC, RP, and INDI) can provide optimal results among all scenarios tested. Here we use the same simulation setup to assess the performance of the multiple imputation estimates under Model (4.1). Multiple imputation (MI) is implemented using R mice. The number of imputations M is set as 50. Tables 4.1 and 4.2 show the results from MI under MCAR and MAR, respectively. For better readability, we include the results from other methods here as well. Clearly, MI yields good results in all cases: for all the estimands considered, both the point and variance estimates have little bias, and the coverage rates are close to the nominal level.

Several extra remarks are given. Under MCAR, estimates from both CC and MI are unbiased for estimating the mean of Y. Yet MI is more efficient and results in a shorter confidence interval ($0.202 < 0.226$). This is because MI uses additional information in X-cases for which Y is missing, and such information is discarded by CC. Under MAR, MI can correct the bias of CC for estimating the mean of Y. Under both MCAR and MAR, CC has the best performance for estimating the regression coefficient of Y on X. MI is almost as efficient as CC as their lengths of confidence intervals are rather close. On the other hand, MI is a clear winner among all missing data methods for estimating the slope coefficient of regressing X on Y, as it corrects the bias and gains extra efficiency compared with other methods.

TABLE 4.1
Example 4.1. Simulation results under MCAR

Method	Bias	SD	$\overline{\text{SE}}$	MSE	CI	COV (%)
			Mean of Y			
BD	0	0.043	0.045	0.00188	0.175	96.1
CC	−0.001	0.057	0.058	0.00322	0.226	94.7
RP	−0.001	0.051	0.040	0.00259	0.157	87.6
MI	0.003	0.051	0.051	0.00260	0.202	95.3
			Slope coefficient of regressing Y on X			
BD	0	0.032	0.032	0.000995	0.124	95.0
CC	−0.001	0.039	0.041	0.00152	0.160	96.0
RP	−0.001	0.039	0.024	0.00152	0.096	78.4
MI	0.003	0.039	0.041	0.00155	0.164	96.2
			Slope coefficient of regressing X on Y			
BD	0	0.016	0.016	0.000251	0.062	95.0
CC	0	0.020	0.020	0.000406	0.080	95.2
INDI	0	0.020	0.024	0.000406	0.095	98.1
RP	0.126	0.019	0.015	0.0162	0.060	0
MI	−0.001	0.018	0.018	0.000327	0.070	94.6

Note: MI: multiple imputation using a normal linear model.

TABLE 4.2
Example 4.1. Simulation results under MAR

Method	Bias	SD	$\overline{\text{SE}}$	MSE	CI	COV (%)
			Mean of Y			
BD	0	0.043	0.045	0.00188	0.175	96.1
CC	−0.350	0.056	0.056	0.1255	0.220	0
RP	−0.001	0.056	0.040	0.00308	0.156	83.0
MI	0.005	0.056	0.055	0.00312	0.218	94.5
			Slope coefficient of regressing Y on X			
BD	0	0.032	0.032	0.000995	0.124	95.0
CC	−0.001	0.045	0.045	0.00204	0.178	95.7
RP	−0.001	0.045	0.024	0.00204	0.095	70.4
MI	0.006	0.045	0.047	0.00207	0.185	95.8
			Slope coefficient of regressing X on Y			
BD	0	0.016	0.016	0.000251	0.062	95.0
CC	−0.045	0.021	0.021	0.00246	0.081	41.5
INDI	−0.045	0.021	0.024	0.00246	0.094	53.1
RP	0.132	0.020	0.015	0.0179	0.060	0
MI	−0.002	0.019	0.018	0.000350	0.070	94.7

Fig. 4.1 shows histograms for deleted Y-values and three sets of imputations from a randomly chosen subset (sample size 250) of one simulation replicate under MAR. The imputed values retain well the marginal distribution of the original missing values. Fig. 4.2 shows scatter plots between X and Y. Again, imputed values capture well the relationship between X and Y, not only for the mean linear relation, but also the joint distribution of the two variables. Readers can compare these plots with those shown for the regression prediction method (Fig. 2.2). In addition, imputed values are also varying across different sets of imputations to preserve the uncertainty of the missing data. Note that multiply imputed values are not identical with the original missing values, as the former are not targeted to recover the latter but are created to yield good statistical inference for the analysis of interest, as demonstrated in Tables 4.1 and 4.2.

Example 4.2. *Multiple imputation of gestational age in birth data using a normal linear regression model*

We illustrate the normal linear regression imputation method using a real-data example. U.S. national natality data are collected by state vital records offices and compiled by NCHS, providing access to a wide variety of information on maternal and infant health characteristics and associated demographics (https://www.cdc.gov/nchs/nvss/births.htm). One variable of major scientific interest is gestational age, which measures the length of pregnancy. Accurate information from this variable in vital records is necessary for determining preterm delivery rates, creating fetal growth charts, and establishing guidelines for monitoring and treating expectant mothers who are most susceptible to preterm labor. Before the 2014 data year, the primary measure used to determine gestational age of the newborn was the interval between the first day of the mother's last normal menses (LMP) and the date of birth. The LMP-based measure of gestational age can be subject to error due to imperfect maternal recall, misidentification of the LMP because of postconception bleeding, delayed ovulation, or intervening early miscarriage (Martin et al. 2003). Parker and Schenker (2007) provided an overview of the data problem of gestational age and proposed using multiple imputation as a potential solution. Starting in 2014, NCHS transitioned to a new standard for estimating the gestational age subject to less error. That is, the obstetric estimate of gestation at delivery replaced the measure based on LMP.

Example 4.2 (and subsequent related examples in this book) is framed to address the issue of incomplete and inaccurate gestational age data where LMP was the major measurement. For the purpose of illustration, we use the 2002 public-use natality data that consists of records from around four million live births. For simplicity, our analytic sample only includes a randomly chosen 10% subset of the original dataset, resulting in a sample size of 40274.

In the 2002 natality file, gestational age is: (a) computed using date of birth of the child and date of LMP; (b) imputed from the LMP date; (c) reported to be the clinical estimate; or (d) reported as unknown when there is insufficient data to impute or no valid clinical estimate. NCHS published imputed weeks

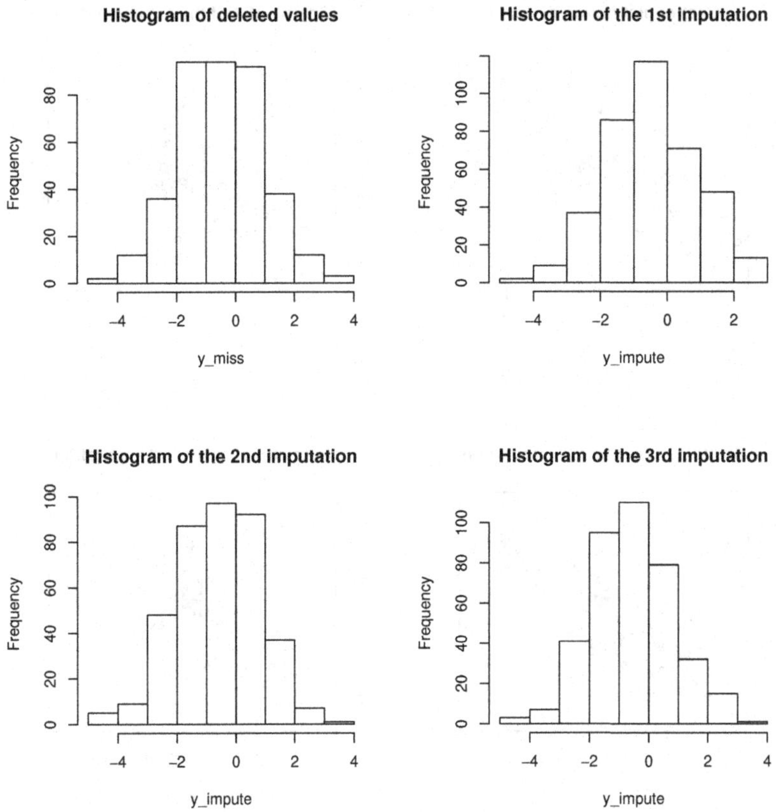

FIGURE 4.1
Example 4.1. Histogram plots of deleted values and multiply imputed values. Top left: deleted values; top right: the 1st imputation; bottom left: the 2nd imputation; bottom right: the 3rd imputation.

of gestation for records missing the day of LMP when there was a valid month and year using a hot-deck method (Taffel et al. 1982) (more detailed discussion about the hot-deck imputation can be found in Section 5.5.1). The clinical estimate was used in three situations: (1) if the LMP date was not reported; (2) when the computed gestational age was outside the reasonable code range (17-47 weeks); or (3) when there were large inconsistencies between gestational age and birth weight. There are around 4.6% of the births in the dataset based on the clinical estimate of gestation. Although LMP-based gestational ages were edited for obvious inconsistences with the infant's plurality and birth weight, reporting problems for this data item persisted, most frequently

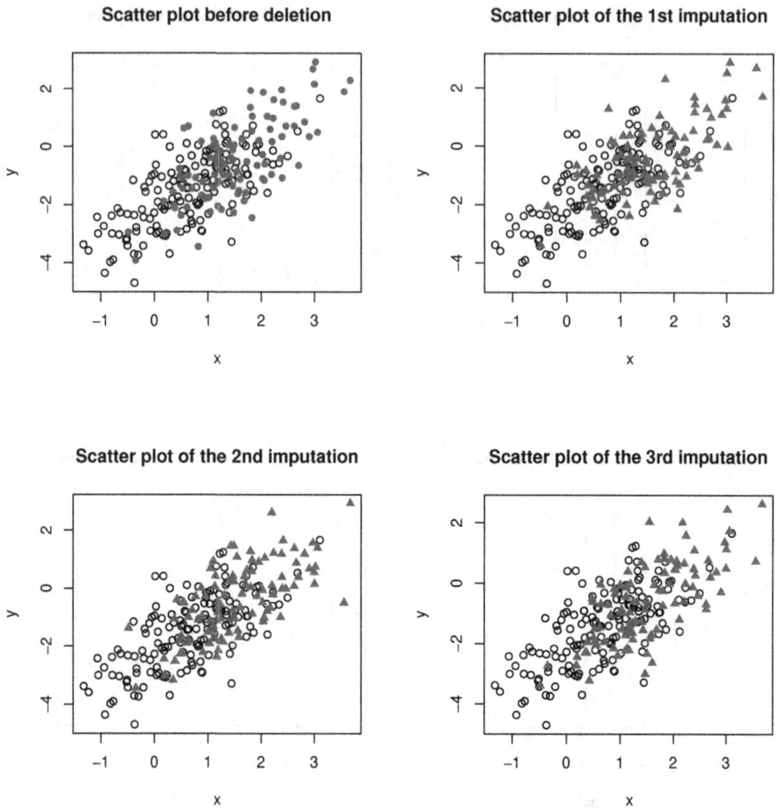

FIGURE 4.2
Example 4.1. Scatter plots of X and Y. Top left: before deletion; top right: the 1st imputation; bottom left: the 2nd imputation; bottom right: the 3rd imputation. Circle: observed values; filled circle: deleted values; filled triangle: imputed values.

among certain subpopulations and births with shorter gestations (Alexander and Allen 1996).

Because of the complexities associated with misspecified or incomplete gestational ages, we set gestational age (unit in weeks), labeled as DGESTAT in the dataset, to missing if the computed length of gestation differed from the clinical estimate by more than two weeks, if it was replaced with the clinical estimate, or if the imputed gestational age was created by the hot-deck method. After these alterations, there were 18.7% missing values for DGESTAT. Note that this process is merely a simplification of the original problem, and the creation of missing DGESTAT is mainly for the illustration of multiple imputation methods in this book. The creation of missing values and subsequent imputation in this dataset does not represent any official NCHS procedure for handling gestational ages in natality files. See Zhang et al. (2014) for a more rigorous approach to the measurement error issue in gestational age based on LMP.

The original natality file consists of around 200 variables, some of which also have missing data. To further simplify the analysis, we used only a limited number of variables that were at least moderately correlated to either DGESTAT and/or its missingness indicator based on some exploratory analyses. The idea of choosing useful predictors in imputation models will be further discussed in later chapters of the book. In this example to frame it as a univariate missing data problem, we removed incomplete cases from these covariates so that all covariates were fully observed. This resulted in a dataset of sample size 24494, including 4142 of cases with missing DGESTAT. A more systematic treatment of multiple incomplete variables in this dataset will be discussed in Chapters 6 and 7. Table 4.3 shows the information of the variables and some descriptive statistics. Among the covariates, DBIRWT (baby's birthweight) has the largest correlation ($r \approx 0.60$ $p < 0.0001$) with DGESTAT.

TABLE 4.3
Example 4.2. Descriptive statistics of the dataset ($n = 24494$)

Variable	Definition	Mean	SD	Range
DGESTAT*	Gestational age (weeks)	38.674	2.081	19-44
DBIRWT	Baby's birth weight (grams)	3325	585.8	227-5670
DMAGE	Age of mother (years)	28.070	5.976	13-49
DFAGE	Age of father (years)	30.600	6.762	15-67
FMAPS	Apgar score	8.929	0.669	0-10
WTGAIN	Weight gain of mother (pounds)	31.070	13.464	0-98
NPREVIS	Total number of prenatal visits	11.764	3.669	0-49
CIGAR	Number of cigarettes per day	0.998	3.749	0-60
DRINK	Number of drinks per week	0.0187	0.411	0-35

Note: SD: standard deviation. * The descriptive statistics of DGESTAT are calculated based on 20352 cases with observed values.

We impute the missing DGESTAT based on Model (4.1) 20 times using SAS PROC MI. We conduct three post-imputation analyses: (1) estimating the mean of DGESTAT; (2) regressing DGESTAT on the covariates; and (3) regressing DBIRWT on DGESTAT and other covariates. Table 4.4 shows the results of the multiple imputation (MI) analysis and complete-case (CC) analysis. The mean estimate of DGESTAT from MI is slightly lower than that from CC (38.64 vs. 38.67). The standard error of the former is also slightly lower than that of the latter (.0140 vs. .0146), indicating some gain of precision by using MI. For the analysis of regressing DGESTAT on the covariates, the results from the two methods are rather similar as expected. If DGESTAT is used as a covariate in the regression model for DBIRWT, there exist moderate differences on coefficient estimates between the two methods. The standard errors from MI are apparently smaller than those from CC, again showing some gain of precision from the former. The pattern of estimates in this real example is consistent with that of the simulation study in Example 4.1.

TABLE 4.4
Example 4.2. Multiple imputation analysis results

Estimand	CC		MI	
Mean of DGESTAT	EST	SE	EST	SE
	38.67	0.0146	38.64	0.0140
Regression for DGESTAT on covariates				
Predictor	EST	SE	EST	SE
Intercept	27.62	0.1746	27.61	0.1706
DBIRWT	0.00209	0.00002037	0.00208	0.00002076
DMAGE	−0.03872	0.00291	−0.03847	0.00294
DFAGE	0.00586	0.00256	0.005664	0.00256
FMAPS	0.5299	0.01766	0.5309	0.01844
WTGAIN	−0.00542	0.000861	−0.00541	0.000870
NPREVIS	0.03581	0.00314	0.03594	0.00307
CIGAR	0.01472	0.00319	0.01428	0.00334
DRINK	−0.00666	0.03472	−0.01106	0.03308
Regression for DBIRWT on DGESTAT and other covariates				
Predictor	EST	SE	EST	SE
Intercept	−3751	67.22	−3886	63.44
DGESTAT	163.08	1.5925	165.55	1.54
DMAGE	10.80	0.8141	10.67	0.7741
DFAGE	−1.4119	0.7166	−1.0638	0.6785
FMAPS	40.55	5.0367	43.98	4.8063
WTGAIN	5.5018	0.2379	5.3954	0.2263
NPREVIS	−0.1952	0.8794	−0.1768	0.8252
CIGAR	−10.8150	0.8899	−10.7636	0.8361
DRINK	2.0136	9.7089	3.9749	7.9001

Note: EST: the estimate for the mean or regression coefficient; SE: the standard error of the estimate.

To run some simple diagnostics, it is always helpful to get a sense of what imputed values "look" like by plotting them. Fig. 4.3 includes histograms of observed and imputed values of DGESTAT. The observed values are somewhat skewed to the left. The imputed data appear to be more symmetrically distributed, and the distribution appears to be a little wider because there are some imputed values that are outside the range of the observed values. Overall, it is often recommended to use plots to examine the distribution of imputed values and compare it with that of observed values, and thus to gain some insights of the adequacy of imputations. This strategy will be applied throughout the book. Chapter 12 will present an in-depth discussion of imputation diagnostics.

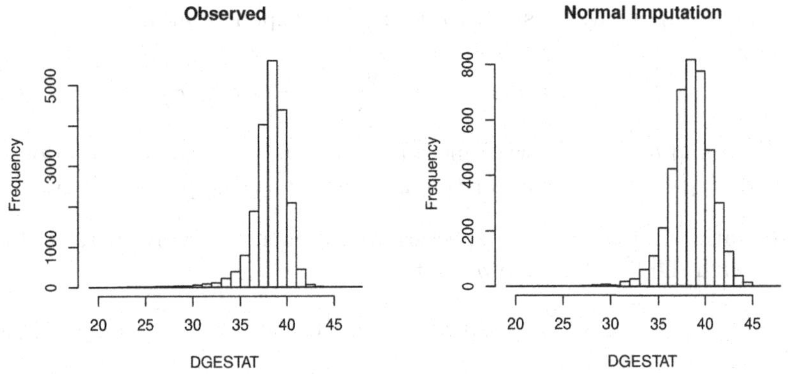

FIGURE 4.3
Example 4.2. Left: the histogram of observed DGESTAT; right: the histogram of imputed DGESTAT from one randomly chosen imputation.

In Example 4.2 we have made quite a few simplifications to the original problem and dataset to make the illustration easier. However, this example indeed implies that real-life missing data problems are often rather complicated and can involve multiple data and subject-matter related issues. Here we only apply a simple normal linear regression model for imputation as a starting point. Multiple questions remain. For example, how good is the normal linear regression imputation model? Can we further improve it? etc. We will continue using the missing data problem of gestational age for illustration as we move to subsequent topics of multiple imputation.

4.3 Imputation for Noncontinuous Data Based on Generalized Linear Models

Arguably, most of the variables encountered in practical data are not in continuous forms. We first introduce some background of generalized linear models (GLM), which are often used to model, impute, and analyze noncontinuous variables. We then discuss several imputation strategies for binary data, which take the basic form of categorical variables. Finally we present some brief reviews on imputation strategies for others types of noncontinuous variables.

4.3.1 Generalized Linear Models

In brief for subject i in the data, a GLM consists of:

(A) Random component: Y_i is generated from an exponential family of distributions:

$$f(Y_i = y_i; \theta_i, \phi) = exp\{\frac{\theta_i y_i - b(\theta_i)}{a(\phi)} + c(y_i; \phi)\}, \tag{4.2}$$

where θ_i and $b(\cdot)$ are known functions that determine the distribution of Y, and $c(y_i; \phi)$ is a known function indexed by a scale parameter ϕ;

(B) Systematic component: the expectation of y_i, denoted as μ_i, is related to x_i through the link function g, that is,

$$\mu_i = E(y_i|x_i) = g^{-1}(x_i\beta), \tag{4.3}$$

where x_i is the i-th row of the covariate matrix X. The relationship between x_i and y_i can be modeled by linking θ_i and μ_i in various ways. For a canonical link g, we have $g(\mu_i) = \theta_i = x_i\beta$.

The normal linear regression of Y on X in Model (4.1) also falls in the framework of GLM. This can be seen by setting $g(\mu_i) = \mu_i = x_i\beta$, the "identity" link function, and $a(\phi) = \phi = \sigma^2$. More importantly, GLM can be used to model variables that are not normally distributed (e.g., categorical variables). The method of maximum likelihood (MLE) is typically used to estimate the unknown parameter β (and ϕ) in GLM. This can be achieved by iterative reweighted least squares. Algorithms fitting GLM models are readily available in statistical software packages. More details about GLM can be found, for example, in McCullagh and Nelder (1989).

4.3.2 Imputation for Binary Data

4.3.2.1 Logistic Regression Model Imputation

For binary data $Y = 1$ or 0, a natural GLM strategy for Y is a logistic regression model:

$$log\frac{f(Y=1)}{1-f(Y=1)} = X\beta, \tag{4.4}$$

or $logit(f(Y=1)) = X\beta$. For case i, Y_i follows a Bernoulli distribution of size 1 and probability of "success" equal to $f(Y_i = 1) = \mu_i = \frac{exp(x_i\beta)}{1+exp(x_i\beta)}$. Model (4.4) is a special case of GLM by setting $g(\mu_i) = log(\mu_i/(1-\mu_i))$ (i.e., the "logit" link) and $a(\phi) = 1$.

Unlike the case of normal linear models in Eq. (4.1), the posterior distribution for β in logistic regression models does not have a closed form, as $f(\beta|Y, X) \propto \prod_{y_i=1} \frac{exp(x_i\beta)}{1+exp(x_i\beta)} \prod_{y_i=0} \frac{1}{1+exp(x_i\beta)}$. Rubin (1987, Section 5.2) proposed an approximate Bayesian imputation algorithm. As outlined in Section 3.4.3, the idea is to approximate the posterior distribution of β using the asymptotic distribution of the MLE: the posterior mean is approximated by the MLE of β, denoted by $\hat{\beta}_{MLE,obs}$; the posterior variance is approximated by the covariance-matrix estimate of $\hat{\beta}_{MLE,obs}$, denoted by $\hat{V}(\hat{\beta}_{MLE,obs})$. Both $\hat{\beta}_{MLE,obs}$ and $\hat{V}(\hat{\beta}_{MLE,obs})$ can be easily obtained by fitting the logistic regression model using cases with both Y and X observed. The imputation algorithm is sketched as follows:

1. Draw β from its approximate posterior distribution as $\beta^{(m)} \sim N(\hat{\beta}_{MLE,obs}, \hat{V}(\hat{\beta}_{MLE,obs}))$.

2. Impute missing Y_i's from independent Bernoulli distributions with the rate of success as $f(Y_{i,mis} = 1) = \frac{exp(x_{i,mis}\beta^{(m)})}{1+exp(x_{i,mis}\beta^{(m)})} = logit^{-1}(x_{i,mis}\beta^{(m)})$. More specifically, for each of the missing cases, calculate $logit^{-1}(x_{i,mis}\beta^{(m)})$, and draw n_{mis} (the number of missing cases) independent uniform (0,1) random numbers, u_i. If $u_{i,mis} > logit^{-1}(x_{i,mis}\beta^{(m)})$, impute $Y_{i,mis} = 0$; otherwise impute $Y_{i,mis} = 1$.

3. Repeat Steps 1 and 2 independently M times from $m = 1, \ldots M$.

The above imputation algorithm appears to be the main algorithm used in major imputation software programs. Nowadays, however, automatic procedures sampling β from its exact posterior distribution are readily available. A lingering question is whether using the latter for imputation offers any advantage.

Example 4.3. *A simulation study comparing the performance of alternative logistic regression imputation algorithms*

We conduct a simple simulation study to compare the performance of alternative logistic regression imputation algorithms including:

1. MLE-MI: approximating the posterior distribution of β using the normal approximation, $N(\hat{\beta}_{MLE,obs}, \hat{V}(\hat{\beta}_{MLE,obs}))$.

2. BOOT-MI: approximating the posterior distribution of β using the bootstrap distribution of $\hat{\beta}_{MLE,obs}$, which is obtained from the nonparametric bootstrap samples of the observed cases. See Section 3.4.3.

3. Bayes-MI: drawing β from its exact posterior distribution assuming a noninformative prior $\pi(\beta) \propto 1$.

We posit a complete-data model: $logit(f(Y = 1)) = \beta_0 + \beta_1 X$, where $X \sim N(1, 1/4)$. We choose the coefficients so that the proportion of $Y = 1$ deceases as β_1 decreases: $\beta_0 = -2$ and $\beta_1 = \{2, 1, 0.5, -1, -2, -3\}$. The complete-data sample sizes include 1000 and 100, and missing values on Y are generated under MCAR with around 20% missingness rate. The simulation is based on 1000 replicates. We focus on two estimands, the marginal mean, $\overline{Y}$, and the slope coefficient of the logistic regression, β_1. Both the MLE-MI and BOOT-MI are implemented by R mice. For Bayes-MI, we obtain the posterior draws of β_1 using R arm (the function Bayesglm). The multiple imputation analyses are based on $M = 20$ imputations.

Table 4.5 shows the main simulation results. Since $\overline{Y}$ can be small, we quantify the bias using the relative bias (RBIAS), which is the ratio of the estimate over the true value in the percentage form. All the methods work similarly well for both the mean and slope coefficient: their biases are close to zero and their MSEs are similar. Coverage rates of the estimates are all around the nominal level and not shown. In the tested scenarios, there is no apparent advantage from Bayes-MI compared with two other methods, suggesting that the approximation of the posterior distribution of β in the latter imputation algorithms work well. The simulation results with complete-data sample size 100 are similar and not shown here.

TABLE 4.5
Example 4.3. Simulation results

Estimand	CC		MLE-MI		BOOT-MI		Bayes-MI	
	RBIAS	MSE	RBIAS	MSE	RBIAS	MSE	RBIAS	MSE
$\overline{Y} = 0.500$	0	0.000292	0	0.000280	0	0.000286	0	0.000285
$\beta_1 = 2$	0.1	0.034	0.7	0.035	0.7	0.034	0.3	0.034
$\overline{Y} = 0.280$	−0.1	0.000257	−0.2	0.000257	−0.3	0.000256	−0.1	0.000258
$\beta_1 = 1$	−0.1	0.027	1.5	0.028	0	0.028	0.5	0.027
$\overline{Y} = 0.185$	−0.1	0.000190	−0.7	0.000196	−0.5	0.000189	−0.7	0.000192
$\beta_1 = 0.5$	0.2	0.034	4.0	0.036	−0.2	0.035	0.7	0.035
$\overline{Y} = 0.0527$	0	0.0000572	−2.0	0.0000564	−0.8	0.0000588	1.5	0.0000589
$\beta_1 = -1$	−0.2	0.111	−0.6	0.113	1.9	0.114	−0.7	0.110
$\overline{Y} = .0280$	0.2	0.0000337	−1.1	0.0000323	−0.7	0.0000335	1.6	0.0000346
$\beta_1 = -2$	0.5	0.241	−0.8	0.242	1.1	0.245	0.9	0.243
$\overline{Y} = 0.0183$	0.2	0.0000243	−0.3	0.0000232	−0.4	0.0000245	1.5	0.0000245
$\beta_1 = -3$	0.7	0.501	−1.3	0.476	1.0	0.516	1.1	0.494

Note: RBIAS: relative bias (%).

When $\beta_1 = -3$ in data generation, warning messages of fitting logistic regression models begin to appear in the simulated data. If β_1 further moves below −3 in data generation so that $\overline{Y}$ is further closer to 0 due to fewer Y = 1s generated, more warning messages appear when fitting the logistic regression. This is the issue of data separation in logistic regression analysis, which will be discussed in Section 4.3.2.4.

4.3.2.2 Discriminant Analysis Imputation

Discriminant analysis is a technique concerned with separating distinct sets of observations and with allocating new observations to previously defined groups. The term "discrimination" in this context can be traced back to Fisher (1938). For example, suppose X is a univariate continuous variable over two populations/classes ($Y = 1$ and 0); then a normal discriminant analysis model is

$$\begin{aligned} f(X|Y=1) &\sim N(\mu_1, \sigma_1^2), \\ f(X|Y=0) &\sim N(\mu_0, \sigma_0^2), \\ Y &\sim Bernoulli(p). \end{aligned} \tag{4.5}$$

Model (4.5) also defines X as a mixture of two normal distributions with a binary grouping indicator Y. Assuming equal costs of misclassification, for $X = x_0$, it can be allocated to population 1 ($Y = 1$) if $-\frac{1}{2}\frac{x_0^2}{1/\sigma_1^2 - 1/\sigma_0^2} + (\mu_1/\sigma_1^2 - \mu_0/\sigma_0^2)x_0 - k >= log(1 - p/p)$, where $k = \frac{1}{2}log\frac{\sigma_1^2}{\sigma_0^2} + \frac{1}{2}(\mu_1^2/\sigma_1^2 - \mu_0^2/\sigma_0^2)$, and can be allocated to population 0 ($Y = 0$) otherwise. More details about discrimination and classification analysis can be found, for example, in Johnson and Wichern (2007, Chapter 11).

Model (4.5) can also be used as a complete-data model for imputing incomplete Y-values. It can be shown that under Model (4.5), we have

$$f(Y = 1|X) = \frac{exp(\alpha_0 + \alpha_1 X + \alpha_2 X^2)}{1 + exp(\alpha_0 + \alpha_1 X + \alpha_2 X^2)}, \tag{4.6}$$

where $\alpha_0 = log(\frac{\sigma_0 p}{\sigma_1(1-p)}) + \frac{\sigma_1^2\mu_0^2 - \sigma_0^2\mu_1^2}{2\sigma_0^2\sigma_1^2}$, $\alpha_1 = \frac{\mu_1\sigma_0^2 - \mu_0\sigma_1^2}{\sigma_0^2\sigma_1^2}$, and $\alpha_2 = \frac{\sigma_1^2 - \sigma_0^2}{2\sigma_0^2\sigma_1^2}$. Assuming prior distribution for parameter $\theta = (\mu_1, \mu_0, \sigma_1^2, \sigma_0^2, p)$ as $\pi(\theta) \propto \sigma_1^{-2}\sigma_0^{-2}$, the complete-data posterior distribution for θ can be summarized as: for $j = 0, 1$ (the group of $Y = 0$ and 1), $f(\sigma_j^2|X, Y) \sim (n_j - 1)s_j^2/\chi^2_{n_j - 1}$, where $(n_j - 1)s_j^2$ is the sum-of-squares of X-values in group j, $f(\mu_j|X, Y, \sigma_j^2) \sim N(\overline{X}_j, \sigma_j^2/n_j)$ (see Example 3.3), and $f(p|Y, X) \sim Beta(n_1 + 1, n_0 + 1)$ (see Eq. (2.13)), where n_j is the number of cases in group j.

For the scenario where Y has some missing values and X is fully observed, a data augmentation (DA) imputation algorithm can be sketched as follows:

1. Begin with an estimate or guess of the parameter θ, say $\theta^{*(t)}$ (where $t = 0$ at the 1st iteration). These estimates can be based on the observed-data statistics.

2. Draw a value of the missing Y based on the Bernoulli distribution defined in Eq. (4.6).

3. Once Y is completed, draw a new value of $\theta^{*(t+1)}$ using the complete-data posterior distribution specified above.

4. Repeat Steps 2 and 3 until the convergence for $(\theta^{*(t)}, Y_{mis}^{*(t)})$ is satisfied, say at $t = T$. Note that imputations for missing Y's can be different across the iterations. The draws of $Y_{mis}^{*(T)}$ constitute the 1st ($m = 1$) set of imputations, $Y_{mis}^{(m)}$.

5. Repeat Steps 1 to 4 independently M times.

This discriminant analysis imputation can be generalized to handle multiple X covariates or a nominal variable Y with more than two categories (i.e., a polytomous variable). Details can be found, for example, in Brand (1999). The discriminant analysis imputation method is implemented in several software packages (e.g., SAS PROC MI and R mice) as an alternative to the logistic regression imputation.

Which method do we use, the logistic regression or discriminant analysis imputation? Van Buuren (2018) commented that the discriminant imputation is inferior to the logistic regression imputation. Here we provide a more detailed discussion on this issue. To narrow the scope, suppose $\sigma_0^2 = \sigma_1^2 = \sigma^2$ under Model (4.5) (i.e., equal variance); then $\alpha_0 = log\frac{p}{1-p} + \frac{\mu_0^2 - \mu_1^2}{2\sigma^2}$, $\alpha_1 = \frac{\mu_1 - \mu_0}{\sigma^2}$, and $\alpha_2 = 0$ in Eq. (4.6), implying a logistic regression model in the form of Eq. (4.4). Our following discussion focuses on this setup.

A key distinction between the two approaches is that the discriminant analysis model assumes a full joint model for Y and X, that is, $f(Y, X)$, whereas the logistic regression model only assumes the conditional model for Y given X, that is, $f(Y|X)$. The former approach is shown to be more efficient than the latter in terms of classification when the joint model, $f(Y, X)$, is correctly specified (Efron 1975). Does this conclusion also apply to the case of imputation?

Example 4.4. *A simulation study comparing the efficiency between the discriminant analysis and logistic regression imputation under a discriminant analysis model*

We conduct a simulation study to compare the performance of the discriminant analysis model imputation (LDA-MI) with that of the logistic regression imputation (LOGIT-MI). We posit a complete-data model under the discriminant analysis: $Y \sim Bernoulli(p)$, $X = 1 + cY + \epsilon$, where $\epsilon \sim N(0, 1)$. We set $p = 0.5$ and $c = 1, 2, 3, 4$. Under this discriminant analysis model, the conditional distribution of Y given X follows a logistic model $logit(f(Y = 1)) = \alpha_0 + \alpha_1 X$, where $\alpha_0 = \frac{-c(2+c)}{2}$ and $\alpha_1 = c$. As c (the logistic regression slope coefficient) increases, X is better separated between $Y = 1$ and $Y = 0$. The complete-data sample size is 1000, and missing values on Y are generated under MCAR with around 40% missingness rate. The simulation is based on 1000 replicates. Both imputation methods are implemented using R mice. We assess both the marginal mean estimate ($\overline{Y}$) and the slope coefficient estimate (α_1) of the logistic regression. The multiple imputation analysis is based on $M = 50$ imputations.

For estimating $\overline{Y}$, both imputation methods yield similar and good results, which are not shown. Table 4.6 presents the simulation results on the logistic regression coefficient. Both methods show little bias. The coverage rates are all around the nominal level and not shown. However, LDA-MI yields more efficient estimates (with lower MSE) than those from LOGIT-MI. The gain of efficiency from the former is more obvious as c increases. Note that when $c = 4$ in data generation, the data separation problem begins to appear in fitting the simulated data. Here the comparative performance on the regression estimate follows a similar pattern identified by Efron (1975), where he showed that the efficiency gain in terms of classification from the discriminant analysis model is more apparent as $\frac{\mu_1-\mu_0}{\sigma}$ increases.

TABLE 4.6
Example 4.4. Simulation results

c	LOGIT-MI		LDA-MI	
	RBIAS	MSE	RBIAS	MSE
1	−0.1	0.000994	0	0.000983
2	0	0.00250	0.1	0.00214
3	−0.1	0.00837	0.3	0.00541
4	−1.5	0.0349	0.6	0.0184

On the other hand, the logistic regression model is usually assumed to be a safer and more robust approach than the discriminant analysis because the former only posits a conditional model, $f(Y|X)$, and does not rely on the marginal distribution assumption of X required in the latter for specifying a joint model, $f(Y,X)$. Suppose $X \sim N(\mu, \sigma^2)$ in Model (4.4); then it can be shown that

$$f(X|Y=1) = \frac{\frac{exp(\beta_0+\beta_1X)}{1+exp(\beta_0+\beta_1X)}\frac{1}{\sqrt{2\pi}\sigma}exp(-\frac{(X-\mu)^2}{2\sigma^2})}{\int\frac{exp(\beta_0+\beta_1X)}{1+exp(\beta_0+\beta_1X)}\frac{1}{\sqrt{2\pi}\sigma}exp(-\frac{(X-\mu)^2}{2\sigma^2})dx},$$

$$f(X|Y=0) = \frac{\frac{1}{1+exp(\beta_0+\beta_1X)}\frac{1}{\sqrt{2\pi}\sigma}exp(-\frac{(X-\mu)^2}{2\sigma^2})}{\int\frac{1}{1+exp(\beta_0+\beta_1X)}\frac{1}{\sqrt{2\pi}\sigma}exp(-\frac{(X-\mu)^2}{2\sigma^2})dx}, \qquad (4.7)$$

which do not have closed forms.

If we use Eq. (4.4) and the additional assumption $X \sim N(\mu, \sigma^2)$ to define a joint model, $f(Y,X)$, and refer to it as a logistic-normal model, then the conditional distribution, $f(X|Y)$, under this joint model does not conform to the normal mixture distribution in Eq. (4.5). In this case, would the discriminant analysis imputation model still work well?

Example 4.5. *A simulation study assessing the robustness of the discriminant analysis imputation under a logistic-normal model*

In this simulation study, we generate X and Y based on a logistic-normal model. We purposefully select values of μ, σ^2, β_0, and β_1 so that $f(Y=1)$ is held at around 0.5. We let β_1 increase (and the corresponding β_0 also has to vary) in different scenarios. The missing data are generated under MCAR with around 40% missingness rate. The simulation is based on 1000 replicates. As in Example 4.4, the estimands of interest include $\overline{Y}$ and the logistic regression coefficient, β_1.

For estimating $\overline{Y}$, the two imputation methods yield comparable and good results (not shown). Table 4.7 shows the simulation results for estimating the logistic regression coefficient. In Scenario 1, results from the two methods are rather similar. In Scenario 2, LDA-MI is more biased (RBIAS=−4.6%) for the point estimate and results in a lower coverage rate (89.8%) than the nominal level. In Scenario 3, LDA-MI yields rather biased estimates (RBIAS=−17.7%) with much higher MSEs as well as lower coverage rates (33.6%) compared with LOGIT-MI.

TABLE 4.7
Example 4.5. Simulation results

Scenario	LOGIT-MI			LDA-MI		
	RBIAS	MSE	COV (%)	RBIAS	MSE	COV (%)
1	0	0.010	95.6	−0.1	0.010	95.5
2	0	0.029	94.4	−4.6	0.029	89.8
3	0	0.079	95.1	−17.7	0.314	33.6

Note: Scenario 1: $\mu = 1.5$, $\sigma^2 = 1.25$, $\beta_0 = -1.5$, $\beta_1 = 1$; Scenario 2: $\mu = 2$, $\sigma^2 = 2$, $\beta_0 = -4$, $\beta_1 = 2$; Scenario 3: $\mu = 2.5$, $\sigma^2 = 13/4$, $\beta_0 = -7.5$, $\beta_1 = 3$.

Simulation assessment results from Examples 4.4 and 4.5 show that the logistic regression imputation model yields valid results as long as the conditional model for Y given X is correctly specified as a logistic regression. It is not necessary to specify the marginal model (distribution) for X for the logistic regression imputation. For the normal discriminant analysis imputation, if the marginal distribution of X is correctly specified (Example 4.4), then it can improve the efficiency of the regression coefficient estimates compared with the logistic regression imputation model. On the other hand, if the marginal distribution of X is misspecified (Example 4.5), then using the normal discriminant analysis imputation method runs the risk of generating suboptimal estimates.

In practice, there can be multiple variables used as X-covariates, and these variables can take different distributional forms. The normality assumptions for the marginal distributions of X-variables in the normal discriminant analysis imputation might not be well satisfied. Therefore, our investigations and conclusions here are consistent with the recommendation from Van Buuren

(2018), that is, the logistic regression imputation method seems to be a more general and valid approach for imputing missing binary (and categorical) variables. Of course when necessary, a comparison analysis can be done to assess whether the results are sensitive between the two methods in practice.

In addition, note that both the logistic-normal and normal discriminant analysis models here are acceptable approaches to jointly modeling one binary variable Y and one continuous variable X. They are based on different factorizations of $f(X, Y)$, the joint distribution. Chapter 6 focuses on the imputation based on a joint modeling of multiple incomplete variables (e.g., both Y and X have missing values) and provides further discussion on related issues.

4.3.2.3 Rounding

A somewhat convenient imputation strategy for binary data is to first use the normal linear regression model (4.1) to carry out the imputation, and then round the (continuous) imputed values to binary numbers using some rules. Rounding appears to be a simple procedure, and yet it can yield suboptimal results. A more principled imputation approach such as that based on the logistic regression model should be preferred. We defer the discussion of the rounding strategy to Section 5.5.3.

4.3.2.4 Data Separation

In the simulation of Example 4.3 we reported that when the logistic regression coefficient (β_1) increases to a rather large value (say $\beta_1 \leq -3$), warning messages about fitting the logistic model begin to appear in software. This is a common issue that can happen in multiple imputation analysis of binary or discrete data, often termed as the data separation or perfect prediction problem (e.g., White et al. 2010 and references therein). For binary data Y, data separation occurs if the strata formed by the covariate X create some cells in which all of the Y-values are either 1 or 0. In this case, the MLE estimate, $\hat{\beta}_{MLE,obs}$, would move to ∞ or $-\infty$, and the associated covariance matrix also would become very large and unstable. The convergence of the estimation is difficult to achieve and thus leads to warning messages in software.

Table 4.8 is an illustrative dataset for data separation in imputation of binary data. In this case, if we fit a logistic regression model, $logit(f(Y = 1)) = \beta_0 + \beta_1 X$, to the complete cases, then there is no finite $\hat{\beta}_{MLE,obs}$. This is simply because we do not observe any 1's for Y when $X = 0$. Without any adjustment, the imputed values could have very different distributions from those of observed: unstable logistic regression estimates and vary large standard errors are supplied to generate unreasonable posterior draws of β, further leading to erratic imputed values. Ideally in this setup, a reasonable imputation method would impute most (if not all) 0's for the m missing cases with $X = 0$. For the n missing cases with $X = 1$, the imputations would retain a similar distribution of 1's and 0's to that of observed cases (i.e., b 1's and c 0's).

In-depth discussions of data separation in imputation of binary variables can be found in White et al. (2010) and Carpenter and Kenward (2013, Section 4.6). Van Buuren (2018, Section 3.6.2) listed several options for dealing with such a problem. Some of the recommended methods are covered in previous examples, such as the bootstrap imputation method and the imputation based on draws of β from its exact posterior distribution (Example 4.3). For the latter method, it is recommended to apply a weakly informative prior distribution (e.g., Cauchy prior distribution) for β to deal with data separation (Gelman et al. 2008). Another well-known method is to add pseudo observations to the data to prevent MLE estimates that are infinite. This strategy is implemented in several software programs (e.g., R mice and SAS PROC MI). In addition, some preliminary simulation study results from Allison (2005) appeared to show that the discriminant analysis imputation can provide better results than the unadjusted logistic regression imputation when the proportion of binary variables is very close to 0 or 1. It is not entirely clear whether this approach can be used to mitigate data separation.

Example 4.6. *A simulation study for assessing the performances of alternative imputation methods for handling data separation*

We conduct a simulation study to assess and compare the performances of alternative imputation methods for handling data separation. These methods include:

1. MLE-MI: the unadjusted logistic regression imputation method approximating the posterior distribution using the asymptotic normal distribution of the MLE estimate.
2. Pseudo-MI: the pseudo observation imputation method.
3. LDA-MI: the discriminant analysis imputation method.
4. BOOT-MI: approximating the posterior distribution using the bootstrap distribution of the MLE estimate.
5. Bayes-MI: drawing β from its exact posterior distribution under a Cauchy distribution prior.

TABLE 4.8
Artificial data for illustrating data separation

X	Y		
	0	1	Missing
0	a	0	m
1	b	c	n

We repeat a simulation design used in White et al. (2010). This study includes four variables: a binary $Y \sim Bernoulli(p = 0.1)$, a binary X with $f(X|Y = 0) \sim Bernoulli(p = 0.8)$, $f(X = 1|Y = 1) = 0$ (i.e., X-values are all 0's if $Y = 1$), a continuous U with $U \sim N(0, 1)$, and a continuous Z with $f(Z|X, Y, U) \sim N(\beta_1 Y + \beta_2 X + \beta_3 U, \sigma^2)$, where $\sigma = 2$. Here data separation happens for Y when $X = 1$. The complete-data sample size is 1000, and missing values on Y are generated under MCAR with around 30% missingness rate. The simulation is based on 1000 replicates. All imputation methods assume a logistic regression model $logit(f(Y = 1)) = \gamma_0 + \gamma_1 X + \gamma_2 U + \gamma_3 Z$. Because of data separation, however, the MLE for γ_1 often results in a rather large negative value.

Estimands in the evaluation include the marginal mean ($\overline{Y}$) and the regression coefficient (β_1) from regressing Z on X, Y, and U. The multiple imputation analyses are based on $M = 30$ imputations. Note that in this setup, the missing Y is a covariate in the regression for Z on Y, X, and U.

Table 4.9 shows the simulation results. Without any adjustment, MLE-MI has a large bias for both the mean and regression estimates. LDA-MI produces a sizable bias (13%) and a low coverage rate (81.5%) for estimating the mean. Its performance for estimating the regression coefficient is more acceptable, despite that the relative bias is still larger than 5%. It seems that applying the discriminant analysis imputation method cannot fully address data separation.

TABLE 4.9
Example 4.6. Simulation results

Estimand	Method	RBIAS (%)	SD	$\overline{SE}$	MSE	COV (%)
	MLE-MI	129.5	0.0126	0.1097	0.0169	100
	Pseudo-MI	1.2	0.0109	0.0109	0.000120	95.9
$\overline{Y}$	LDA-MI	13.0	0.0127	0.0122	0.000330	81.5
	BOOT-MI	−0.1	0.0110	0.0107	0.000121	94.7
	Bayes-MI	0.9	0.0109	0.0109	0.000120	94.7
	MLE-MI	−36.9	0.165	0.351	0.164	88.0
	Pseudo-MI	0.5	0.236	0.226	0.056	94.9
β_1	LDA-MI	−5.4	0.228	0.217	0.055	93.1
	BOOT-MI	0.5	0.240	0.226	0.057	95.0
	Bayes-MI	−0.9	0.236	0.226	0.056	95.1

All other three imputation methods (i.e., Pseudo-MI, BOOT-MI, and Bayes-MI) work similarly well in handling data separation. Their biases are small and their coverage rates are around the nominal level. Specifically, Fig. 4.4 shows 1000 posterior draws of γ_1 in Bayes-MI from a randomly selected simulation replicate. The use of weakly informative prior has stabilized the posterior distributions of γ_1 to an approximately normal distribution centered around -7, a large yet manageable coefficient value. On the other hand, the corresponding MLE estimate in MLE-MI is −22 and its variance estimate is around 2200, both of which are extremely large and can result in bad imputations and estimates.

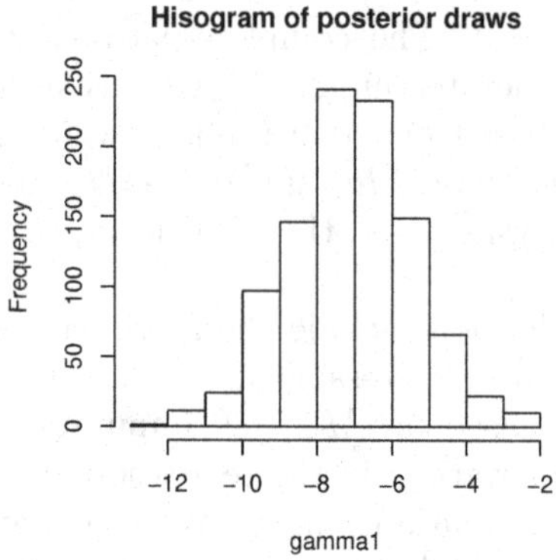

FIGURE 4.4
Example 4.6. The histogram of 1000 posterior draws of γ_1 in Bayes-MI from a simulation replicate.

In addition to using appropriate adjustments in imputation with data separation, it is important for researchers and practitioners to carefully review descriptive statistics in advance to identify any possible extremely unbalanced distributions of binary and categorical variables and thus foresee and prepare for such problems.

4.3.3 Imputation for Nonbinary Categorical Data

The logistic regression model for binary data can be easily extended to accommodate categorical variables (nominal and ordinal variables) with more than two categories. Nominal categorical variables do not have an internal order. An example is the variable identifying subjects' racial/ethnic groups such as "Non-Hispanic White", "Non-Hispanic Black", "Non-Hispanic Others", and "Hispanic". Ordinal categorical variables have internal orders. An example is the variable for measuring subjects' self-rated health status such as "Poor", "Fair", "Good", "Very Good", and "Excellent".

A multinomial logistic regression model can be used to impute missing nominal variables. An ordered logistic regression model or proportional odds model can be used to impute missing ordinal variables. All these (generalized) logistic regression models fall into the framework of GLM. For simplicity we do not provide mathematical details of these models, which can be found, for example, in Agresti (2002; 2010). The imputation algorithms used for binary logistic models can be used to impute nonbinary categorical data in a similar manner. We use MLE-MI as an example. Suppose $\hat{\beta}_{MLE,obs}$ is the regression coefficient estimate and $\hat{V}(\hat{\beta}_{MLE,obs})$ is the covariance matrix estimate in generalized logistic regressions using completed cases, the imputation step includes drawing β from its approximate posterior distribution, $N(\hat{\beta}_{MLE,obs}, \hat{V}(\hat{\beta}_{MLE,obs}))$. Once β is drawn, it is straightforward to draw missing values in Y using fitted probabilities, which is a function of β and covariates X under the generalized logistic regression models. These imputation algorithms are implemented in major software packages. Imputations based on the bootstrap method (BOOT-MI) and exact posterior distribution (Bayes-MI) can be applied as well.

Example 4.7. *Multiple imputation of satisfaction level of health care in survey data*

Zhou et al. (2017) conducted a multiple imputation study of nonbinary categorical missing data. For illustration, we use the dataset in their application, which is a subset of 2013 Behavioral Risk Factor Surveillance System (BRFSS) public-use data. Established in 1984 and managed by CDC, BRFSS is a nation-wide system of health-related telephone surveys that annually collects state data about U.S. residents regarding their health-related risk and preventive behaviors, health conditions, and information about health services. More information about BRFSS can be found in (https://www.cdc.gov/brfss).

In this analysis, we focus on the satisfaction level with health care received for the Hispanic subpopulation who were unable to work and had annual household income less than 15000 dollars. From the public-use BRFSS data system, a subset of 1430 participants was selected with fully observed data of potentially associated covariates. More specifically, the missing variable of interest (Y) consists of three categories: 1=Very satisfied, 2=Somewhat satisfied, and 3=Not at all satisfied. In the working dataset, we treat those participants who answered "Don't know/not sure", "Not applicable", "Refused" all as missing data. This results in 363 cases with missing satisfaction level, counting around 25% of the original sample.

There are many variables in the original BRFSS data. For simplicity, we conduct some exploratory analyses to select six fully observed covariates (X) including: age group, sex, education level, health care coverage, having delayed medical care, and self-rated general health status. Table 4.10 shows some descriptive statistics of the dataset. Again for simplicity, we treat age group, education level, and general health status as continuous variables in this example. Doing so can still retain the general association between these variables with the missing variable and missingness indicator. It can also avoid possible data separation as some of the groups formed by these variables are rather small.

We run a logistic regression for the missingness indicator of Y on these covariates. Table 4.11 shows the results. Most of the regression coefficients are significant at the 0.05 level (p-values not shown), suggesting that the missingness is not completely at random. Note that the delay of health care is the strongest predictor of the missingness. This can also be seen from simple statistics: for the cases with "Yes", the rate of the missingness is $351/(351+416) \approx 45.8\%$; yet for the cases with "No", the rate of missingness is only $12/(12+651) \approx 1.8\%$.

In addition, we run a multinomial logistic regression of Y on X using observed cases, treating "Very satisfied" as the comparison group. Table 4.12 shows the regression estimates. Most notably, females are much less likely to answer "Not at all satisfied" vs. "Very satisfied". Participants with higher education level are more likely to answer "Not at all satisfied" vs. "Very satisfied". Participants without health care coverage are much more likely to answer "Not at all satisfied" vs. "Very satisfied". Participants with delayed health care are much more likely to answer "Not at all satisfied" or "Somewhat satisfied" vs. "Very satisfied". Participants with lower self-rated health

TABLE 4.10
Example 4.7. Descriptive statistics of the dataset ($n = 1430$)

Variable	Distribution
Satisfaction of health care received*	1=Very satisfied (58.5%), 2=Somewhat satisfied (33.5%), 3=Not at all (8.1%)
Age group	1=18-24 yrs (2.2%), 2=25-34 yrs (7.5%), 3=35-44 yrs (9.9%), 4=45-54 yrs (24.6%), 5=55-64 yrs (33.4%), 6=65 yrs and above (22.5%)
Sex	1=Male (36.6%), 2=Female (63.4%)
Education	1=Less than high school (43.6%), 2=High school graduate (31.4%), 3=Attended college (16.6%), 4=College graduate (8.3%)
Health care coverage	1=Yes (88.4%), 2=No (11.6%)
Delayed health care	1=Yes (53.6%), 0=No (46.4%)
Self-rated health status	1=Excellent (2.7%), 2=Very good (5.5%) 3=Good (18.5%), 4=Fair (40.3%), 5=Poor (32.9%)

Note: * based on 1067 observed values.

TABLE 4.11
Example 4.7. A logistic regression analysis of the missingness indicator

Predictor	EST	SE
Intercept	−3.499	0.620
Age group	0.156	0.059
Sex	−0.219	0.150
Education level	−0.214	0.077
Health care coverage	0.460	0.214
Delayed health care	3.988	0.305
Self-rated health status	−0.271	0.075

status are more likely to answer "Somewhat satisfied" vs. "Very satisfied". The directions of these associations make general sense.

TABLE 4.12
Example 4.7. A multinomial logistic regression analysis using complete cases

	Somewhat satisfied		Not at all satisfied	
Predictor	EST	SE	EST	SE
Intercept	−1.501	0.551	−3.773	0.941
Age group	−0.098	0.058	−0.185	0.100
Sex	−0.226	0.143	−0.839	0.246
Education level	0.137	0.070	0.353	0.120
Health care coverage	0.382	0.237	1.216	0.330
Delayed health care	0.651	0.142	1.553	0.260
Self rated health status	0.207	0.075	0.264	0.138

Note: The comparison group in the logistic regression is "Very satisfied" .

We apply a multinomial logistic regression imputation with $M = 50$ using R mice. Table 4.13 presents the mean estimates of Y from the multiply imputed (MI) data and those from the complete-case (CC) analysis . Compared with CC, MI increases the estimates for the category of "Not all satisfied" (9.6% vs. 8.1%) or "Somewhat satisfied" (34.9% vs. 33.5%) and decreases the estimate for the category of "Very satisfied" (55.5% vs. 58.5%). Going back to Tables 4.10 and 4.12, this can be partially explained by the fact that cases with delayed health care are far more likely to have missing responses than those without the delay, plus the former cases are also much more likely to answer "Not at all satisfied" or "Somewhat satisfied". Therefore we expect that the imputed values, by picking up this covariate effect (and others), are more likely to fall into these two categories. This example shows that it is important to run some exploratory analyses before imputation to better understand the direction and pattern of multiple imputation analysis results.

TABLE 4.13
Example 4.7. Mean estimates of care satisfaction

Care Satisfaction	CC		MI	
	EST (%)	SE (%)	EST (%)	SE (%)
Very	58.5	1.51	55.5	1.44
Somewhat	33.5	1.44	34.9	1.38
Not at all	8.1	0.83	9.6	0.96

We also apply a proportional odds imputation model using R mice. Such a model explicitly accounts for the internal order of the satisfaction level from the survey question. The mean estimates are similar to those under the multinomial logistic regression imputation. Again, the proportional odds model identifies the delay of health care as a strong predictor in a similar way as in the multinomial logistic model. The corresponding imputations have picked up this covariate effect as well. Details are not shown.

4.3.4 Imputation for Other Types of Data

There exist variables other than continuous and categorical forms. One of them is count data. Examples of count data include the non-negative number of certain events in a defined period. Count data can be fitted by Poisson regressions as a special case of GLM. Briefly speaking, by setting $g(\mu_i) = log(\mu_i)$, the "log" link, and $a(\phi) = 1$ in Eqs. (4.2) and (4.3), then Y_i follows a Poisson distribution with mean $\mu_i = X_i\beta$. The negative binomial model (i.e., by using a negative binomial link function in the GLM modeling) can be used to account for overdispersion (i.e., extra variability) in count data. Sometimes real studies may encounter an excessive amount of zeros for count data, then zero-inflated extensions of both the Poisson model and negative binomial model can be used to improve the model fit.

All these types of models are special cases of GLM, and therefore the corresponding imputation algorithms for count data can be devised and executed in a similar way to those of logistic regression models. Some algorithms have been implemented in software packages. For example, an option for imputing count data using Poisson models is available in IVEware (Raghunathan et al. 2001), and an option using the negative binomial model is available in STATA ICE (Royston 2009). Researchers and practitioners need to check the features and options of specific imputation programs that they use.

Another type of data is referred to as the mixed type, also called semicontinuous variables. Such variables have a high probability mass at one value (often number zero) and a continuous distribution over the remaining values. A classic example is the medical cost/spending variable within a defined period, say annual medical spending. This variable typically has a high probability at zero from healthy subjects who do not see medical providers. It often has a highly skewed unimodal distribution from remaining subjects who have positive medical spending. Note that semicontinuous data are still treated as continuous from the continuous part, yet count data are generally considered as discrete.

A well established class of statistical models for semicontinuous data is the two-part model (Duan et al. 1983). The first part is to model the probability of the variable being the point mass vs. positive values. Conditional on being a positive value from the first part, the second part is to model its continuous distribution from positive values. Typically a logistic regression model can be used in the first part, and a normal linear regression model can be used in the second part. The imputation algorithms for both models, as we discussed before, can thus be combined to impute a semicontinuous variable. For example, an option for imputing mixed type of data is available in IVEware (Raghunathan et al. 2001).

Beyond GLM, there exist alternative imputation models for count and semicontinuous data in the literature. For example, a nonparametric transformation (e.g., quantile normal scores) has been demonstrated to perform well for imputing data containing mixed variables with many zeros (Nevalainen et al. 2009). Yu et al. (2007) and Vink et al. (2014) showed that semicontinuous variables can also be well imputed by predictive mean matching methods, which will be covered with more details in Chapter 5.

4.4 Imputation for a Missing Covariate in a Regression Analysis

From the perspective of imputation, the missing variable Y is the outcome/response of the imputation model, and other variables are treated as covariates/predictors in the imputation model. On the other hand, the imputed variable can be used as a predictor or covariate in a regression analysis using completed data, where the regression outcome variable is one of the fully observed X-variables. For instance, the multiply imputed family income variable in NHIS data (Example 1.2) is often used as a key demographic covariate in regression analyses for health outcomes. See also Examples 4.1, 4.2, and 4.6. In general, if the main purpose of a multiple imputation project is to construct completed datasets for a broad variety of analyses, then the imputed variable(s) are expected to act as either the response variable or covariate in these analyses, depending on the goal of scientific investigations.

Suppose we know that the missing variable is going to be included as a covariate in a planned regression analysis. What are the key issues in the imputation? We use the following example to initiate some discussion.

Example 4.8. *A missing continuous covariate problem in a linear regression with two covariates*

Suppose we have three continuous variables, X, Y, and Z. Variable X has some missing data, and both Y and Z are fully observed. We plan to run a normal linear regression analysis of Y on X and Z as follows:

$$Y = \beta_0 + \beta_1 X + \beta_2 Z + \epsilon_{Y|X,Z}, \tag{4.8}$$

where $\epsilon_{Y|X,Z} \sim N(0, \sigma^2)$. Suppose the missingness of X is MAR. The key to building an appropriate imputation model for X is to figure out the posterior predictive distribution of missing X-values, which involves the conditional distribution $f(X|Y,Z)$. By the Bayes rule, $f(X|Y,Z) \propto f(X,Y,Z) = f(Y|X,Z)f(X|Z)$. The first component, $f(Y|X,Z)$, is already specified by the analysis model in Eq. (4.8). The second component, $f(X|Z)$, can be generally referred to as the distribution of the covariate (given the other covariate Z). However, since $f(X|Z)$ might not be the primary analysis interest, it cannot be readily determined by the analysis model alone as in Eq. (4.8). Typically this component needs to be specified by the imputer.

For illustration, we can posit a normal linear regression model for X on Z as

$$X = \alpha_0 + \alpha_1 Z + \epsilon_{X|Z}, \tag{4.9}$$

where $\epsilon_{X|Z} \sim N(0, \tau^2)$.

Combining Eqs. (4.8) and (4.9), it can be shown that the conditional distribution $f(X|Y, Z)$ is also a normal distribution in the form of a linear regression. Therefore, the correct imputation model for X is

$$X = \gamma_0 + \gamma_1 Y + \gamma_2 Z + \epsilon_{X|Y,Z}, \tag{4.10}$$

where $\epsilon_{X|Y,Z} \sim N(0, \omega^2)$. We can use the normal linear regression model to impute the missing X-values treating Y and Z as fully observed covariates in the imputation model.

In summary, to impute a missing covariate for a targeted analysis, we need to build an imputation model that characterizes the conditional distribution of the missing covariate given both the outcome and other covariates in the analysis (e.g., $f(X|Y, Z)$ in Example 4.8). Deriving this model can follow the Bayes rule and typically requires two pieces of information: one is embedded in the analysis model (already known), and the other is embedded in the conditional distribution of the missing covariate given other covariates. However, we often need to specify a model for the second part.

For many practitioners, using the outcome variable of the planned analysis to impute a missing covariate and subsequently estimating the association between the same covariate and outcome variable may appear to artificially strengthen (i.e., overestimate) the original association. Therefore, there often exists a resistance to including the outcome variable in imputing the missing covariate. The imputation model then would only include other covariates as imputation predictors. However, this concern and practice are not right. Ignoring the outcome from the imputation of the covariate can typically attenuate the estimated relation between them towards zero when the imputed values are used in the analysis. It can also bias the estimates for the association between other covariates with the outcome.

Example 4.9. *The impact of excluding the outcome variable from the imputation of a missing continuous covariate in a linear regression*

We use the setup in Example 4.8 to gain some insights into the consequences of excluding Y in imputing missing X-values. Based on Models (4.8) and (4.9), $f(Y|Z)$ can be expressed as

$$Y = \beta_0 + \alpha_0\beta_1 + (\alpha_1\beta_1 + \beta_2)Z + \epsilon_{Y|Z}, \tag{4.11}$$

where $\epsilon_{Y|Z} \sim N(0, \beta_1^2\tau^2 + \sigma^2)$. For simplicity, suppose that the missingness of X is MCAR (with probability p_{mis}) and let R denote the response indicator. For cases with observed X-values, $f(Y|X_{obs}, Z, R = 1) = N(\beta_0 + \beta_1 X + \beta_2 Z, \sigma^2)$. Assume that we impute X using Model (4.9) (by ignoring Y) so that $f(X_{imp}|Z, R = 0) \approx N(\alpha_0 + \alpha_1 Z, \tau^2)$, where X_{imp} denote the imputed values of X. Therefore $f(Y|X_{imp}, Z, R = 0) = \frac{f(Y,X_{imp},Z,R=0)}{f(X_{imp},Z,R=0)} = \frac{f(X_{imp}|Y,Z,R=0)}{f(X_{imp}|Z,R=0)} f(Y|Z, R = 0) = f(Y|Z, R = 0) = f(Y|Z)$. Here $f(X_{imp}|Y, Z, R = 0) = f(X_{imp}|Z, R = 0)$ because the imputation model excludes Y. By Eq. (4.11), $f(Y|X_{imp}, Z, R = 0) = N(\beta_0 + \alpha_0\beta_1 + (\alpha_1\beta_1 + \beta_2)Z, \beta_1^2\tau^2 + \sigma^2)$. Therefore, the conditional distribution of Y given (completed) X and Z is a mixture of two normal distributions with different means and variances. If we regress Y on X and Z, then the regression coefficients can be expressed as

$$
\begin{aligned}
E(Y|X, Z) &= E(E(Y|X, Z, R)) \\
&= (1 - p_{mis})(\beta_0 + \beta_1 X + \beta_2 Z) + p_{mis}(\beta_0 + \alpha_0\beta_1 + (\alpha_1\beta_1 + \beta_2)Z) \\
&= \beta_0 + p_{mis}\alpha_0\beta_1 + (1 - p_{mis})\beta_1 X + (\beta_2 + p_{mis}\alpha_1\beta_1)Z. \qquad (4.12)
\end{aligned}
$$

Comparing Eq. (4.12) with Eq. (4.8), all the regression coefficients become biased after we exclude Y in the imputation of X. The slope coefficient for X changes from β_1 to $(1 - p_{mis})\beta_1$ (being attenuated), and the slope coefficient for Z also changes from β_2 to $\beta_2 + p_{mis}\alpha_1\beta_1$ to compensate for the attenuation.

We conduct a simple simulation study to provide some numerical ideas. We generate X, Y, and Z assuming $Z \sim N(0, 1)$, $X = 1 + Z + \epsilon_{X|Z}$, and $Y = -2 + X + Z + \epsilon_{Y|X,Z}$, where both $\epsilon_{X|Z}$ and $\epsilon_{Y|X,Z} \sim N(0, 1)$. We generate missing values on X under MCAR with around 30% missingness rate. The completed-data sample size is 1000, and the simulation is based on 1000 replicates. We assess three estimands: the mean of X ($\overline{X}$) and the coefficients for regressing Y on X and Z (β_1 and β_2). We consider two imputation methods for X: one includes both Y and Z as predictors (MI including Y), and the other only includes Z as the predictor (MI excluding Y). The multiple imputation analyses are based on $M = 50$ imputations.

Table 4.14 shows the results from the simulation study. For estimating $\overline{X}$, both imputation methods yield estimates with little bias and around nominal coverage rates under MCAR. They are more efficient than the complete-case (CC) method, judged by comparing their respective MSEs. In addition, MI including Y is even more efficient than MI excluding Y because the former uses extra information from Y.

For the regression coefficient β_1, MI excluding Y yields attenuated estimates, for which the relative bias is around negative 30%. For the other regression coefficient β_2, a large positive bias is also produced. It can be verified that the magnitude of biases match with the theoretical derivations in Eq. (4.12). These estimates also have large MSEs and bad coverage rates (0 in this case). Overall, the performance of MI excluding Y is even worse than that of CC. On the other hand, MI including Y produces estimates with little bias and good coverage rates. Speaking of precision, MSEs of this correct imputation method are still larger than those of BD, exhibiting no erratic behavior of overstating the association between Y and X.

Example 4.10. *The impact of birth weight in imputing gestational age*
In Example 4.2, we included an analysis of regressing the baby's birth weight (DBIRWT) on the mother's gestational age (DGESTAT) and other variables. Here the missing variable DGESTAT is a predictor in this analysis. The correct imputation strategy for DGESTAT used in Example 4.2 includes DBIRWT as a predictor. In this example, we apply an incorrect imputation method by excluding DBIRWT from the predictors. Table 4.15 compares results between two methods. For estimating the mean of DGESTAT, the point estimates are little different, and the standard error increases slightly when DBIRWT is excluded from the imputation. However, the regression coefficient estimates from the two imputation methods are rather different. Most notably the coefficient estimate for DGESTAT is considerably smaller when DBIRWT is excluded from the imputation (135.26 vs. 165.55), showing an attenuation of the association between DGESTAT and DBIRWT. Excluding DBIRWT from the imputation also changes other coefficient estimates in a sizable manner. In addition, the standard errors increase, showing a loss of precision when DBIRWT is excluded from the imputation.

Including the outcome variable in the imputation model for the missing predictor has been emphasized multiple times in the past literature (e.g., Little 1992; Moons et al. 2006). In certain cases, if the analysis model and/or the distribution of covariate is nonlinear or complicated, the exact model

TABLE 4.14
Example 4.9. Simulation results

Estimand	Method	RBIAS	SD	$\overline{\text{SE}}$	MSE	COV (%)
	BD	0	0.043	0.045	0.00189	96.1
$\overline{X}$	CC	−0.1	0.053	0.053	0.00276	95.0
	MI including Y	0	0.046	0.047	0.00207	95.8
	MI excluding Y	−0.4	0.048	0.050	0.00230	95.8
	BD	0	0.032	0.032	0.000998	94.7
β_1	CC	0.1	0.038	0.038	0.00147	94.5
	MI including Y	0	0.036	0.035	0.00129	93.4
	MI excluding Y	−30.3	0.030	0.050	0.0926	0
	BD	0	0.046	0.045	0.00211	94.0
β_2	CC	−0.1	0.056	0.054	0.00309	94.1
	MI including Y	0	0.052	0.049	0.00271	92.7
	MI excluding Y	30.4	0.048	0.065	0.0945	0

for imputing the missing covariate can be complicated and might need to be approximated for practical purposes. To illustrate this point a little further, we still use the setup of running a linear regression model for Y on an incomplete X and a fully observed Z. Recall that the imputation task needs a specification of the model $f(X|Y,Z)$. If we would like to fully account for the targeted analysis, we can derive the model by the Bayes rule: $f(X|Y,Z) \propto f(X,Y,Z) = f(Y|X,Z)f(X|Z)$, where $f(Y|X,Z)$ is the analysis model and $f(X|Z)$ models the covariate distribution. However, this is not the only strategy: the specification can fully ignore such theoretical derivation and instead can be determined by convenience or by some exploratory analysis on the relationship between X and Y as well as Z from the actual data. We can refer to this strategy as the direct imputation approach, aiming to approximate the exact model of $f(X|Y,Z)$.

Example 4.11. *A missing binary covariate problem in a linear regression with two covariates*

Suppose we are interested in running a linear regression for Y on a binary covariate X and continuous covariate Z, where X has some missing values. The analysis model, $f(Y|X,Z)$, can be expressed as $Y = \beta_0 + \beta_1 X + \beta_2 Z + \epsilon_{Y|X,Z}$, where $\epsilon_{Y|X,Z} \sim N(0,\sigma^2)$. Now since X is binary, it is natural to specify a logistic regression to characterize the distribution, $f(X|Z)$, as $logit(f(X=1|Z)) = \alpha_0 + \alpha_1 Z$. Under these specifications, it can be shown that $f(X=1|Y,Z) \propto \frac{exp(-\frac{(Y-\beta_0-\beta_1-\beta_2 Z)^2}{2\sigma^2}+\alpha_0+\alpha_1 Z)}{1+exp(\alpha_0+\alpha_1 Z)}$, and $f(X=0|Y,Z) \propto \frac{exp(-\frac{(Y-\beta_0-\beta_2 Z)^2}{2\sigma^2})}{1+exp(\alpha_0+\alpha_1 Z)}$. Although both the analysis model and covariate distribution are simple, the

TABLE 4.15
Example 4.10. Multiple imputation analysis results

Estimand	MI including DBIRWAT		MI excluding DBIRWT	
Mean of DGESTAT	EST	SE	EST	SE
	38.64	0.0140	38.66	0.0144
Regression for DBIRWT on DGESTAT and other covariates				
Predictor	EST	SE	EST	SE
Intercept	−3886	63.44	−3009	69.99
DGESTAT	165.55	1.54	135.26	1.82
DMAGE	10.67	0.7741	9.87	0.8389
DFAGE	−1.0638	0.6785	−0.8101	0.7313
FMAPS	43.98	4.8063	75.18	5.3636
WTGAIN	5.3954	0.2263	5.6090	0.2450
NPREVIS	−0.1768	0.8252	1.6242	0.8840
CIGAR	−10.7636	0.8361	−11.241	0.8675
DRINK	3.9749	7.9001	5.6047	8.4352

exact imputation model is not straightforward, as $f(X|Y, Z)$ does not have a closed form. However, it can be seen that Y is still included in the conditional distribution and thus is needed for imputing X.

On the other hand, we can directly posit a logistic regression imputation model for X as $logit(f(X = 1|Y, Z)) = \gamma_0 + \gamma_1 Y + \gamma_2 Z$. It can be verified that $f(X|Y, Z)$ under this logistic regression model does not match with that derived above by the Bayes rule. However, the direct imputation approach might work well in many practical cases even if it can be viewed as an approximation strategy.

The existence of two imputation strategies in Example 4.11, one based on the derivation under the joint model, $f(X, Y, Z)$, and the other based on a direct specification of $f(X|Y, Z)$, implies two broad strategies, namely the joint modeling approach and the fully conditional specification approach. Such a distinction is more salient when multiple incomplete variables are to be imputed. See Chapters 6 and 7 for more details.

4.5 Summary

Univariate missing data are the basic form of missing data problems. For a missing variable Y with fully observed covariates X and assuming ignorable missingness, the key to imputation is to identify a complete-data model, $f(Y|X, \theta)$, which characterizes well the relation between Y and X using parameter θ. In this chapter we consider and discuss imputation models on the basis of GLM. Classic examples include normal linear regression models for a continuous Y-variable and logistic regression models for a binary Y-variable and other categorical variables. The multiple imputation algorithms center on drawing missing Y-values from their posterior predictive distributions under the assumed model. They can be based on draws of parameters from the exact posterior distributions or their approximations. In many cases, there is little practical difference between the two.

On the other hand, the GLM family includes multiple options for the same type of variables. This is especially the case for categorical variables as there exist alternative modeling strategies for the same categorical variable (Agresti 2002; 2010). A simple example is to choose between the logistic and probit links for a binary variable. For a probit regression model, we can set $g(\mu_i) = \Phi^{-1}(\mu_i)$, where $\Phi(\cdot)$ is the cumulative distribution function for the standard normal distribution. And its inverse function, $\Phi^{-1}(\cdot)$, is called the probit function. More research is needed to assess and compare the performances of imputation methods under alternative modeling options in GLM for the same type of categorical variables (e.g., Wu et al. 2015; Van der Palm et al. 2016).

As shown in some of the examples, if Y is the response variable, a complete-case analysis for estimating the regression relation (i.e., regression coefficients in normal linear or logistic models) between Y and X is a valid method. To see that, let R be the response indicator of Y, $f(Y|X, R = 1) = \frac{f(Y,X,R=1)}{f(X,R=1)} = \frac{f(R=1|Y,X)f(Y,X)}{f(R=1|X)f(X)} = \frac{f(R=1|Y,X)}{f(R=1|X)}f(Y|X)$. When $f(R = 1|Y, X) = f(R = 1|X)$, that is, the missingness of Y is MAR (including MCAR as a special case), $f(Y|X, R = 1) = f(Y|X)$, implying that using complete cases can obtain the same conditional distribution (i.e, the regression coefficients) as the complete data. Multiple imputation estimates for these coefficients are asymptotically equivalent to complete-case estimates because missing Y-values contribute no information to estimating the coefficients. However, by using the information embedded in X-values for which Y is missing, the advantage of applying multiple imputation is to yield good estimates for other quantities, such as the mean of Y.

On the other hand, if Y is a covariate in a targeted regression analysis, then using complete cases for the regression is not necessarily a good method. For simplicity, suppose the fully observed X is the outcome of this regression. Under MAR, $f(X|Y, R = 1) = \frac{f(R=1|X)}{f(R=1|Y)}f(X|Y)$. However, unless the missingness of Y is MCAR, $f(R = 1|X) \neq f(R = 1|Y)$ so $f(X|Y, R = 1) \neq f(X|Y)$ in general. As we have shown, however, using multiple imputations that include X (the response variable of the regression model) as the predictor is a good missing data strategy.

5

Multiple Imputation for Univariate Missing Data: Robust Methods

5.1 Overview

"All models are wrong, but some are useful." George E. P. Box's famous statement also applies for multiple imputation modeling. In Chapter 4, we presented some basic parametric imputation models targeted for data with one missing variable. In some cases, these models might need to be adjusted to better account for various complex features of real data. For instance, recall that in Example 1.1, the family income variable in NHIS is highly skewed. In Example 4.2, the default unit of the gestational age variable is weeks so that the imputed values are preferred to be integers for data consistency. In addition, compared with parameter methods, semiparametric or nonparametric imputation models might render additional robustness against model misspecifications. Moreover in these examples, often multiple covariates are included in the imputation model. Is there any practical guideline for choosing covariates/predictors to be included in imputation?

Chapter 5 presents some strategies for improving the robustness and flexibility of imputation models. Section 5.2 discusses the use of transformation in imputation. Section 5.3 considers applying smoothing regression techniques in imputation models. Section 5.4 extends the normal linear imputation model to account for data with bounds. Section 5.5 briefly reviews a commonly used robust imputation strategy, predictive mean matching. Section 5.6 introduces an important principle in constructing imputation models, the inclusive imputation strategy. Section 5.7 provides a summary.

5.2 Data Transformation

5.2.1 Transforming or Not?

Many natural phenomena do not follow the normal law. Therefore it is not uncommon that analysts have to deal with data with substantial

DOI: 10.1201/9780429156397-5

nonnormality, such as continuous variables exhibiting large skewness and/or kurtosis. In typical statistical analysis and model fitting, a transformation for continuous variables is often applied to make the normality assumption more plausible or to stabilize the variance. It is therefore tempting to embed transformation in multiple imputation analysis if the incomplete, continuous variable Y deviates significantly from normality. Some relevant recommendations can be found from the literature (e.g., Schafer and Olsen 1998; Schafer and Graham 2002; Allison 2001; Raghunathan et al. 2001). Practical procedures can be summarized as follows:

1. Transform the missing variable to approximate normality using commonly used transformations, such as the square-root or cubic-root transformation.

2. Draw multiple imputations based on a model for the missing variable on the transformed scale.

3. Back-transform the imputed values to the original scale.

4. Conduct post-imputation analyses on the original scale.

However, the use of transformation might be more subtle than it appears to be. We first look at the simplest case, a univariate missing data problem without any covariate. Demirtas et al. (2008) conducted a simulation study. Their results suggested that for estimating the marginal mean and variance, the performance of normality-based imputation is overall rather robust under a variety of nonnormal distributions for continuous data. This pattern might be justified by the law of large numbers. On the other hand, if the estimand of interest concerns the shape of the marginal distribution, imputing nonnormal data using normal models might yield suboptimal results. He and Raghunathan (2006) considered the proportion of the sample with values below (or above) various cut-off points. Their simulation study demonstrated that for nonnormal continuous data, imputations from a normal model can behave badly for estimating these proportions.

We then look at the case when covariates X are included. We consider a generalized version of the normal linear model as follows:

$$Y = X\beta + \epsilon, \tag{5.1}$$

where $\epsilon \sim f_\epsilon(\epsilon)$ with $E(\epsilon) = 0$ and $Var(\epsilon) = \sigma^2$.

Compared with the normal linear model, Model (5.1) has a more general distributional assumption for the error terms, ϵ. Eq. (5.1) suggests that the marginal distribution of Y can be viewed as the sum of the distribution from $X\beta$ and that from ϵ. If both $X\beta$ and ϵ come from normal distributions, then Y is also marginally distributed as a normal distribution. Suppose that the distribution of $X\beta$ deviates from normality and ϵ is still distributed as a normal variable. Further suppose that the former dominates over the latter in terms of variability; then the marginal distribution of Y can deviate considerably from normality. However, the normal linear assumption still holds in this case, and therefore the corresponding imputation procedure can yield valid results. No transformation of Y is necessary. In summary, as long as regression error terms follow independent homogenous normal distributions in Model (5.1), the normal model-based imputation can be conducted for missing Y whether Y is marginally normal or not.

What if ϵ is not normally distributed? In this case, the marginal distribution of Y is unlikely to be normal, either. The imputation procedure outlined in Section 4.2 draws residuals of missing cases from a normal distribution. Some evidence from the literature (e.g., Rubin and Schenker 1986; Schenker and Taylor 1996; He and Raghunathan 2009) suggested that using this procedure would still yield reasonably good estimates for the marginal mean of Y and regression coefficient β, especially when the variation of ϵ is relatively smaller compared with that of $X\beta$. That is, Model (5.1) has a relatively large regression R^2.

There also exist methods that directly take account of the nonnormal error distributions without using the transformation. For univariate incomplete nonnormal data, Rubin and Schenker (1986) proposed an adjustment procedure that draws residuals of missing cases from those of observed cases to retain the actual distribution of error terms. He and Raghunathan (2009) extended this approach to the regression setting in Model (5.1).

On the other hand, applying a transformation to Y under Model (5.1) might run the risk of distorting the conditional linear relationship. For example, suppose a transformation $T(Y)$ results in a better normality for the marginal distribution of Y. Yet $E(Y|X) = X\beta$ does not necessarily imply $E(T(Y)|X)$ is still linear on X. Therefore conducting imputation assuming a normal linear model on the transformed scale can lead to invalid results (Von Hippel 2013).

The above reasoning suggests that we should first explore the relationship between Y and X. If on the original scale the conditional linear relationship, $E(Y|X) = X\beta$, holds well, then there is little need to apply transformation for Y even if its marginal distribution deviates from normality. If we find some evidence of nonnormality for the error distribution, we should either use the normal linear imputation model or apply some adjustments (e.g., predictive mean matching). We caution against transforming Y mechanically.

5.2.2 How to Apply Transformation in Multiple Imputation

With covariates, we believe that the main purpose of using a transformation in imputation is to identify some transformation $T(Y;\lambda)$, where λ is the transformation parameter and usually unknown, so that the conditional linear modeling assumption might hold well on the transformed scale. A frequently used transformation technique in statistics is the Box-Cox transformation (Box and Cox 1964). To facilitate the discussion, we consider a normal linear model on the Box-Cox transformed scale:

$$\begin{aligned} \frac{Y^\lambda - 1}{\lambda} &= X\beta + \epsilon, \lambda \neq 0, \\ log(Y) &= X\beta + \epsilon, \lambda = 0, \end{aligned} \tag{5.2}$$

where $\epsilon \sim N(0, \sigma^2)$.

There are several alternative imputation methods for missing Y under Model (5.2). For illustration, we can consider an approximate Bayesian method. Assuming $\pi(\lambda) \propto 1$ (i.e., noninformative prior distribution), the posterior distribution of λ does not have a closed form. We approximate it by $N(\hat{\lambda}_{ML}, \hat{\sigma}^2_{\hat{\lambda}_{ML}})$, where $\hat{\lambda}_{ML}$ and $\hat{\sigma}^2_{\hat{\lambda}_{ML}}$ are the MLE of λ and its variance estimate from the complete cases, respectively. Both estimates can be obtained using R car (the powerTransform function). The imputation algorithm is sketched as follows:

1. Draw λ from $N(\hat{\lambda}_{ML}, \hat{\sigma}^2_{\hat{\lambda}_{ML}})$.
2. Obtain $Y^{(\lambda)} = \frac{Y^\lambda - 1}{\lambda}$ (data at the transformed scale) and impute the corresponding missing values using the normal linear imputation method (Section 4.2). Transform the imputations to the original scale.
3. Repeat Steps 1 and 2 independently M times from $m = 1, \ldots M$.

Example 5.1. *A simulation study for assessing the performance of using transformation in multiple imputation*

We consider a complete-data model by setting $\lambda = 1/3$ (i.e., the cubic-root transformation), $X \sim N(10, 4)$, $\beta_0 = 1$, $\beta_1 = 1$, and $\epsilon \sim N(0, 4)$. The resulting Y is skewed to the right (Fig. 5.1). The missing data in Y are generated by a moderate MAR mechanism: $logit(f(R = 0|X)) = 36 - 4X$, where X is fully observed. This results in an increasing probability of missingness with a smaller X and overall around 30% missing cases. The complete-data sample size is 1000, and the simulation consists of 1000 replicates. The imputation methods include the normal linear model imputation at the original scale (i.e., without the transformation) and transformation-based imputation using the above procedure.

In the evaluation, we include the marginal mean of Y ($\overline{Y}$) as well as the proportions of Y that are less than the bottom 5%-tiles of the population ($Pr(Y < Y_{.05})$). Table 5.1 shows the simulation results. Since the missingness is MAR, the complete-case (CC) method yields biased results for both the mean and proportion estimates. Because smaller X-values tend to have more missing Y-values, the normal imputation treats the distribution as a symmetric one and tends to impute Y-values smaller than expected, resulting in a negative bias for the mean estimate (RBIAS=-3.2%) and a low coverage rate (74.5%). Since the imputed values are concentrated on the lower end of the Y-distribution, they also result in a positive bias for the proportion estimate (RBIAS=150.5%). On the other hand, the transformation-based imputation performs well for both the mean and proportion estimates because the method captures the original distribution of Y well. A slight wrinkle is that the coverage rate for the proportion estimate is larger than the nominal level (98.5%), largely due to the fact that the variance is somewhat overestimated ($\overline{\text{SE}} = 0.0132 > \text{SD} = 0.0103$). This is caused by imputation uncongeniality (Meng 1994). An in-depth explanation of this phenomenon can be found in Example 14.2.

TABLE 5.1
Example 5.1. Simulation results

		Mean of Y				
Method	RBIAS (%)	SD	$\overline{\text{SE}}$	MSE	CI	COV (%)
BD	0	2.049	2.101	4.195	8.237	96.1
CC	18.6	2.481	2.517	456.8	9.868	0
Normal Imputation	−3.2	2.465	2.644	18.99	10.41	74.5
Transformation	0.1	2.333	2.523	5.461	9.949	96.6
	Proportion of Y less than the bottom 5%-tile					
BD	0	0.00678	0.00688	0.0000459	0.027	94.5
CC	−83.8	0.00339	0.00334	0.001765	0.013	0
Normal Imputation	150.5	0.0127	0.0158	0.005822	0.063	0
Transformation	−1.1	0.0103	0.0132	0.000106	0.053	98.5

Fig. 5.1 shows histograms of Y-values using one randomly chosen simulation replicate. Completed data using the transformation-based imputation yield a similar distribution to that of the original data. Fig. 5.2 shows the scatter plot of X and Y as well as the predicted lines from the linear fit and LOWESS curve before missing values are deleted, where LOWESS stands for locally weighted estimated scatter plot smoothing. Note that in this setup, the fit for the two models (at the original and transformed scale) is little different: their regression R^2 are both close to 0.5. However, the transformation-based imputation does show some merits judging by the simulation study.

To conduct transformation-based imputation, we can also draw λ from its exact posterior distribution using some advanced Bayesian sampling algorithms or approximate the posterior distribution using the bootstrap distribution of $\hat{\lambda}_{ML}$. The idea is similar to strategies discussed for the logistic regression imputation of binary data (Example 4.3). All these methods are proper multiple imputation approaches since they account for the uncertainty of model parameters including the transformation parameter λ.

Example 5.2. *A cubic-root transformation used in the NHIS income multiple imputation project*

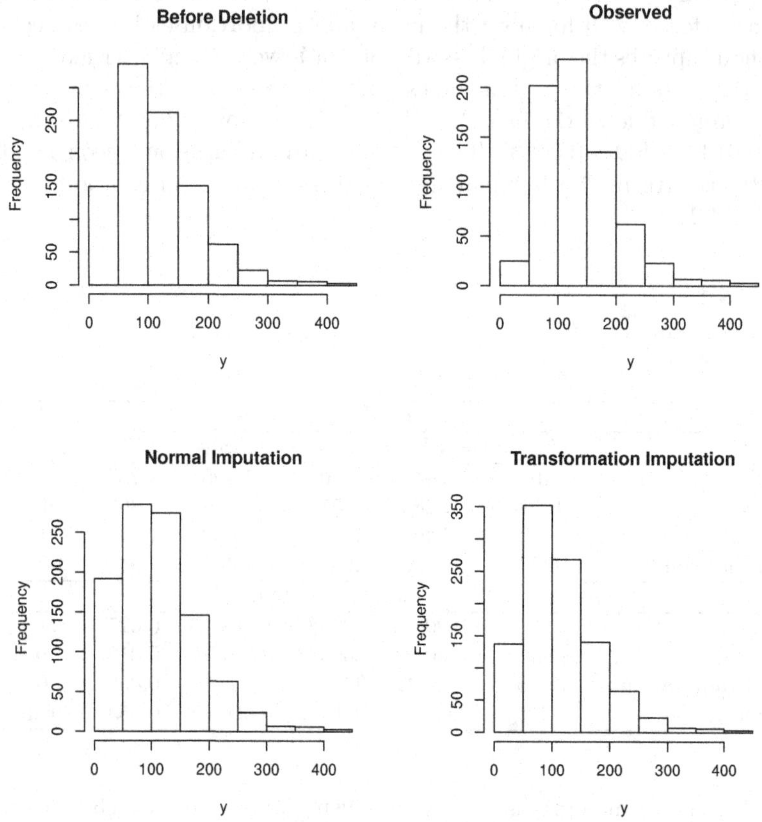

FIGURE 5.1
Example 5.1. Histograms of Y-values from one simulation replicate. Top left: before deletion; top right: observed values; bottom left: completed data under the normal model from one imputation; bottom right: completed data under the transformation model from one imputation.

In Example 1.2 we introduced the example on the NHIS income imputation project. To handle the high skewness of the income variable, a cubic-root transformation has been applied in the imputation. From the literature, the cubic root (i.e., $\lambda = 1/3$) is similar to the transformation (to the power .375) found by Paulin and Sweet (1996) to be optimal in modeling income data from the Consumer Expenditure Survey of the U.S. Bureau of Labor Statistics. For illustration, Fig. 5.3 plots the cubic-root transformed income from the observed values in the NHIS 2016 public-use data. The cubic-root transformation improves the normality of the data, reducing their right skewness and making the distribution more symmetric. Because of the top-coding, the normality is not that obvious on the right tail of the distribution.

We mainly use the Box-Cox transformation to illustrate the main ideas. Other types of transformation can be used in imputation analysis. For example, He and Raghunathan (2006) applied Tukey's gh-transformation to impute missing data with strong skewness and/or kurtosis. It is of further research

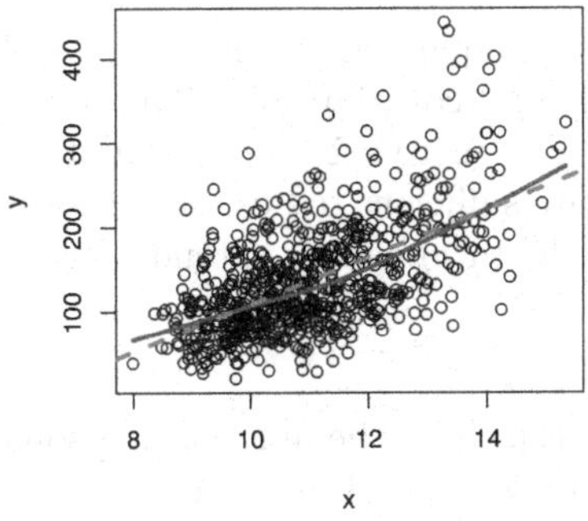

FIGURE 5.2
Example 5.1. Scatter plot of X and Y from one simulation replicate. Dashed line: linear fit; solid line: LOWESS fit.

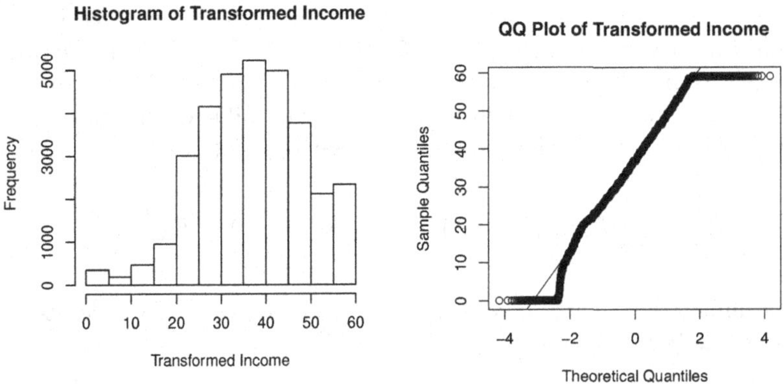

FIGURE 5.3
Example 5.2. Left: the histogram of cubic-root transformed family income from the NHIS 2016 public-use data; right: the QQ plot.

interest to devise strategies identifying which transformation family is more appropriate for actual data by comparing the performance of imputation estimates under different transformation families.

5.3 Imputation Based on Smoothing Methods

5.3.1 Basic Idea

Smoothing regression techniques can be used to model complex relationships between the outcome Y and covariates X. This strategy can also be used to impute missing data in Y if there exists some doubt that regular models (e.g., normal linear models) are sufficient to describe these relationships for data at hand. For example, Model (4.1) can be extended to

$$y_i = f(x_i) + \epsilon_i, \tag{5.3}$$

where $\epsilon_i \sim N(0, \sigma^2)$. In Eq. (5.3), the function f is some unspecified "smooth" function that needs to be estimated from $\{x_i, y_i\}$, $i = 1, \ldots n$.

For illustration, we assume that X is a continuous variable. One major approach to smoothing is via local polynomial fitting (smoother). At point x, the smoothing is obtained by fitting the pth-degree polynomial model,

$$E(y_i) = \beta_0 + \beta_1(x_i - x) + \ldots + \beta_p(x_i - x)^p,$$

using weighted least squares with kernel weights $K(\frac{x_i - x}{\lambda})$. The kernel function $K(Z)$ is usually taken to be a symmetric positive function with $K(Z)$ decreasing as $|Z|$ increases. For example, one popular choice of $K(Z)$ is the standard normal density, that is, $K(\frac{x_i-x}{\lambda}) = exp(-\frac{1}{2}\frac{(x_i-x)^2}{\lambda^2})$. The parameter $\lambda > 0$ is the smoothing parameter for local polynomial smoothers and is usually referred to as the bandwidth parameter. The value of the curve estimate at data point x is the height of the fit $\hat{\beta}_0$, where $\hat{\beta} = (\hat{\beta}_0, \ldots, \hat{\beta}_p)^t$ minimizes

$$\sum_{i=1}^{n} \{y_i - \beta_0 - \beta_1(x_i - x) - \ldots - \beta_p(x_i - x)^p\}^2 K(\frac{x_i - x}{\lambda}).$$

Assuming the invertibility of $X_c^t W_x X_c$, standard weighted least-squares theory leads to the solution:

$$\hat{\beta} = (X_c^t W_x X_c)^{-1} X_c^t W_x Y,$$

where $X_c = \begin{bmatrix} 1 & x_1 - x & \ldots & (x_1 - x)^p \\ 1 & x_2 - x & \ldots & (x_2 - x)^p \\ & & \ldots & \\ 1 & x_n - x & \ldots & (x_n - x)^p \end{bmatrix}$ is an $n \times (p + 1)$ design matrix,

$W_x = diag\{K(\frac{x_1-x}{\lambda}), \ldots, K(\frac{x_n-x}{\lambda})\}$ is an $n \times n$ design matrix of weights, and Y is the vector of y-values. Since the estimator of $f(x) = E(y|x)$ is the intercept coefficient, we obtain

$$\hat{f}(x; p, b) = e_1^t (X_c^t W_x X_c)^{-1} X_c^t W_x Y,$$

where e_1 is the $(p+1) \times 1$ vector having 1 in the first entry and 0 elsewhere.

The case $p = 0$ results in the *Nadaraya-Watson* (Nadaraya 1964; Watson 1964) estimator:

$$\hat{f}(x; 0, b) = \frac{\sum_{i=1}^{n} K(\frac{x_i-x}{\lambda}) y_i}{\sum_{i=1}^{n} K(\frac{x_i-x}{\lambda})},$$

often referred to as the "local constant" estimator.

Aerts et al. (2002) proposed a local multiple imputation approach based on the kernel regression techniques. They used two strategies to generate missing values. One is nonparametric and imputes a missing y_j-value (index j is for the cases with missing y) by drawing from the cumulative distribution function $\sum_{i=1}^{n_{obs}} w_{ji}(x_j, x_i) I(y_i <= y)$, where $w_{ji} = \frac{K(\frac{x_j-x_i}{\lambda})}{\sum_{i=1}^{n_{obs}} K(\frac{x_j-x_i}{\lambda})}$ denote the positive, normalized kernel weights, $I(\cdot)$ is the identity function, and index i is for the observed cases. That is, y_j is replaced by a random draw from observed cases, $\{y_i\}_{i=1}^{n_{obs}}$, with probabilities $\{w_{ji}\}_{i=1}^{n_{obs}}$. Intuitively, this can be thought of as a selection of the k nearest neighbors in which $k = n_{obs}$ and with unequal selection probabilities defined by $\{w_{ji}\}_{i=1}^{n_{obs}}$. Another strategy is semiparametric, imputing y_j by drawing from $N(E(y_j|X_c), \sigma^2(y_j|X_c))$, in which both the mean and variance estimates can be obtained from the aforementioned local polynomial regression fitting. For both strategies, the bootstrap resampling needs to be applied to the observed data for each imputation to ensure the multiple imputation is proper .

Example 5.3. *Multiple imputation using kernel smoothers*

We conduct a simple simulation study to demonstrate the idea of imputing missing values using random draws from observed cases, with the probability of drawing weighted by the kernel function (i.e., the nonparametric method). We consider a complete-data model $Y = -2 + X + X^2 + \epsilon$, where $X \sim N(0, 1)$ and $\epsilon \sim N(0, 1)$. We impose a MCAR mechanism to generate around 30% missing values in Y. For the kernel smoothing imputation method (KMI), we use the kernel function $K(\frac{x_i-x}{\lambda}) = exp(-\frac{1}{2}\frac{(x_i-x)^2}{\lambda^2})$. For the bandwidth parameter, we explore a range of values by setting $\lambda = (0.2, 0.1, 0.05, 0.01)$. A normal linear model imputation (NMI) is also applied, which has a misspecified model by excluding the X^2 term from the predictors. The complete-data sample size is set to be 1000, and the simulation study includes 1000 replicates. The estimands of interest include the mean of Y ($\overline{Y}$) and both the linear and quadratic coefficients of regressing Y on X. All multiple imputation analyses are based on $M = 50$ imputations.

Table 5.2 shows the simulation results. Due to MCAR, CC yields good results for all estimates. NMI yields unbiased estimates for $\overline{Y}$. However, the regression coefficients for both linear and quadratic effects in NMI are severely biased due to the model misspecification. KMI yields little bias for estimating $\overline{Y}$, and there exist some minor differences across the choices of λ. Compared with NMI, KMI considerably reduces the bias for estimating the regression coefficients. Across different λ's, their relative biases are all below 5%. However, choosing $\lambda = 0.1$ and 0.05 seem to yield slightly better results than the two other choices: the biases and MSE's are smaller based on these two bandwidth parameters, and the coverage rates are somewhat closer to the nominal level.

TABLE 5.2
Example 5.3. Simulation results

			Mean of Y			
Method	RBIAS (%)	SD	SE	MSE	CI	COV (%)
BD	0	0.109	0.110	0.012	0.429	94.4
CC	−0.2	0.129	0.131	0.017	0.512	94.7
NMI	0.2	0.115	0.115	0.013	0.452	94.6
KMI ($\lambda = 0.2$)	−1.4	0.115	0.115	0.013	0.452	94.6
KMI ($\lambda = 0.1$)	−0.6	0.112	0.111	0.012	0.436	94.5
KMI ($\lambda = 0.05$)	−0.4	0.112	0.111	0.012	0.436	94.4
KMI ($\lambda = 0.01$)	−0.7	0.112	0.111	0.013	0.435	94.4
			Linear regression coefficient			
BD	0	0.055	0.055	0.0030	0.216	95.1
CC	−0.2	0.065	0.066	0.0043	0.259	95.9
NMI	60.0	0.142	0.096	0.382	0.379	0
KMI ($\lambda = 0.2$)	2.4	0.071	0.072	0.0056	0.282	95.3
KMI ($\lambda = 0.1$)	1.9	0.072	0.068	0.0056	0.267	93.5
KMI ($\lambda = 0.05$)	1.9	0.073	0.067	0.0056	0.263	92.6
KMI ($\lambda = 0.01$)	3.6	0.100	0.074	0.0113	0.293	88.1
			Quadratic regression coefficient			
BD	0	0.022	0.023	0.00049	0.088	94.8
CC	0	0.027	0.027	0.00071	0.106	95.6
NMI	−30.0	0.058	0.038	0.094	0.152	0
KMI ($\lambda = 0.2$)	−3.2	0.033	0.031	0.00211	0.124	83.9
KMI ($\lambda = 0.1$)	−1.7	0.033	0.029	0.00139	0.114	89.7
KMI ($\lambda = 0.05$)	−1.3	0.034	0.028	0.00130	0.111	90.0
KMI ($\lambda = 0.01$)	−2.6	0.060	0.034	0.00426	0.136	84.7

To better understand the idea of the nearest-neighbor selection in KMI, Fig. 5.4 plots the selection probabilities for a particular missing data point from one simulation replicate. The X-value for the missing Y is −0.2059, the sampling probabilities (normalized w_{ji}'s) are determined by applying the kernel function (setting $\lambda = 0.1$) to the X-values of all cases with observed Y-values. It is apparent that observed Y-values from X-values that are close to -0.2059 are more likely to be selected as the imputation for the missing Y-value (left panel). These Y-values are between -5 and 0 (right panel). This can be verified by noting that $E(Y|X = -0.2059) = -2 + X + X^2 = -1.75$, and adding random error from $N(0, 1)$ to -1.75 would make that range. As expected, the selection probabilities for the Y-values corresponding to X-values that are far away from -0.2059 quickly decrease to 0. Fig. 5.5 shows scatter plots between X and Y from a randomly chosen subset (sample size 250) of one simulation replicate. Imputed values from NMI follow a linear relation between X and Y, which is obtained from a misspecified model by excluding the quadratic effect. On the other hand, imputed values from KMI directly use observed Y-values from the neighboring X-values and better capture the quadratic relation from the observed data. The idea of selecting from the nearest neighbors in KMI is also closely related to the predictive mean matching imputation (Section 5.5).

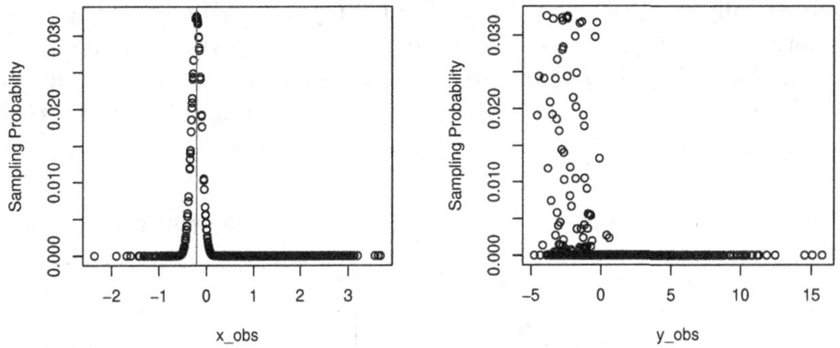

FIGURE 5.4
Example 5.3. Sampling selection probabilities for a missing Y-value when $X = -0.2059$ (the vertical line) in KMI ($\lambda = 0.1$). Left: over the range of X_{obs}; right: over the range of Y_{obs}.

In general, however, choosing the optimal smoothing (or bandwidth) parameter is not a trivial issue in applying smoothing regression methods. A smaller bandwidth parameter (i.e., less smooth) would put more sampling weights to the nearest neighbors and therefore reduce the bias of estimates. However, the associated variance would increase because this nearest neighbor could be used repeatedly multiple times in the imputation. On the other hand, a larger bandwidth parameter (i.e., more smooth) could select a neighbor far away and thus incur more bias. However, since more data are likely to be reused, such estimates tend to have less variability. This is the phenomenon of bias-variance trade-off in data smoothing. In the context of multiple imputation, the impact of the smoothing parameter can also be affected by the distribution of missingness. That is, where do missing data occur in the distribution of X? In practice, it is advised to apply a range of smoothing parameters for comparison and decision.

5.3.2 Practical Use

There exist many other well developed smoothing techniques in statistics. For example, $f(X)$ in Model (5.3) can also be estimated by penalized smoothing splines (Eilers and Marx 1996). Interested readers can, for example, refer to Ruppert, Wand, and Carroll (2003) for an introductory overview of nonparametric or semiparametric regressions. In principle, these smoothing techniques can be used to build imputation models using the similar idea to kernel smoothing. Model (5.3) can be extended to include multiple X-covariates, for example, via generalized additive models (Hastie and Tibshirani 1990). However, with a large number of X-variables, applying smoothing techniques to multivariate data is likely to encounter the "curse of dimensionality" that could drastically increase the modeling and computational complexity.

Sometimes it might be more feasible to apply smoothing methods to a key X-variable in imputation analysis. For longitudinal data having nonlinear temporal patterns, it might be useful to apply smoothing techniques to the

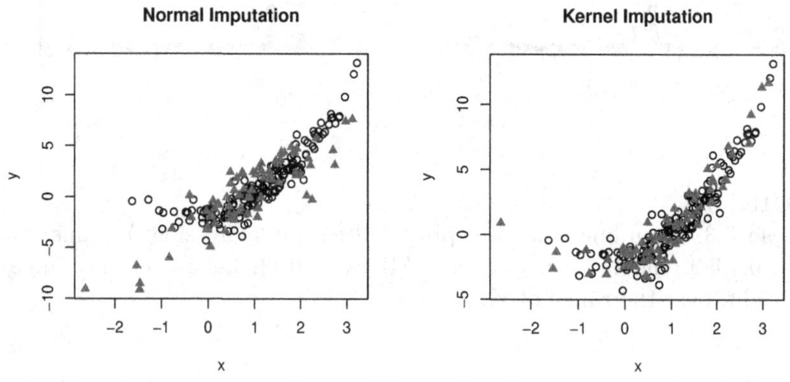

FIGURE 5.5
Example 5.3. Scatter plots of X and Y from one simulation replicate. Left: NMI; right: KMI ($\lambda = 0.1$). Circle: observed values; filled triangle: imputed values.

continuous time factor t. For example and briefly speaking, we may consider a model such as $y_i = f(t) + x_i\beta + \epsilon_i$, where x_i denote covariates that are time-invariant for subject i. He et al. (2011) proposed a multiple imputation approach that models longitudinal response profiles as smooth functions of time using cubic smoothing splines (Example 9.4).

For cross-sectional data, we might apply smoothing methods to a univariate construct that captures the major information from all X-variables. For example, Zhang and Little (2009) considered a robust imputation model, in which a penalized spline is applied to the propensity score derived from all covariates. The propensity score can be viewed as a univariate summary of the information from multiple covariates. Hsu et al. (2016) proposed a robust imputation approach, in which a kernel smoothing technique is applied to a combination of the linear predictive score and propensity score, both of which summarize information from covariates. A related discussion can be found in Section 5.6.

5.4 Adjustments for Continuous Data with Range Restrictions

Lots of continuous data are subject to certain bounds. For example, income or spending data are always nonnegative. Proportions or rates are between 0 and 1. The normal linear regression imputation generates imputed values with no restrictions. A practical adjustment strategy is to restrict the imputed values to meet the specified range. In principle, this can be done using the idea of rejection sampling. That is, keep drawing the imputed value until it falls into the specified range. However, this idea can be less feasible if the number of missing values is large: there is no guarantee how many iterations of the rejection sampling are needed to ensure all imputed values fall into the range.

Some imputation programs (e.g., SAS PROC MI and IVEware) include automatic options for generating imputations with range restrictions, which is convenient for practitioners. In addition, on the basis of the normal linear model imputation, it is often easier to draw values from a truncated normal distribution. For illustration, suppose we would like to draw values from a normal distribution $N(\mu, \sigma^2)$ truncated at a and b, where $a < b$. This can be achieved by the inverse function method. We first draw u from a uniform distribution, and then let $y^* = \sigma\Phi^{-1}((\Phi(b) - \Phi(a))u + \Phi(a)) + \mu$. Here a can be $-\infty$ ($\Phi(a) = 0$) or b can be ∞ ($\Phi(b) = 1$), where $\Phi(\cdot)$ is the cumulative distribution function of the standard normal distribution. In the normal linear model imputation algorithm (Section 4.2), for case i of the m-th imputation, we set $\mu = x_{i,mis}\beta^{(m)}$ and $\sigma^2 = \sigma^{2(m)}$ and draw the truncated values as y_i^*. Note that the restrictions (truncations) can be varying for different cases, that is, a_i and b_i instead of a and b.

Note that the aforementioned adjustment procedure is somewhat post hoc since it is based on a model that does not incorporate range restrictions in the estimation. In the case of handling missing data from truncated normal distributions, a principled complete-data model should be, for example, $y_i = x_i\beta + \epsilon_i$, where $\epsilon_i \sim N(0, \sigma^2)$ and $y_i \in (a_i, b_i)$, for $i = 1, \ldots n$. The corresponding imputation algorithm would be different from the adjustment procedure discussed here. In Chapter 8 we will discuss some related ideas in the context of survival data: y_i's are censored event times in that context.

5.5 Predictive Mean Matching

5.5.1 Hot-Deck Imputation

While imputations are random numbers generated from some models, they might be different from actual observed data in terms of certain features (e.g., units, ranges, distributions). It is natural to think of using only observed values as imputations to mitigate these differences. In Example 5.3, the kernel smoothing-based imputations are taken directly from observed values. In this section we focus on predictive mean matching (PMM), which is a multiple imputation procedure that also effectively uses observed values to impute/replace missing values. The idea of PMM originates from hot-deck imputation, which had been a popular approach to handling item nonresponse problems in organizations which manage and release large scale surveys.

In brief, hot-deck imputation is a strategy for imputing missing data in which each missing value is replaced with an observed response from a similar unit (Kalton and Kasprzyk 1986). In this context, the nonrespondent is called the "recipient", and the respondent from which the value is used for imputation is called the "donor". Donor pools, also referred to as imputation classes or adjustment cells by survey researchers, are formed based on auxiliary variables (i.e., covariates) that are observed for both donors and recipients. Suppose X-covariates are available for imputing a variable Y; a key to hot-deck imputation is to classify responding and nonresponding units into imputation cells/classes based on X-values (Brick and Kalton 1996). To create these cells, continuous covariates can be categorized. Imputation is then carried out by randomly picking a donor to replace each nonrespondent within each cell.

Historically, the term "hot-deck" comes from the use of computer punch cards for data storage, and refers to the deck of cards for donors available for a nonrespondent. The deck was "hot" since it was currently being processed, as opposed to the "cold deck" which refers to using pre-processed data as the donors, that is, data from a previous data collection or a different dataset. One of the earliest applications of hot-deck imputation was for item nonresponses in the Income Supplement of the Current Population Survey which dated back to 1947 (Ono and Miller 1969). A comprehensive review of hot-deck imputation can be found in Andridge and Little (2010).

5.5.2 Basic Idea and Procedure

PMM (e.g., Rubin 1986; Little 1988; Heitjan and Little 1991) is an extension of hot-deck imputation. It uses an explicit model (rather than adjustment cells) to create proper multiple imputations. For illustration, consider the normal linear model in Eq. (4.1), the key components of a PMM procedure (e.g., Schenker and Taylor 1996) is sketched as follows:

1. Draw β and σ^2 from their posterior distributions: $\sigma^{2(m)} \sim SSE_{obs}/\chi^2_{n_{obs}-p-1}$ and $\beta^{(m)} \sim N(\hat{\beta}_{obs}, (X^T_{obs}X_{obs})^{-1}\sigma^{2(m)})$.
2. Define the predictive mean of an incomplete observation j to be $\hat{y}_j = x_j\beta^{(m)}$. For each incomplete case j, PMM draws an observation randomly from a set (say r) of observed cases i's (i.e., the donor set) having predictive means $\hat{y}_i = x_i\hat{\beta}$ close ("close" will be discussed later) to that of the incomplete case $\hat{y}_j$, where $\hat{\beta}$ is the least-squares estimate of β using complete cases. The value of y from the selected observed case is then used to replace the missing value.
3. Repeat Steps 1 and 2 independently M times from $m = 1, \ldots M$.

Step 1 of the PMM procedure draws β and σ^2 from their posterior distributions and ensures the multiple imputation is proper. This is identical to the step of drawing parameters from the normal linear model. Let i index the observed cases and j index the incomplete cases; Step 2 essentially defines a function $d(j, i)$ to measure the distance between subject j and i based on x_j and x_i. Here the distance function, $d(j, i) = |\hat{y}_j - \hat{y}_i|$, is the difference from the predictive means under the model. Intuitively, the smaller the $d(j, i)$, the donor i is more similar to recipient j. Values from observed cases with smaller distances are more likely to be chosen to replace the missing value. This is achieved by randomly selecting one out of the r closest donors.

Fig. 5.6 sketches the idea of PMM, plotting the distance function $d(j, i)$ over the range of X and observed Y-values. We use the same setup as in Fig. 5.4 (Section 5.3), where the X-value for the missing Y is -0.2059. As expected, the left panel shows that the distance function achieve minimum values near $X = -0.2059$. The right panel plots the distance function, which is limited to be less than 0.2 for the purpose of presentation, over the possible donors of observed Y-values. The Y-value for the smallest distance is -1.5162, which is very close to $E(Y|X = -0.2059) = -1.75$. These donors (17 of them) range between -5 to 0, similar to the case for the kernel-based imputation.

In general, PMM can be robust against some model misspecifications. Schenker and Taylor (1996) considered a variety of misspecifications of the normal linear model. Modifications of models include nonlinear relationships between Y and X, interactions among X-variables, and nonnormal error distributions. Their simulation study demonstrated overall satisfactory performance of PMM, yet the normal linear imputation method can break down in some cases.

Example 5.4. *A simulation study for PMM*

We conduct a simulation study to demonstrate the robustness of PMM. We consider two complete-data models (scenarios) based on the simulation design from Schenker and Taylor (1996): Scenario I: $Y = 10 + X_1 + X_2 + \epsilon$; Scenario II: $Y^{1/4} = X_1 + X_2 + \epsilon$, where $X_1 \sim N(5, 1)$, $X_2 \sim N(5, 1)$, and $\epsilon \sim N(0, 1)$ in both scenarios. Note that the normal linear modeling assumption is violated in Scenario II. We generate around 30% missing cases in Y assuming MCAR. The complete-data sample size is set to be 1000, and the simulation study includes 1000 replicates. The estimands of interest include the marginal mean of Y ($\overline{Y}$) and the proportion of Y less than the bottom 5%-tile of the population ($Pr(Y < Y_{.05})$). PMM is implemented using R mice based on five donors (i.e., $r = 5$). All multiple imputation analyses are based on $M = 30$ imputations.

Table 5.3 summarizes the simulation results. In Scenario I, the normal imputation method (NMI) performs well because the underlying assumption holds well. PMM performs very closely to NMI with a slight increase of MSEs. In Scenario II, both imputation methods perform well for estimating the mean of Y. However, for estimating the proportion, NMI clearly breaks down with a large bias (RBIAS close to 50%) and a low coverage rate (28.6%). On the

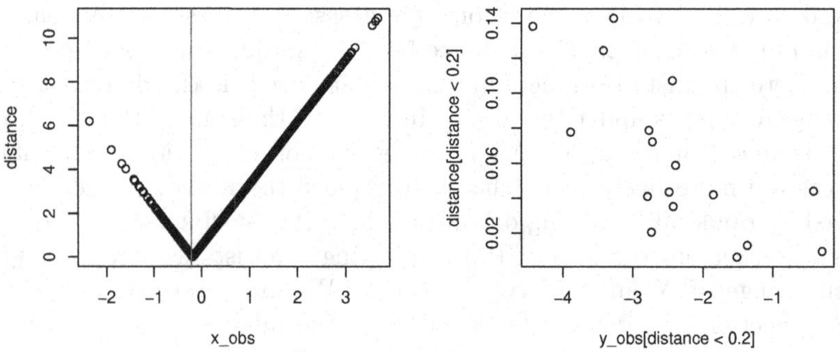

FIGURE 5.6
A sketch of the idea behind PMM. Distance functions for a missing Y-value when $X = -0.2059$ (the vertical line) in PMM. Left: over the range of X_{obs}; right: over the range of Y_{obs} for cases whose distance is less than 0.2.

other hand, PMM performs well with little bias and a coverage rate close to the nominal level.

TABLE 5.3
Example 5.4. Simulation results

Method	RBIAS (%)	SD	$\overline{\text{SE}}$	MSE	CI	COV (%)
Scenario I: mean of Y						
BD	0	0.053	0.055	0.00276	0.214	95.1
CC	0	0.065	0.065	0.00428	0.257	95.0
NMI	0	0.058	0.058	0.00332	0.226	94.4
PMM	0	0.058	0.058	0.00341	0.229	95.0
Scenario I: $Pr(Y < Y_{.05})$						
BD	0	0.0072	0.0069	0.0000521	0.027	92.8
CC	0.2	0.0086	0.0082	0.0000747	0.032	92.6
NMI	0.3	0.0071	0.0076	0.0000506	0.030	95.9
PMM	−0.3	0.0082	0.0076	0.0000674	0.030	91.8
Scenario II: mean of Y						
BD	0	242	253	58450	992	96.0
CC	0	300	302	90214	1185	95.2
NMI	0.1	270	269	72840	1055	94.7
PMM	−0.1	269	273	72491	1071	94.4
Scenario II: $Pr(Y < Y_{.05})$						
BD	0	0.0072	0.0069	0.0000521	0.027	92.8
CC	0.2	0.0086	0.0082	0.0000747	0.032	92.6
NMI	45.2	0.0072	0.0095	0.000561	0.037	28.6
PMM	−0.2	0.0082	0.0076	0.0000666	0.030	92.2

Again, we use graphics to illustrate the difference between two imputation methods. Fig. 5.7 shows the histogram of Y-values using a subset (sample size 250) of one simulation replicate in Scenario II. The original distribution from the complete data is skewed to the right. NMI generates imputed values that move the distribution of completed data to be more symmetric. As a result, some negative values are imputed despite that original values are all positive. Yet completed data from PMM retain the shape of the original distribution well. Fig. 5.8 shows scatter plots between $X_1 + X_2$ and Y from a randomly chosen replicate, including $Y_{.05}$ as the cut-off line. The imputed values from NMI are scattered around a linear regression line between $X_1 + X_2$ and Y. The imputed values from PMM are observed Y-values from close donors, and therefore better capture the underlying nonlinear relation. If we focus on the lower left corner of the scatter plots, it is quite obvious that some of the imputations from NMI are stretched to an even lower region than they should have been (e.g., $Y < 0$) to fit the imposed linear regression relationship. This results in a large, positive bias for the estimate of $Pr(Y < Y_{.05})$, as can be seen in the simulation results. Such deviations are somewhat avoided in PMM because only similar, observed Y values are used as imputations.

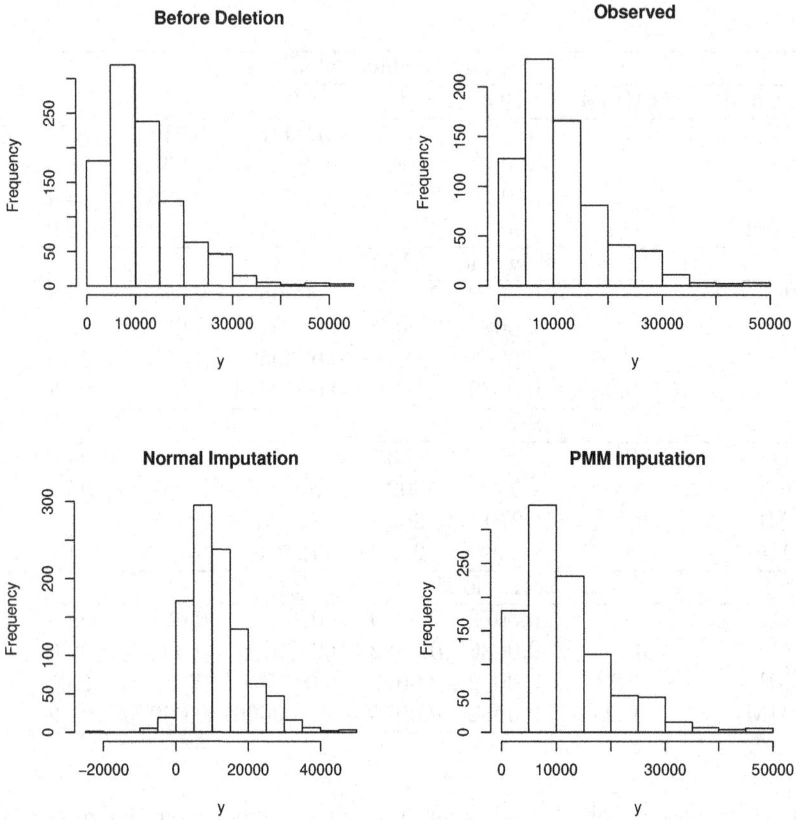

FIGURE 5.7
Example 5.4. Scenario II: histograms of Y-values from one simulation replicate. Top left: before deletion; top right: observed values; bottom left: completed data from NMI of one imputation; bottom right: completed data from PMM of one imputation.

Example 5.5. *Using PMM to impute gestational age*

In Example 4.2, we introduced the missing data problem of gestational age (DGESTAT). There we started with a normal linear model-based imputation (NMI). A lingering question is that whether this method is good enough. Here we go one step further and explore additional imputation methods. Since birth weight (DBIRWT) has the strongest association with DGESTAT, we plot DGESTAT against DBIRWT using complete cases and identify a nonlinear relationship between them (the 1st row of Fig. 5.10). This pattern suggests some room for improving the imputation model.

We then consider three additional imputation methods: (a) using PMM including the original covariates (PMM); (b) adding a quadratic term of DBIRWT to the covariates and applying the normal imputation model (NMI-QUAD); (c) applying the PMM version of (b) (PMM-QUAD). All imputation methods are conducted 20 times. Fig. 5.9 shows histograms from observed and imputed values from one randomly chosen completed dataset. The distribution from the observed DGESTAT is somewhat left skewed. Imputed values from NMI appear to more symmetrically distributed. Compared with those from NMI, imputed values from PMM, NMI-QUAD, and PMM-QUAD seem to retain the shape of the original distribution somewhat better. In addition, imputed values from NMI and NMI-QUAD spread somewhat more widely than those from PMM and PMM-QUAD. This is because the former methods can generate imputations outside the range from the observed data.

Fig. 5.10 shows the corresponding scatter plots of DBIRWT and DGESTAT. Not only is the nonlinear relation apparent, but there also exist some outliers (extreme values) of DGESTAT at different values of DBIRWT, which deviate from the smooth boundary of the bivariate distribution. Clearly, imputations from NMI fail to capture the nonlinear relation. Imputations from PMM have some apparent improvement, especially for the region with larger

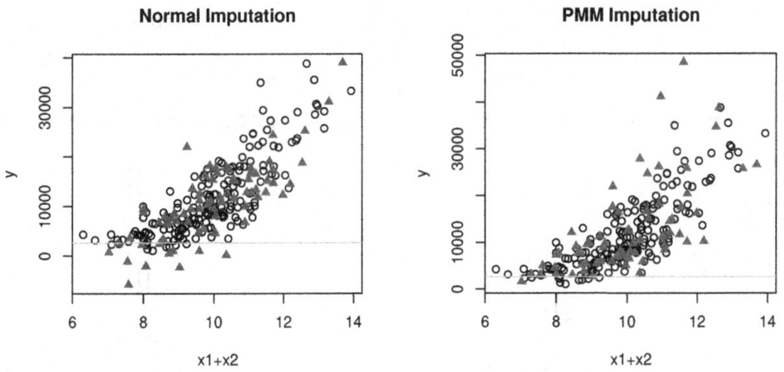

FIGURE 5.8
Example 5.4. Scenario II: scatter plots of $X_1 + X_2$ and Y from one simulation replicate. Left: completed data from NMI; right: completed data from PMM. Circle: observed values; filled triangle: imputed values. The horizonal line is $Y_{.05}$.

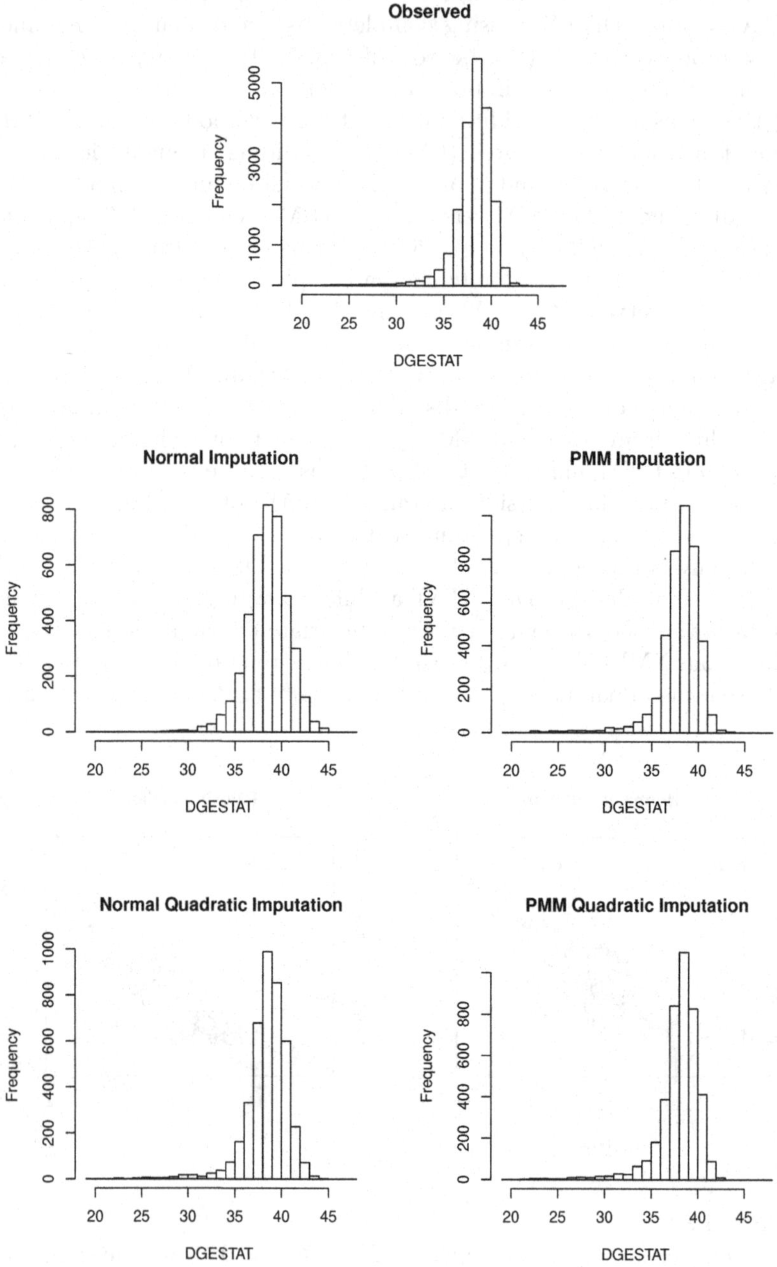

FIGURE 5.9
Example 5.5. Histograms of DGESTAT from both observed (the 1st row) and imputed values (the 2nd and 3rd rows). The latter are from one randomly chosen set of imputations.

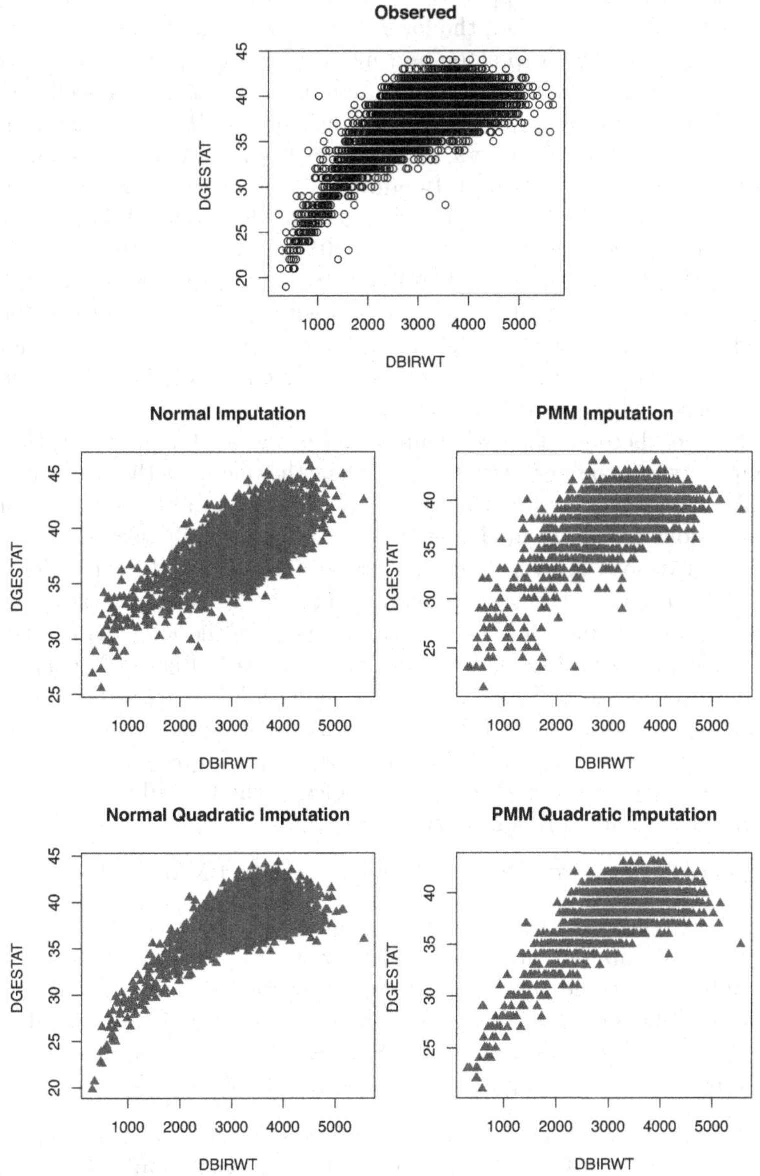

FIGURE 5.10
Example 5.5. Scatter plots of DBIRWT and DGESTAT from both observed (the 1st row) and imputed values (the 2nd and 3rd rows), which are from one randomly chosen set of imputations. Circle: observed values; filled triangle: imputed values.

DBIRWT values at the upper right corner. However, for the region with smaller DBIRWT values (i.e., the lower left corner), the imputed values seem to fail to capture the steep slope and instead tend to spread around. Imputations from NMI-QUAD capture the nonlinear relation very well, yet the distribution appears to be a little over-smoothed as there are few imputed values that appear to be outliers. Imputations from PMM-QUAD capture the nonlinear relation similarly well. In addition, there exist a few more imputed values that look like outliers in a way similar to the observed data.

Based on the graphs (Figs. 5.9 and 5.10), imputed values for DGESTAT from NMI-QUAD and PMM-QUAD seem to match well with the observed data in terms of the distributional relation with DBIRWT. Between the two methods, it appear that PMM-QUAD also retain the "roughness" of the original data slightly better. This is due to the fact that PMM-QUAD directly uses the observed values for imputation.

Note that there exist more complicated proposals for modeling the relationship between the gestational age and birth weight in the literature. For example, Schwartz et al. (2010) introduced a Bayesian finite mixture of bivariate regressions to jointly model gestational age and birth weight to estimate the effect of various covariates and risk factors (e.g. sex and birth order of the baby, maternal age, race, and smoking). The use of such bivariate mixture models can also account for the nonnormal shape of the joint distribution for birth weight and gestational age, which can be seen in Fig. 5.10. Zhang et al. (2014) also proposed a Bayesian mixture model for birth weight conditional on reported gestational age groups to identify implausible/misreported gestational ages. For the purpose of illustrating multiple imputation, however, we model the relation between the two variables in the typical linear regression modeling framework throughout this book.

In general, there are several advantages of using PMM:

1. Similar to the traditional hot-deck imputation, a major advantage of PMM is that the imputed values are always within the range of observed data, as opposed to some purely model-based imputation methods that can generate numbers outside the range. The unit of imputed values is also consistent with that of observed data. Such a property is desirable if the main purpose of multiple imputation is to release public-use data.

2. PMM exhibits certain robustness against some model misspecifications, as suggested by a large body of literature (e.g., Van Buuren 2018, Section 3.4). The key to its robustness is that the method uses observed values from cases with similar covariates to replace missing values, where the similarity is quantified by distance functions. The use of distance functions plays a pivotal role in guarding against misspecifications of exact model forms.

3. PMM can be viewed as an intermediate approach between the parametric imputation and the traditional hot-deck method based on adjustment cells. With a fixed sample size and when the dimension of X increases, creating

adjustment cells can be clumsy (e.g., running into sparse cells), yet using the predicted value based on some parametric models to define the donor set can be more convenient and feasible.

4. When the traditional hot-deck procedure is used to create multiply imputed datasets, and the same donor pool is used for a respondent for all M datasets, the method is not a proper multiple imputation procedure. Since the creation of adjustment cells (i.e., an implicit model) is fixed in the imputation process, the uncertainty of such a model is not propagated and thus can lead to the underestimation of the variance (Rubin 1987, Section 4.3). In typical PMM procedures, the issue of "improperness" has been addressed by including a posterior step so that the distance metric and donor set vary across M sets of imputations.

5.5.3 Predictive Mean Matching for Noncontinuous Data

The idea of PMM started with continuous data. However, similar ideas might be applied to other types of data. We explore the use of PMM for binary data. Before the logistic imputation model for a binary variable (0 or 1) is widely implemented, a seemingly straightforward strategy is to impute the binary variable using a normal linear model and then round off imputed values to either 0 or 1 using some rules. More generally, Schafer (1997) suggested rounding off continuous imputed values in categorical data to the nearest category to preserve the distributional properties as fully as possible. Interestingly, the strategy of rounding can be viewed as a primitive nearest-neighbor PMM for binary data. However in rounding, the predicted means of missing cases in the normal linear model are not compared with those of observed cases. In simple rounding, for example, the former are directly compared with observed values (i.e., 0 and 1), and one of the closer donors is chosen as the imputation.

However, Horton et al. (2003) showed that simple rounding may introduce bias for the mean estimate of the binary variable. Allison (2005) confirmed this pattern using a simulation study. The bias caused by simple rounding is also shown to exist for ordinal variables with more than two categories (Xia and Yang 2016). A couple of methods have been proposed to mitigate the bias associated with rounding (e.g., Bernaards et al. 2007; Yucel et al. 2008, 2011; Demirtas 2009, 2010). However, the bias cannot be completely removed after applying these corrections. Therefore, rounding is generally not recommended for imputing categorical variables (Van Buuren 2018).

To implement PMM for binary data, we need to devise appropriate distance functions based on some working imputation models. Specifically, the distance functions can take the form: $d(j,i) = |\hat{y}_j - \hat{y}_i = x_j\beta^* - x_i\hat{\beta}|$, where β^* and $\hat{\beta}$ are the posterior draw and maximum likelihood (MLE) estimate of the regression coefficients from the working model, and j and i index the missing and observed cases, respectively. Interestingly, we can try either the normal linear model or logistic regression model as the working imputation model.

Example 5.6. *A simulation study for exploring PMM for binary data*

We conduct a simple simulation study to gauge the behavior of the aforementioned two PMM methods (PMM-linear and PMM-logit). We consider two complete-data models: Scenario I (used in Example 4.3): $logit(f(Y = 1)) = -2 + X$; Scenario II: $logit(f(Y = 1)) = -2 + X + 0.5X^2$, where $X \sim N(1, 1/4)$ in both scenarios. In both PMM methods, the working imputation model only includes X as the predictor and excludes X^2 so it is misspecified in Scenario II. In addition, the working model in PMM-linear is on a misspecified scale because it attempts to fit a linear regression model to the binary Y-variable. We generate around 30% missing cases in Y assuming MCAR . The complete-data sample size is 1000, and the simulation includes 1000 replicates. The estimands of interest include the mean of Y ($\overline{Y}$) and the logistic regression coefficients for the main effect X. We also consider two other imputation methods: one uses the logistic regression model (LOGIT) and only includes X as the predictor, which is misspecified in Scenario II; the other simply rounds off imputed values based on a normal linear model (ROUND). All multiple imputation analyses are based on $M = 30$ imputations.

Table 5.4 presents the simulation results. In Scenario I, LOGIT works well as expected. ROUND produces the largest bias for estimating $\overline{Y}$ (RBIAS=3.6%) and the regression coefficient (RBIAS=−9.4%). Both PMM methods also perform well, despite that PMM-linear is based on a misspecified working model. In Scenario II, ROUND still yields the largest bias for estimating $\overline{Y}$. All other imputation methods have little bias. For estimating the regression coefficients, both LOGIT and ROUND yield large biases. On the other hand, both PMM methods mitigate the relative bias to around 2%.

This simulation study demonstrates that for binary data, using the PMM strategy might have certain potential to guard against some model misspecifications. In addition, the similar performance between two PMM methods suggests that forming the metrics of predictive means (i.e., distance function) might be less dependent on the scale of the missing variable.

Example 5.7. *A simulation study for PMM with data separation*

Can PMM deal with the issue of data separation (Section 4.3.2.4) for categorical data? We use an identical simulation design in Example 4.6 to gain some insights. The imputation methods assessed here include the rounding and two PMM methods. Table 5.5 shows the results. PMM-logit clearly breaks down because it still relies on the MLE estimate of β from the logistic regression fitting to form the distance function. When data separation occurs, both the estimate and its posterior draw are unstable, and therefore the matching and donor selection processes based on the distance function can behave erratically. On the other hand, PMM-linear performs the best with small biases and coverage rates close to the nominal level. Although the working model used in PMM-linear is misspecified, the corresponding regression estimates and posterior draws are more stable, which can lead to a better performance in this setting. Again, its good performance is due to the use

TABLE 5.4
Example 5.6. Simulation results

Method	RBIAS (%)	SD	$\overline{\text{SE}}$	MSE	CI	COV (%)
Scenario I: mean of Y						
BD	0	0.014	0.014	0.000207	0.056	94.9
CC	0	0.017	0.017	0.000292	0.066	94.8
LOGIT	0.6	0.017	0.017	0.000291	0.066	95.2
ROUND	3.6	0.017	0.016	0.000384	0.065	89.8
PMM-linear	0.1	0.017	0.017	0.000296	0.067	94.6
PMM-logit	0	0.017	0.017	0.000292	0.067	95.3
Scenario I: logistic regression coefficient on X						
BD	0	0.148	0.152	0.022	0.596	96.5
CC	0	0.177	0.182	0.031	0.713	96.0
LOGIT	1.6	0.177	0.186	0.032	0.733	96.0
ROUND	−9.4	0.160	0.177	0.034	0.697	94.8
PMM-linear	−0.3	0.179	0.182	0.032	0.718	95.1
PMM-logit	−0.4	0.179	0.183	0.032	0.723	96.1
Scenario II: mean of Y						
BD	0	0.016	0.016	0.000253	0.061	95.4
CC	0	0.019	0.019	0.000350	0.073	94.9
LOGIT	0.2	0.018	0.018	0.000325	0.072	94.8
ROUND	1.2	0.017	0.018	0.000316	0.070	94.8
PMM-linear	0	0.018	0.018	0.000334	0.072	94.9
PMM-logit	0	0.018	0.018	0.000330	0.072	95.6
Scenario II: logistic regression coefficient on X						
BD	0	0.574	0.577	0.328	2.262	95.5
CC	0.6	0.682	0.693	0.465	2.715	95.6
LOGIT	31.4	0.526	0.674	0.376	2.650	97.8
ROUND	19.5	0.466	0.652	0.255	2.567	99.0
PMM-linear	2.6	0.684	0.663	0.469	2.609	94.4
PMM-logit	2.4	0.683	0.663	0.466	2.609	94.8

of a reasonable distance function. In addition, ROUND can be thought of as a primitive PMM-linear method so its performance, although not optimal, is still better than PMM-logit and the original MLE-MI method in Example 4.6.

Besides binary variables, a few studies (e.g., Yu et al. 2007; Vink et al. 2014) have shown that PMM may have good potential for imputing semicontinuous variables. More research is needed to apply PMM to a wide variety of variables and understand its performance.

5.5.4 Additional Discussion

PMM is widely available in imputation software programs. However, a couple of subtle issues exist in practical implementation. One of them is specifying the distance function between missing and observed cases. Since the predictive mean for case i can be generally written as $x_i\beta$, there exist a couple of options specifying the distance function, depending upon whether the MLE or posterior draw is used for β. Past literature has shown the optimal option is to use $x_j\beta^* - x_i\hat{\beta}$, where β^* and $\hat{\beta}$ are the posterior draw and MLE, and j and i index the missing and observed cases, respectively. Detailed discussions can be found in Morris et al. (2014) and Van Burren (2018, Section 3.4). In addition, it is recommended to standardize covariates in the imputation because many distance functions can be sensitive to scales of covariates.

The second issue is forming the donor pool and making the selection, similar to that in hot-deck imputation (Andridge and Little 2010). Following the algorithm presented in Section 5.5.2, the essential question is how to determine r, the number of donors to sample from the donor pool. The easiest choice is to set $r = 1$, which always chooses the nearest neighbor. Intuitively, using the nearest donor might be the most accurate choice, leading to the least bias for the estimates based on imputed datasets. However, when setting $r = 1$ or a small value, the same donor might be selected over and over again. As a result, there might exist considerable correlation for the imputed cases both within and between imputations. This can result in higher variability of point estimates in repeated sampling. Setting r to a high value alleviates the duplication problem, but might incur more bias since the likelihood of suboptimal matches increases when a replacement is randomly chosen from a large number of possible donors. For instance, if we let $r \rightarrow n_{obs}$, then PMM is like randomly selecting one of any observed cases for imputation, obviously discarding the information embedded in X that is predictive of Y.

The bias-variance trade-off rising from the choice of r (i.e., as r increases, the bias increases yet variance decreases) is similar to the issue of choosing the

TABLE 5.5
Example 5.7. Simulation results

Estimand	Method	RBIAS (%)	SD	SE	MSE	COV (%)
$\overline{Y}$	ROUND	-0.4	0.013	0.011	0.000167	90.8
	PMM-linear	0.2	0.011	0.012	0.000123	95.6
	PMM-logit	83.3	0.032	0.072	0.000797	99.9
β_1	ROUND	−10.4	0.216	0.228	0.058	93.5
	PMM-linear	−1.3	0.242	0.229	0.058	94.7
	PMM-logit	−32.2	0.169	0.329	0.133	88.6

optimal smoothing parameter in the context of nonparametric/semiparametric regressions (Section 5.3). Schenker and Taylor (1996) derived an adaptive scheme for choosing the optimal r. Siddique and Belin (2008a) sampled one donor from all observed cases with a probability that depends on $d(j,i)$, which is similar to the use of kernel weights in selecting imputations (Section 5.3). However, practitioners might be more used to a fixed r in the implementation. Past research (e.g., Schenker and Taylor 1996; Morris et al. 2014) suggested that reasonable choices of r might range from 3 to 20. When the sample size increases, larger values of r can also be considered. The default value for r in software packages implementing PMM typically ranges from 5 to 10. We recommend that practitioners test different choices of r for comparison and decision.

Third, PMM is not a panacea for every difficult missing data problem. Morris et al. (2014) suggested that this approach might not work well with few donors in the vicinity of an incomplete case. Donor sparsity is expected when there is a large proportion of missing data for certain regions of covariates or when missing data occur in the tails of distributions. In addition, PMM might not work well if the distance has the same sign for all donors in the pool (e.g., predictive means from donors are all larger than that of the missing case). As for all other multiple imputation methods, it is important to conduct some imputation diagnostics after applying PMM (Chapter 12).

PMM can have some extensions. For example, Schenker and Taylor (1996) also proposed a local residual draw (LRD) method for imputation. Unlike PMM, after identifying the possible donor, LRD does not directly use the observed value from that donor. Instead, it still uses the prediction from the working model (e.g., a normal linear model), yet adds a residual from that donor. That is, $y_j^* = x_j\beta^* + r^*$, where $r^* = y_i - x_i\hat{\beta}$ is the residual from the randomly selected donor y_i. The aim is to adjust for the local lack of fit of the regression model in the imputations. The method can be used if the error distributions deviate from normality.

Unlike PMM, LRD might generate values outside the range of observed data. It can be viewed as an intermediate between the parametric model-based imputation and PMM. Simulation studies (Schenker and Taylor 1996; Morris et al. 2014) seemed to suggest some differences between PMM and LRD in terms of their performances. Compared with PMM, relatively little research has been devoted to study LRD. For example, it is not immediately clear how this method can be applied to other types of variables such as binary data. Although LRD is not emphasized in this book, further investigations are warranted.

5.6 Inclusive Imputation Strategy

5.6.1 Basic Idea

In previous sections we have discussed a few strategies that aim to robustify parametric imputation models. All these strategies, as well as the methods discussed in Chapter 4, are built on the assumption that multiple covariates are already selected as predictors for imputation. In some practical cases, however, it is not immediately clear which variables are needed in imputation. That is, how should we select predictors used for imputation?

Naturally, all the variables involved in completed-data analyses need to be included as predictors. For example, to impute a missing covariate in a regression analysis, the response variable and other covariates for this analysis are to be included in the imputation model (Section 4.4). In addition, all other variables that are associated with either the missing variable or its missingness indicator are to be included. This general and important idea is termed the "inclusive imputation strategy" following Collins et al. (2001).

Example 5.8. *A simulation study for the inclusive imputation strategy*

We use a simple simulation study to demonstrate the utility of the inclusive imputation strategy. We continue using the simulation setup in Examples 2.5 and 4.1. Suppose the complete-data analysis includes both Y and X, where $Y = -2 + X + \epsilon$, where $X \sim N(1,1)$ and $\epsilon \sim N(0,1)$. We also consider an auxiliary variable Z, which is not relevant to the analysis and yet can be correlated to either Y or its response indicator R. We consider the following four scenarios:

I. Z is unrelated to both Y and R: $Z \sim N(1,3)$; MCAR for Y.

II. Z is correlated with Y yet not correlated with R: $Z = 2 + Y + \epsilon$, where $\epsilon \sim N(0,1)$; MCAR for Y.

III. Z is unrelated to Y yet correlated with R: $Z \sim N(1,3)$; $logit(f(R = 0|X,Y,Z)) = -1.7 + Z$.

IV. Z is related to both Y and R: $Z = 2+Y+\epsilon$, where $\epsilon \sim N(0,1)$; $logit(f(R = 0|X,Y,Z)) = -1.7 + Z$.

What is the missingness mechanism in these scenarios given that the auxiliary variable Z is included? Since the completed-data analysis only concerns X and Y (i.e., Z is not involved), we should examine $f(R = 0|X,Y)$, which can be expressed as

$$f(R=0|X,Y) = \int f(R=0|X,Y,Z)f(Z|X,Y)dZ.$$

In Scenarios I and II when $f(R = 0|X,Y,Z)$ is a constant C, $f(R = 0|X,Y) =$

$\int Cf(Z|X,Y)dZ = C$, implying MCAR. In Scenario III, $f(R = 0|X,Y) = \int logit^{-1}(-1.7 + Z)N_Z(1,3)dZ$ is again some constant without deriving the actual integral, implying MCAR. In Scenario IV, however, $f(R = 0|X,Y) = \int logit^{-1}(-1.7 + Z)N_Z(2 + Y, 1)dZ$ is a function of Y, which implies MNAR because Y is incomplete. In all scenarios, the missingness rate is around 40% in the simulation.

In Scenarios II and IV, it can be shown that the complete-data model of $f(Y|X,Z)$ is $Y = -2 + X/2 + Z/2 + \epsilon$, where $X \sim N(1,1)$, $Z \sim N(1,3)$ and $\epsilon \sim N(0,1/2)$. We apply two imputation approaches. One is the inclusive approach (inclusive MI), imputing Y based on a linear model including both X and Z as predictors. The other is the exclusive (or restrictive) approach (exclusive MI), imputing Y based on a linear model only including X as the predictor. The estimands of interest include the mean of Y ($\overline{Y}$) and the slope coefficient of regressing Y on X. The complete-data sample size is 1000, and the simulation includes 1000 replicates.

Table 5.6 shows the simulation results. When the auxiliary variable Z is unrelated to either Y or R in Scenario I, including it in the imputation essentially adds some noise to the estimates. There is little difference between the inclusive and exclusive MI methods. When Z is related to Y yet not R in Scenario II, adding Z to the imputation model improves the efficiency of both the mean and regression estimates. Specifically, the standard deviations of estimates from the inclusive MI method are less than those from the exclusive MI method, resulting in tighter confidence intervals. When Z is unrelated to Y yet related to R in Scenario III, adding Z to the imputation model changes little for estimating the regression coefficient. However, the mean estimate becomes less efficient with a larger standard deviation and wider confidence interval. Finally, when Z is related to both Y and R in Scenario IV, both the CC and exclusive MI methods yield biased point estimates and low coverage rates. This phenomenon happens because the missingness of Y is not MAR any more by only conditioning on X in this scenario. Only the inclusive MI method yields good estimates with little bias and coverage rates close to the nominal level. Results in Scenario IV is also a counter argument for the proposed procedure "Multiple Imputation, Then Deletion Method for Dealing With Missing Outcome Data" by Von Hippel (2007), which corresponds to using CC. See also comments from Sullivan et al. (2015).

This simulation study suggests that including auxiliary variables as predictors in imputation brings sound benefits. Although auxiliary variables are not particularly relevant to the analysis by the design, they have potential to improve the performance of the multiple imputation analysis. Applying an inclusive strategy might make it less likely to omit an important cause of missingness (e.g., Scenario IV). That is, it can make the MAR assumption more plausible in the dataset for imputation. It might also lead to noticeable gains in terms of increased efficiency (e.g., Scenarios II and IV) and reduced bias (Scenario IV). The reduced efficiency in estimating the mean in Scenario III is only a minor cost. Collins et al. (2001) conducted more comprehensive

TABLE 5.6
Example 5.8. Simulation results

Method	Bias	SD	$\overline{\text{SE}}$	MSE	CI	COV (%)
Scenario I: Z unrelated to both Y and R						
Mean of Y						
BD	0	0.043	0.045	0.00188	0.175	96.1
CC	−0.003	0.058	0.058	0.00335	0.226	94.6
Exclusive MI	0.002	0.051	0.051	0.00260	0.202	95.7
Inclusive MI	−0.004	0.051	0.051	0.00262	0.203	95.7
Slope coefficient of regressing Y on X						
BD	0	0.032	0.032	0.000995	0.124	95.0
CC	−0.002	0.041	0.041	0.00165	0.160	95.0
Exclusive MI	0.002	0.041	0.041	0.00166	0.163	95.6
Inclusive MI	−0.002	0.041	0.041	0.00166	0.163	94.6
Scenario II: Z related to Y and unrelated to R						
Mean of Y						
CC	−0.003	0.058	0.058	0.00335	0.226	94.6
Exclusive MI	0.002	0.051	0.051	0.00260	0.202	95.7
Inclusive MI	−0.002	0.047	0.048	0.00224	0.190	95.8
Slope coefficient of regressing Y on X						
CC	−0.002	0.041	0.041	0.00165	0.160	95.0
Exclusive MI	0.002	0.041	0.041	0.00166	0.163	95.6
Inclusive MI	−0.001	0.037	0.037	0.00136	0.145	94.7
Scenario III: Z unrelated to Y and related to R						
Mean of Y						
CC	−0.002	0.058	0.057	0.00335	0.224	94.6
Exclusive MI	0.003	0.051	0.051	0.00262	0.200	95.2
Inclusive MI	−0.005	0.056	0.056	0.00319	0.221	95.4
Slope coefficient of regressing Y on X						
CC	0	0.041	0.040	0.00164	0.158	94.4
Exclusive MI	0.004	0.041	0.041	0.00167	0.161	94.5
Inclusive MI	0	0.041	0.041	0.00167	0.161	95.1
Scenario IV: Z related to both Y and R						
Mean of Y						
CC	−0.512	0.052	0.051	0.2654	0.200	0
Exclusive MI	−0.278	0.048	0.049	0.0798	0.191	0
Inclusive MI	−0.003	0.050	0.051	0.00254	0.199	95.7
Slope coefficient of regressing Y on X						
CC	−0.108	0.040	0.040	0.01326	0.158	21.9
Exclusive MI	−0.103	0.040	0.041	0.01211	0.162	29.7
Inclusive MI	−0.002	0.038	0.038	0.00143	0.150	95.6

simulation studies to exhibit the benefits of applying the inclusive strategy for missing data analysis.

How do we identify predictors that are correlated to either the missing variable or the missingness probability? A liberal approach is to use as many variables as possible. This sounds straightforward yet might not be very practical if too many variables are included in a large dataset. More practically, some exploratory analyses prior to imputation can be conducted to identify variables that are significantly associated with the missing variable or the missingness probability. These variables can then be used as predictors for imputation.

5.6.2 Dual Modeling Strategy

Besides imputation, the idea of using variables that are either related to the missing variable or missingness indicator (i.e., nonresponse mechanism) is very general for missing data analysis. In this book we refer to this strategy in the context of multiple imputation as the dual modeling strategy (Jolani et al. 2013). We provide some general background before discussing more specifics.

5.6.2.1 Propensity Score

In Chapter 2 we introduced the concept of missingness mechanisms. Under MAR, the missingness of Y is unrelated to the actual values of Y after conditioning on fully observed covariates X. That is, $f(R=1|Y,X)=f(R=1|X)$. Let $g(X)=f(R=1|X)$, a scalar function of X that the missingness is based on, then $g(X)$ is termed as the (response) propensity score (Rosenbaum and Rubin 1983). The (response or nonresponse) propensity score is arguably one of the most important statistics in missing data analysis and has important properties. For example, the MAR assumption is generally equivalent to the assumption of conditional independence of Y and R given X or $g(X)$:

$$f(Y,R|g(X))=f(Y|g(X))f(R|g(X)), \tag{5.4}$$

which further implies that

$$f(Y,R=1|g(X))=f(Y,R=0|g(X))=f(Y|g(X)). \tag{5.5}$$

Eq. (5.5) suggests that conditional on the propensity score, Y_{obs} and Y_{mis} have identical (i.e., balanced) distributions, which is the so-called balance property of the propensity score (Rosenbaum and Rubin 1983).

In addition, although X is typically multi-dimensional, $g(X)$ is a scalar function and is therefore much easier to handle for practical analysis. Likewise, specifying a model for Y conditional on $g(X)$ is much easier than specifying a full-scale model for Y conditional on all X-variables. In many cases, $g(X)$ might be viewed as a scalar summary of the information contained in all X-variables. Therefore, it has been recommended to include $g(X)$ or variables that strongly affect $g(X)$ (i.e., the variables that are predictive of the nonresponse propensity score) as covariates for imputation (Lavori et al. 1995; Van Buuren et al. 1999). This is consistent with the inclusive imputation strategy.

5.6.2.2 Calibration Estimation and Doubly Robust Estimation

Why can we try the dual modeling strategy in multiple imputation? From a general and theoretical perspective, the use of propensity score is a key component in doubly robust (DR) estimation in missing data analysis. The idea of DR estimation can be briefly explained through the calibration estimator (CE). The earliest literature about CE can be traced back to the generalized regression estimator (Cassel et al. 1976) for survey nonresponse. Suppose μ is the population mean of a sample of Y which is not fully observed. Assuming MAR, CE uses two working models based on the fully observed covariates X: one model predicts the missing values (i.e, the outcome or completed-data model), and the other predicts the missingness probabilities (i.e., the propensity model). Specifically, the estimator is a result of expressing the mean of Y as a sum of prediction and inverse probability weighted errors as

$$\mu = E[E(Y|X)] + E[R\frac{Y - E(Y|X)}{g(X)}]. \tag{5.6}$$

For a sample of n cases, Eq. (5.6) leads to the following estimator:

$$\hat{\mu}_{CE} = \frac{\sum_{i=1}^{n} \hat{y}_i}{n} + \frac{\sum_{i=1}^{n} r_i w_i \hat{\epsilon}_i}{n}. \tag{5.7}$$

In Eq. (5.7), $\hat{y}_i$ is the prediction of case i based on the outcome regression model for $E(Y|X)$, r_i is the response indicator, $w_i = 1/\hat{g}(x_i)$ is the inverse of the estimated probabilities of being observed (i.e., response or propensity weights) based on the nonresponse model $f(R = 1|X)$, and $\hat{\epsilon}_i = \hat{y}_i - y_i$. If the outcome model is correctly specified, then the second term converges to 0 and $\hat{\mu}_{CE}$ converges to μ. If the propensity model is correctly specified, then it can be shown that the second term consistently removes any bias that may be associated with the first term, and hence $\hat{\mu}_{CE}$ still converges to μ. As a result, $\hat{\mu}_{CE}$ is consistent if either of the two models is correctly specified, a property of "double robustness".

Eq. (5.7) had been extended to more general settings of missing data estimation including regression (e.g., Robins et al. 1994) and repeated measurement data (e.g., Robins et al. 1995). These research proposed a class of augmented orthogonal inverse probability-weighted estimators, which extend weighted estimating equation estimates by extracting information from incomplete cases. Therefore, this class of estimators, also known as semiparametric estimators, have better efficiency than the typical inverse probability weighted estimators that completely discard information from incomplete cases. On the other hand, the semiparametric estimators are generally not as efficient as the maximum likelihood estimators obtained using only the correctly specified outcome model. More importantly, the DR property of the semiparametric estimators is that, if either (a) the prediction/outcome model is correctly specified or (b) the response/propensity model on which the weight is based is correctly specified, then the estimate is consistent. However, when both the prediction and response models are misspecified, using the semiparametric estimators does not necessarily have any edge over using either model alone. Some reviews about this topic can be found in Carpenter et al. (2006) and Kang and Schafer (2007). More theoretical discussions can be found in Tsiatis (2006).

5.6.2.3 Imputation Methods

Applying the dual modeling strategy might robustify the imputation models. Example 5.8 already demonstrated this effect in a simple simulation study. The property of DR estimators also provides some theoretical justifications. Now suppose these predictors are identified; we discuss a few strategies to form the appropriate imputation model.

Let X denote the variables correlated with the incomplete variable Y and Z denote the variables correlated with the nonresponse of Y. Variables X and Z might have some overlap. Suppose Y is continuous; the goal of the dual modeling strategy is to form a model as

$$Y = f(X, Z; \theta) + \epsilon, \tag{5.8}$$

where $f(\cdot)$ is some mean response function parameterized by θ and ϵ is the error component. If Y is not continuous, we might consider the GLM framework.

The simplest strategy is to apply the normal linear model for Eq. (5.8), treating variables in X and Z as linear predictors as in Example 5.8. On the other hand, the information contained in Z-variables might be summarized by the corresponding propensity score estimate $\hat{g}(Z) = \hat{f}(R = 1|Z)$, its inverse $1/\hat{g}(Z)$ (i.e., the nonresponse weights), or strata formed using $\hat{g}(Z)$. These quantities can also enter Model (5.8) as linear predictors. However, Jolani et al. (2013) reviewed several of such strategies and concluded that as long as the relevant variables are taken into the imputation model, how the propensity score variables (Z) are included in Model (5.8) has only a minor effect on the quality of the imputations.

Nonparametric or semiparametric adjustments of Model (5.8) have also been proposed to make the models less restrictive. One adjustment approach is PMM (Section 5.5).

Example 5.9. *A dual modeling imputation using PMM*

Long et al. (2012) proposed a dual modeling strategy using PMM. They considered two working models, one regressing Y on X (i.e., the complete-data model) and the other regressing R on Z (i.e., the logistic propensity score model). Briefly speaking, suppose the models are $Y = X\beta + \epsilon$ and $logit(f(R = 1)) = Z\alpha$. Based on the model fit, two predictive scores (after standardization) can be obtained as $S_X = X\hat{\beta}$ and $S_Z = Z\hat{\alpha}$. This strategy summarizes the multidimensional structure of the fully observed variables into a two-dimensional summary score. The hope is that this two-dimensional summary score contains most, if not all, information from X and Z.

A PMM method can be executed based on the two predictive scores. Specifically, a distance function is defined as

$$d(j,i) = \sqrt{w_X(S_X(j) - S_X(i))^2 + w_Z(S_Z(j) - S_Z(i))^2},$$

where j and i index the missing and observed cases, respectively. Here w_X and w_Z are two positive weights used to control the relative contribution from the two predictive scores, where $w_X + w_Z = 1$. To make the multiple imputation proper, for each imputation, $\hat{\beta}$ and $\hat{\alpha}$ can take the estimates from a bootstrap sample of observed data. This idea can also be applied to impute censored event times in the context of survival analysis (Section 8.2.3).

Both theoretical arguments and simulation study show that this PMM-based dual modeling strategy is robust for estimating the mean of Y if either the complete-data model or propensity score model is correctly specified. Since this procedure is relatively simple to program, it has been extended to a variety of scenarios (e.g., Hsu et al. 2014; Zhou et al. 2017).

Smoothing techniques introduced in Section 5.3 have also been used in the dual modeling strategy. A series of studies (e.g., Little and An 2004; An and Little 2008; Zhang and Little 2009, 2011) considered modeling the propensity score using a penalized smoothing spline and treated the spline estimate as a covariate in Model (5.8). Theoretical arguments and simulation studies show that this strategy is doubly robust for estimating the mean of Y.

Example 5.10. *Applying the dual modeling strategy to impute gestational age*

The dual modeling imputation strategy might have nice theoretical properties as the hope is to use the propensity score model to rescue the situation if the complete-data model is misspecified. However, a practical issue is that the propensity score model can also be subject to possible misspecifications. Example 5.5 shows that adding a quadratic term of DBIRWT (i.e., NMI-QUAD) can considerably improve the imputation model compared with NMI in terms of capturing the nonlinear relation between DBIRWT and DGESTAT. Here we explore whether including a propensity score model can improve upon NMI as well.

We fit a propensity score model for the response indicator of DGESTAT using all the main effects of the predictors. Among these, DBIRWT, DBAGE, and CIGAR are highly significant predictors ($p < 0.0001$), and DRINK is marginally significant ($p = 0.083$). Fig. 5.11 shows the scatter plot of the estimated response propensity score $\hat{g}(X)$ and its inverse ($\hat{W}(X) = 1/\hat{g}(X)$) over DGESTAT. We smooth the response weights by top coding them at 1.5. Unlike DBIRWT, the nonlinear relation between $\hat{g}(X)$ or $\hat{W}(X)$ and DGESTAT is not obvious.

We then consider two imputation models for DGESTAT based on Eq. (5.8): one includes all the original covariates X plus the response weight $\hat{W}(X)$ (PMI); the other includes X, $\hat{W}(X)$, and the interactions between $\hat{W}(X)$ and X (PMI-INT). Both methods include additional imputation predictors derived from the propensity score model.

Fig. 5.12 shows the scatter plots between DBIRWT and imputed DGESTAT-values from the two models using one randomly chosen set of imputations. Fig. 5.12 can be compared with Fig. 5.10. Compared with NMI, it seems that PMI has incorporated slightly more nonlinearity for the relation between DBIRWT and DGESTAT, more obviously at the lower-left corner of the distribution. The nonlinear relation is more apparent from PMI-INT, yet still less obvious than from the imputation models that explicitly include the quadratic term of DBIRWT (e.g., NMI-QUAD and PMM-QUAD in Example 5.5). Contrasts among different imputation methods can also be seen from some model-fitting statistics. For example, we regress DGESTAT on DBIRWT, its squared term, and other covariates using one completed dataset and calculate R^2-statistics based on different methods and show them here: NMI (adjusted R^2=0.503), PMI (adjusted R^2 = 0.509), PMI-INT (adjusted R^2 = 0.529), and NMI-QUAD (adjusted R^2 = 0.553). These R^2-statistics clearly show that there is some improvement by including the propensity score variable to the misspecified NMI model, yet the improvement cannot match with that from using NMI-QUAD.

It might be helpful to include the information from the propensity score model to improve the imputations if the initial complete-data model speci-

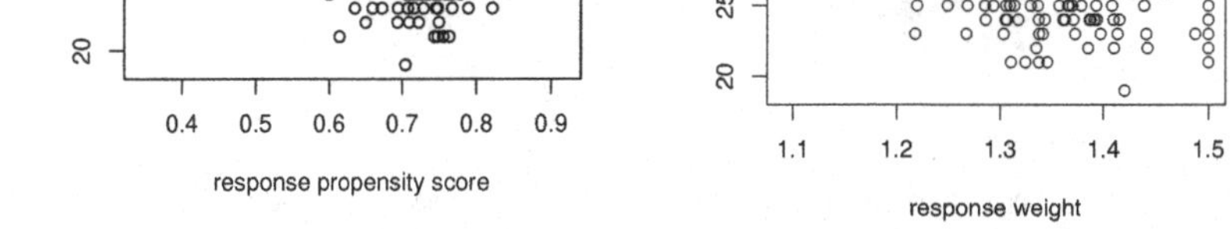

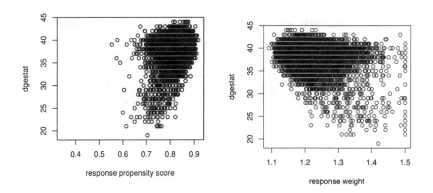

FIGURE 5.11
Example 5.10. Scatter plots of propensity function estimates and DGESTAT. Left: response propensity $\hat{g}(X)$ vs. DGESTAT; right: response weight $\hat{W}(X) = 1/\hat{g}(X)$ vs. DGESTAT.

fication is not optimal. However, this requires a correct specification of the propensity model. In Example 5.10, we did not explore whether the propensity model used can be further improved. However, investing in the propensity model might not be as efficient as directly improving the complete-data model (e.g., by directly adding a quadratic term of DBIRWT as in NMI-QUAD or using PMM as in PMM-QUAD in Example 5.5).

5.7 Summary

In this chapter we have discussed several strategies that aim to improve the quality of imputations so that they conform better to complex features of real data. For example, some of the methods (e.g., PMM) introduced in this chapter are semiparametric, which somewhat yet clearly deviate from the parametric, fully-Bayesian models emphasized in Chapter 4. These additional choices of imputation modeling strategies again show the flexibility of the multiple imputation approach to missing data problems.

The research on robustifying imputation models is a fast-growing field, and therefore the topics presented in this book are surely not exhaustive. Besides the techniques introduced here, in the current era of machine learning and data science, it is expected that many of the techniques from predictive analytics can be used for model building in imputation. For example, Burgette and Reiter (2010) is a classic paper on using the regression tree method for

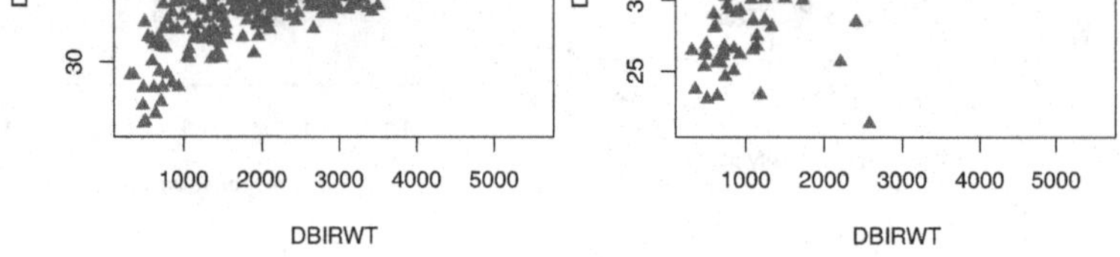

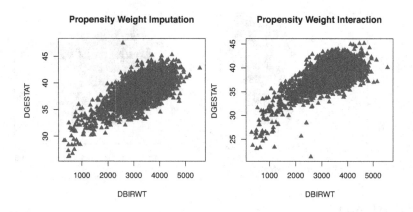

FIGURE 5.12
Example 5.10. Scatter plots of DBIRWT and imputed values of DGESTAT from one randomly chosen set of imputations. Left: PMI; right: PMI-INT.

multiple imputation. Van Buuren (2018, Chapter 3) listed more literature on this topic. Note that predictive approaches in general need to be adapted to produce proper multiple imputations because they generate predictive means and not posterior predictive draws.

Besides all the robust modeling techniques, the importance of the inclusive imputation strategy cannot be emphasized more for producing high-quality imputations. In many cases, it is fair to say that including sufficient predictors in imputation is perhaps more important than applying sophisticated modeling techniques yet ignoring important predictors. That is, "variables" are more important than "modeling forms". For example, the model can include both main effects as well as interactions among variables.

In Example 5.8, the advantage of including the auxiliary variable Z in the imputation is clear. However, one related question is, if the imputation model includes both X and Z, yet the analysis model only includes X (say regressing Y on X), is this a principled imputation procedure? The answer is yes. However, the apparent inconsistency between the imputation model and analysis procedure there often occurs in other settings and is generally referred to as "imputation uncongeniality" (Meng 1994). A more focused discussion of this topic can be found in Section 14.2.

Finally, in all examples illustrated in Chapters 4 and 5, the variables (both the incomplete Y and fully observed X) are assumed to be collected in one single dataset or data source. This is a natural setup of the univariate missing data problem (Table 1.2) yet does not include all possible scenarios. For example, we can have one dataset A that has both X and Y observed, and another dataset B that has X observed yet Y missing. An apparent solution is to use the model $f(Y|X)$ built up from dataset A and then impute the missing Y-values in dataset B, using methods introduced in Chapters 4 and 5. However, we need to make sure that the variables from the two datasets are measuring the same constructs. We also need to assume that the model posited for $f(Y|X)$ is consistent between two data sources. That is, the model is "transportable" (Carroll et al. 2006). The use of multiple imputation in this context is a strategy for "combining information from multiple data sources" (Schenker and Raghunathan 2007). Related examples and discussions will be given in Chapter 11.

6

Multiple Imputation for Multivariate Missing Data: The Joint Modeling Approach

6.1 Introduction

Chapters 4 and 5 focus on univariate missing data problems with fully observed covariates. In practice, however, missing data often occur on multiple variables that are typically related to each other in a dataset. In principle, it is optimal to impute multiple incomplete variables jointly, that is, impute them while incorporating their relations. Chapters 6 and 7 are devoted to multivariate missing data problems in cross-sectional data. Chapter 6 discusses the joint modeling (JM) approach. The term "joint modeling" was first coined by Van Buuren (2007), although some of the major work had been developed before (e.g., Schafer 1997).

In summary, suppose variables of interest are $Y = (Y_1, Y_2, \ldots Y_p)$, $(p > 1)$; each variable has some missing values. Suppose X are some fully-observed covariates. Assuming MAR for Y-variables, JM starts by specifying a parametric multivariate density, $f(Y, X|\theta)$, for complete data Y, covariate X, and parameter θ. Given this specification of the complete-data model and appropriate prior distributions $\pi(\theta)$ for θ, in principle one can use the data augmentation (DA) algorithm (Section 3.4.1) to generate multiple imputations as draws from $f(Y_{mis}|Y_{obs}, X)$, the posterior predictive distribution of missing data. However in many cases, especially for complex models and general missingness patterns, implementing the DA algorithm is not trivial. Researchers and practitioners often need some special software programs for assistance.

This chapter presents several JM strategies, largely determined by the type and pattern of missing variables. Section 6.2 discusses the scenario where the missing data pattern is monotone. Assuming a general missingness pattern, Section 6.3 deals with the case where all variables are continuous, Section 6.4 focuses on the case where all the variables are categorical, and Section 6.5 handles a mixture of both continuous and categorical variables. Section 6.6 discusses the problem in which the outcome variable and covariates are incomplete in a targeted regression analysis. Section 6.7 provides a summary.

DOI: 10.1201/9780429156397-6

6.2 Imputation for Monotone Missing Data

In monotone missing data (Table 1.5), $Y_{j+1}, \ldots, Y_p$ are all missing for cases where Y_j is missing, $j = 1, \ldots p-1$. The complete-data model (conditional on X-covariates), $f(Y|X, \theta)$, can be expressed as

$$f(Y|X,\theta) = f(Y_1|X,\theta_1) \times f(Y_2|Y_1,X,\theta_2) \ldots \times f(Y_p|Y_1,\ldots,Y_{p-1},X,\theta_p), \tag{6.1}$$

where θ_1, ..., θ_p index the parameter of each of the p conditional distributions.

Eq. (6.1) factorizes the $f(Y|X, \theta)$ into p univariate conditional distributions (models). The posterior predictive distribution of missing values in Y_j is determined by $f(Y_j|Y_1, \ldots, Y_{j-1}, X, \theta_j)$. The imputation can be conducted based on each of the conditional models, starting from Y_1 till Y_p. Major components of the algorithm are sketched as follows:

1. Impute missing values in Y_1 based on the model $f(Y_1|X, \theta_1)$.

2. Impute missing values in Y_2 based on the model $f(Y_2|Y_1, X, \theta_2)$. Note that Y_1 becomes completed after Step 1.

   ⋮

p Impute missing values in Y_p based on the model $f(Y_p|Y_1, \ldots, Y_{p-1}, X, \theta_p)$. Note that Y_1 through Y_{p-1} become all completed from the previous $p-1$ steps.

$p+1$ Repeat Steps 1 to p independently M times.

In each of the steps, we can use the methods targeted for univariate missing data problems (Chapters 4 and 5) since the variables being conditioned are fully completed from the previous step. For monotone missing data, drawing $Y_{1,mis}$ only needs to be based on $f(Y_1|X, \theta_1)$ because the corresponding values in Y_2, ..., Y_p are also missing, contributing no information to $Y_{1,mis}$. Similarly, drawing $Y_{2,mis}$ only needs to be based on $f(Y_2|Y_1, X, \theta_2)$ since the corresponding values in Y_3, ..., Y_p are also missing, contributing no information to $Y_{2,mis}$, and so on. To produce one set of the multiple imputations, the algorithm only needs to be run for one pass and has no need for iteration among the variables.

For data with a general missingness pattern, although the complete-data model can still be factorized as in Eq. (6.1), the imputation cannot be conducted in a similar manner. This is because, for example, the posterior predictive distribution of missing values from Y_1 is not solely determined by $f(Y_1|X, \theta_1)$, but also from other Y-variables which have observed values where Y_1 is missing. JM strategies for imputing data with a general missing pattern are to be presented in subsequent sections. In practice, however, if the missing data pattern is not monotone yet very close to it, it had been suggested to intentionally remove some of the observed cases to make the pattern monotone for imputation (e.g., Little and Rubin 2020).

6.3 Multivariate Continuous Data

6.3.1 Multivariate Normal Models

For simplicity, we now use Y to denote all variables including both the incomplete and fully observed ones. If all Y-variables are continuous, a classic JM method is based on a multivariate normal model for the complete data, that is, $Y \sim N(\mu, \Sigma)$, where μ is a $p \times 1$ vector and Σ is a $p \times p$ covariance matrix. Let $\theta = (\mu, \Sigma)$ and to start a Bayesian imputation process, a prior $\pi(\theta)$ is imposed for θ (e.g., a flat prior for μ and an inverse-Wishart prior for Σ), then the DA algorithm can be executed for multiple imputation. It iterates between the posterior step (P-step) and imputation step (I-step): the former draws model parameters given completed data, and the latter imputes missing values given parameter draws and makes the data completed.

Example 6.1. *The data augmentation imputation steps for a bivariate normal sample subject to missing data*

Suppose $Y = (Y_1, Y_2)$ follows a bivariate normal distribution. The conditional distribution of Y_i given Y_j $(i, j \in (1, 2), j \neq i)$ is a univariate normal linear regression model :

$$f(Y_1|Y_2) = N(\beta_{01} + \beta_{11}Y_2, \tau_1^2), \quad (6.2)$$
$$f(Y_2|Y_1) = N(\beta_{02} + \beta_{12}Y_1, \tau_2^2). \quad (6.3)$$

Suppose $\mu = (\mu_1, \mu_2)^t$, $\Sigma = \begin{bmatrix} \sigma_1^2 & \sigma_{12} \\ \sigma_{12} & \sigma_2^2 \end{bmatrix}$, and let $\theta = (\mu, \Sigma)$. In linear regression models (6.2) and (6.3), the parameters (β's and τ^2's) are reparameterizations of θ: $\beta_{01} = \mu_1 - \beta_{11}\mu_2$, $\beta_{11} = \sigma_{12}/\sigma_2^2$, $\tau_1^2 = \sigma_1^2 - \sigma_{12}^2/\sigma_2^2$, $\beta_{02} = \mu_2 - \beta_{12}\mu_1$, $\beta_{12} = \sigma_{12}/\sigma_1^2$, and $\tau_2^2 = \sigma_2^2 - \sigma_{12}^2/\sigma_1^2$.

After imposing a diffuse prior for θ such as $\pi(\mu, \Sigma) \propto |\Sigma|^{-1}$, major components of the imputation algorithm can be sketched as follows:

1. Begin with an estimate or guess of the parameter θ, say $\theta^{*(t)}$ (where $t = 0$ at the 1st iteration). Reparametrize $\theta^{*(t)}$ to $\beta^{*(t)}$'s and $\tau^{2*(t)}$'s using the above formula.

2. I-step: for missing cases in Y_1 where their values are observed in Y_2, imputing them using Model (6.2) with parameter draws (i.e., $\beta_{01}^{*(t)}$, $\beta_{11}^{*(t)}$, and $\tau_1^{2*(t)}$) from Step 1; impute missing cases in Y_2 in a similar way using

Model (6.3); if both variables are missing, they can be generated from $N_2(\mu^{*(t)}, \Sigma^{*(t)})$.

3. P-step: once Y_1 and Y_2 are completed from Step 2, draw $\mu^{*(t+1)}$ and $\Sigma^{*(t+1)}$ as: $\Sigma^{*(t+1)}/(n-1) \sim Inverse - Wishart(S^{(t+1)}, n-1)$, and $\mu^{*(t+1)} \sim N_2(\overline{Y}^{(t+1)}, \Sigma^{*(t+1)}/n)$, where $\overline{Y}^{(t+1)}$ and $S^{(t+1)}$ are the sample mean and covariance matrix from completed data $Y^{(t+1)}$, respectively; transform $\theta^{*(t+1)}$ to $\beta^{*(t+1)}$'s and $\tau^{2*(t+1)}$'s.

4. Repeat Steps 2 and 3 until the convergence for $(\theta^{*(t)}, Y_{mis}^{*(t)})$ is satisfied, say at $t = T$. The draws of $Y_{mis}^{*(T)}$ constitute the 1st ($m = 1$) set of imputations, $Y_{mis}^{(m)}$.

5. Repeat Steps 1 to 4 independently M times.

JM under a bivariate normal model can be generalized to a multivariate normal model with p-dimensions. We first provide a quick review of the partition of a multivariate normal distribution. If we partition the p-variate Y into two parts, Y_1 and Y_2, with dimensions p_1 and p_2 where $p = p_1 + p_2$, then $Y = (Y_1, Y_2)^t \sim N((\mu_1, \mu_2)^t, \begin{pmatrix} \Sigma_{11} & \Sigma_{12} \\ \Sigma_{21} & \Sigma_{22} \end{pmatrix})$, where μ_1 and μ_2 are $p_1 \times 1$ and $p_2 \times 1$ vectors, respectively, and Σ_{ij} are covariance matrices with dimensions $p_i \times p_j$ for $i, j = 1, 2$. The conditional distribution of Y_2 given Y_1 is a p_2-variate normal distribution: $f(Y_2|Y_1) \sim N(\mu_{2|1}, \Sigma_{2|1})$, where $\mu_{2|1} = \mu_2 + \Sigma_{21}\Sigma_{11}^{-1}(Y_1 - \mu_1)$ is the predicted mean from the p_2-variate linear regression of Y_2 on Y_1, and $\Sigma_{2|1} = \Sigma_{22} - \Sigma_{21}\Sigma_{11}^{-1}\Sigma_{12}$ is the residual covariance matrix from that regression.

To carry out JM for more than two variables, we need to partition subjects into groups according to which variables have missing values. Then for the group with Y_2-variables missing and Y_1-variables observed, their imputations can be generated using the multivariate linear regression model characterized by $f(Y_2|Y_1)$ as above. The I-step of the DA algorithm goes through all these groups and is then followed by the P-step. For simplicity, here we do not present all the details of the algorithm. Interested readers can refer to, for example, Schafer (1997) and Little and Rubin (2020), which also summarized the likelihood-based estimation approach to θ for incomplete Y. These methods have been widely implemented in imputation packages such as SAS PROC MI (MCMC option) and R norm.

6.3.2 Models for Nonnormal Continuous Data

For multivariate continuous variables that exhibit strong skewness and/or heavy tails, one convenient JM strategy is to apply some transformations to make the normality assumptions more plausible and then conduct the multivariate normal model imputation on the transformed scale.

Another effective strategy for nonnormal continuous data is to use mixture models. Mixture models allow for flexible joint modeling, as they can reflect complex distributional and dependence structures automatically. Finite mixtures of normal distributions (McLachlan and Peel 2000) are a powerful tool for statistical modeling in a wide variety of situations. For example, Fraley and Raftery (2002) and Marron and Wand (1992) showed that many probability distributions may be well approximated by finite mixture models. In addition, Priebe (1994) showed that with 10000 observations, a lognormal density may be well approximated by a mixture of 30 normal components.

To sketch the idea, Let $Y = (Y_1, \ldots, Y_n)$ comprise n complete observations, where each Y_i is a p-dimensional vector. Suppose that each individual belongs to exactly one of K latent mixture components (groups or classes). For $i = 1, \ldots, n$, let $Z_i \in \{1, \ldots, K\}$ indicate the component of individual i, and let $\pi_k = Pr(Z_i = k)$. Assume that $\pi = (\pi_1, \ldots, \pi_K)$ is the same for all individuals. Within any component k, suppose that the p variables follow a component-specific multivariate normal distribution with mean μ_k and variance Σ_k. Let $\theta = (\mu, \Sigma, \pi)$, where $\mu = (\mu_1, \ldots, \mu_K)$ and $\Sigma = (\Sigma_1, \ldots, \Sigma_K)$. The finite mixture model can be expressed as

$$\begin{aligned} Y_i|Z_i, \mu, \Sigma &\sim N(\mu_{Z_i}, \Sigma_{Z_i}), && (6.4) \\ Z_i|\pi &\sim Multinomial(\pi_1, \ldots, \pi_K). && (6.5) \end{aligned}$$

Marginalizing over Z_i's, this mixture model is equivalent to

$$f(Y_i|\theta) = \sum_{k=1}^{K} \pi_k N(\mu_k, \Sigma_k). \quad (6.6)$$

To complete a Bayesian specification of the model, common prior distributions for θ can be imposed. For example, we can specify: $\pi(\mu_k|\Sigma_k) \sim N(\mu_0, \tau^{-1}\Sigma_k)$ and $\pi(\Sigma_k) \sim Inverse-Wishart(m, \Lambda)$ $(k = 1, \ldots K)$ for given prior mean vector μ_0, prior scalar precision parameter τ, and prior degrees of freedom m for prior covariance matrix Λ (Gelman et al. 2013). In addition, $\pi(\pi_1, \ldots, \pi_K) \sim Dirichlet(1, \ldots, 1)$, the Dirichlet distribution.

As a mixture of multivariate normal distributions, the model is flexible enough to capture distributional features like skewness and nonlinear relations that a single multivariate normal distribution would fail to encode. For example, when $K = 1$, the normal mixture model is the typical normal model. Setting $K = 2$ and $\mu_1 = \mu_2$ in Models (6.4) and (6.5), data with outliers can be viewed having risen from two classes: one contains the majority of normal values and the other comprises extreme values/outliers that have inflated variances. This is also referred to as the contaminated normal model in Little and Rubin (2020, Chapter 12). In addition, although K is often fixed and preset, it can be treated as unknowns and obtained in a data-driven manner via advanced Bayesian methods. This option further expanded the flexibility and utility of normal mixture models, especially for high-dimensional data (Section 14.4).

A major distinction between the mixture model and the discriminant analysis model (Section 4.3.2.2) is that here the grouping indicator Z_i's are not observed and probabilistic. This feature makes the model fitting and imputation rather complicated. Specific code or computational routines are often required. Applications of normal mixture models to missing data imputations can be found, for example, in Elliott and Stettler (2007), Böhning et al. (2007), and Kim et al. (2014). A simple illustration with the aid of WinBUGS will be given in Example 6.9.

In the same spirit as normal mixture models, another JM strategy for nonnormal data is to impose models that can accommodate nonnormal features of data using extra parameters (i.e., in addition to μ and Σ) and include normal models as special cases. For example, Liu (1995) developed imputation models assuming a multivariate t-family. It is well known that in the t-family, the degrees of freedom ν control the tail behavior of the distribution; as $\nu \rightarrow \infty$, the t-distribution converges to a normal distribution. He and Raghunathan (2012) considered a multivariate extension of the gh family (Tukey 1977), which is a transformation of a standard normal variable to accommodate different skewness and elongation of the distribution from nonnormal variables, and the transformation is controlled by several unknown parameters. In addition to contaminated multivariate normal models, Little and Rubin (2020, Chapter 12) provided some examples on the family of weighted multivariate normal models. Also similar to normal mixture models, a technical challenge of using these models for imputation is that the posterior distributions of parameters can be rather complex. The corresponding DA algorithm might need specific code or need to be run with the aid of Bayesian software packages. Sometimes approximations of the DA algorithm (e.g., Section 3.4.3) may be considered.

Example 6.2. *Imputation for hospital performance data*

To illustrate the idea of using transformation and nonnormal distribution families in imputation, we present an example involving some hospital performance data used in He and Raghunathan (2012). The dataset contains performance measures from 5021 U.S. hospitals for care delivered to patients over 65 years old from October, 2005 to September, 2006. We focus on three measures: "Use Beta Blocker at arrival for patients with heart attack" (AMI), "Left ventricular systolic dysfunction assessment for patients with heart failure" (CHF), and "Adult smoking cessation advice/counseling for patients with pneumonia" (CAP). These measures are widely used as hospital quality indicators by the health services research community. Compliance rates are calculated as the ratio between the number of patients who are provided with those procedures (i.e., the numerator) among those who are eligible (i.e., the denominator) within the defined period. Note that some patients might have contradictions to certain clinical procedures. The proportion measures range between 0 and 1, and a hospital with higher compliance rates might demonstrate better performances in implementing these standard procedures. A main analytic goal for the hospital performance data is to determine the top performers, as such information can guide patients, employers, and insurance plans to make better choices of hospitals from which they would receive or buy services. For example, a hospital might be classified as a top performer if its compliance proportion for a measure (e.g., AMI) is greater than 90% or some other threshold.

The missingness rates of the dataset are 25.5%, 20.6%, and 19.5% for AMI, CHF, and CAP, respectively. Fig. 6.1 (the top row) shows that marginal distributions of performance measures are skewed to the left. In addition, Fig. 6.2 (the left panel) shows that there exist strong, positive correlations among three measures, suggesting that it is necessary to impute the three variables jointly.

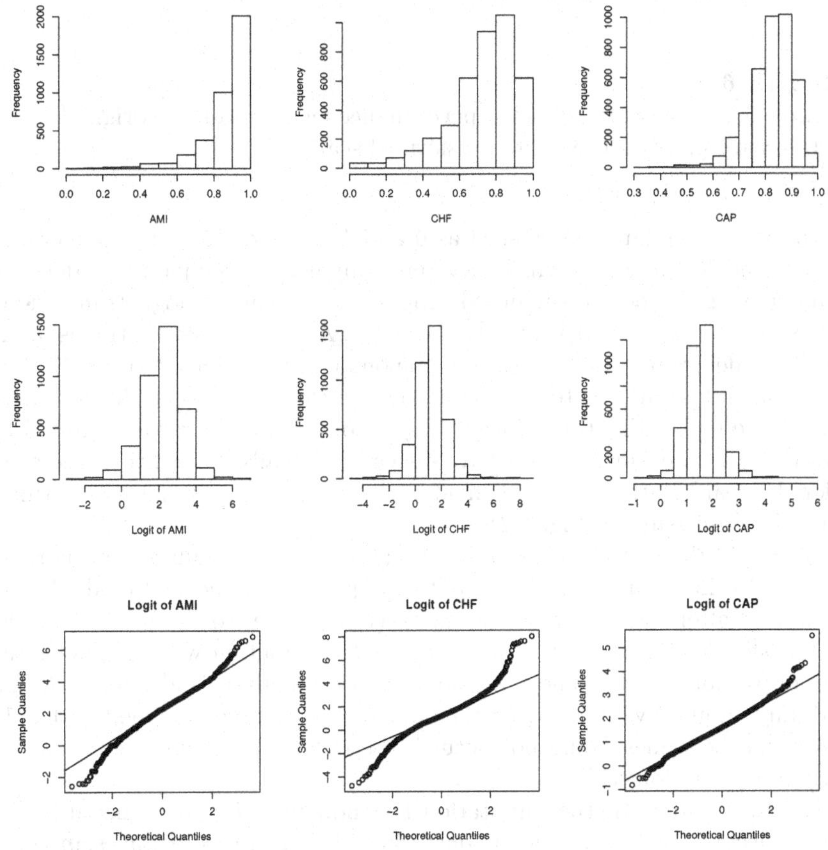

FIGURE 6.1
Example 6.2. Top: histograms of performance measures on the original scale; middle: histograms on the logit-transformed scale; bottom: QQ plots on the logit-transformed scale.

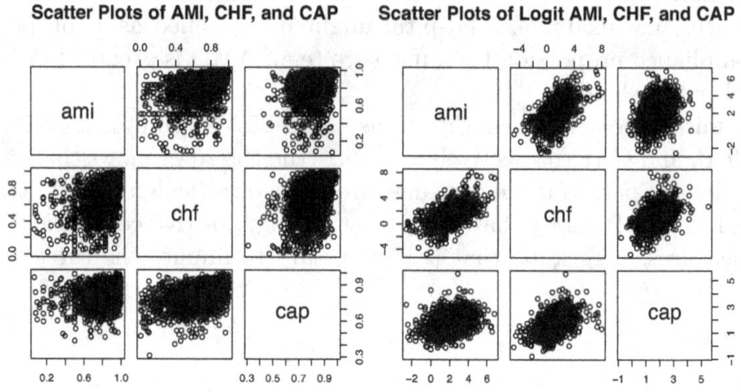

FIGURE 6.2
Example 6.2. Left: scatter plots of performance measures on the original scale; right: scatter plots on the logit-transformed scale.

To avoid boundary values such as 0 and 1, we add 0.5 to the numerator and 1 to the denominator, which is a commonly used technique for smoothing proportions. Let p be the smoothed proportion; we apply a logit transformation, $Y = log(p)/(1 - log(p))$, to the measures. A logit transformation is often used to model proportions arising from binomial distributions. Marginally, the transformed data appear to be symmetric and closer to normal distributions (the middle row of Fig. 6.1). However, QQ plots (the bottom row of Fig. 6.1) show that the transformed data have rather heavy tails. In addition, the conditional linear relations seem to hold well for the data on the transformed scale (the right panel of Fig. 6.2).

We apply three imputation methods: (1) applying a multivariate normal model-based imputation for p's (NMI); (2) applying the normal model-based imputation after the logit-transformation (LMI); (3) modeling and imputing the transformed-data assuming a trivariate t-distribution with unknown degrees of freedom (TMI). The complete-case (CC) analysis is also applied. For NMI, any imputed values outside the boundary are simply assigned as 0 or 1. However, such values would not occur if the imputation is conducted on the logit-transformation scale.

For NMI and LMI, the imputation is conducted $M = 50$ times using R norm. Although we do not focus on technical details of the Bayesian computation in this book, here we plot posterior samples of parameters from the imputation model for illustration. Fig. 6.3 shows the trace/history plots of the parameter draws from the DA algorithm in LMI. The MCMC chain started with the MLE estimate of the model and then ran for 2000 iterations. A burn-in period of 1000 iterations was used. By visual inspection, the posterior draws appear to achieve the convergence (i.e., stationary distributions) since there exist no apparent irregular patterns for these trace plots. Formal convergence statistics can be calculated (e.g., Gelman et al. 2013) and are omitted here. In practice it is important to assess the convergence of posterior draws of parameters before using imputations from the DA algorithm.

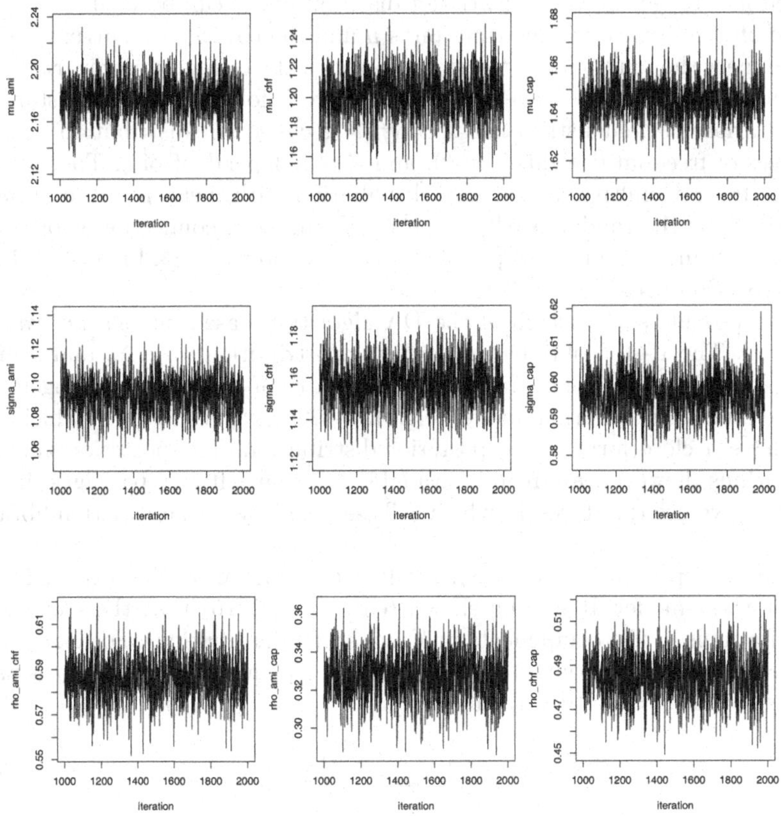

FIGURE 6.3
Example 6.2. History plots of the trivariate normal model (on the logit-transformed scale) parameters against iterations from the DA algorithm in LMI. The parameters include the means, standard deviations, and correlations of the variables.

TMI is more complicated because it involves a multivariate t-model. We provide additional specifics here. The conventional version of the multivariate t distribution $Y \sim t_p(\mu, \Sigma, \nu)$, with location μ, scale matrix Σ, and degrees of freedom ν, has the probability density function as

$$f(Y)=\frac{\Gamma((\nu+p)/2)}{\Gamma(\nu/2)(\nu\pi)^{p/2}|\Sigma|^{1/2}}(1+\nu^{-1}(Y-\mu)^t\Sigma^{-1}(Y-\mu))^{-(\nu+p)/2}, \quad (6.7)$$

where $\Gamma(z)=\int_0^{\infty}x^{z-1}e^{-x}dx$ is the gamma function for a positive value z. When $p=1$, $f(Y)$ is the univariate t-distribution. It can be verified that the mean of a multivariate t-distribution is μ and the covariance matrix is $\frac{\nu}{\nu-2}\Sigma$. To connect the multivariate t-distribution with the ordinary multivariate normal model, we have $Y=\mu+\Sigma^{1/2}Z/\sqrt{q}$, where Z follows a p-variate standard multivariate normal distribution, $q\sim\chi^2_{\nu}/\nu$ (a chi-square distribution with the degrees of freedom ν divided by ν), and Z is independent of q. Therefore, Y differs from a multivariate normal distribution with mean μ and covariance matrix Σ by the random scaling factor $\sqrt{q}$, and such connection is often the basis of estimating unknown parameters from t-models (e.g., Little and Rubin 2020, Section 8.4.2).

We provide some sketch of the DA algorithm based on the multivariate t-model. The P-step obtains the posterior distribution of μ, Σ, and ν given completed data Y. This is more complicated than the corresponding P-step in the multivariate normal model (e.g., Liu 1995), especially for ν which does not have a close form of the posterior distribution. The posterior sampling algorithms, however, are readily available in some software packages. In this example we use the R miscF, which calls several Bayesian estimation libraries in R.

The I-step imputes the missing values conditional on the observed cases under the t-model. It can be shown (e.g., Ding 2016) that the conditional distributions of the components in Y are still t-distributions. For example, if we partition the p-variate Y into two parts, Y_1 and Y_2, with dimensions p_1 and p_2 where $p=p_1+p_2$, then

$$f(Y_2|Y_1)\sim t_{p_2}(\mu_{2|1},\frac{\nu+d_1}{\nu+p_1}\Sigma_{2|1},\nu+p_1), \quad (6.8)$$

where $\mu_{2|1}=\mu_2+\Sigma_{21}\Sigma_{11}^{-1}(Y_1-\mu_1)$ and $\Sigma_{2|1}=\Sigma_{22}-\Sigma_{21}\Sigma_{11}^{-1}\Sigma_{12}$ (the predicted mean and residual covariance matrix as in the case of a multivariate normal distribution), and $d_1=(Y_1-\mu_1)^t\Sigma_{11}^{-1}(Y_1-\mu_1)$ is the squared Mahalanobis distance of Y_1 from μ_1 with scale matrix Σ_{11}. In summary, the conditional distribution of the multivariate t-distribution is very similar to that of the multivariate normal distribution. The conditional location parameter is the linear regression of Y_2 on Y_1. The conditional scale matrix is $\Sigma_{2|1}$ multiplied by the factor $(\nu+d_1)/(\nu+p_1)$. The conditional degrees of freedom increase to $\nu+p_1$. Therefore, random number generating functions based on multivariate or univariate t-distributions can be used for imputation (e.g., R mvtnorm).

For TMI, Fig. 6.4 plots the posterior draws of ν. The left panel is the trace/history plot of 1000 iterations, suggesting the convergence of the DA algorithm. The right panel is the corresponding histogram plot, showing that the posterior mean of ν is around 5.5. Such an estimate is consistent with the heavy tails shown in Fig. 6.1 (the bottom row). For simplicity, we do not display posterior draws of other model parameters. We collect 50 sets of imputations for every 20 iterations of the DA algorithm.

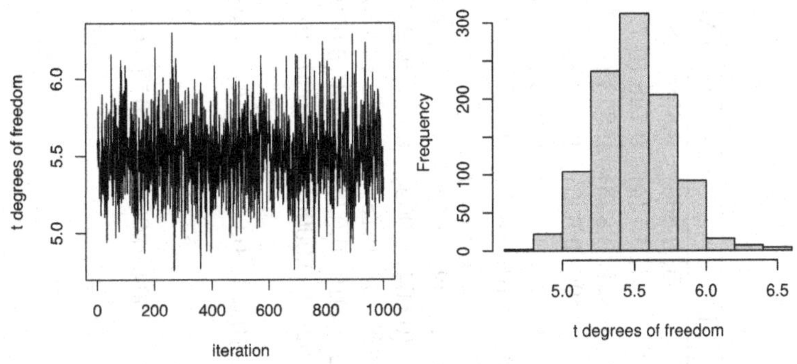

FIGURE 6.4
Example 6.2. Left: the history plot of the posterior draws for the degrees of freedom ν from the trivariate-t model (on the logit-transformed scale) against iterations from the DA algorithm in TMI; right: the histogram of the posterior draws of ν.

Table 6.1 shows point estimates and their 95% confidence intervals (in parentheses) for the means of each measure and proportions of hospitals having compliance rates greater than some arbitrary cut-off points. Some proportion estimates are computed from multiple measures jointly. For some tail probabilities (e.g., greater than 99%), the lower bound of the symmetrical confidence interval is set as 0 if it is negative (for simplicity). In general, the results for estimating means of these measures are similar among different missing data methods. However, there appear to exist some differences for the proportions across different methods. For example, there is a considerable difference between NMI and the other two imputation methods for estimating $Pr(AMI > 90\%)$. The distribution of AMI is highly skewed to the left, and the imputed values from the normal model would move to the center of the distribution and result in an apparently lower proportion estimate. In addition, TMI in general produces slightly higher proportion estimates for tail probabilities than LMI (e.g., $Pr(AMI > 99\%)$), which is expected under the assumption of the t-model with heavy tails.

TABLE 6.1
Example 6.2. Analysis results

Estimands	CC	NMI	LMI	TMI
$E(AMI)$	0.866 (0.862, 0.870)	0.857 (0.853, 0.861)	0.859 (0.855, 0.863)	0.858 (0.854, 0.863)
$E(CHF)$	0.730 (0.725, 0.735)	0.728 (0.722, 0.733)	0.728 (0.722, 0.734)	0.728 (0.723, 0.734)
$E(CAP)$	0.823 (0.821, 0.826)	0.823 (0.821, 0.826)	0.823 (0.821, 0.826)	0.823 (0.820, 0.825)
$Pr(AMI > 90\%)$	0.537 (0.521, 0.553)	0.486 (0.470, 0.501)	0.514 (0.499, 0.530)	0.518 (0.503, 0.534)
$Pr(AMI > 95\%)$	0.238 (0.224, 0.251)	0.232 (0.219, 0.245)	0.231 (0.218, 0.245)	0.233 (0.219, 0.246)
$Pr(AMI > 99\%)$	0.014 (0.010, 0.017)	0.046 (0.038, 0.053)	0.013 (0.010, 0.017)	0.017 (0.013, 0.021)
$Pr(CHF > 90\%)$	0.155 (0.144, 0.166)	0.159 (0.148, 0.170)	0.163 (0.151, 0.174)	0.157 (0.146, 0.168)
$Pr(CHF > 95\%)$	0.053 (0.046, 0.060)	0.066 (0.058, 0.074)	0.056 (0.049, 0.063)	0.055 (0.048, 0.062)
$Pr(CHF > 99\%)$	0.008 (0.005, 0.011)	0.022 (0.017, 0.027)	0.007 (0.004, 0.009)	0.008 (0.005, 0.011)
$Pr(CAP > 90\%)$	0.167 (0.156, 0.179)	0.169 (0.158, 0.180)	0.169 (0.158, 0.181)	0.165 (0.153, 0.176)
$Pr(CAP > 95\%)$	0.023 (0.019, 0.028)	0.031 (0.025, 0.036)	0.022 (0.017, 0.026)	0.023 (0.019, 0.028)
$Pr(CAP > 99\%)$	0.0002 (0, 0.0007)	0.004 (0.002, 0.007)	0.0002 (0, 0.0006)	0.0004 (0, 0.001)
$Pr(AMI > 90\%, CHF > 90\%, CAP > 90\%)$	0.053 (0.046, 0.060)	0.055 (0.048, 0.062)	0.058 (0.050, 0.065)	0.056 (0.049, 0.063)
$Pr(AMI > 95\%, CHF > 95\%, CAP > 95\%)$	0.006 (0.004, 0.009)	0.008 (0.005, 0.011)	0.006 (0.004, 0.009)	0.007 (0.004, 0.009)

Note: Estimates and their 95% confidence intervals (in parentheses).

6.4 Multivariate Categorical Data

Interestingly, it might be fair to say that lots of the variables collected in real data are in categorical forms. For example, in survey data, it might be more convenient to ask or answer survey questions through multiple categories. However, compared with continuous variables, it is often harder to model and impute multivariate categorical data. This section briefly reviews a few alternative approaches.

6.4.1 Log-Linear Models

A classic approach to imputing multiple categorical variables is based on a contingency table formed by these variables. Missing data in contingency tables can be modeled using log-linear models (e.g., Little and Rubin 2020, Chapter 13). Briefly, suppose all p Y-variables are categorical. They can be arranged as a p-dimensional contingency table, with C cells defined by joint levels of the variables. The entries in the table are counts $n_{ij,\ldots,u}$, where $n_{ij,\ldots,u}$ is the number of observed counts in the cell with $Y_1 = i$, $Y_2 = j$, ..., $Y_p = u$, where $i = 1, \ldots, I$, $j = 1, \ldots, J$, ..., and $u = 1, \ldots U$. The target quantities for the contingency table modeling is the cell probability $p_{ij,\ldots,u}$, which can be modeled through a log-linear model, for example, as

$$log(p_{ij,\ldots,u}) = \beta_0 + \beta_i^{(1)} + \beta_j^{(2)} + \ldots + \beta_u^{(p)}. \tag{6.9}$$

Constraints are needed to estimate these β's uniquely, and a common set of constraints is to have them sum up to zero across different levels of each marginal, that is, $\sum_{i=1}^{I} \beta_i^{(1)} = \ldots = \sum_{u=1}^{U} \beta_u^{(p)} = 0$.

Note that Model (6.9) only includes main effects from each of the p variables (i.e., factors). Higher-order interactions among these effects such as between Y_1 and Y_2 can be parameterized by including $\beta_{ij}^{(12)}$ in the model. On the other hand, some of the coefficients (either main effects or interactions) can be set as 0 for simplification. Thus there are many possible specifications of log-linear models. Additional discussions about log-linear models can be found, for example, in Agresti (2002).

Missing values in the contingency table framework result in partially classified counts. Schafer (1997) presented Bayesian DA imputation algorithms under the log-linear models. In summary, the posterior step of the DA algorithm draws sets of log-linear model parameters from Dirichlet conditional distributions given current values of the other log-linear model parameters and the imputed data; the imputation step of the DA algorithm allocates each partially classified count into the set of possible cells as draws from a multinomial distribution, with conditional probabilities calculated from the posterior step. These algorithms are implemented, for example, by R cat.

As the number of categorical variables increases, the level of all the possible joint classifications in the contingency table increases. As a result, even for complete data with sufficient sample sizes, some of the cells might have zero count or small numbers (i.e., sparse cells) and can create a problem in estimating these cell probabilities. This is similar to data separation in logistic regression (Section 4.3.2.4). Many log-linear models used in practice only keep main effects in order to avoid small sample size and corresponding estimation issues when possible interactions are included. However, it is important to assess and defend such specification when only main-effect models are implemented. In general, the imputation strategy works better if the number of variables is small to moderate (e.g., 3 to 5). With a large number of categorical variables, alternative modeling strategies might need to be considered.

6.4.2 Latent Variable Models

An increasingly popular approach is to use latent variable models to impute categorical variables. To illustrate the basic idea, suppose Y_j's, $j = 1, \ldots p$ are all binary variables. We first define a standard normal latent variable Z_j such that $Y_j = 1$ if and only if $Z_j > 0$ and $Y_j = 0$ otherwise. Then Y_j's can be modeled as

$$\begin{aligned} Y_j &= 1 \text{ if } Z_j > 0 \\ &= 0 \text{ if } Z_j <= 0, j = 1, \ldots p, \\ (Z_1, \ldots Z_p) &\sim N_p(\mu, \Omega), \end{aligned} \tag{6.10}$$

where $\mu = (\mu_1, \ldots \mu_p)$ is the mean vector, and Ω is the correlation matrix for p-variate standard normal variables (i.e., Z's), capturing the tetrachoric correlation among p binary variables of Y. Model (6.10) is also referred to as a multivariate probit model.

Note that Z's are all unobserved. Once Z's are determined, the corresponding Y's can be generated. Under Model (6.10), we need to draw/impute Z's as well to impute missing Y-values. A Bayesian imputation algorithm can be developed under commonly used priors for μ and Ω. Details are omitted due to the complexity. A key element of the imputation algorithm involves drawing Z's from truncated normal distributions given Y's (e.g., Albert and Chib 1993).

The idea of the latent modeling approach can also be extended to imputing ordinal and nominal variables. For example, suppose Y is an ordinal variable with K categories $K >= 3$. Then we can express Y as

$$\begin{aligned} Y &= 1 \text{ if } Z \in (-\infty, \tau_1] \\ &= \vdots \\ &= k \text{ if } Z \in (\tau_{k-1}, \tau_k] \\ &= K \text{ if } Z \in (\tau_{K-1}, \infty), \end{aligned} \tag{6.11}$$

where $\tau_1, \ldots, \tau_{K-1}$ represent the $K - 1$ threshold parameters that separate the standard normal distribution Z into K segments. With multiple variables, latent variable Z's from different Y's can be linked together using a standard multivariate normal distribution as in Model (6.10).

In general, the modeling and imputation algorithms for latent normal models are rather sophisticated. See Carpenter and Kenward (2013, Sections 4.3 and 5.2) and references therein for more details. Their imputation algorithms are implemented by R jomo. Similar imputation models have been implemented in Mplus in the framework of structure equation models (https://www.statmodel.com/; e.g., Asparaouhov and Muthen 2010). In addition, Demirtas and Hedeker (2007; 2008) used the idea of the multivariate normal model imputation by converting categorical variables to normal variables, although in a less rigorous way than latent normal models.

With a large number of categorical variables, Vermunt et al. (2008) proposed to use latent class models to impute categorical data. This will be introduced in the topic of imputation for high-dimensional data (Section 14.4).

Example 6.3. *Imputing multivariate incomplete binary data using log-linear and latent normal models*

We go back to the dataset used in Example 4.7, in which the variable measuring the subject's satisfaction of care received (CARERCVD) has around 25% missing cases. We consider a binary version of this variable: CARERCVD_B=1: Very Satisfied; CARERCVD_B=0: Somewhat satisfied/not satisfied at all. We also choose three other variables and make them binary if they are not in the original form. These variables are all related to health and health care including the self-rating of the health status (SELFHEALTH_B=1: Excellent/Very Good/Good; SELFHEALTH_B=0: Fair/Poor), having health insurance coverage (COVERAGE=1: Yes; COVERAGE=0: No), and having received the delayed medical care (DELAYM=1: Yes; DELAYM=0: No). We randomly delete 10% cases from each of the three variables so that all four variables have some missing values. After these modifications, removing subjects with any missing variable would delete around 46% of the subjects. The working dataset now includes the four incomplete binary variables.

We use both the log-linear modeling and latent variable modeling approaches to the multiple imputation analysis. For the log-linear modeling approach, we consider two candidate models: one only includes main effects from the four variables (MMI); the other is the saturated log-linear model, including both the main effects and all the two-way, three-way, and four-way interactions (SMI). Both MMI and SMI are implemented by R cat. For the latent modeling approach, we assume Model (6.10) for these four binary variables (PMI) and use R jomo to execute the imputation. The multiple imputation analysis is based on $M = 50$ sets of imputations.

For the analysis, we focus on the distribution of CARERCVD_B, including its overall mean and means in subgroups formed by the other three variables. Table 6.2 shows the results. Results from all methods show similar patterns to those identified in Example 4.7: subjects were more likely to be very satisfied when they had higher self-rated health status, had health care coverage, or did not receive delayed health care. In general, CC yields estimates with reduced precisions (i.e., wider confidence intervals) compared with the imputation methods. This is especially the case for some subgroup mean estimates (e.g., when $S = 1, C = 1, D = 1$). The saturated log-linear model used in SMI is more general, and thus might be preferred for the purpose of imputation. The main-effects log-linear model used in MMI includes fewer parameters and therefore sometimes yield estimates with shorter confidence intervals (e.g., when $S = 1, C = 0, D = 1$). Note that point estimates from MMI and SMI are somewhat different as well. In this case, however, since we know that the four variables are associated with each other, MMI might be suboptimal because it assumes mutual independence among the variables.

As for the results from PMI, they are closer overall to those from SMI. This is expected because the multivariate probit model captures tetrachoric pairwise correlations, which can be viewed in a similar way as two-way interactions in the log-linear modeling framework. The posterior mean estimates for the latent-variable model are $\mu = (-0.136, 0.621, -1.189, -0.085)$ and $\Omega = \begin{pmatrix} 1, 0.137, 0.198, -0.358 \\ 0.137, 1, -0.108, -0.080 \\ 0.198, -0.108, 1, -0.179 \\ -0.358, -0.080, -0.179, 1 \end{pmatrix}$. The correlation matrix Ω is consistent with the pattern from the data. For example, $\Omega[1,4] = -0.358$, showing subjects experiencing delayed medical care were less likely to be very satisfied.

A remaining question is which imputation model we would choose. This is related to imputation diagnostics and will be discussed in Example 12.6.

6.5 Mixed Categorical and Continuous Variables

6.5.1 One Continuous Variable and One Binary Variable

It is common that both continuous and categorical variables are subject to missing data and need imputation. We first illustrate some basic modeling ideas for one binary variable Y and one continuous variable X. The joint distribution for Y and X, $f(Y, X)$, for the complete data can be factorized as either $f(Y, X) = f(Y|X)f(X)$ or $f(Y, X) = f(X|Y)f(Y)$.

We first consider the factorization $f(Y, X) = f(X)f(Y|X)$. A possible joint modeling specification is

$$\begin{aligned} X &\sim N(\mu, \sigma^2), \\ f(Y|X) &\sim Bernoulli(p(X)). \end{aligned} \tag{6.12}$$

Let $p(X) = \frac{exp(\beta_0+\beta_1 X)}{1+exp(\beta_0+\beta_1 X)}$; then Model (6.12) specifies a logistic regression

TABLE 6.2
Example 6.3. Analysis results

Mean of Y	CC	MMI	SMI	PMI
Overall	0.585 (0.555, 0.614)	0.585 (0.556, 0.615)	0.556 (0.523, 0.588)	0.557 (0.526, 0.587)
$S = 1, C = 1, D = 1$	0.600 (0.469, 0.731)	0.572 (0.477, 0.668)	0.559 (0.438, 0.680)	0.557 (0.453, 0.660)
$S = 1, C = 1, D = 0$	0.788 (0.718, 0.858)	0.760 (0.691, 0.830)	0.786 (0.717, 0.855)	0.773 (0.705, 0.842)
$S = 1, C = 0, D = 1$	0.385 (0.109, 0.660)	0.475 (0.254, 0.695)	0.319 (0.076, 0.563)	0.385 (0.177, 0.593)
$S = 1, C = 0, D = 0$	0.375 (0.130, 0.620)	0.406 (0.174, 0.639)	0.327 (0.113, 0.541)	0.435 (0.207, 0.663)
$S = 0, C = 1, D = 1$	0.461 (0.393, 0.529)	0.506 (0.453, 0.560)	0.444 (0.385, 0.503)	0.447 (0.388, 0.506)
$S = 0, C = 1, D = 0$	0.631 (0.578, 0.685)	0.641 (0.593, 0.688)	0.646 (0.598, 0.694)	0.648 (0.601, 0.696)
$S = 0, C = 0, D = 1$	0.348 (0.149, 0.547)	0.487 (0.339, 0.634)	0.348 (0.182, 0.514)	0.330 (0.183, 0.477)
$S = 0, C = 0, D = 0$	0.762 (0.575, 0.949)	0.668 (0.492, 0.845)	0.756 (0.587, 0.925)	0.671 (0.491, 0.852)

Note: Y: CARERCVD_B; S: SELFHEALTH_B; C:COVERAGE; D:DELAYM.

for Y on X as $logit(f(Y=1|X)) = \beta_0 + \beta_1 X$. This is referred to as a logistic-normal model (Section 4.3.2.2).

We can also consider the factorization $f(Y,X) = f(Y)f(X|Y)$. A possible joint modeling specification is

$$\begin{aligned} f(Y) &\sim Bernoulli(p), \\ f(X|Y=l) &\sim N(\mu_l, \sigma_l^2), \end{aligned} \tag{6.13}$$

where $l = 0, 1$. Note that Eq. (6.13) defines a normal discriminant analysis, classifying X into two subgroups according to Y. The marginal distribution of X is a mixture of two normal distributions.

In Section 4.3.2, we discussed these two models for imputing incomplete Y, assuming X is fully observed. They are the logistic regression and discriminant analysis imputation methods. In addition, there (Eq. (4.5)) we also showed that in general, the joint models, $f(Y,X)$, implied by Models (6.12) and (6.13) are not equivalent. In addition, even within the same factorization, different specifications can be used. For example, we can consider a simpler version of Eq. (6.13) by making the variance equal across X (i.e., $\sigma_1^2 = \sigma_2^2$). Then the conditional distribution of Y given X can be expressed as a logistic regression model. Yet the two joint models (Eqs. (6.12) and (6.13)) are still not equivalent.

Even for only two variables, this simple illustration shows that there exist alterative ways of specifying the joint model, and these models are unlikely to be equivalent to each other. In subsequent sections, we introduce several commonly used strategies for specifying joint models and imputing a mixture of continuous and categorical variables.

6.5.2 General Location Models

Let $X = (X_1, \ldots, X_p)$ denote the collection of p continuous variables and $Y = (Y_1, \ldots, Y_q)$ the collection of q categorical variables. We consider the factorization $f(Y,X) = f(Y)f(X|Y)$ using the general location model (GLOM), originally proposed by Olkin and Tate (1961). Specifically, Let I_j be the number of the levels of the j-th categorical covariate Y_j, $j = 1, \ldots, q$. Then the cross-classification on the components of Y-variables creates a contingency table with $L = \prod_{j=1}^{q} I_j$ cells. Let l index the cell in this contingency table, where $l = 1, \ldots, L$. Let W be the $L \times 1$ vector recording the number in each cell. Under a GLOM, W follows a multinomial distribution with p_l as the cell probability for the lth cell; the continuous variables X have a multivariate normal distribution within each cell of the contingency table, with mean μ_l and $p \times p$ covariance matrix Σ_l. It is often assumed that Σ_l's are all equivalent.

Eq. (6.13) is the simplest GLOM with $p = 1$ and $q = 1$. In many cases for $q > 1$, the contingency table formed by Y-variables can be more effectively modeled using log-linear models to achieve model parsimony (Section 6.4). In that sense, GLOM essentially combines the fitting for multivariate normal models and for log-linear models. Among the categorical variables Y, we can fit a log-linear model. At each level of the contingency tables, we can fit linear regression models for one continuous X-variable given others based on the decomposition of a multivariate normal distribution into conditional univariate normal distributions (Section 6.3).

GLOM has been used as a popular modeling strategy for multivariate missing-data problems with a general missingness pattern. Briefly speaking, the imputation algorithm combines that for multivariate normal models with that for log-linear models (Sections 6.3 and 6.4). Additional complexity arises from the conditional distribution of categorical variables given continuous variables, $f(Y|X)$, under the GLOM. Schafer (1997) and Little and Rubin (2020) provided details on these algorithms, which have been implemented by R mix.

A notable application of the multiple imputation analysis based on GLOM is the National Health and Nutrition Examination Survey imputation project (e.g., Ezzati-Rice et al. 1995; Schafer 2001) See also Example 10.3. On the other hand, Belin et al. (1999) pointed out some limitations of GLOM in a case study, revealing considerable differences between imputed values and follow-up data for the initial nonrespondents.

6.5.3 Latent Variable Models

Since categorical variables can be modeled as normal variables on the latent scale (Section 6.4.2), it is natural to connect the latent normal variables with continuous variables to form an extended multivariate normal model. Continuing with the model shown in Eq. (6.10), suppose now there exist q binary Y-variables and p continuous X-variables; this extension can be expressed as

$$\begin{aligned} Y_j &= 1 \text{ if } Z_j > 0 \\ &= 0 \text{ if } Z_j <= 0, j = 1, \ldots q, \\ (Z_1, \ldots Z_q, X_1, \ldots, X_p) &\sim N_{p+q}(\mu_{ZX}, \Omega_{ZX}), \end{aligned} \tag{6.14}$$

where μ_{ZX} is a $(p+q) \times 1$ mean vector, and Ω_{ZX} is the $(p+q) \times (p+q)$ covariance matrix used to capture the correlation among all the variables. It should be noted that since Z's are standard normal variables, the specification of Ω_{ZX} needs to account for that.

The idea of Model (6.14) can be extended to include ordinal and nominal variables. For example, Zhang et al. (2015) modeled multivariate incomplete data assuming multivariate normal linear regression models for multivariate continuous measures, multivariate probit models for correlated ordinal measures, and multivariate multinomial probit models for multivariate nominal measures. Carpenter and Kenward (2013) provided more discussion on the use of latent variable models for imputing a mixture of normal and categorical variables. The corresponding Bayesian imputation algorithms are sophisticated, involving complex MCMC sampling procedures. These imputation algorithms have been implemented through R jomo.

Example 6.4. *Imputing multiple incomplete variables in the gestational age dataset*

We continue using the gestational age dataset (Examples 4.2, 5.5, and 5.10) for illustration. In previous examples, we focused on a subset that only has a single missing variable, DGESTAT, with several fully observed continuous covariates. This subset has a reduced sample size. We now consider a working dataset with the full sample size ($n = 40274$) that includes two continuous variables, DGESTAT and DBIRWT, and three categorical variables: MRACE (mother's race: White/Black/Others), FRACE (father's race: White/Black/Others), and MEDUC (mother's education: 0-8 years/9-11 years/12 years/13-15 years/>= 16 years). In addition to the missing DGESTAT, MRACE is fully observed, DBIRWT has little missing data, MEDUC has around 1.4% missing cases, and FRACE has around 15% missing cases. Table 6.3 shows some of the descriptive statistics of these variables. We impose around 10% missing cases randomly to each of the variables, DBIRWT, MRACE, and MEDUC, so that none of the variables in the working dataset is fully observed. After these modifications, removing subjects with any missing variable would delete around 49% of the cases.

TABLE 6.3

Example 6.4. Descriptive statistics of the dataset ($n = 40274$)

Variable	Definition	n_{obs}	Statistics
DGESTAT	Gestational age in weeks	32748	Mean (38.65), SD (2.246), Range (18-47)
DBIRWT	Baby's birth weight in grams	40240	Mean (3297), SD (604.9), Range (227-6039)
MRACE	Mother's race	40274	White (79.1%), Black (14.5%), Others (6.3%)
FRACE	Father's race	34626	White (81.4%), Black (12.5%), Others (6.1%)
MEDUC	Mother's education	39721	0-8 yrs (6.0%), 9-11 yrs (15.3%), 12 yrs (31.0%), 13-15 yrs (21.8%), 16 yrs and above (25.9%)

We apply the GLOM-based imputation procedure for missing variables using R mix. We first assume a saturated log-linear model for the categorical variables, namely FRACE, MRACE, and MEDUC. This creates $3 \times 3 \times 5 = 45$ cross-classification cells in the continency table and somehow brings in the issue of sparse cells for complete cases. We then encounter a problem of running the program. Therefore we collapse the first and second categories of MEDUC into one (i.e., 0-11 years) to avoid this issue. The corresponding contingency table has $L = 3 \times 3 \times 4 = 36$ cells.

Conditional on each of the 36 cells of the contingency table, the GLOM model would assume that DGESTAT and DBIRWT follow a bivariate normal model. This is equivalent to assuming a conditional linear regression relationship between DBIRWT and DGESTAT. However, as previously shown (Example 5.5), there exists a clearly nonlinear relationship between the two variables. To address this issue here, we apply a square-root transformation of DBIRWT (DBIRWT_SQRT) and instead assume that DBIRWT_SQRT and DGESTAT follow a bivariate normal model conditional on the cells of the contingency table. Fig. 6.5 shows the histogram plots of DGESTAT, DBIRWT_SQRT, and their scatter plots. The relation between DBIRWT_SQRT and DGESTAT appears to be closer to a linear one. We conduct a GLOM-based imputation using DBIRWT_SQRT (SQRTMI). For comparison, we apply a GLOM-based imputation using DBIRWT at the original scale (LINMI).

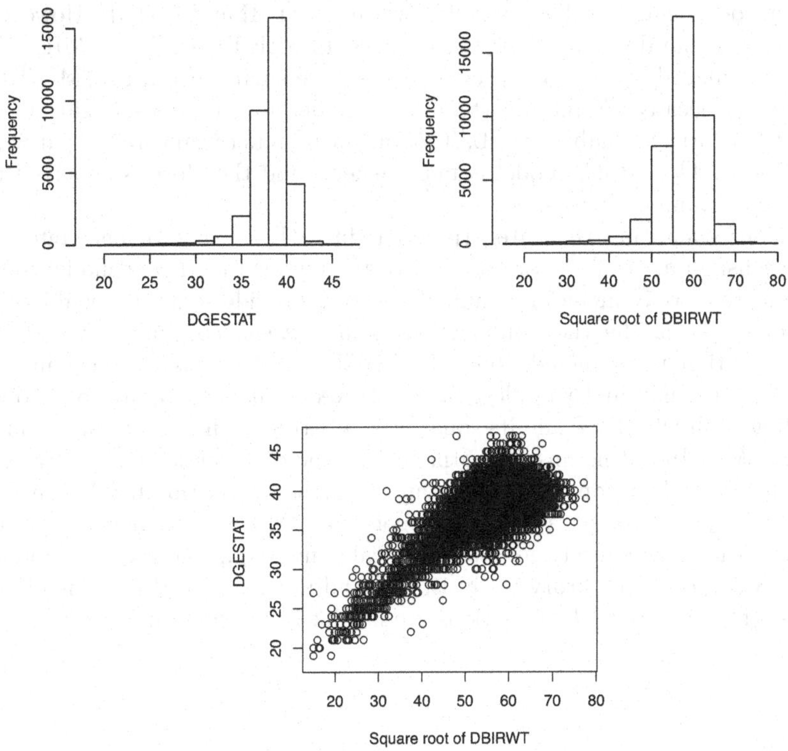

FIGURE 6.5
Example 6.4. Top: histograms of DGESTAT and the square root of DBIRWT; bottom: the scatter plot of the two variables.

We also apply an imputation using the latent variable model (LATMI) for DGESTAT, DBIRWT_SQRT, FRACE, MRACE, and MEDUC. The latter three are treated as nominal categorical variables. R jomo is used for the imputation.

We consider a post-imputation analysis of regressing DGESTAT using other variables in the dataset as predictors. Table 6.4 shows the results based on $M = 50$ imputations. Besides DBIRWT and its squared term, the magnitudes of some other regression coefficients are considerably different between CC and multiple imputation methods. The standard errors of CC are generally larger than those of the imputation estimates, reflecting the loss of precision due to removing cases with observed information. For LINMI, its coefficient estimate on the squared term of DBIRWT is clearly attenuated compared with that of CC and two other imputation methods (SQRTMI and LATMI). This is expected because the former method failed to capture the nonlinear relation between DBIRWT and DGRSTAT in the missing cases. Regression coefficients are close overall between SQRTMI and LATMI. However, there exist some differences for the coefficients with FRACE and MRACE, and the standard errors for these coefficients are smaller from LATMI. This is expected because SQRTMI attempts to model all the cross-classifications among categorical variables, yet LATMI only attempts to capture the pairwise correlations. The former model is more general and therefore is expected to have larger standard errors.

Besides collapsing the categories to reduce the chance of encountering sparse cells, an alternative strategy is to use simplified log-linear model specifications (e.g., only including main effects) and provide some reasonable justifications. To handle the nonlinear relation between continuous variables, another option is to include both DBIRWT and its squared term in the GLOM. Note that in Example 5.5, we showed that the imputation model including both DBIRWT and its square has a better fit than that only including DBIRWT. Imputing missing variables and their interactions, however, can be challenging. This topic will be discussed further in Section 6.6.3.2. In Example 6.4 and in general, applying a generic GLOM (or any other models) to real data will almost surely get into practical issues (e.g., sparse cells or model unfit). Based on exploratory analysis, some adjustments need to be made to address these issues, and the options are not likely to be unique.

6.6 Missing Outcome and Covariates in a Regression Analysis

6.6.1 General Strategy

In preceding sections, the rationale behind modeling of multivariate incomplete data focuses on using multivariate distributions/models to accommodate different types (i.e., continuous and/or categorical) of variables. These impu-

TABLE 6.4
Example 6.4. Analysis results

Predictor	CC	LINMI	SQRTMI	LATMI
Intercept	19.9	22.6	21.7	21.9
	(0.16)	(0.15)	(0.13)	(0.13)
DBIRWT	9.95×10^{-3}	8.24×10^{-3}	8.77×10^{-3}	8.62×10^{-3}
	(9.88×10^{-5})	(9.32×10^{-5})	(8.20×10^{-5})	(8.23×10^{-5})
DBIRWT	-1.25×10^{-6}	-9.74×10^{-7}	-1.05×10^{-6}	-1.02×10^{-6}
square	(1.57×10^{-8})	(1.46×10^{-8})	(1.33×10^{-8})	(1.33×10^{-8})
MRACE black	0.0859	0.263	0.242	0.116
	(0.0643)	(0.0742)	(0.0734)	(0.0626)
MRACE others	0.129	−.0996	−.0604	0.0816
	(0.0683)	(0.0898)	(0.0699)	(0.0411)
FRACE black	0.0696	−0.130	−0.0923	0.103
	(0.067)	(0.0904)	(0.0840)	(0.0645)
FRACE others	−0.0167	0.194	0.161	0.0364
	(0.064)	(0.0884)	(0.0685)	(0.0433)
MEDUC high school	−0.161	−0.206	−0.201	−0.199
	(0.0323)	(0.0278)	(0.0276)	(0.0283)
MEDUC some college	−0.236	−0.296	−0.288	−0.279
	(0.0339)	(0.0302)	(0.0289)	(0.0288)
MEDUC college	−0.215	−0.273	−0.269	−0.275
and above	(0.0322)	(0.0278)	(0.0285)	(0.0273)

Note: Regression coefficient estimates and standard errors (in parentheses) from all missing data methods. The reference group is White for MRACE, White for FRACE, and less than high school for MEDUC.

tation models, in general, are not targeted to accommodating specific analyses of imputed data (also termed as "substantive models" in some literature, see Carpenter and Kenward 2013), although in some cases conditional univariate distributions under these models do conform to commonly used analysis models (e.g., linear regression or logistic regression models). This section presents some JM strategies that are purposefully devised to accommodate a targeted regression analysis.

Suppose the targeted analysis is to relate an outcome variable Y with some predictors X. The analysis model can be expressed as $f(Y|X;\theta_{Y|X})$, where the estimand of interest is the parameter $\theta_{Y|X}$. For example, in the analysis of running a normal linear regression for Y on X, $Y = X\beta + \epsilon$, where $\epsilon \sim N(0,\sigma^2)$, $\theta_{Y|X} = (\beta, \sigma^2)$.

Some univariate missing data problems discussed in Chapters 4 and 5 fall into scenarios where either Y or one of the X-variables is incomplete. We now sketch the principled JM strategy for a more general setup. Suppose Y and/or some of the X-variables have missing values and let $Y = (Y_{mis}, Y_{obs})$ and $X = (X_{mis}, X_{obs})$. A joint model for complete data, $f(Y, X|\theta)$, needs to be specified. To purposefully account for the targeted analysis, the complete-data model can be expressed as

$$f(Y, X|\theta) = f(Y|X, \theta_{Y|X})f(X|\theta_X), \tag{6.15}$$

where $\theta = (\theta_{Y|X}, \theta_X)$ is the parameter governing the complete-data model. Here $\theta_{Y|X}$ characterizes the relationship between Y and X, and θ_X is the parameter for the distribution of X-covariates. In Eq. (6.15), the first term, $f(Y|X, \theta_{Y|X})$, can be simply referred to as the "analysis model" (or substantive model), and the second term, $f(X|\theta_X)$, can be referred to as the "covariate model" or "covariate distribution". The targeted analysis already determines the form of the analysis model. However in general, it does not provide any information about the covariate model. Some exploratory analysis of data, coupled with possibly subjective input, can be used to determine the distribution of the covariates. Finally, from a Bayesian modeling perspective, adequate prior distributions $\pi(\theta)$ need to be specified to complete the model specification.

Under Model (6.15) and assuming $\pi(\theta) \propto \pi(\theta_{Y|X})\pi(\theta_X)$, posterior predictive distributions of missing values can be derived and sampled to generate multiple imputations. We sketch the basic idea of the DA algorithm. The posterior step of the DA algorithm includes:

1. Draw $\theta_{Y|X}$ from $f(\theta_{Y|X}|Y, X)\pi(\theta_{Y|X}) \propto f(Y|X, \theta_{Y|X})\pi(\theta_{Y|X})$.
2. Draw θ_X from $f(\theta_X|Y, X)\pi(\theta_X) \propto f(X|\theta_X)\pi(\theta_X)$.

The imputation step of the DA algorithm includes:

1. Draw Y_{mis} from $f(Y_{mis}|X, \theta_{Y|X})$.
2. Draw X_{mis} from

$$f(X_{mis}|Y, X, \theta) \propto f(Y_{obs}, Y_{mis}|X_{obs}, X_{mis}, \theta_{Y|X})f(X_{mis}|X_{obs}, \theta_X).$$

Upon the convergence of the DA algorithm after iterations between the posterior and imputation steps, draws of Y_{mis} and X_{mis} constitute the imputations. The whole process is independently repeated M times to obtain multiple imputations. Note that the imputation of X_{mis} has to be conditional on both Y and other covariates in X. This is consistent with the approach to imputing a single missing covariate X conditional on Y in a regression analysis (Section 4.4). Simply imputing X_{mis} from $f(X_{mis}|X_{obs}, \theta_X)$ that excludes Y is incorrect.

Example 6.5. *Missing data in a logistic regression with two continuous predictors*

Suppose Y is a binary variable, $X = (X_1, X_2)$ contains two continuous covariates, and all three variables have missing values. The analysis of interest is to run a logistic regression for Y on X_1 and X_2. Following the idea in Eq. (6.15), the joint model for the three variables, $f(Y, X_1, X_2|\theta)$, can be decomposed as:

1 Analysis model: $f(Y|X_1, X_2, \theta_{Y|X})$ is determined by the logistic regression model: $logit(f(Y = 1|X_1, X_2)) = \beta_0 + \beta_1 X_1 + \beta_2 X_2$; here $\theta_{Y|X} = (\beta_0, \beta_1, \beta_2)$.

2 Covariate model: for illustration, suppose $(X_1, X_2) \sim N_2(\mu, \Sigma)$ (a bivariate normal model); here $\theta_X = (\mu, \Sigma)$.

Assuming noninformative prior distributions for θ and letting the complete-sample size be n, we now sketch major steps of the DA algorithm. The posterior steps include:

1. Draw $\beta^* = (\beta_0^*, \beta_1^*, \beta_2^*)$ from $f(\beta|Y, X) \propto \prod_{y_i=1} \frac{exp(X_i\beta)}{1+exp(X_i\beta)} \prod_{y_i=0} \frac{1}{1+exp(X_i\beta)}$, where X_i is the i-th row of the X-covariate matrix.

2. Draw μ^* and Σ^* from $\Sigma/(n-1) \sim Inverse - Wishart(S_X, n-1)$, and $\mu \sim N_2(\overline{X}, \Sigma/n)$, where $\overline{X}$ and S_X are the sample mean and covariance matrix from completed data X, respectively.

For missing case i, the imputation steps include

1. Draw missing Y_i from a Bernoulli distribution, $f(Y_i = 1) = \frac{exp(x_i\beta^*)}{1+exp(x_i\beta^*)}$

2. Draw missing X_{1i} from the distribution,

$$f(X_{1i}|Y_i, X_{2i}, \theta) \propto f(Y_i|X_i, \beta) f(X_{1i}|X_{2i}, \mu, \Sigma)$$

$$\propto [\frac{exp(X_i\beta^*)}{1+exp(X_i\beta^*)}]^{I(Y_i=1)} [\frac{1}{1+exp(X_i\beta^*)}]^{1-I(Y_i=1)} N(\gamma_{01}^* + \gamma_{11}^* X_{2i}, \tau_1^{*2}),$$

where $N(\gamma_{01} + \gamma_{11}X_{2i}, \tau_1^2)$ is the conditional normal distribution of X_{1i} given X_{2i} under a bivariate normal model for X, for which γ_{01}^*, γ_{11}^*, and τ_1^{*2} can be obtained by reparameterizing μ^* and Σ^* (e.g., Example 6.1).

3. Draw missing X_{2i} from the distribution $f(X_{2i}|Y_i, X_{1i}, \theta)$, which can be formed in a similar way as from $f(X_{1i}|Y_i, X_{2i})$.

Even in this simple example, the conditional distribution for β and posterior predictive distribution of X_{mis} have no closed forms. We can certainly write our own programming code. In some cases, however, we might be able to take advantage of existing Bayesian analysis software packages to implement the imputation (Section 6.6.3).

6.6.2 Conditional Modeling Framework

Given various forms of targeted analyses and variables involved, the complete-data model in Eq. (6.15) has to be formulated case by case. Since the analysis model is already determined by the targeted analysis, the modeling task is often to specify the covariate model. One strategy is to specify a multivariate model/distribution for all X-covariates. This can be done using the strategies introduced in Sections 6.3-6.5 (e.g., multivariate normal models, log-linear models, general location models, etc.)

However, if some of the X-variables are fully observed, it is sufficient to specify the conditional models of X_{mis} given X_{obs}. This strategy can be referred to as the conditional modeling framework, a term coined by Ibrahim et al. (2001). Note that in Model (6.15), $f(X|\theta_X)$ can be further expressed as $f(X_{mis}|X_{obs}, \theta_A) f(X_{obs}|\theta_B)$, where θ_A parameterizes the distribution of X_{mis} conditional on X_{obs}, and θ_B parameterizes the marginal distribution of X_{obs}. However, for the purpose of imputation, we only need to specify $f(X_{mis}|X_{obs}, \theta_A)$. The specification of $f(X_{obs}|\theta_B)$ is totally unnecessary because X_{obs} is fully conditioned when deriving the posterior predictive distribution of missing data. That is, $f(X_{mis}|X_{obs}, Y, \theta) \propto f(Y|X, \theta_{Y|X}) f(X_{mis}|X_{obs}, \theta_A)$. In addition, specifying only $f(X_{mis}|X_{obs}, Y, \theta_A)$ can be more general and robust than specifying $f(X|\theta_X)$ because the former avoids making the assumption for the marginal distribution of the fully observed X_{obs}.

In Example 6.5, since both X_1 and X_2 are assumed to be incomplete, specifying a bivariate normal model for X_1 and X_2 is necessary. However, now suppose only X_1 contains some missing values and X_2 is fully observed. To model the covariates, we only need to specify $f(X_1|X_2)$. If X_2 is highly skewed, for example, then it can be more adequate to apply a normal linear regression model: $X_1 = \gamma_{01} + \gamma_{11} X_2 + \epsilon$, where $\epsilon \sim N(0, \tau_1^2)$, than assuming a bivariate normal model for both X_1 and X_2.

In many cases, modeling $f(X_{mis}|X_{obs}, \theta_A)$ might require specifying a series of one-dimensional conditional distributions. Suppose X_{mis} consists of p variables, then

$$\begin{aligned} f(X_{mis}|X_{obs}, \theta_A) &= f(X_{mis,p}|X_{mis,1}, \ldots, X_{mis,p-1}, X_{obs}, \theta_{A,p}) \\ &\times f(X_{mis,p-1}|X_{mis,1}, \ldots, X_{mis,p-2}, X_{obs}, \theta_{A,p-1}) \\ &\times \ldots \\ &\times f(X_{mis,2}|X_{mis,1}, X_{obs}, \theta_{A,2}) \\ &\times f(X_{mis,1}|X_{obs}, \theta_{A,1}), \end{aligned} \tag{6.16}$$

where $\theta_{A,j}$ parameterizes the j-th conditional distribution, $j = 1, \ldots p$, and $\theta_A = (\theta_{A,1}, \ldots, \theta_{A,p})$.

In practice, these univariate conditional distributions can be specified using typical regression models such as the normal linear regression model and logistic regression models discussed in Chapter 4.

6.6.3 Using WinBUGS

6.6.3.1 Background

We have presented the general JM framework for missing data problems in a targeted analysis. However, the imputation algorithms can be rather sophisticated even for a simple analysis such as in Example 6.5. In general, most of the imputation software packages (e.g., R mice, R norm, SAS PROC MI, etc.) are not designed to tune the imputation to targeted analyses.

In Chapter 2, we mentioned that there exist multiple software packages for conducting Bayesian data analysis, which provide user-friendly coding environments to execute complicated MCMC sampling steps under commonly used and structured analysis models. Some of the packages offer options of drawing missing values in variables by treating missing data as parameters. Therefore, we may use these packages to execute the imputation algorithm.

In this book, we use WinBUGS (https://www.mrc-bsu.cam.ac.uk/software/bugs) to illustrate the idea. In the past two decades or so, WinBUGS (and related software packages) had been a very popular programming language for research and education in applied Bayesian analysis (e.g., Lunn et al. 2000; Lunn et al. 2009). In brief, WinBUGS is a statistical software package for Bayesian analysis using MCMC methods. It is based on the BUGS (Bayesian inference Using Gibbs Sampling) project started in 1989. WinBUGS is a freeware that can be run under Microsoft Windows. The last version of WinBUGS was version 1.4.3, released in August 2007. WinBUGS 1.4.3 remains available as a stable version for routine use, but is no longer being developed. Further development was picked up by OpenBUGS, an open-source version of the package (http://www.openbugs.net/w/FrontPage). More recently, R nimble is a Bayesian analysis package that can write and implement algorithms that operate on models written in BUGS, with many expanded capabilities.

The WinBUGS language is based on the notion of directed acyclic graph models (i.e., layered conditional models) and has employed multiple MCMC sampling algorithms. Both the data elements (random variables) and parameters can be considered as stochastic nodes, for which their posterior (predictive) distributions can be drawn under the specified model. For parameters, their priors need to be specified by users.

WinBUGS has been used in Bayesian analyses with missing data in the literature. For example, Erler et al. (2016) used a fully Bayesian approach to missing covariates in longitudinal studies, and Hemming and Hutton (2012) conducted Bayesian sensitivity analyses for missing covariates in a Cox proportional hazards model for survival data. Lunn et al. (2013, Chapter 9) provided a few examples of using WinBUGS to fit Bayesian models with missing data in either the outcome or covariates. Although these applications focus on parameter estimates, the imputations for the missing variables under a well defined joint model can also be automatically generated by the WinBUGS software. We illustrate this strategy in the book.

Example 6.6. *Imputation of missing data in a logistic regression with two continuous predictors using WinBUGS*

To impute the missing data following Model (6.15), both the analysis model and the covariate model need to be programmed. We illustrate this process using the setup in Example 6.5. The main WinBUGS program syntax is included as follows:

```
model
{       # beginning of the model
# complete-data model
for( i in 1 : M ) {
# analysis model
y_miss[i]~ dbern(p[i])
logit(p[i]) <-beta0+beta1*x1_miss[i]+beta2*x2_miss[i]
# covariate model
x1_miss[i] ~ dnorm(mu, tau)
x2_miss[i] ~ dnorm(theta[i], psi)
theta[i] <- gamma0+gamma1*x1_miss[i]
}
# prior for complete-model parameters
beta0 ~ dnorm(0.0, 1.0E-3)
beta1 ~ dnorm(0.0, 1.0E-3)
beta2 ~ dnorm(0.0, 1.0E-3)
mu ~ dnorm(0.0, 1.0E-3)
gamma0 ~ dnorm(0.0, 1.0E-3)
gamma1 ~ dnorm(0.0, 1.0E-3)
tau ~ dgamma(1.0E-3, 1.0E-3)
psi ~ dgamma(1.0E-3, 1.0E-3)
} # end of the model
# Data
list(M = 1000)
# Initial values
list(beta0=-1, beta1=0.5, beta2=0.5; mu=0,
gamma0=0, gamma1=0, tau=1, psi=1)
```

In this sample code, the three variables are named as y_miss, x1_miss, and x2_miss, respectively. The imputed values are also stored in these variables (i.e. they are treated as parameters in WinBUGS) when they need to be fetched. The dataset has $M = 1000$ observations and the data input and model specification are executed in a loop:

```
for( i in 1 : M ) {
}
```

In the analysis model, the three logistic regression coefficients are named as beta0, beta1, and beta2, respectively. The analysis model is executed through:

```
y_miss[i] ~ dbern(p[i])
logit(p[i]) <-beta0+beta1*x1_miss[i]+beta2*x2_miss[i]
```

The covariate model is executed through:

```
x1_miss[i] ~ dnorm(mu, tau)
x2_miss[i] ~ dnorm(theta[i], psi)
theta[i] <- gamma0+gamma1*x1_miss[i]
```

Here we decompose the bivariate normal model for (X_1, X_2) into the product of two univariate distributions: one is a normal model for X_1, and the other is a normal linear regression model for X_2 given X_1. The reason for doing this is that if we had specified X_1 and X_2 directly as a bivariate normal distribution, then WinBUGS would have set both X_1 and X_2 as missing if one of them is missing. This issue might be addressed in the more advanced variant of WinBUGS such as R nimble. However in general, if any variable has missing data, it is recommended to specify its distribution using a univariate model (conditional on other variables) rather than specifying it with other variables jointly as multivariate models. This strategy also follows the conditional modeling framework discussed in Section 6.6.2.

The prior distributions for the model parameters as well as their initial values are specified through:

```
beta0 ~ dnorm(0.0, 1.0E-3)
beta1 ~ dnorm(0.0, 1.0E-3)
beta2 ~ dnorm(0.0, 1.0E-3)
mu ~ dnorm(0.0, 1.0E-3)
gamma0 ~ dnorm(0.0, 1.0E-3)
gamma1 ~ dnorm(0.0, 1.0E-3)
tau ~ dgamma(1.0E-3, 1.0E-3)
psi ~ dgamma(1.0E-3, 1.0E-3)
list(beta0=-1, beta1=0.5, beta2=0.5; mu=0,
gamma0=0, gamma1=0, tau=1, psi=1)
```

These prior distributions are common diffuse and proper prior distributions for logistic and linear regression model parameters. In addition, we find that it is often unnecessary to specify initial values for missing data.

The output of the WinBUGS program includes posterior draws for both the parameters and imputations of three variables (i.e., y_miss, x1_miss, and x2_miss). It is more convenient to call WinBUGS within the platform of major statistical software packages (e.g., R) to process the imputations. In this book, we use R R2WinBUGS to call WinBUGS from R. But all these programs can also be executed in R nimble without directly calling WinBUGS.

Example 6.7. *A simulation study on imputing missing data in a logistic regression with two continuous predictors*

We run a simulation to assess the performance of the JM imputation (via WinBUGS) in Examples 6.5 and 6.6. The complete-data model is: $logit(f(Y = 1|X_1, X_2)) = -1 + X_1/2 + X_2/2$, and $(X_1, X_2) \sim N_2((0,0), \begin{pmatrix} 1, 1/2 \\ 1/2, 1 \end{pmatrix})$. We then randomly remove around 15% of cases for each of the three variables so that the missing values follow MCAR. The complete-data sample size is 1000, and the number of simulations is 1000. We also apply the general location model-based imputation (GLOM), assuming that the two continuous X variables follow a bivariate normal distribution at each level of Y with possibly different mean vectors and yet the identical covariance matrices. Here

GLOM can be viewed as a linear discriminant analysis model with two X-variables classified by Y. Note that the imputation model under the GLOM is somewhat misspecified when compared with the data-generating model. In addition, GLOM is not specifically devised to accommodate the logistic regression analysis. The multiple imputation estimates are based on $M = 50$ imputations. For JM, we let the Bayesian imputation algorithm run 5000 iterations as the burn-in period, and then run it for another 5000 iterations. We take a single set of imputations for every 100 iterations. The estimands of interest include the mean of Y ($\overline{Y}$) and the slope coefficient of X_1 when running a logistic regression of Y on X_1 and X_2.

Fig. 6.6 shows the trace/history plots of the parameter draws based on 5000 iterations from one simulation replicate. By visual inspection, the posterior samples appear to achieve the convergence after the burn-in period of 5000 iterations.

Table 6.5 shows the simulation results. Because of MCAR, CC yields estimates with little bias and coverage rates close to the nominal level. Estimates from both imputation methods are also good. For the logistic regression coefficient, both imputation methods gain precision (e.g., smaller MSEs) compared with CC because the former use extra information from subjects with missing values. Note that in spite of the model misspecification in GLOM, its results are similarly good to those in JM.

TABLE 6.5
Example 6.7. Simulation results

Method	RBIAS (%)	SD	$\overline{\text{SE}}$	MSE	CI	COV (%)
Mean of Y						
BD	0	0.014	0.014	0.000201	0.057	95.2
CC	−0.1	0.015	0.016	0.000240	0.061	95.5
JM	−0.1	0.015	0.016	0.000237	0.061	95.3
GLOM	−0.2	0.015	0.016	0.000239	0.061	95.6
Logistic regression coefficient of X_1						
BD	0	0.092	0.089	0.00847	0.350	94.3
CC	0.2	0.117	0.114	0.0136	0.448	94.3
JM	0.2	0.110	0.107	0.0122	0.422	94.6
GLOM	−0.3	0.109	0.107	0.0118	0.420	94.4

6.6.3.2 Missing Interactions and Squared Terms of Covariates in Regression Analysis

In many cases, the interaction effects of covariates (including the squared term of a covariate) cannot be simply ignored in a targeted regression analysis. This often creates additional complexity if the covariate has missing values

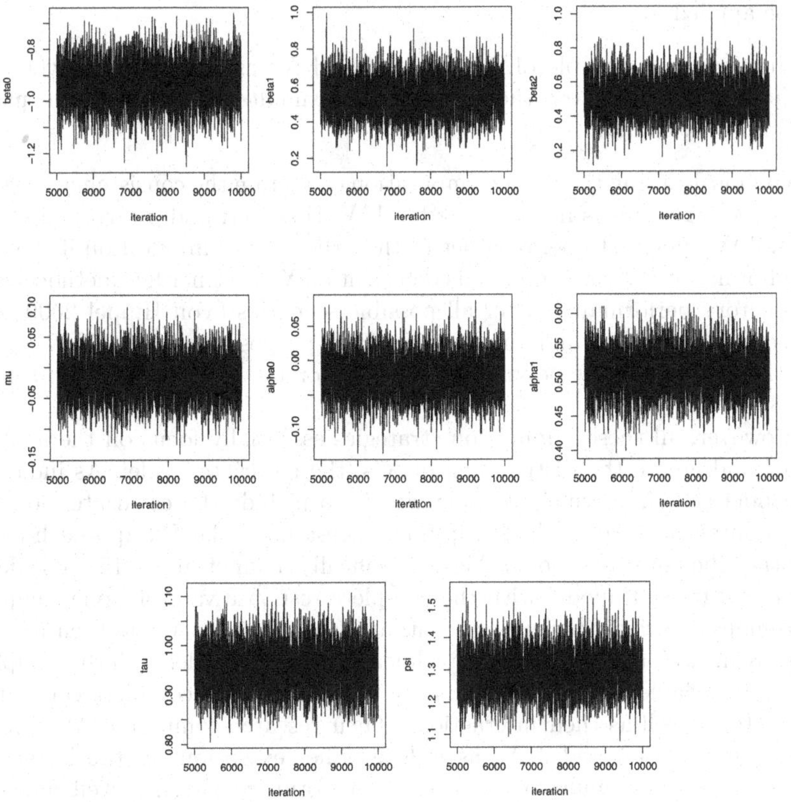

FIGURE 6.6
Example 6.7. History plots of the posterior draws for the complete-data model parameters against iterations of MCMC from one simulation replicate.

and needs imputation. For example, if an interaction between X_1 and X_2 is included as an extra regressor in Example 6.5, the conditional distributions involved in the DA algorithm become even more complicated.

Imputation for missing covariates with squared terms or interactions has received attention in recent literature. For illustration, suppose that there is only one X-covariate in the regression and the analysis model is a linear regression model $Y = \beta_0 + \beta_1 X + \beta_2 X^2 + \epsilon$, where $\epsilon \sim N(0, \sigma^2)$ and X has some missing values. Possible imputation strategies include:

1. Passive imputation: X is imputed under a linear regression model for X given Y, and the corresponding X^2 is obtained passively by taking the

square of imputed X; a PMM version of the passive imputation can also be applied.

2. Just another variable (JAV): treat X and X^2 as two unrelated variables (say let $Z = X^2$) and then impute them simultaneously in a joint model for Y, X, and Z.

Obviously, using the passive imputation can retain the consistency between X and X^2. Yet this is not the case for JAV. However, under certain assumptions, JAV appears to perform better than the passive imputation if the estimand of interest is the regression coefficient of X^2. Yet neither method yields satisfactory performances over all possible scenarios (Von Hippel 2009; Seaman et al. 2012). Vink and Van Buuren (2013) proposed to impute $\beta_1 X + \beta_2 X^2$ together, termed as the polynomial combination method. See also Van Burren (2018, Section 6.4.2).

However, all these imputation strategies essentially focus on the analysis model and ignore the distribution of X's, the covariate model. As indicated by Model (6.15), a principled approach is to include the covariate model in the imputation. Kim et al. (2015; 2018) investigated the JM approach incorporating the covariate model. More specifically, Kim et al. (2015) considered a linear regression model with the complete response variable, one complete continuous covariate, one missing continuous covariate, as well as their interactions. Kim et al. (2018) considered a linear regression model with the complete response variable, one complete binary covariate, one incomplete continuous covariate, as well as their interactions. Their research compared JM with the passive imputation and JAV, as well as other extensions of the latter two methods. In their simulation studies, JM performs consistently well across all tested scenarios and is the best approach among alternative methods.

Example 6.8. *A logistic regression with a missing covariate and its squared terms*

The challenge of applying JM to missing data problems with regression interactions is that the posterior predictive distribution of the missing covariate could be rather complicated. In the settings considered in Kim et al. (2015; 2018), the posterior predictive distribution of the missing continuous covariate is still normal so that it is relatively easy to program. Here we consider a logistic regression model, $logit(f(Y = 1|X)) = \beta_0 + \beta_1 X + \beta_2 X^2$, where $X \sim N(\mu, \sigma^2)$. In this case, the posterior predictive distribution of the missing X-values does not have a closed form.

Similar to Example 6.6, we can use WinBUGS to draw imputations for missing X-values. This would save practitioners from deriving complicated conditional distributions and programming the imputation algorithm. For simplicity, we omit the detailed WinBUGS code here. We run a simulation study to assess its performance. The simulation design is similar to one used in Seaman et al. (2012). In the completed-data model, we set $X \sim N(2, 1)$, $\beta_0 = -3/2$, $\beta_1 = 1/3$, and $\beta_2 = 1/6$. We set Y as fully observed and generate missing values for X based on $logit(f(R = 1)) = 2 - 2Y$, which follows an MAR mechanism. This produces around 30% of the missing cases in X. The complete-data sample size is 2000, and the number of simulations is 1000. In the evaluation, the estimand of interest is the slope coefficient of X^2 when running a logistic regression for Y on X and X^2.

In addition to JM, we consider two other imputation methods. One is the passive imputation method: impute X by assuming X given Y is a normal linear regression model and then square the imputed X to get the corresponding X^2. The other is JAV: let $Z = X^2$ and then JAV is based on a general location model for Y, X, and Z. All imputation methods are conducted $M = 50$ times.

Table 6.6 shows the simulation results. Since the probability of being missing in X is only related to Y, CC provides satisfactory results. This is consistent with the fact that valid inferences can be obtained from case-control study data using ordinary logistic regression where Y is treated as the outcome (Prentice and Pyke 1979). Neither the passive nor JAV method performs well, as both of them generate large biases. Results from these two methods are consistent with those reported from Seaman et al. (2012). On the other hand, JM (via WinBUGS) performs well with little bias and coverage rates close to the nominal level.

TABLE 6.6
Example 6.8. Simulation results

Method	RBIAS (%)	SD	$\overline{\text{SE}}$	MSE	CI	COV (%)
Coefficient of X^2 from regressing Y on X and X^2						
BD	0	0.048	0.048	0.00227	0.187	94.8
CC	−1.2	0.058	0.060	0.00340	0.234	95.4
JM	−1.5	0.058	0.059	0.00338	0.234	95.6
Passive	−31.8	0.041	0.056	0.00442	0.221	93.5
JAV	60.7	0.052	0.057	0.0128	0.225	58.5

6.6.3.3 Imputation Using Flexible Distributions

Besides the GLM-type models, WinBUGS and its variants have a wide variety of options of distributions (e.g., t-distributions) and models that might be effectively used for imputation. For example, in Section 6.3.2 we briefly introduced the use of normal mixture models for imputing continuous data that deviate from normal distributions. Relevant imputation algorithms can be run through WinBUGS. We provide a simple illustration here.

Example 6.9. *Imputation of birth weight using a mixture model*

Zhang et al. (2014) applied Bayesian mixture models to public-use U.S. birth weight data (from 2001 to 2008) to identify gestational ages (DGESTAT) that are subject to misreporting or misspecifications. This problem was mentioned in Example 4.2, where we assigned some suspicious values of DGESTAT to be missing for the purpose of illustrating multiple imputation. In Zhang et al. (2014), they attempted to separate the problematic DGESTAT from normal ones by modeling the corresponding birth weight variable (DBIRWT). Fig. 6.7 (top left panel) includes the distribution of DBIRWT for reported DGESTAT=30 weeks from 2001. It clearly shows that the data follow a mixture of two distributions (i.e., from two groups). The idea behind the mixture modeling is that subjects whose DBIRWT falls into one group might have their DGESTAT misspecified. It is likely that the problematic group has a larger mean of DBIRWT.

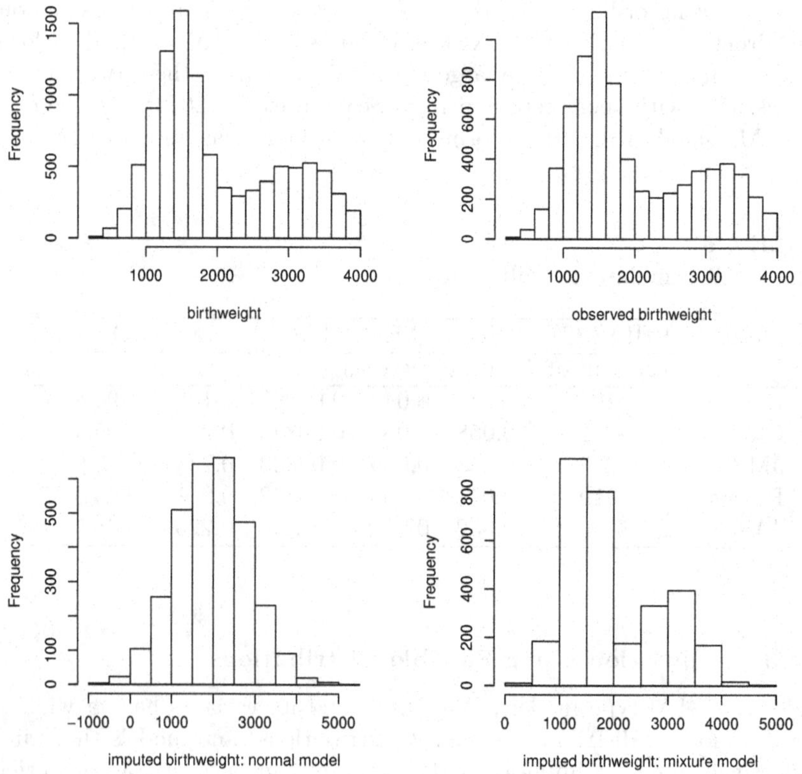

FIGURE 6.7
Example 6.9. Histograms of DBIRWT. Top left: complete data; top right: observed values after removing around 30% of the cases; bottom left: imputed values using a univariate normal model; bottom right: imputed values using a normal mixture model. The imputed values are from one set of imputations.

To showcase the use of normal mixture models for imputation, we take a working dataset that is a subset of the data used in Zhang et al. (2014). The original sample size is 10196, and we randomly remove around 30% of the cases, resulting in an incomplete dataset with sample size 7167. We impute the missing DBIRWT-values using a univariate normal mixture model with two groups (i.e., group 1 and 2):

$$Y_i|Z_i, \mu_1, \mu_2, \sigma_1, \sigma_2 \sim N(\mu_{z_i}, \sigma^2_{z_i}), \quad (6.17)$$

$$Z_i|\pi_1 \sim 1 + Bernoulli(\pi_1), \quad (6.18)$$

where Z_i takes 1 or 2 for subject i.

We apply typical prior distributions to model parameters. The imputation algorithm is run using WinBUGS, and we include some sample code here. In order to better identify the mixture components, we reparameterize $\lambda_2 = \lambda_1 + \theta$ where λ_1 and λ_2 are the means of the two groups (i.e., μ_1 and μ_2 in Eqs. (6.17) and (6.18)) and $\theta > 0$. In addition, we sort the data into ascending order, and fix $T_1 = 1$ (the minimum DBIRWT=260) and fix $T_N = 2$ (the maximum DBIRWT=3997), where T is the variable for the unknown grouping indicator Z_i's. For other subjects, the T-values are input as missing values to WinBUGS. Doing this can prevent the labels of the two mixture components from switching back and forth as the MCMC sampler progresses.

```
model
{
for( i in 1 : N ) {
y_miss[i] ~ dnorm(mu[i], tau[i])
mu[i] <- lambda[T[i]]
tau[i] <- gamma[T[i]]
T[i] ~ dcat(P[])
}
P[1:2] ~ ddirch(alpha[])
theta ~ dnorm(0.0, 1.0E-6) I(0.0, )
lambda[2] <- lambda[1] + theta
lambda[1] ~ dnorm(0.0, 1.0E-6)
mean1 <- lambda[1]
mean2 <- lambda[2]
P1 <- P[1]
sigma1 ~ dunif(0.01,10000)
gamma[1]<- 1/(sigma1*sigma1)
sigma2 ~ dunif(0.01,10000)
gamma[2]<- 1/(sigma2*sigma2)
}
```

Table 6.7 shows results from the Bayesian model fitting using both complete data and observed cases (i.e., after removing missing cases). The posterior estimates are obtained from 1000 iterations after setting the burn-in period as 10000 and the thinning period as 10 iterations, respectively. The model-fitting results are consistent with the visual inspection of the data in Fig. 6.7. The major component (P_1 is around 2/3) of the sample has the mean of DBIRWT close to 1400 grams, while the minor component ($1-P_1$ is around 1/3) of the sample has the mean of DBIRWT close to 3000 grams. In addition, DBIRWT from subjects in the minor group is more spread out, as $\sigma_2 > \sigma_1$ from the estimates. As expected, posterior mean estimates from using the observed cases are similar to those before deletion under MCAR. The model fitting results here are also similar to those from Zhang et al. (2014), which used more years of data.

TABLE 6.7
Example 6.9. Bayesian model fitting results

Parameter	Complete data	Observed values
μ_1	1438 (5.282)	1432 (6.501)
σ_1	363 (4.199)	361.2 (5.101)
μ_2	3080 (10.82)	3077 (13.06)
σ_2	470.6 (8.192)	468.6 (10.04)
P_1	0.658 (0.00536)	0.660 (0.00650)

Note: Posterior means (standard deviations) based on 1000 iterations of the posterior draws, after running the chain for a burn-in period of 10000 MCMC iterations and a thinning interval of 10 iterations.

More importantly, the imputations under the mixture model faithfully adhere to the distribution of the original data, which display the two groups and are shown in Fig. 6.7 (bottom right panel). On the other hand, imputations from a univariate normal model (e.g., using the method introduced in Example 3.1) clearly breaks from the original distribution (bottom left panel in Fig. 6.7), and some negative values are also produced.

6.7 Summary

In this chapter we have discussed a few common JM strategies for specifying joint models to impute multivariate incomplete data. All these methods are based on a fully Bayesian framework that involves specifying a complete-data model and prior distribution, deriving the posterior (predictive) distributions for the model parameter and missing values, and drawing missing values using the DA algorithm. With a general missingness pattern, models discussed in Sections 6.3-6.5 are mainly driven by the types of variables in the data. In real applications, some of the models (e.g., the log-linear models for categorical variables) might need to be refined for model parsimony. However, the model refining steps are usually not automatic and need to be based on exploratory analysis.

When a specific analysis is planned for data at hand, in principle both the analysis model and covariate model should be considered in the JM imputation. The implementation is often not simple for general practitioners. In addition, the commonly used imputation software packages are mostly designed for general purposes and not targeted to specific analyses. Therefore, we propose to use Bayesian analysis software to implement the analysis-targeted imputation. In this book we list a few examples using WinBUGS. Based on our limited knowledge, other Bayesian analysis software packages (e.g., OpenBUGS, SAS PROC MCMC, R rstan, R nimble) might have similar capacities. More research is warranted to explore their utilities and limitations.

A related question is, if the Bayesian estimate for the analysis model (e.g., $\theta_{Y|X}$ in Eq. (6.15)) can be obtained directly via the software in the presence of incomplete data, why bother doing multiple imputation analyses of Y and X using completed data? There are two considerations. First, if both the analysis and covariate models are correctly specified, then the multiple imputation-based estimates are asymptotically equivalent to the Bayesian estimate (Section 3.2). The former might be slightly less efficient when the number of imputations is moderate. Second, as pointed out before, having multiply imputed datasets allows more flexibility. For example, besides estimating $\theta_{Y|X}$, additional analyses can also be applied to the completed data.

7

Multiple Imputation for Multivariate Missing Data: The Fully Conditional Specification Approach

7.1 Introduction

The JM strategy discussed in Chapter 6 has a well justified statistical foundation for multivariate missing data problems. However in some practical settings, using JM is often not straightforward. One issue is the practical complexity associated with many Bayesian multivariate models, although some of this can be addressed by using special software packages (e.g., WinBUGS). More importantly, some data exhibit complicated data features and structures. For instance, large and complex survey data often have many variables that might include skip patterns, restrictions, and bounds. In these cases, it can be difficult, if not impossible, to specify a reasonable multivariate distribution as the complete-data model to accommodate all these features. Even if such a model existed, the imputation algorithm would not be simple.

Oftentimes, a more practical and effective approach is to impute multivariate missing data using the fully conditional specification (FCS) strategy. In a nutshell, FCS operates on a variable-by-variable basis. It specifies the complete-data model as a collection of univariate models for each of the incomplete variables conditional on other variables. The imputations are created in an iterative fashion. Since its birth, FCS has significantly boosted the utility of multiple imputation analysis for practical missing data problems. Its success builds on implementations of this strategy in multiple software packages such as SAS PROC MI FCS option, R mice, IVEware, and STATA ICE.

The term "Fully Conditional Specification" was originally coined by Van Buuren et al. (2006) and Van Buuren (2007) as an alternative strategy to the joint modeling approach. Two other popular terms in the literature are "SRMI" (sequential regression multiple imputation) and "MICE" (multiple imputation by chained equations). There exist two excellent books on multiple imputation analysis, Van Buuren (2018) and Raghunathan et al. (2018), that devote a major part of their contents to FCS and illustrate it using R mice and IVEware, respectively.

DOI: 10.1201/9780429156397-7

In Chapter 7 we provide some general discussion of FCS. Section 7.2 presents the basic idea. Section 7.3 discusses specifications of conditional models in FCS. Section 7.4 discusses how FCS can be used to handle some of common complex data features. Section 7.5 discusses issues related to the practical implementation of FCS. Section 7.6 discusses some of the theoretical implications from FCS. Section 7.7 illustrates FCS using a real example. Section 7.8 provides a summary.

7.2 Basic Idea

First we quickly recap the idea behind JM. Suppose the data consist of p variables $Y = (Y_1, \ldots, Y_p)$, each of which has some missing values. With a general missingness pattern, we specify a complete-data model, $f(Y|\theta)$ with a prior distribution $\pi(\theta)$. In a sketchy manner, the DA algorithm draws Y_{mis} and θ by alternating between the following steps:

1. Posterior step: $\theta \sim f(\theta|Y_{obs}, Y_{mis})$.

2. Imputation step: $Y_{mis} \sim f(Y_{mis}|Y_{obs}, \theta)$.

The imputation step can be typically decomposed into p-steps. In each step, we impute missing values in Y_j $(j = 1, \ldots, p)$ from its conditional distribution,

$$Y_j \sim f(Y_j|Y_{-j}, \theta_j), \tag{7.1}$$

where $Y_{-j} = (Y_1, \ldots, Y_{j-1}, Y_{j+1}, \ldots Y_p)$ includes the rest of the variables in the data by excluding Y_j. In Eq. (7.1), $f(Y_j|Y_{-j}, \theta_j)$ is determined by the joint model, $f(Y|\theta)\pi(\theta)$, where θ_j is a function of θ. See Examples 6.1 and 6.5.

In contrast to JM, FCS does not start from the explicit multivariate model, $f(Y|\theta)\pi(\theta)$, to derive models specified in Eq. (7.1). Instead, FCS directly specifies the collection of conditional models in Eq. (7.1). If a Bayesian paradigm is taken, then prior distributions for θ_j's are also assigned. These specified conditional models are then used to impute missing values in Y_j's sequentially. By doing so, FCS transforms a p-dimensional, multivariate missing data problem into p one-dimensional, univariate missing data problems. This idea is somewhat similar to that behind the Gibbs sampling that draws one parameter (conditional on others) at a time sequentially to obtain the posterior draws of all the model parameters (Section 2.3.3). Arguably FCS enables the creation of more flexible multivariate models because it is often easier to specify univariate models directly. The apparent flexibility is not without price, however, because it attempts to define a multivariate joint model $f(Y_1, \ldots, Y_p|\theta)$ by a series of specified conditional models $f(Y_j|Y_{-j}, \theta_j)$. Related issues will be discussed in Section 7.6.

Example 4.11 provided a contrast between JM and FCS for one missing covariate. Note that the idea of specifying conditional models is also generally used in JM, such as in monotone missing data settings (Section 6.2) and in handling missing covariates in a targeted analysis by the conditional modeling framework (Section 6.6). However, distinction can be seen in the variables that are conditioned on between the two strategies. For FCS with a general missingness pattern, all other $p-1$ variables are conditioned on in the specification of each univariate conditional model. For JM used in the two above scenarios, the variables to be conditioned on gradually expand and do not include all $p-1$ variables until the last specification.

7.3 Specification of Conditional Models

In FCS, there exists no unique rule for specifying conditional models, $f(Y_j|Y_{-j},\theta_j)$. Since the modeling task focuses on a collection of univariate missing data problems, methods and ideas discussed in Chapters 4 and 5 can be considered and applied.

In Chapter 4 we discussed using GLM-type parametric models for univariate missing data problems. A commonly used strategy for specifying $f(Y_j|Y_{-j},\theta_j)$ is to apply a GLM model including each of Y_{-j}'s as a main-effect predictor. For example, if Y_j is continuous, we can use a normal linear model such as

$$Y_j = \beta_{0j}+\beta_{1j}Y_1+\beta_{2j}Y_2,\ldots,+\beta_{j-1,j}Y_{j-1}+\beta_{j+1,j}Y_{j+1}+\ldots+\beta_{pj}Y_p+\epsilon_j, \quad (7.2)$$

where $\epsilon_j \sim N(0,\sigma_j^2)$.

If Y_j is noncontinuous (e.g., categorical variable), then the mean function of the GLM model can be specified as

$$g_j(E(Y_j)) = \beta_{0j}+\beta_{1j}Y_1+\beta_{2j}Y_2,\ldots,+\beta_{j-1,j}Y_{j-1}+\beta_{j+1,j}Y_{j+1}+\ldots+\beta_{pj}Y_p, \quad (7.3)$$

where g_j is an appropriate link function (e.g., the logistic link for a binary Y_j.)

Table 7.1 lists some possible specifications for conditional models used in FCS following some standard practices in statistical modeling. They include both the parametric, GLM-type models and additional adjustments discussed in Chapters 4 and 5. Note that PMM is straightforward to implement for univariate models in FCS. On the other hand, it is not immediately clear how to implement PMM directly for a multivariate joint model in JM. This list is certainly not exhaustive and can only serve as a starting point, as additional modeling strategies exist and new ones are being developed for univariate missing data problems.

TABLE 7.1
Common specifications of conditional models in FCS

Type of Variables	Model
Continuous	Linear regression, PMM, Transformation
Binary	Logistic/Probit regression, Discriminant analysis, PMM
Ordinal	Proportional odds regression, Discriminant analysis, PMM
Nominal	Multinomial logistic regression, Discriminant analysis
Count	Poisson regression, PMM
Semicontinuous	Two-part model, PMM

It is common to include variables in Y_{-j} simply as main-effect predictors, such as in Eqs. (7.2) and (7.3). However in some cases, interactions among predictors cannot be simply ignored (Section 6.6.3) and thus can be included in the conditional models.

Example 7.1. *FCS specifications for a binary variable and two continuous variables*

We consider a setup with a binary variable $Y_1 = 1$ (or 0) and two continuous variables Y_2 and Y_3. To apply FCS, we need to specify three univariate conditional models: $f(Y_j|Y_{-j}, \theta_j)$, $j = 1, \ldots, 3$. There exist multiple options. Following the idea of Eqs. (7.2) and (7.3), we may consider

$$\begin{aligned} logit(f(Y_1 = 1)|Y_2, Y_3, \alpha_0, \alpha_1, \alpha_2) &= \alpha_0 + \alpha_1 Y_2 + \alpha_2 Y_3, & (7.4) \\ f(Y_2|Y_1, Y_3, \beta_0, \beta_1, \beta_2, \tau^2) &= N(\beta_0 + \beta_1 Y_1 + \beta_2 Y_3, \tau^2), & (7.5) \\ f(Y_3|Y_1, Y_2, \gamma_0, \gamma_1, \gamma_2, \sigma^2) &= N(\gamma_0 + \gamma_1 Y_1 + \gamma_2 Y_2, \sigma^2). & (7.6) \end{aligned}$$

Eq. (7.4) specifies a logistic regression for Y_1 on Y_2 and Y_3, Eq. (7.5) specifies a normal linear regression for Y_2 on Y_1 and Y_3, and Eq. (7.6) specifies a normal linear regression for Y_3 on Y_1 and Y_2. Other specifications can be applied, depending on the feature of the data and planned completed-data analysis. For example, we might include interactions in the predictors of these models if there exists some strong evidence that these interactions are significant predictors. In some cases, if the data have a multilevel structure (Chapter 9), this feature can also be incorporated in conditional models. More elaborations will follow in subsequent sections.

7.4 Handling Complex Data Features

Large datasets from surveys or administrative databases often have complex structure and features. For example, values of some variables are bounded or fall in certain ranges. In addition, structure missingness due to not applicable or skip patterns are rather frequent in large surveys. Compared with JM, it is relatively easier to handle these complexities using FCS.

7.4.1 Data Subject to Bounds or Restricted Ranges

In general, one can easily specify a model $f(Y_j|Y_{-j}, \theta_j)$ if Y_j is bounded or has a restricted range. For Y_j with categorical values, the GLM-type models for categorical variables can automatically preserve the range of the imputations. For Y_j with continuous values (e.g., income has to be nonnegative), there exist a couple of strategies. One is to adjust the imputed values by drawing them from a truncated normal distribution with a specified range (Section 5.4). The other is to use PMM so that imputed values can only be chosen from observed cases and thus automatically fall into the desired ranges or bounds.

7.4.2 Data Subject to Skips

In survey data, some variables have skip patterns (i.e., "not applicable" answers). We briefly introduced this problem in Example 1.5. We provide more discussion here.

Example 7.2. *Ideas to handle skip patterns in imputing tobacco-use variables from RANDS I*

As stated in Example 1.5, health surveys might have multiple questions on tobacco use. In Example 2.1, we introduced a probability panel-based web survey of health variables (RANDS). The following lists a subset of questions implemented in RANDS I with some simplifications for the purpose of presentation.

1. SMKEV: These next questions are about cigarette smoking. Have you smoked at least 100 cigarettes in your entire life? Answers: 1 Yes; 2 No; 9 Don't Know.

   Skip: (If code 1 SMKEV continue, otherwise skip to SMKAY)

2. SMKNOW: How often do you now smoke cigarettes? Every day, some days or not at all? Answers: 1 Every Day; 2 Some Days; 3 Not at All; 9 Don't Know.

   Skip: (If code 3 continue, if code 1 or 2 skip to CIGQTRY, if code 9 or blank skip to a question after the questions on tobacco use).

3. SMKQTNO: How long has it been since you quit smoking cigarettes? Answers: Enter number of times for Months and Years Ago.

4. CIGQTYR: During the past 12 months, have you stopped smoking for more than one day because you were trying to quit smoking? Answers: 1 Yes; 2 No; 9 Don't Know.

   Skip: (All in CIGQTRY skip to a question after the questions on tobacco use).

5. SMKANY: Have you ever smoked a cigarette even one time? Answers: 1 Yes; 2 No; 9 Don't Know. Programmer: (Only ask SMKANY of those who were code 2, 9, or blank in SMKEV).

The above questions are intended to measure various aspects of tobacco use. For simplicity, our discussion focuses on the first two questions, SMKEV and SMKNOW, and the skip between the two variables. In a perfect scenario, that is, there is no "don't know" (or missing value) answered for the two variables, then SMKEV and SMKNOW can be used to group the population. One of the possible grouping, labeled as SMOKER, is:

1. Current smokers (SMOKER=1): have smoked at least 100 cigarettes in their lifetime (SMKEV=1) and still currently smoke every day or on some days (SMKNOW=1 or 2).

2. Former smokers (SMOKER=2): have smoked at least 100 cigarettes in their lifetime (SMKEV=1) but currently do not smoke at all (SMKNOW=3).

3. Nonsmokers (SMOKER=3): have never smoked at least 100 cigarettes in their lifetime (SMKEV=2 and thus is skipped for SMKNOW).

Therefore, the skip can be used to stratify the population. However, with missing values in both SMKEV and SMKNOW, SMOKER can also have missing values. Using RANDS I data ($n = 2304$) for illustration, a simple frequency table for SMOKER is in Table 7.2. There are two types of missing values for SMOKER (=8 or 9). When SMKEV is 1 yet SMKNOW is missing, the participant is surely not a nonsmoker, yet it is unclear if he/she is a current or former smoker. When SMKEV is missing, this person can be in any of the three groups, current smokers, former smokers, or nonsmokers.

TABLE 7.2
Example 7.2. Frequency table of SMOKER in RANDS I

SMOKER value	Definition	Frequency
1	Current smoker: SMKEV = 1 and SMKNOW = 1 or 2	297/2304
2	Former smoker: SMKEV = 1 and SMKNOW = 3	711/2304
3	Nonsmoker: SMKEV = 2 and SMKNOW=Not applicable	1278/2304
8	Smoker but not sure current or former: SMKEV = 1 and SMKNOW = 9	3/2304
9	Not clear: SMKEV=9	15/2304

For the purpose of imputation, if we directly impute SMOKER, then it is not straightforward to distinguish between the two types of missing values (i.e., SMOKER=8 and SMOKER=9). It is more convenient to impute SMKEV and SMKNOW separately. For illustration, let X denote the rest of the variables in the dataset, which are not included in the questions about tobacco use. The conditional models for SMKEV and SMKNOW can be specified as follows:

$$log(\frac{f(SMKEV=1)}{f(SMKEV=2)}) = X\beta_1, \quad (7.7)$$

$$log(\frac{f(SMKNOW=1)}{f(SMKNOW=3)}|SMKEV=1) = X\beta_2, \quad (7.8)$$

$$log(\frac{f(SMKNOW=2)}{f(SMKNOW=3)}|SMKEV=1) = X\beta_3, \quad (7.9)$$

where β_1-β_3 are the regression coefficients. Eq. (7.7) specifies a logistic regression imputation model for SMKEV. Since SMKNOW has three categories, Eqs. (7.8) and (7.9) specify a multinomial logistic regression imputation model conditional on SMKEV=1. Therefore, the imputation for SMKNOW is dependent on the imputed values for SMKEV, as SMKNOW is skipped for the cases whose value for SMKEV is 2.

In some practical scenarios, despite that the skip switch is on, there might exist cases who provide conflicting answers, that is, answering 2 for SMKEV and answering from 1 to 3 for SMKNOW. When this occurs, data editing steps are needed to resolve the inconsistency before carrying out model-based imputations (Section 10.5.1).

In addition to imputing missing variables with skips, another issue is to use them as predictors for imputing other variables. In RANDS I data, for example, there is a variable (CBRCHYR) asking the participant, "Have you ever been told by a doctor or other health professional that you had chronic bronchitis?" The answers include: 1 Yes; 2 No; 9 Don't Know. Suppose we are interested in including variables on tobacco use as predictors for imputing the missing data in CBRCHYR (i.e., cases whose answer is 9). A simple strategy is to include SMOKER, a composite variable from SMKEV and SMKNOW as a predictor. Although SMOKER should be completed after we impute SMKEV and SMKNOW, the variable-forming step needs to be inserted in the iterative FCS algorithm. A more general modeling specification is to impute CBRCHYR separately by the three groups of SMOKER. For example, among former smokers (SMOKER=2), we can also include the time of quitting smoking (SMKQTNO) as another predictor, which is not available in both current smokers (SMOKER=1) and nonsmokers (SMOKER=3).

In summary, the key to imputing variables with skip patterns is to figure out sensible groups that can be formed by these variables and associated skip patterns. The grouping can be largely determined by the need of analysis and subject-matter knowledge. In our example, classifying the population into groups for the current, former, and never smokers (nonsmokers) is a conventional method. The imputation models might need to be specified separately by these groups. On the other hand, although grouping is important in dealing with skip patterns, there does not exist universal grouping strategies for different variables in different data. The imputation analysis needs to be conducted on a case-by-case manner.

To our knowledge, it might be difficult to have a fully automatic method to handle skip patterns. He et al. (2010) used a strategy that first treats all the skips as "imputable" missing data and then reassigns these values back to skips after the imputation is finished. This procedure is straightforward to implement yet it might distort the logic relationships between variables that contain skips. For example, as pointed out in Eqs. (7.7)-(7.9), SMKNOW is not a legitimate predictor for SMKEV in Eq. (7.7), and SMKEV cannot be simply used as a predictor for SMKNOW in Eqs. (7.8) and (7.9).

Certain imputation software packages allow some automatic options for handling data with restricted values and skip patterns. For example, in IVEware, the statement "restrict" allows the imputer to restrict the imputation to be conducted only for a subgroup of the sample, which can be used to handle skips in some cases such as in Eqs. (7.8) and (7.9); the statement "bounds" allows the imputed values to fall in specified ranges. These features are helpful for practitioners. In addition to using these software features whenever possible, it is important to have a good understanding of the structure of variables and scientific implications of associated analyses before setting up the imputation models. For example, Abmann et al. (2017) provided very specific steps and reasoning in handling skip patterns for imputing the incomplete income variables in the National Educational Panel Survey from Germany.

7.5 Implementation

7.5.1 General Algorithm

For each of the Y_j's ($j = 1, \ldots, p$), we first specify the conditional model $f(Y_j|Y_{-j}, \theta_j)$ and the associated priors for θ_j. We then obtain the posterior predictive distribution of missing values, $f(Y_{j,mis}|Y_{j,obs}, Y_{-j}, \theta_j)$, and the posterior distribution of the parameter, $f(\theta_j|Y_{j,obs}, Y_{-j})$. A general imputation algorithm for FCS can be sketched as follows:

1. For each j, fill in starting imputations $Y_{j,mis}^{*(0)}$ (e.g., random draws from $Y_{j,obs}$).

2. Define $Y_{-j}^{*(t)} = (Y_1^{*(t)}, \ldots, Y_{j-1}^{*(t)}, Y_{j+1}^{*(t-1)}, \ldots, Y_p^{*(t-1)})$ as the currently completed data except Y_j.

3. Draw $\theta_j^{*(t)} \sim f(\theta_j|Y_{j,obs}, Y_{-j}^{*(t)})$.

4. Draw imputations $Y_j^{*(t)} \sim f(Y_{j,mis}|Y_{j,obs}, Y_{-j}^{*(t)}, \theta_j^{*(t)})$.

5. Repeat Steps 2-4 for $j = 1, \ldots p$ to cycle through all variables.

6. Repeat Steps 2-5 for $t = 1, \ldots T$, where T is the number of iterations run for a single set of imputations.

9. Repeat Steps 1-6 independently M times to create multiple imputations.

One main issue is how many iterations, T, are needed. Theoretically, it is difficult to define the concept of "convergence" for parameter draws, $\{\theta_j\}_{j=1}^{p}$, in the FCS algorithm. The related discussion can be found in Section 7.6. In practice, however, a reasonable choice is $T \geq 20$ with a small or moderate proportion of missing values. Of course, T can be increased as needed. In addition, it is recommended to assess the "convergence" of the FCS imputation chain by examining the behavior of common statistics (e.g., mean, standard deviation, correlation) of the completed data across the iterations of the algorithm.

In addition, the initial values for $Y_{j,mis}^{*(0)}$ do not seem to matter when T is large enough. As for the sequence of arranging Y_j's in the imputation process, there seems to exist little evidence that it matters. A typical rule is to start with the variable with the least number of missing values and then go over the variables arranged by the order of increasing amount of missingness rate. This rule appears to mimic the situation of monotone missing data (Section 5.2). See also Van Buuren (2018, Section 4.5) for more discussions about various properties of the algorithm.

7.5.2 Software

There exist multiple software packages automating the FCS imputation algorithm under various models. The commonly used ones include, but are not limited to, SAS PROC MI FCS, R mice, IVEware, and STATA ICE. In this book we do not offer detailed discussions of specific features of these packages, as relevant references are widely available. For example, Van Buuren (2018) and Raghunathan et al. (2018) provided many examples using R mice and IVEware, respectively. On the other hand, it is expected that these packages are being upgraded periodically to incorporate new modeling options and data tools.

FCS imputation programs include multiple options for specifying conditional models (e.g., Table 7.1). They also include options to accommodate some practical data issues. For example, in Section 7.3 we mentioned that some packages (e.g., IVEware) can help in handling skip patterns. Another main issue is related to which variable would be included as a predictor in imputation. For imputing Y_j, in theory all variables in Y_{-j} (and their interactions) can be included as predictors. When the number of variables in the dataset is large (say on the scale of hundreds), including all these variables in the conditional model might not be practical. First, there might only exist a few variables in Y_{-j} that are effectively related to Y_j or the missingness of Y_j, and therefore are worth being included in the imputation model. Having other variables as predictors would essentially propagate additional noise to imputations. Second, having all variables in the imputation model might increase computational burden. To address this issue, practitioners can manually drop some predictors when specifying conditional models, justified by both the statistical and subject-matter evidence. For example, R mice allows the user to specify which predictors to be included in the predictorMatrix parameter of the MICE function. In addition, if such a process is desired to be somewhat automatic, some conventional variable selection procedures can be applied when processing conditional models. For example, in IVEware, the statement "minrsqd" executes a variable selection procedure using R^2-statistics of the fitted regressions based on the conditional models.

7.5.2.1 Using WinBUGS

Practitioners can also program the FCS algorithm on their own so that model specifications do not have to be limited to those used in the aforementioned packages. Again, the challenge is to program the possibly sophisticated posterior and imputation steps for each of the univariate conditional models, $f(Y_j|Y_{-j}, \theta_j)$. In Chapter 6 we briefly introduced the use of WinBUGS for implementing JM. The advantage of WinBUGS is that it can automatically draw parameters and impute missing values under a well specified Bayesian missing data model. Can WinBUGS be used to implement FCS? Our limited exploratory research appears to show that this is plausible.

In WinBUGS, it is straightforward to code the imputation step for each of the univariate conditional models, that is, imputing missing values in Y_j conditional on Y_{-j} and θ_j. However, from WinBUGS's perspective, just coding this step will also imply that any missing values from Y_{-j} are affected by Y_j simultaneously, which is not desired because any missing values in Y_{-j} should be imputed in other conditional models from different steps. We can achieve this by using the cut() function in WinBUGS to stop the feedback from Y_j to Y_{-j}. It creates a copy, say $Y^*_{-j} = Y_{-j}$, so that Y^*_{-j} is used to impute Y_j yet at the same time, the information from Y_j is not looped back to impute Y_{-j}.

Example 7.3. *A bivariate normal model with missing data in both variables*

We consider the setup of bivariate normal models (Examples 2.5 and 4.1), which includes two continuous variables Y and X. Assuming that both X and Y have missing values, one imputation strategy is to use JM directly (Example 6.1). We can also apply FCS by assuming that $f(Y|X)$ is a normal linear model and vice versa for $f(X|Y)$. We attach some sample WinBUGS code here, ignoring the part for prior distributions and initial values.

```
# The JM approach
for( i in 1 : M ) {
y_miss[i] ~ dnorm(mu_y[i], tau_y)
```

```
mu_y[i]<-beta0+beta1*x_miss[i]
x_miss[i] ~ dnorm(mu, tau_x)
}
# Wrong code for the FCS approach
for( i in 1 : M ) {
y_miss[i] ~ dnorm(mu_y[i], tau_y)
mu_y[i]<-beta0+beta1*x_miss[i]
x_miss[i] ~ dnorm(mu_x[i], tau_x)
mu_x[i]<-alpha0+alpha1*y_miss[i]
}
# Correct code for the FCS approach using the cut()
function for( i in 1 : M ) {
y_miss[i] ~ dnorm(mu_y[i], tau_y)
mu_y[i]<-beta0+beta1*x_miss_star[i]
x_miss_star[i] <- cut(x_miss[i])
x_miss[i] ~ dnorm(mu_x[i], tau_x)
mu_x[i]<-alpha0+alpha1*y_miss_star[i]
y_miss_star[i] <- cut(y_miss[i])
}
```

In the WinBUGS code, y_miss and x_miss are used to hold variables Y and X, respectively. In JM, the WinBUGS code specifies $X \sim N(\mu_X, \sigma_X^2)$ and $f(Y|X) \sim N(\beta_0 + \beta_1 X, \sigma_Y^2)$, where $\sigma_X^2 = 1/\tau_X$ and $\sigma_Y^2 = 1/\tau_Y$.

In FCS, the conditional models are specified as $f(Y|X) \sim N(\beta_0+\beta_1 X, \sigma_Y^2)$ and $f(X|Y) \sim N(\alpha_0 + \alpha_1 Y, \sigma_X^2)$. However, as we stated before, coding them directly will not achieve the desired results. For example, the information from Y flows back to X in the WinBUGS code

```
mu_y[i]<-beta0+beta1*x_miss[i]
```

To prevent this from happening, we create a copy $X^* = X$, and let $f(Y|X^*) \sim N(\beta_0 + \beta_1 X^*, \sigma_Y^2)$, stopping the feedback from Y to X. This is achieved by coding

```
mu_y[i]<-beta0+beta1*x_miss_star[i]
x_miss_star[i] <- cut(x_miss[i])
```

To fetch the imputed values, we can use either X or X^*. Similar tricks can be applied to Y and Y^* as well. This technique might be extended to include more variables and models. A related discussion of using the cut function in general Bayesian models can be found in Plummer (2015).

We run a simple simulation study to assess the performance of FCS executed by WinBUGS. We use the setup in Examples 2.5 and 4.1. The complete-data model is: $X \sim N(1, 1)$, and $Y = -2 + X + \epsilon$, where $\epsilon \sim N(0, 1)$. We randomly delete 20% of cases for both Y and X so that the missing values are MCAR. We apply three imputation methods: (a) JM implemented by R norm; (b) FCS implemented by R mice; (c) FCS implemented by WinBUGS (using the cut() function). All the methods are conducted $M = 50$ times.

The estimands of interest include the slope coefficient of regressing Y on X and regressing X on Y, respectively. The complete-data sample size is set as $n = 1000$, and the simulation includes 1000 replicates.

Table 7.3 shows the simulation results. The three imputation methods yield very similar results with little bias and coverage rates close to the nominal level. All of them gain some efficiency over CC as expected. On the other hand, using the wrong WinBUGS code (without the cut() function) would lead to incorrect results (details not shown).

TABLE 7.3
Example 7.3. Simulation results

Method	RBIAS (%)	SD	$\overline{\text{SE}}$	MSE	CI	COV (%)
		Slope coefficient of regressing Y on X				
BD	0	0.032	0.032	0.000995	0.124	95.0
CC	0	0.040	0.040	0.00163	0.155	94.4
JM	−0.1	0.039	0.038	0.00149	0.148	93.5
FCS (R mice)	−0.1	0.039	0.037	0.00150	0.147	93.5
FCS (WinBUGS)	−0.1	0.039	0.037	0.00149	0.145	93.1
		Slope coefficient of regressing X on Y				
BD	0	0.016	0.016	0.000251	0.062	95.0
CC	0.1	0.019	0.020	0.000380	0.078	95.6
JM	0	0.019	0.019	0.000347	0.074	95.8
FCS (R mice)	0	0.019	0.019	0.000348	0.074	94.9
FCS (WinBUGS)	0	0.019	0.018	0.000345	0.073	95.3

Based on our limited experience, we are not aware of any previous literature on using WinBUGS (or related Bayesian analysis software packages) to execute FCS. The custom has always been to use WinBUGS to specify and estimate a well defined joint Bayesian model. However, an apparent advantage of using WinBUGS for FCS is that it can accommodate a wide variety of Bayesian analysis models, possibly more than those available in the routine imputation software packages. Of course, since the original goal of WinBUGS is not for imputation, practitioners need to be more specific about the models and write them using the WinBUGS code. When the number of variables increases, the code is expected to be more clumsy. Yet this might be improved by the more recent variant of WinBUGS such as R nimble. In general we believe that WinBUGS provides an extra option of implementing FCS.

7.6 Subtle Issues

7.6.1 Compatibility

The practical flexibility of FCS does not come without a price. A major theoretical concern for this strategy is the issue of compatibility. In practical terms, two conditional distributions $f(Y_1|Y_2)$ and $f(Y_2|Y_1)$ are compatible if a joint distribution $f(Y_1, Y_2)$ exists and has $f(Y_1|Y_2)$ and $f(Y_2|Y_1)$ as its conditional distributions (e.g., Van Buuren 2018). In the context of FCS, the issue is that having well specified conditional models, $f(Y_j|Y_{-j}, \theta_j)$'s, does not necessarily imply the existence of a joint model for all Y_j's. This is in contrast to JM, which first specifies the joint model $f(Y_1, \ldots, Y_p|\theta)$ and then derives the conditional models under it. We use a few examples to illustrate this point.

Example 7.4. *The FCS specifications for multivariate normal data*

Let Y denote the collection of p continuous variables; we now consider the classic example of multivariate normal models (Section 6.2). For JM, if $f(Y|\theta)$ follows a multivariate normal distribution, then $f(Y_j|Y_{-j}, \theta_j)$'s are all univariate normal distributions, which can be specified as linear regression models with normal error distributions. For FCS, if $f(Y_j|Y_{-j}, \theta_j)$'s are specified as linear regression models with only main-effect predictors and constant normal error distributions (e.g., Eq. (7.2)), then these conditional specifications imply that the joint distribution of all Y-variables follows a multivariate normal model. These conditional linear regression models are deemed as compatible.

In Example 7.3, we have shown that the multiple imputation analysis results from both JM and FCS are very close under a bivariate normal model. A subtle difference between JM and FCS is the number of parameters used. In JM, the number of parameters (for the mean vector and covariance matrix of the p-variate normal model) is $p + \frac{p(p-1)}{2} = \frac{p(p+1)}{2}$. In FCS, the number of parameters (for p linear regression models) is $p(p+1)$. Although they are different, with a large sample size relative to the number of variables, the efficiency of estimates from the two methods are expected to be similar.

Example 7.5. *Three binary variables*

Consider three binary variables Y_1-Y_3 that can take 1 or 0. For JM, a log-linear model can be specified for them with all the main effects and two-way interactions, lacking the three-way interaction. This joint model would imply that a conditional distribution, $f(Y_1|Y_2, Y_3, \theta_1)$, can be expressed as a logistic regression model with only main-effect predictors: $logit(f(Y_1 = 1)) = \theta_{10} + \theta_{11}Y_2 + \theta_{12}Y_3$. Similar results hold for $f(Y_2|Y_1, Y_3, \theta_2)$ and $f(Y_3|Y_1, Y_2, \theta_3)$. For FCS, if we specify the conditional distributions for Y_1, Y_2, and Y_3 (given the other two variables) as these logistic regressions; then it is essentially the case that we are specifying a log-linear model without the three-way interaction for Y_1-Y_3. These logistic regression models with only main effects are deemed compatible.

Furthermore, if the three-way interaction is included in the log-linear model for Y_1-Y_3, then the conditional distribution for each variable can be expressed as a logistic regression with both main-effect predictors and their interactions. For example, $logit(f(Y_1 = 1)) = \theta_{10} + \theta_{12}Y_2 + \theta_{13}Y_3 + \theta_{14}Y_2Y_3$. For FCS, if we specify the conditional models as such logistic regressions, then these models are also deemed compatible.

However, if the interaction term is included in some conditional model specifications but not in others, it is not clear whether the conditional models are compatible with each other. For instance, consider the following specifications of a possible FCS imputation:

$$logit(f(Y_1 = 1)) = \theta_{10} + \theta_{11}Y_2 + \theta_{12}Y_3 + \theta_{13}Y_2Y_3, \quad (7.10)$$

$$logit(f(Y_2 = 1)) = \theta_{20} + \theta_{21}Y_1 + \theta_{22}Y_3, \quad (7.11)$$

$$logit(f(Y_3 = 1)) = \theta_{30} + \theta_{31}Y_1 + \theta_{32}Y_2. \quad (7.12)$$

Does there exist a joint model for Y_1-Y_3 so that the conditional distributions under this model match with Eqs. (7.10)-(7.12)? As reasoned before, a log-linear model for three binary variables without the three-way interaction would imply $\theta_{13} = 0$ in Eq. (7.10). On the other hand, a log-linear model with the three-way interaction would imply all the logistic models shall include the interaction term between the two predictors. That is, we need to add $\theta_{23}Y_1Y_3$ to the right hand side of Eq. (7.11) and $\theta_{33}Y_1Y_2$ to the right hand side of Eq. (7.12). Therefore, the current specifications do not appear to be compatible with each other. That is, there might not exist a well defined joint model based on Eqs. (7.10)-(7.12).

In general, it might not be simple to verify the compatibility of arbitrarily specified conditional models except for some simple cases. Therefore in practice, it is rather likely that many of the specifications of conditional models in FCS do not imply a well defined joint distribution/model for the variables involved. In theory, if the incompatibility exists, we cannot simply assume that iterations of parameter estimates (from the FCS imputation algorithm) would behave exactly as the Gibbs sampling algorithm derived under a well defined joint model. As a result, the imputations (under a theoretically misspecified model when such incompatibility exists) are not technically coherent with the correct posterior predictive distribution of the missing data.

However, past research suggests that possible incompatibility of the FCS imputation models usually does not matter much in practice, especially in scenarios where the conditional model specifications match with the analysis models (e.g., Van Buuren 2018). For instance, in Example 7.5, if the analysis model is to run a logistic regression of Y_1 using Y_2 and Y_3, as well as their interaction as predictors (corresponding to Eq. (7.10)), then it is well justified to use this specification in FCS with less concern about the compatibility of all conditional models. Similarly, we might modify (or keep) the specifications of imputation models for Y_2 and Y_3 to match the analysis models for them, respectively. In addition, if the missing variable is used as a covariate in the analysis model, the information from the analysis model is also recommended to be incorporated in the specification of the conditional model for the missing variable in FCS (Bartlett et al. 2015).

In-depth discussions about the statistical theory related to the issue of compatibility can be found in Arnold et al. (1999). Van Buuren (2006) initiated the discussion in the context of FCS. Van Buuren (2018, Section 4.5.3) provides a nice overview. Some recent literature can be found in Hughes et al. (2014); Liu et al. (2014); and Zhu and Raghunathan (2015). Omitting technical details, the relevant research concluded that the distributions from which the FCS and JM approaches draw the missing values (i.e., the imputation distributions) are asymptotically the same when the conditional models used in FCS are compatible with a joint model. Seaman and Hughes (2018) further showed that the asymptotic equivalence of imputation distributions does not imply the two methods would yield asymptotically equally efficient inference about the parameters of the model of interest. Using a restricted general location model as the study setting, they concluded that in general FCS is only slightly less efficient than the corresponding JM, yet the former can be more robust if the joint model is misspecified in JM.

As shown in Section 6.2, when the missingness pattern is monotone, a theoretically valid JM approach is to specify, for each variable with missing values, a conditional distribution given the variables with fewer or the same number of missing values and sequentially draw imputations from these distributions. There are some proposals to combine the idea of FCS and monotone imputation strategies to mitigate the possible incompatibility of the former. For example Rubin (2003) suggested first using FCS to impute missing values that destroy the monotone pattern and then conduct JM based on created monotone missing data. See also Baccini et al. (2010) and Li et al. (2014).

7.6.2 Performance under Model Misspecifications

In many cases, FCS imputation models might not be compatible, or they are compatible yet the implied joint model might not match with the true complete-data model. Here we vaguely term these situations as "misspecified models." How well would misspecified FCS models work and what are their performances? We use a few examples to gain some insights.

Example 7.6. *A binary variable and a continuous variable*

In Section 6.5.1, we considered a missing data problem for a binary variable Y (=1 or 0) and a continuous variable X. For simplicity, we assume that the true complete-data model is a logistic-normal model

$$logit(f(Y=1)) = \beta_0 + \beta_1 X, \tag{7.13}$$
$$X \sim N(\mu, \sigma^2). \tag{7.14}$$

We consider the FCS imputation method. It is convenient to specify $f(Y|X)$ as a logistic regression model matching with Eq. (7.13). For $f(X|Y)$, it is also convenient to use a linear regression model as $f(X|Y) \sim N(\alpha_0 + \alpha_1 Y, \tau^2)$. By doing so marginally X can be considered as a mixture of normals with unequal means and identical variances. These two specifications are compatible because they imply a general location model for Y and X . However, the general location model is generally not identical to the logistic-normal model as shown in Section 6.5.1.

Under Models (7.13)-(7.14), it can be shown that (also Eq. (4.5)):

$$\begin{aligned} f(X|Y=1) &= \frac{\frac{exp(\beta_0+\beta_1 X)}{1+exp(\beta_0+\beta_1 X)} \frac{1}{\sqrt{2\pi}\sigma} exp(-\frac{(X-\mu)^2}{2\sigma^2})}{\int \frac{exp(\beta_0+\beta_1 X)}{1+exp(\beta_0+\beta_1 X)} \frac{1}{\sqrt{2\pi}\sigma} exp(-\frac{(X-\mu)^2}{2\sigma^2}) dx}, \\ f(X|Y=0) &= \frac{\frac{1}{1+exp(\beta_0+\beta_1 X)} \frac{1}{\sqrt{2\pi}\sigma} exp(-\frac{(X-\mu)^2}{2\sigma^2})}{\int \frac{1}{1+exp(\beta_0+\beta_1 X)} \frac{1}{\sqrt{2\pi}\sigma} exp(-\frac{(X-\mu)^2}{2\sigma^2}) dx}, \end{aligned} \tag{7.15}$$

which verifies that the exact conditional distribution of X given Y is not a normal linear model as specified in FCS.

However, can $f(X|Y=1)$ or $f(X|Y=0)$ in Eq. (7.15) be approximated well by a normal distribution in the FCS specification? To see that, take the logarithm of $f(X|Y=1)$ and obtain

$$log(f(X|Y=1)) = \beta_0 + \beta_1 X - log(1+exp(\beta_0+\beta_1 X)) - \frac{(X-\mu)^2}{2\sigma^2} + C, \tag{7.16}$$

where C is some constant which does not depend on X. If $log(1 + exp(\beta_0 + \beta_1 X))$ can be well approximated by a quadratic function of X, then Eq. (7.16) can be well approximated by a quadratic function of X, which would imply that $f(X|Y=1)$ can be well approximated by a normal distribution.

For simplicity, we consider a quadratic Taylor approximation of $log(1 + exp(\beta_0 + \beta_1 X))$ at $X = \mu$. After some algebra, we obtain that

$$\begin{aligned} log(1+exp(\beta_0+\beta_1 X)) &\approx log(1+exp(\beta_0+\beta_1\mu)) + \frac{\beta_1 exp(\beta_0+\beta_1\mu)}{1+exp(\beta_0+\beta_1\mu)}(X-\mu) \\ &+ \frac{\beta_1^2 exp(\beta_0+\beta_1\mu)}{2(1+exp(\beta_0+\beta_1\mu))^2}(X-\mu)^2 \end{aligned} \tag{7.17}$$

By plugging this Eq. (7.17) into Eq. (7.16), we obtain

$$log(f(X|Y=1)) \approx AX^2 + BX + C, \tag{7.18}$$

where

$$A = \frac{(1+exp(\beta_0+\beta_1\mu))^2 + \beta_1^2 exp(\beta_0+\beta_1\mu)\sigma^2}{2(1+exp(\beta_0+\beta_1\mu))^2\sigma^2},$$

$$B = \frac{-2[\beta_1(1+exp(\beta_0+\beta_1\mu))\sigma^2 + \beta_1^2 exp(\beta_0+\beta_1\mu)\mu\sigma^2 + (1+exp(\beta_0+\beta_1\mu))^2\mu]}{2(1+exp(\beta_0+\beta_1\mu))^2\sigma^2},$$

and C is some constant which does not depend on X.

Eq. (7.18) implies that $f(X|Y=1) \sim N(\mu_1, \tau^2)$, where

$$\mu_1 = \frac{\beta_1(1+exp(\beta_0+\beta_1\mu))\sigma^2 + \beta_1^2 exp(\beta_0+\beta_1\mu)\mu\sigma^2 + (1+exp(\beta_0+\beta_1\mu))^2\mu}{(1+exp(\beta_0+\beta_1\mu))^2 + \beta_1^2 exp(\beta_0+\beta_1\mu)\sigma^2},$$

and

$$\tau^2 = \frac{2(1+exp(\beta_0+\beta_1\mu))^2\sigma^2}{(1+exp(\beta_0+\beta_1\mu))^2 + \beta_1^2 exp(\beta_0+\beta_1\mu)\sigma^2}.$$

We can apply the same idea to $f(X|Y=0)$. After some algebra, we obtain that $f(X|Y=0) \sim N(\mu_0, \tau^2)$, where

$$\mu_0 = \frac{\beta_1^2 exp(\beta_0+\beta_1\mu)\mu\sigma^2 + (1+exp(\beta_0+\beta_1\mu))^2\mu}{(1+exp(\beta_0+\beta_1\mu))^2 + \beta_1^2 exp(\beta_0+\beta_1\mu)\sigma^2} - \frac{exp(\beta_0+\beta_1\mu)(1+exp(\beta_0+\beta_1\mu))\beta_1\sigma^2}{(1+exp(\beta_0+\beta_1\mu))^2 + \beta_1^2 exp(\beta_0+\beta_1\mu)\sigma^2}$$

Note that if we set $\beta_1 = 0$, then $\mu_1 = \mu_0 = \mu$ and $\tau^2 = \sigma^2$, which describes the joint model in the special case where X and Y are independent.

The quadratic approximation to $log(1 + exp(\beta_0 + \beta_1 X))$ provides some theoretical insights behind using the normal distribution to fit $f(X|Y)$ in the FCS specification. To have an idea of how well the approximation would work, we randomly generate 10000 X-samples from $N(0,1)$ and fit a quadratic curve on the corresponding $log(1+exp(\beta_0+\beta_1 X))$ using a range of β's. Fig. 7.1 shows the corresponding results. When β_1 is close to 0, the quadratic approximation works very well. As β_1 moves away from 0, the approximation becomes worse as expected. However, even when β_1 increases to 3, the overall approximation by the quadratic curve is reasonable for a wide range of X's (e.g., -2 to 2). In addition, the component $\beta_0 + \beta_1 X - \frac{(X-\mu)^2}{2\sigma^2}$ in Eq. (7.16), which is itself a quadratic function of X, can help improve the overall quadratic approximation to $log(f(X|Y=1))$. In actual model fitting, the FCS imputation algorithm is expected to find the best quadratic fit of $log(f(X|Y=1))$ and $log(f(X|Y=0))$ based on the data to fit the two normal linear models.

Example 7.7. *A simulation study for a binary variable and two continuous variables*

We consider a more general scenario than that in Example 7.6, including one binary variable Y and two continuous variables $X = (X_1, X_2)$. Similar to Example 6.7, the complete-data model is $logit(f(Y=1)) = -3 + 1.5X_1 + 1.5X_2$, and $(X_1, X_2) \sim N_2((0,0), \begin{pmatrix} 1, 0.5 \\ 0.5, 1 \end{pmatrix})$. Different from Example 6.7, the logistic regression coefficients from $f(Y|X)$ are larger here, as β's change from 0.5 to 1.5.

To generate missing data, we randomly remove around 15% of the cases for all three variables so that missing values are MCAR. The complete-data sample size is 1000, and the simulations include 1000 replicates. As in Example 6.7, the estimands of interests include the mean of Y ($\overline{Y}$) and the slope coefficient of X_1 when running a logistic regression for Y on X_1 and X_2. In addition to JM and GLOM considered in Example 6.7, we consider an FCS imputation here. The conditional specification for $f(Y|X_1, X_2)$ is the logistic regression model. The conditional specification for $f(X_l|Y, X_m)$ $(l, m = 1, 2)$ is a normal linear regression including only main-effect predictors. FCS is implemented using R mice. All imputation methods are conducted $M = 50$ times.

Table 7.4 shows the simulation results. For the mean estimand, all three imputation methods yield similar and satisfactory results. For estimating the logistic regression coefficient, JM performs well as expected. GLOM yields a considerable bias (|RBIAS| close to 7%), and its coverage rate (91.8%) is somewhat lower than the nominal level. As noted in Example 6.7, the model assumed in GLOM is misspecified. With increased β's (compared with those in Example 6.7), such misspecification starts to generate worse estimates in GLOM. The conditional models in FCS are also misspecified, as we reasoned in Example 7.6. However, the corresponding results are reasonably good: the bias is tolerable (|RBIAS| $< 5\%$), and the coverage rate (93.4%) is close to the nominal level.

Example 7.8. *A simulation study for a binary variable and a continuous variable, with a quadratic relationship in the logistic regression model*

We go back to the problem that includes a binary Y-variable and a con-

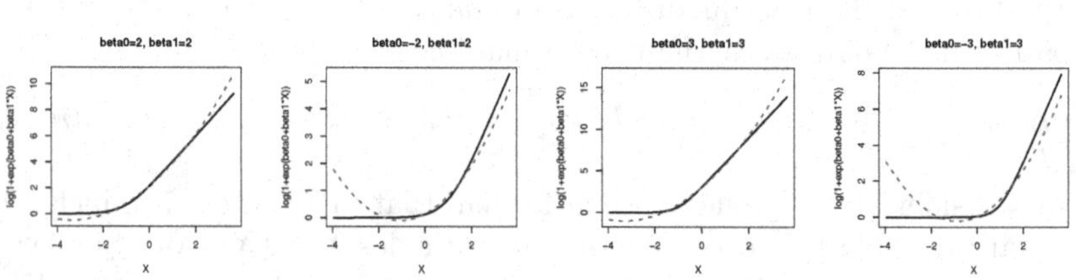

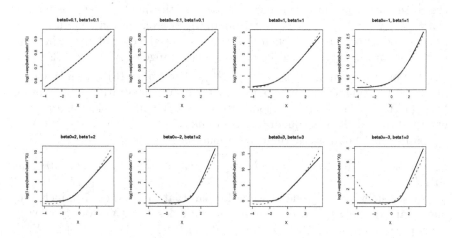

FIGURE 7.1
Example 7.6. Quadratic approximations of $log(1+exp(\beta_0+\beta_1 X))$ with selected values of β_0 and β_1. Solid line: $log(1+exp(\beta_0+\beta_1 X))$; dashed line: its quadratic approximation.

tinuous X-variable. Suppose that the data are distributed as

$$logit(f(Y=1)) = \beta_0 + \beta_1 X + \beta_2 X^2, \quad (7.19)$$
$$X \sim N(\mu, \sigma^2), \quad (7.20)$$

which are more complicated than the model specified in Eqs. (7.13) and (7.14). In Example 6.8, WinBUGS was used to impute missing X when Y is fully observed to illustrate the idea of imputing quadratic terms via JM.

Suppose now both X and Y have missing values; we can also apply FCS for imputation. For $f(Y|X)$, we can specify a logistic regression model with a quadratic term as in Eq. (7.19). For $f(X|Y)$ we can consider two specifications : (a) $f(X|Y=l) \sim N(\mu_l, \sigma^2)$ and (b) $f(X|Y=l) \sim N(\mu_l, \sigma_l^2)$, where $l = 0, 1$. In both specifications, X is assumed to be a mixture of normal distributions. In specification (a), the two normal distributions are assumed to have possibly different means yet equal variances. In specification (b), the two normal distributions are allowed to have both different means and variances. The model for X in the latter is more general than in the former.

As discussed in Section 6.5, we use WinBUGS to implement FCS. We attach some sample code here, ignoring the part for prior distributions and initial values. Note the trick of cut() is again used here to stop the feedback in fitting conditional models.

```
model
{
# complete-data model
for( i in 1 : M ) {
# specification of f(Y|X)
```

TABLE 7.4
Example 7.7. Simulation results

Method	RBIAS (%)	SD	SE	MSE	CI	COV (%)
			Mean of Y			
BD	0	0.011	0.011	0.000132	0.047	94.8
CC	−0.2	0.013	0.013	0.000159	0.051	95.3
JM	−0.2	0.012	0.013	0.000149	0.049	95.2
GLOM	−0.2	0.012	0.013	0.000151	0.049	94.8
FCS	0	0.012	0.013	0.000150	0.049	95.1
			Logistic regression coefficient of X_1			
BD	0	0.161	0.163	0.0259	0.634	94.6
CC	0.4	0.211	0.208	0.0444	0.815	95.1
JM	0.9	0.197	0.195	0.039	0.766	94.7
GLOM	−7.2	0.172	0.187	0.0413	0.738	91.8
FCS	−4.7	0.180	0.189	0.0372	0.745	93.4

```
y2_miss[i] ~ dbern(p[i])
logit(p[i])<-beta0+beta1*x_miss_star[i]
+beta2*x2_miss_star[i]
x2_miss_star[i] <- pow(x_miss_star[i],2)
x_miss_star[i] <- cut(x_miss[i])

# For FCS, specification (a) of f(X|Y)
#       indi[i] <- y2_miss_star[i]+1
#       x_miss[i] ~ dnorm(mu[indi[i]], tau)
#       y2_miss_star[i] <- cut(y2_miss[i])

# For FCS, specification (b) of f(X|Y)
#       indi[i] <- y2_miss_star[i]+1
#       x_miss[i] ~ dnorm(mu[indi[i]], tau[indi[i]])
#       y2_miss_star[i] <- cut(y2_miss[i])
}
}
```

We use a simulation study to assess the performance of all imputation methods based on $M = 50$. We randomly delete 20% of the missing cases in both X and Y so that missing values are MCAR. The complete-data sample size is 1000, and simulation consists of 1000 replicates. The estimand of interest is the coefficient of the quadratic term, β_2, in Eq. (7.19).

Table 7.5 shows the simulation results. As expected, JM performs well because the complete-data model is correctly specified. The results from both FCS methods show some biases because specifications of $f(X|Y)$ do not match with the true $f(X|Y)$ derived under Eqs. (7.19) and (7.20). Estimates from FCS (b) seem to be better than those from the FCS (a) with a smaller bias. One technical explanation for this difference is that having a quadratic term on X^2 in the logistic model of $f(Y|X)$ would lead to different variances for $f(X|Y = l)$ $(l = 0, 1)$ if we apply the quadratic Taylor approximation as in Example 7.6. This is verified by a numerical evaluation using a very large sample size $n = 1000000$ (e.g., $E(X|Y = 1) = 2.42$, $Var(X|Y = 1) = 0.91$, $E(X|Y = 0) = 1.60$, and $Var(X|Y = 0) = 0.76$). Overall, this example suggests that results from FCS are not immune to model misspecifications. In addition, besides trying to match the specification of conditional models with the analysis model (e.g., for $f(Y|X)$), it is desirable to use more general specifications for other conditional models (e.g., for $f(X|Y)$, FCS (b) is better than FCS (a)).

In Section 6.3 we stated that it is common to specify FCS conditional models for statistical convenience. Based on examples and discussion here, a more thoughtful strategy might be to specify conditional models using information from some plausible joint models. In certain cases, we might be able to posit a plausible joint model, $f(Y|\theta)$, for complete data. Under this joint model, the exact conditional models, $f(Y_j|Y_{-j}, \theta_j)$'s, might be complicated and do not have closed forms. However, we might try to approximate them using somewhat simpler regression models. If the approximation is reasonable, then results of the corresponding FCS imputation are expected to be close to those of JM under $f(Y|\theta)$. Example 7.6 shows the technique of approximating the logarithm of conditional distributions by quadratic functions. This technique will be illustrated again in more complex situations (e.g., Section 8.3.3). See also Raghunathan et al. (2018, Section 1.10.3) for similar arguments.

In spite of aforementioned theoretical issues, the general performance of FCS in real data is satisfactory, which is supported by a large body of literature (Van Buuren 2018). For example, Kropko et al. (2014) assessed the relative performance of JM and FCS when data consist of continuous, binary, ordinal, and nominal variables. The JM approach is based on the multivariate normal model. Implementations of such a strategy typically assume that categories of discrete variables are probabilistically constructed from continuous values. Their simulations suggested that FCS is more accurate than JM whenever the data include categorical variables. Other comparisons between the two methods can be found, for example, in Van Buuren (2007), Yu et al. (2007), and Lee and Carlin (2010).

7.7 A Practical Example

We provide a real-data example of using FCS here, although many examples can be found in the literature.

Example 7.9. *FCS imputation for multiple incomplete variables in the gestational age dataset*

We continue using the gestational age dataset (Examples 4.2, 5.5, 5.10, and 6.4) for illustration. Similar to Example 6.4, we now consider a subset with the full sample size ($n = 40274$) including 33 variables covering a wide range of

TABLE 7.5
Example 7.8. Simulation results

Method	RBIAS (%)	SD	$\overline{\text{SE}}$	MSE	CI	COV (%)
Coefficient of X^2 from regressing Y on X and X^2						
BD	0	0.067	0.068	0.00445	0.266	95.6
CC	−0.9	0.084	0.085	0.00701	0.334	95.3
JM	−0.7	0.084	0.085	0.00705	0.336	95.4
FCS (a)	−20.6	0.068	0.082	0.00574	0.322	97.0
FCS (b)	−8.6	0.078	0.082	0.00632	0.324	95.9

demographic and clinical characteristics for newborns and their parents. Table 7.6 shows the definition of the variables and their respective missingness rates.

Note that not all the variables have missing values, as fully observed ones can be used as predictors for imputation. In addition, for the simplicity of illustration, some of the variables in the original vital data file are summed into one category to be used in the imputation model. These summary variables include: the total number of medical risk factors (MEDRISK), the total number of obstetric procedures (OBSTETRIC), the total number of complications of labor or delivery (LABCOMP), the total number of abnormal conditions of the newborn (NEWBORN), and the total number of congenital anomalies (CONGNTL). Variables in this example have different types including continuous (e.g., gestational age, birth weight, age of mother), categorical (e.g., race of mother, marital status), count (e.g., number of prenatal visits), and mixed variables (e.g., number of cigarettes per day, number of drinks per week).

This dataset does not have any skip pattern. However, there exist some natural bounds for certain variables that must be accounted for. For example, the imputed values for DMAGE must be greater than 10 and less than 54 and those for DFAGE must be greater than 10.

We use SAS PROC MI FCS option to implement the imputation. We first tried unadjusted logistic regression imputation models for all categorical variables and encountered data separation. Therefore we implemented the pseudo observation method (Section 4.3.2.4) in the logistic regression imputation method, which stabilizes the estimation. In addition, some of the ordinal categorical variables have highly skewed distributions. Taking NLBNL (number of living births now living) as an example, its category ranges from 0 to 13. Most (95.9%) of the observations fall in categories 0-3, and there are only a few observations at the higher end of the distribution (e.g., only 0.2% of the cases are in categories greater than 8). Imputing it using a proportional odds model is apparently not optimal. On the other hand, imputing it using a multinomial logistic regression model would bring more data separation issues because of the sparse data in the top categories. Given its ordered nature, a predictive mean matching (PMM) imputation might better preserve the distribution and is used here.

After some exploratory analyses and trial-and-error steps, our procedure can be summarized as:

1. Apply pseudo observation logistic regression imputation method for nominal variables with not too many categories (e.g., ≤ 4).

2. Apply PMM for ordinal categorical variables with many categories and highly skewed distributions.

3. Apply PMM for continuous, mixed, and count variables. PMM can automatically preserve the range of imputed values to be consistent with that of observed data.

We include some sample SAS code here:

TABLE 7.6
Example 7.9. Variable Information

Variable	Definition	Missingness rate (%)
	Categorical variable	
RESTATUS	Resident status	0
PLDEL3	Place of delivery	0
REGNRES	Region of residence	0
CITRSPOP	Population size of city of residence	0
CNTRSPOP	Population size of county of residence	0
METROES	Metropolitan	0
CSEX	Sex	0
DPLURAL	Plurality	0
MRACE3	Race of Mother	0
DMAR	Marital status of mother	0
MEDUC6	Education of mother	1.37
MPLBIRR	Place of birth of mother	0.21
ADEQUACY	Adequacy of care	4.01
MPRE5	Mother prenatal care began	1.95
DFRACE4	Race of father	14.02
DELMETH5	Method of delivery	0.63
NLBNL	Number of live births, now living	0.20
NLBND	Number of live births, now dead	0.27
NOTERM	Number of other terminations	0.30
MEDRISK[1]	Total number of medical risks	0.83
NEWBORN[2]	Total number of newborn complications	0.81
LABCOMP[3]	Total number of labor complications	0.60
OBSTETRIC[4]	Total number of abnormal conditions	0.48
CONGNTL[5]	Total number of congenital anomalies	0.51
	Continuous variable	
DMAGE	Age of mother	0
DBIRWT	Birth weight of child (gram)	0.08
DFAGE	Age of father	13.25
DGESTAT	Gestational age	18.7
FMAPS	Apgar score	22.76
WTGAIN	Weight gain (lb)	18.60
	Count variable	
NPREVIS	Total number of prenatal visits	2.57
	Mixed variable	
CIGAR	Number of cigarettes/day	14.79
DRINK	Number of drinks/week	13.94

Note: 1. MEDRISK: medical risk variables include anemia, cardiac disease, acute or chronic lung disease, etc.; 2. NEWBORN: newborn complications include anemia, birth injury, fetal alcohol, etc.; 3. LABCOMP: labor complications include febrile, meconium, premature rupture of membrane, etc.; 4. OBSTETRIC: obstetric procedures include amniocentesis, electronic fetal monitor, induction of labor, etc.; 5. CONGNTL: congenital anomalies include anencephalus, spina bifida, hydrocephalus, microcephalus, etc.

```
proc mi data=sbirth_missing
out=sbirth_impute nimpute=20;
class restatus pldel3 regnres citrspop metrores
cntrspop mrace3 meduc6 dmar mplbirr adequacy mpre5
frace4 csex dplural delmeth5;
fcs nbiter=20 logistic (pldel3 / likelihood=augment);
fcs nbiter=20 logistic (meduc6 / link=glogit
likelihood=augment);
fcs nbiter=20 logistic (mplbirr / likelihood=augment);
fcs nbiter=20 logistic (adequacy / link=glogit
likelihood=augment);
fcs nbiter=20 logistic (mpre5 / link=glogit
likelihood=augment);
fcs nbiter=20 logistic (frace4 / link=glogit
likelihood=augment);
fcs nbiter=20 logistic (delmeth5 / link=glogit
likelihood=augment);
fcs nbiter=20 regpmm (nlbnl);
fcs nbiter=20 regpmm (nlbnd);
fcs nbiter=20 regpmm (noterm);
fcs nbiter=20 regpmm (medicalrisk);
fcs nbiter=20 regpmm (obstetric);
fcs nbiter=20 regpmm (congntl);
fcs nbiter=20 regpmm (newborn);
fcs nbiter=20 regpmm (labcomp);
fcs nbiter=20 regpmm (dbirwt);
fcs nbiter=20 regpmm (dfage);
fcs nbiter=20 regpmm (fmaps);
fcs nbiter=20 regpmm (wtgain);
fcs nbiter=20 regpmm (nprevis);
fcs nbiter=20 regpmm (cigar);
fcs nbiter=20 regpmm (drink);
fcs nbiter=20 regpmm (dgestat);

var restatus pldel3 regnres citrspop metrores cntrspop
mrace3 meduc6 dmar mplbirr adequacy mpre5 frace4 csex
dplural delmeth5 nlbnl nlbnd noterm medicalrisk obstetric
congntl newborn labcomp dbirwt dmage dfage fmaps wtgain
dgestat nprevis cigar drink;
run;
```

We provide a few remarks about the SAS code. For example, a pseudo observation logistic regression imputation is used for MEDUC6, a categorical variable with more than two categories, through the following code:

```
fcs nbiter=20 logistic (meduc6 / link=glogit
```

```
likelihood=augment);
```

A PMM imputation is used for NLBNL through the following code:

```
fcs nbiter=20 regpmm (nlbnl);
```

In addition, the "DETAILS" option can be used in these statements to display the regression coefficients in the regression model used in each imputation. The FCS imputation consists of 20 iterations and the imputation is conducted $M = 20$ times. For each missing variable, all the rest of the variables in the dataset are used as main-effect predictors.

To have a quick assessment and diagnostics of the quality of imputations, we compare the distribution of observed values with that of completed data. Tables 7.7 and 7.8 show the frequency of categorical variables from both the observed values and one randomly chosen completed dataset. Missing values are those labeled as "unknown" or "not stated". It is worth noting that for variables with a large number of categories and highly skewed distributions (Table 7.8), the distribution of completed values matches well with those of observed values. For these variables, most (if not all) of the imputed cases go to the categories with higher frequency. Table 7.9 shows the means and standard deviations of noncategorical variables for both observed and completed data. The results are mostly similar.

TABLE 7.7
Example 7.9. Categorical variables I: observed vs. completed data frequency

Variable	Category	Definition	Observed	Completed
PLDEL3	1	In hospital	39909	39911
	2	Not in a hospital	363	363
	Missing	Unknown/not stated	2	
MEDUC6	1	0-8 yrs	2397	2451
	2	9-11 yrs	6061	6168
	3	12 yrs	12324	12515
	4	13-15 yrs	8651	8751
	5	>= 16 yrs	10288	10389
	Missing	Not stated	553	
MPLBIRR	1	Native born	30712	30774
	2	Foreign born	9476	9500
	Missing	Unknown/not stated	86	
ADEQUACY	1	Adequate	29464	30578
	2	Intermediate	7179	7520
	3	Inadequate	2018	2176
	Missing	Unknown	1613	
MPRE5	1	1st trimester	33040	33654
	2	2nd trimester	5017	5136
	3	3rd trimester	1056	1092
	4	No prenatal care	374	392
	Missing	Unknown/not stated	787	
DFRACE4	1	White	28186	31379
	2	Other	2107	2321
	3	Black	4333	6574
	Missing	Unknown/not stated	5648	
DELMETH5	1	Vaginal	28974	29101
	2	Vaginal birth after previous C-section	589	591
	3	Primary C-section	6328	6383
	4	Repeated C-section	4130	4199
	Missing	Not stated	253	

Our model/code here can be best viewed as a starting point. It is likely that the imputation model can be further refined or improved. For example, to better account for the nonlinear relation between DGESTAT and DBIRWT,

TABLE 7.8
Example 7.9. Categorical variables II: observed vs. completed data frequency

Variable	Definition	Category	Observed	Completed
NLBNL	Number of live births, now living	0	16166	16195
		1	13303	13331
		2	6574	6586
		3	2585	2594
		4	894	894
		5	349	350
		6	155	156
		7	86	86
		8	31	31
		9	21	21
		10	14	14
		11	8	8
		12	5	6
		13	2	2
		Missing	81	
NLBND	Number of live births, now dead	0	39515	39624
		1	549	550
		2	73	73
		3	15	15
		4	7	7
		5	1	1
		6	1	1
		9	3	3
		Missing	110	
NOTERM	Number of other terminations	0	30572	30669
		1	6463	6481
		2	2062	2067
		3	682	683
		4	226	226
		5	76	76
		6	44	44
		7	12	12
		8	9	9
		9	3	4
		10	3	3
		Missing	121	
MEDRISK	Total number of medical risks	0	27767	28029
		1	9721	9783
		2	1975	1983
		3	399	402
		4	58	58
		5	15	15
		6	4	4
		Missing	335	
OBSTETRIC	Total number of abnormal conditions	0	2816	2843
		1	8062	8111
		2	17251	17322
		3	9686	9721
		4	2063	2073
		5	193	193
		6	11	11
		Missing	192	
CONGNTL	Total number of congenital anomalies	0	39276	39877
		1	341	348
		2	38	40
		3	6	6
		4	3	3
		Missing	610	
NEWBORN	Total number of newborn complications	0	36994	37300
		1	2545	2563
		2	341	343
		3	60	60
		4	8	8
		Missing	326	
LABCOMP	Total number of labor complications	0	27159	27312
		1	10118	10194
		2	2307	2317
		3	368	369
		4	68	68
		5	12	12
		6	2	2
		Missing	240	

we might replace DBIRWT with its square root in the model (Example 6.4), although the use of PMM has already accounted for some of the nonlinear relation (Example 5.5). For simplicity, we do not present further imputation diagnostics and refining steps. Some of these techniques will be discussed more systematically in Chapter 12.

7.8 Summary

FCS imputes multivariate data based on conditional modeling specifications for each incomplete variable. The idea is straightforward, and the implementation is usually simpler than JM in practice. In general, in spite of some theoretical concerns (e.g., incompatibility and model misspecification), performance of FCS in real data is surprisingly satisfactory. FCS has been readily applied to large-scale, complex datasets that are vital to many important scientific research studies (e.g., Schenker et al. 2006; He et al. 2010). Some of the relevant examples in the context of survey data will be discussed in Chapter 10. In our opinion, this will continue to be the case in the future as more research is conducted to advance the ideas and software packages for FCS. Practitioners are generally encouraged to use FCS whenever necessary, and yet be mindful about its theoretical limitations.

Topics covered in Chapters 4-7 focus on missing data problems in regular, cross-sectional settings. In subsequent chapters we will discuss imputation strategies targeted to data arising from more specialized study designs (e.g.,

TABLE 7.9
Example 7.9. Continuous/mixed/count variables: observed vs. completed data statistics

Variable	Observed Mean	Observed SD	Completed Mean	Completed SD
DBIRWT	3297.12	604.87	3296.37	606.51
DMAGE	27.35	6.2	27.35	6.2
DFAGE	30.49	6.85	29.95	6.96
DGESTAT	38.65	2.25	38.61	2.32
FMAPS	8.91	0.75	8.91	0.74
WTGAIN	30.89	13.8	30.87	13.9
NPREVIS	11.55	3.99	11.54	4.00
CIGAR	1.08	3.88	1.02	3.79
DRINK	0.04	0.71	0.04	0.70

survival data and longitudinal data). However there exist a few general principles for multiple imputation that can be summarized as follows:

1. Imputation should be conditioned on observed data to reduce the bias due to nonresponse, improve precision, and preserve the association between missing and observed variables.

2. Imputation model should be multivariate and inclusive to preserve the associations among variables.

3. A balance needs to be sought between model generality and model parsimony/practicality.

4. Imputed values should be draws from the predictive distribution rather than mean predictions to provide valid variance estimates for a wide range of estimands based on multiple imputation combining rules.

5. Multiple imputations should be proper to incorporate the uncertainty of the model parameter estimates.

8

Multiple Imputation in Survival Data Analysis

8.1 Introduction

In this chapter, we focus on censored data that are partially observed, and discuss the use of multiple imputation in survival analysis. The outcome of survival analysis consists of the occurrence of the event of interest and time to event (also often termed as "failure time"), which is subject to censoring. For example in many medical studies, a participant is followed until an event (e.g., death) occurs or he/she exits the study. The latter situation is termed as "right censoring", and the individual is still free of event (e.g., being alive) at the censoring time when he/she leaves the study. Besides right censoring, left or interval censoring can also occur. In the latter case, the event is only known to occur within a time interval. Interval censoring typically happens in screening studies for the occurrence of certain diseases, in which each participant is screened periodically (e.g. every 6 months). As will be discussed later, the notion of censoring is somewhat similar to missingness because the information regarding time to the event is not fully observed.

Survival analysis is a fundamental area of statistics, and there exists a large volume of literature (e.g., Kalbfleisch and Prentice 2002 and references therein). Extensive statistical methods have been developed for analyzing survival data and can be classified into nonparametric, semiparametric, and parametric methods. For nonparametric methods, for example, the Kaplan-Meier (KM) estimator (Kaplan and Meier 1958) can be used to estimate the marginal survival function, and a log-rank test can be used to compare two marginal survival functions. Cox's proportional hazard regression (Cox 1972) can be used to relate the hazard rate with risk factors, which is considered as a semiparametric method since it only requires specifying the relationship between the hazard rate and risk factors through an exponential function and leaves the baseline hazard function unspecified. Parametric survival models such as accelerated failure time (AFT) models can be used to predict failure time and identify the variables predictive of failure time by fully specifying the failure time distribution. All these established methods rely on the assumption of independent censoring, that is, censoring time is independent of failure time,

DOI: 10.1201/9780429156397-8

or dependent censoring, that is, the dependence between censoring time and failure time can be explained by some fully observed covariates.

Conceptually, censored data can also be considered as being similar to missing data (Heitjan and Rubin 1991; Heitjan 1993; 1994). Therefore, methods for handling missing data such as multiple imputation can be used to recover information for censored observations. Although most survival analysis methods do not explicitly frame censored observations as missing data, there is an increasing use of multiple imputation to either simplify the survival analysis approach or relax the assumption of censoring mechanisms behind some of the analyses. In addition, multiple imputation can be used as a convenient and effective strategy to handle missing covariates in survival analysis.

This chapter is organized as follows. Section 8.2 first links censoring mechanisms to missingness mechanisms through the concept of coarsened data, which provides a theoretical basis for using multiple imputation to handle censored data. It then discusses imputation strategies for censored event times. Section 8.3 discusses imputation strategies for missing covariates in survival analyses. Section 8.4 provides a summary.

8.2 Imputation for Censored Event Times

8.2.1 Theoretical Basis

Censoring makes survival data unique and complicates data analysis. In addition, censoring induces partially observed event-time data for subjects who have not experienced the event of interest before they are lost to follow-up or before the study ends. Analysis without accounting for censoring could produce biased results in estimating the survival function. Heitjan and Rubin (1991) and Heitjan (1993; 1994) referred to censored data as one type of coarsened data and established a general system of describing the random mechanism leading to coarsened data, which include missing data, censored data, grouped or heaped data as special cases. For example, a condition of "coarsened at random" is required for the likelihood-based inference to be valid ignoring the process leading to the coarsened data. Little and Rubin (2020, Section 6.4) provided a concise summary of the concepts. The general notion of coarsened data provides the theoretical basis for one to treat censored data in a similar way as missing data and then use multiple imputation to handle censored data in survival analysis.

We use right-censored data to illustrate the idea of coarsened data and the associated mechanisms. To illustrate the idea of right censoring, Fig. 8.1 shows the time-to-event for five hypothetical subjects in a follow-up study. The time-to-event for subjects 1,2,4, and 5 are 7,5,4.5, and 2, respectively. Subject 3 is alive until time (t) which is 8. After that, his/her status is unknown due to some reason (e.g., the study ended at $t = 8$ or subject 3 dropped out of the study at $t = 8$). So subject 3's censoring time is 8, and his/her event time has to be greater than 8 yet remains unknown or missing.

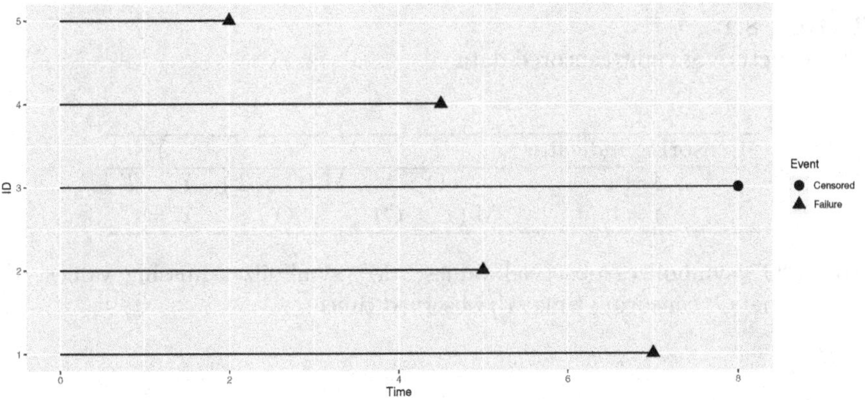

FIGURE 8.1
An illustrative plot of right-censored data.

To lay out some formal notations, let T denote the event time, C denote the censoring time, and the censoring indicator be $\delta = I(T <= C)$. Let $Y = min(T, C)$, which is the observed time from the data. We assume that there are n independent observations of Y_i's and δ_i's ($i = 1, \ldots n$). Because of the right censoring, we observe the event time when $\delta = 1$ and observe the censoring time when $\delta = 0$. That is, $Y_i = T_i$ if $\delta_i = 1$ and $Y_i = C_i$ otherwise. For the latter cases, we know their unobserved $T_i \in (C_i, \infty)$. However, importantly, the complete data would be T_i and C_i for $i = 1, \ldots, n$, that is, every subject would have both the event time and censoring time observed. Therefore for a subject whose T_i is observed, his/her C_i is unobserved and $C_i \in (T_i, \infty)$. Here δ_i's can also be viewed as the missingness indicator variable (or the more general term, coarseness indicator) in the context of right-censored data. Table 8.1 shows the general structure of the right-censored data. Fig. 8.2 plots the hypothetical, complete data including unobserved event time and censoring time for subjects in Fig. 8.1: the time-to-censoring for subjects 1,2,4,5 are 9, 6, 7.5, 3, respectively, and the time-to-event for subject 2 is 10.

We also let $X = (X_1, X_2, \ldots, X_p)$ denote the fully observed, time-independent auxiliary variables/covariates. In the context of censored data, Carpenter and Kenward (2013, Chap. 8) developed three notions of censoring mechanisms that are parallel to MCAR, MAR, and MNAR. In practical terms, they can be described as follows:

1. Survival data are said to be censored completely at random (CCAR) when the distribution of time to censoring is completely independent of the

TABLE 8.1
The structure of right-censored data

Censoring indicator	T	C	Y
$\delta = 1$	O	M ($C > T$)	$Y = T$
$\delta = 0$	M ($T > C$)	O	$Y = C$

Note: "O" symbolizes observed values; "M" symbolizes missing values. T: event time; C: censoring time; Y: observed time.

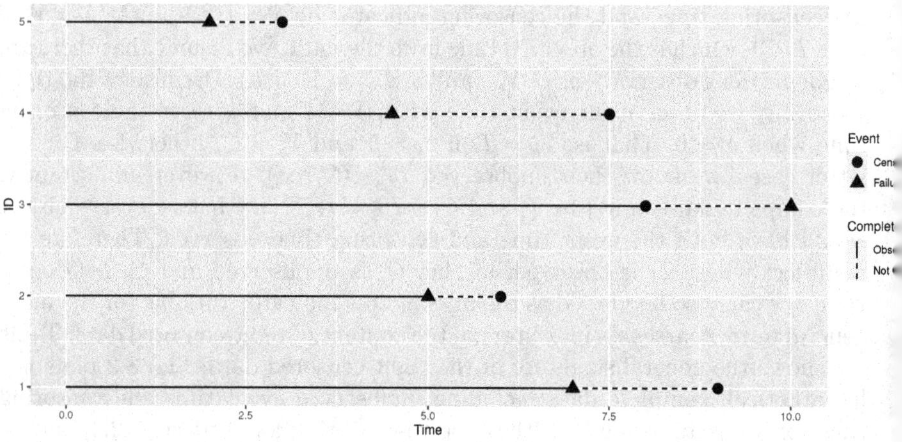

FIGURE 8.2
An illustrative plot of right-censored data with hypothetically complete information

survival process under investigation. This notion is similar to "independent censoring".

2. Survival data are said to be censored at random (CAR) if conditional on the auxiliary variables in the survival model, the censoring process is independent of the survival time. This notion is similar to "dependent censoring". Note that CCAR is a special case of CAR.

3. Survival data are said to be censored not at random (CNAR) if, even conditional on covariates, the censoring process is still dependent on the survival time. This notion is similar to "informative censoring".

Both CCAR and CAR are similar to "noninformative" or "ignorable" censoring, while CNAR is similar to "informative" censoring in the literature for survival analysis. Additional discussions of the terminology can be found, for example, in Jackson et al. (2014) and Atkinson (2019). Following the recommendations in Carpenter and Kenward (2013), we believe that clearly stating these assumptions is important when applying multiple imputation to censored data, similar to clearly stating the missingness mechanism assumptions when applying multiple imputation to conventional missing data. This chapter assumes CAR behind all the multiple imputation analyses discussed.

8.2.2 Parametric Imputation

Although both T and C are partially observed, most survival analyses focus on T, the event time, to infer about subjects' survival distributions. Taking the problem of right-censoring, the imputation task is to find a probable event time T^* longer than the observed censoring time C under a specified model. To sketch the idea, the observed-data likelihood function for a parametric survival model can be expressed as

$$\prod_{\delta_i=1} f(t_i|x_i,\theta) \prod_{\delta_i=0} \int_{c_i}^{\infty} f(t_i|x_i,\theta)dt, \tag{8.1}$$

where $f(t|\theta)$ is the density distribution function of the event time T governed by parameter θ, t_i and c_i are the observed event time or censoring time for subject i, x_i is the associated time-invariant covariate, and $i = 1, \ldots n$.

Once the model (e.g., Eq. (8.1)) for predicting event time is specified, the model can be estimated and then used to impute T^* for each of the censored observations. In principle, a wide variety of parametric survival analysis models can be used for imputation. A few examples are provided in this section.

Example 8.1. *Imputation of event times under an AFT model*

An AFT model (e.g., Kalbfleisch and Prentice 2002) uses a linear model to predict log event time, i.e. $log(T)$ using

$$log(T) = X\beta + \sigma W, \tag{8.2}$$

where β (a $(p+1) \times 1$ vector) parameterizes regression coefficients for X-covariates including the intercept, σ is a scale parameter, and W is the error term. Under AFT models, the role of covariates is to alter the rate at which an individual proceeds along the time axis (i.e., accelerate or decelerate the time to event). The expression in Model (8.2) is somewhat similar to that of a typical normal linear regression model. The AFT model is characterized by specifying that the logarithm of a failure time be linearly related with covariates. It sometimes can be more appealing and easier to interpret than Cox's proportional hazard regression model because of its rather direct physical interpretation (Wei 1992). Literature on estimation for AFT models can be found, for example, in Buckley and James (1991).

To illustrate the idea of imputation, we further assume that $W \sim N(0,1)$ so that the error follows a normal distribution, and thus Model (8.2) is an AFT model with log-normal error terms. Assuming noninformative prior distributions to σ and β, the DA algorithm can be sketched as follows:

1. Let $Y = log(T)$. Initialize y_i for $\delta_i = 0$ (e.g., set $y_i = log(c_i)$) so that T and Y become completed.

2. Draw β and σ^2 as follows. Let $\hat{\beta} = (X^tX)^{-1}X^tY$, where Y is now completed from Step 1.

$$f(\beta^*|Y,X,\sigma) \sim N(\hat{\beta},(X^tX)^{-1}\sigma^{*2}),$$

   and

$$f(\sigma^{*2}|Y,X,\beta) \sim SSE/\chi^2_{n-p-1},$$

   where $SSE = \sum_{i=1}^{n}(y_i - x_i\beta^*)^2$

3. Impute y_i^* for $\delta_i = 0$ as $y_i^* \sim N(x_i\beta^*, \sigma^{*2})$ subject to the constraint that $y_i^* \geq log(c_i)$. This could be done by drawing y_i^* from a truncated normal distribution: $y_i^* \sim \sigma^*\Phi^{-1}(Uniform(0,1)(1-\Phi(\frac{log(c_i)-x_i\beta^*}{\sigma^*}))+\Phi(\frac{log(c_i)-x_i\beta^*}{\sigma^*}))+x_i\beta^*$

4. Iterate between Steps 2 and 3 until the convergence of parameter draws for β and σ. Take the last draws of y_i^* (hence $t_i^* = exp(y_i^*)$) as imputed event time for the censored observations.

5. Repeat Steps 1 - 4 independently M times to obtain multiple imputations.

The above DA algorithm appears to be similar to that for imputing the missing outcome variable under a linear regression model (Section 4.2). Yet there are a few important distinctions. One is that if Y is missing (not censored) under the linear regression model, then the imputation algorithm does not have to be iterative. In that setting, parameters β and σ^2 can be drawn directly from the model using cases with only observed Y-values. This is because under MAR, all the information about these parameters are contained in the complete cases only. However, if Y includes censoring time under the

AFT model, then the censored observations (i.e., for which $\delta_i = 0$) also contain information about β and σ^2. Therefore drawing β and σ^2 from cases with only observed event time (i.e., for which $\delta_i = 1$) is insufficient, and an iterative DA algorithm is necessary. In addition, here y_i^* is drawn from a truncated normal distribution due to the right censoring. For simplicity, here we also assume an unlimited study period so that the event can happen to every participant, sooner or later. More realistically, an upper-limit for T, T_{max}, can be set so that the imputed event time for y_i would fall in between $log(c_i)$ and $log(T_{max})$. In a more general sense, such an imputation algorithm can also be used to handle the missing data problems where the missing values have to fall in certain ranges, for which a post-hoc solution was provided in Section 5.4.

Example 8.2. *A simulation study for the imputation of event times under an AFT Model*

We conduct a simple simulation study to assess the performance of the imputation method in Example 8.1. Specifically, we generate event time from $log(T) = -1 + X + \epsilon_T$, where $X \sim uniform(0, 1)$ and $\epsilon_T \sim N(0, 1/9)$. We consider two types of censoring mechanisms, CCAR and CAR. For CCAR, we generate $C \sim exp(1)$, which is completely independent of T. For CAR (but not CCAR), we generate $log(C) = -1 + X + \epsilon_C$, where ϵ_C is independent of ϵ_T. That is, conditional on X, T and C are independent. However in CAR, T and C are marginally correlated with a correlation coefficient of around 0.4.

The simulation consists of 1000 replicates, in which the complete sample size (before censoring is applied) is fixed at 1000. Applying both censoring mechanisms would yield a censoring rate around 50%. The estimand of interest is the probability of the survival time greater than the population median, that is, $Pr(T >= T_{50}) = 0.5$, where T_{50} is estimated as the average of the median of the survival time (before censoring is applied) across 1000 replicates. Specifically, $T_{50} = 0.6069$ in this example. We consider three methods, applying the KM estimator to the data before censoring (FO), to the observed data (PO), and to the multiply imputed data using the algorithm in Example 8.1 (PMI) based on 50 imputations with 200 iterations of each imputation. The standard error estimates of the survival probability are derived based on Greenwood's formula (Greenwood 1926).

Table 8.2 shows the simulation results. Under CCAR, all methods yield estimates with little bias as expected. Interestingly, PMI is more efficient than PO as indicated by MSE. The length of the CI is narrower in the former, yet the coverage is even larger than the nominal level. The advantage of the PMI over PO here shows one phenomenon of imputation uncongeniality (Meng 1994), that is, superefficiency (Rubin 1996), which will be discussed in detail in Section 14.2. Briefly speaking, this is because the PMI estimate is based on imputations generated from a correctly specified model using information from covariates, yet the PO estimate is based on a nonparametric estimator (i.e., KM estimator) which does not incorporate covariates into estimation.

Under CAR, the PO estimate is positively biased (RBIAS $\approx$ 15%) with a much larger MSE and a low coverage rate. This phenomenon is similar to the classic missing data problem in regression models: if the missingness is random but not completely at random, then the marginal mean estimate of the incomplete variable from the complete-case analysis is likely to be biased (Example 4.1). In this case, since the censoring time is positively correlated with X, the censoring event is more likely to happen for longer survival time marginally. This would result in a positively biased median survival estimate based on the KM estimator, which corresponds to the survival probability estimate greater than 0.5. On the contrary, the PMI estimate corrects the bias by incorporating X into the imputation model to induce CCAR conditional on X.

An early application of multiply imputing the event time based on an AFT model can be found in Taylor et al. (1990). Interestingly, imputing censored regression data using Model (8.2) was used as an early example of illustrating the DA algorithm (Wei and Tanner 1991), which is a general imputation algorithm for missing data problems. In more complicated scenarios, nonnormal error distributions for W (e.g., log-Weibull and log-logistic) can be considered in Model (8.2). In these cases, the conditional distributions of β and σ might not have closed forms, and more advanced Bayesian sampling techniques are needed to run the DA algorithm (e.g., Cho and Schenker 1999). Some of the models can be implemented using WinBUGS, as will be shown next.

Example 8.3. *Imputation of event times under a Weibull proportional hazards model*

The Weibull proportional hazards model is one of the most popular parametric survival models due to its flexible hazard function. It is also an AFT model with extreme-value error terms. Under a Weibull model, the density function for event time T for subject i can be expressed as

$$f(t_i|x_i) = re^{x_i\beta}t_i^{r-1}exp(-e^{x_i\beta t_i^r}), \tag{8.3}$$

TABLE 8.2
Example 8.2. Simulation results

			CCAR			
Method	RBIAS (%)	SD	$\overline{\text{SE}}$	MSE	CI	COV (%)
FO	0	0.016	0.016	0.000249	0.062	95.2
PO	0.1	0.019	0.020	0.000367	0.079	96.8
PMI	0.1	0.016	0.018	0.000259	0.072	97.3
			CAR			
FO	0	0.016	0.016	0.000249	0.062	95.2
PO	15.1	0.019	0.018	0.00605	0.072	17.0
PMI	0.1	0.016	0.018	0.000259	0.072	97.4

where t_i is the event time of an individual with covariate vector x_i and β is a vector of unknown regression coefficients, and r is the shape parameter of the survival function. In survival data, the hazard is the probability that a subject has an event at time t, conditional on the subject having survived to that time. Algebraically that is, $h(t) = \frac{f(t)}{S(t)}$, where $S(t) = Pr(T \geq t) = 1 - F(t) = 1 - \int_0^t f(u)du$ is the survival function. In a Weibull model, the baseline hazard function, $h_0(t)$, has the form:

$$h_0(t_i) = rt_i^{r-1}. \tag{8.4}$$

Setting $\mu_i = e^{x_i\beta}$ gives the parameterization $T_i \sim Weibull(r, \mu_i)$. For censored observations, the survival distribution is a truncated Weibull distribution, with a lower bound corresponding to the censoring time.

We consider using a Weibull proportional hazards model to impute event times. The density function in Eq. (8.3) is complicated, so devising the DA algorithm is not simple. In this example we briefly introduce the use of WinBUGS to generate imputations. The WinBUGS example book (Volume 1, Example 16, http://www.mrc-bsu.cam.ac.uk/wp-content/uploads/WinBUGS_Vol1.pdf) includes an example of fitting a Bayesian Weibull regression model titled under "Mice: Weibull regression." Dellaportas and Smith(1993) analyzed this dataset measuring photocarcinogenicity in four groups, each containing 20 mice, whose survival time had been recorded, along with whether they died or were censored at that time. A portion of the data, giving survival times in weeks, is shown in Table 8.3. This dataset contains 15 censored time points.

TABLE 8.3
Example 8.3. A schematic portion of data

Mouse	Irradiated control	Vehicle control	Test substance	Positive control
1	12	32	22	27
...	...	...	...	...
18	*40	30	24	12
19	31	37	37	17
20	36	27	29	26

Note: * indicates a censoring time.

Sample code is included in the following. Details on how to set up a Bayesian estimation for the Weibull regression model can be found in the WinBUGS example book. Note that the event time data matrix t[i,j] would set censoring time as missing (NA) for the purpose of imputation, while the censoring time data matrix t.cen[i,j] would set event time as 0. The I(a,b) construct in WinBUGS can be used to limit the stochastic quantity lying in the interval (a, b), which is an effective tool for handling censored or bounded data. Although the original program is aimed to estimate the regression coefficient and median survival time, we can request the program to also output the node t, which will impute survival times for each censored observation. To distinguish the imputations from the observed data in terms of the notation, we assign the former a new name "*t_impute*": *t_impute*$[i, j]$ denotes imputed event time for the jth mouse in the ith group.

```
model
{
for(i in 1 : M) {
for(j in 1 : N) {
t[i, j] ~ dweib(r, mu[i])I(t.cen[i, j],)
t_impute[i,j] <- t[i,j]
}
mu[i] <- exp(beta[i])
beta[i] ~ dnorm(0.0, 0.001)
median[i] <- pow(log(2) * exp(-beta[i]), 1/r)
}
r ~ dexp(0.001)
veh.control <- beta[2] - beta[1]
test.sub <- beta[3] - beta[1]
pos.control <- beta[4] - beta[1]
}
```

After we run a data augmentation chain for 10000 iterations with a burn-in of 5000 iterations, we take draws for every 100 iterations to create 50 sets of imputations. Table 8.4 shows some summary statistics for imputed event times, which are larger than the censoring times as expected.

8.2.3 Semiparametric Imputation

If the parametric modeling assumption of the event time distribution is questionable, then a widely used alternative is the Cox proportional hazards (Cox) model:

$$h(t|X) = h_0(t)exp(X\beta), \tag{8.5}$$

where $h(t)$ is the hazard function at $T = t$ and $h_0(t)$ is the corresponding baseline hazard function. In the Cox model, the relationship between time-independent covariates X and hazard function is fully specified but the baseline hazard function is unspecified. Hence, the Cox model is considered as a semiparametric model and expected to be more robust to model misspecifications than a parametric model.

Imputing event times for censored observations using the Cox model is possible but the baseline hazard function needs to be specified. An example can be found in Faucett et al. (2002), where they considered a joint modeling approach to AIDS clinical trial data. They predicted the trajectory of CD4 counts, which can be used as time-dependent predictors for the event time, and then imputed the censored event times based on the Cox model. They considered a step function on a predefined set of time intervals as the baseline hazard function. The procedure is rather complicated, and details are not shown.

Instead of imputing an event time for a censored observation based on an explicit distribution, an alternative strategy is to replace it from the collection of observed event times (i.e., t_i's for which $\delta_i = 1$). This follows the idea of the predictive mean matching (PMM) imputation (Section 5.5). Recall that in the typical settings of PMM or hot-deck imputation, adjustment cells are created to select donors that are similar to the incomplete cases with respect to either auxiliary variables or some distance measures. In the context of survival data, the adjustment cell is termed as "risk set" (Taylor et al. 2002; Hsu et al. 2006; 2007; 2015; Hsu and Taylor 2009). In the absence of auxiliary variables and assuming CCAR, the risk set for a censored subject i (with censoring time c_i) includes all individuals still at risk, i.e. who survive longer than time c_i (Taylor et al. 2002). In the presence of auxiliary variables, the risk set can be defined using the information from these covariates.

More specifically, Hsu et al. (2006) proposed imputation procedures, termed as a nearest-neighbor multiple imputation (NNMI), which can be summarized as follows:

1. Fit two working Cox models, one for event time and the other for censoring

TABLE 8.4
Example 8.3. Summary statistics of multiple imputations of censored event times

Censored event time	Censoring time	Mean	SD	25%-tile	50%-tile	75%-tile
t_impute[1,18]	40	43.6	3.0	41.1	42.9	45.2
t_impute[2,6]	40	49.6	8.9	42.5	46.4	53.5
t_impute[2,7]	40	48.7	7.7	43.0	46.4	52.1
t_impute[2,11]	40	48.7	8.0	42.7	45.7	52.3
t_impute[2,13]	40	48.9	7.3	43.9	46.9	54.0
t_impute[2,14]	40	48.2	7.1	43.1	45.6	49.7
t_impute[2,15]	40	49.0	7.4	42.6	46.5	54.0
t_impute[2,16]	40	50.6	8.7	44.0	48.9	54.5
t_impute[3,3]	10	29.5	10.9	22.6	27.8	37.6
t_impute[3,13]	24	33.4	7.4	28.4	31.4	36.0
t_impute[3,15]	40	44.4	3.4	41.5	43.3	46.6
t_impute[3,16]	40	44.4	3.6	41.9	43.4	46.6
t_impute[4,8]	20	27.7	5.2	23.7	27.3	31.2
t_impute[4,13]	29	33.4	3.8	30.6	33.0	34.6
t_impute[4,14]	10	20.6	6.9	15.1	19.8	25.6

time. For the latter model, the observed event time is considered to be right censored for the censoring time (Section 8.2.1). The idea of fitting two working models follows the dual modeling strategy in which both the outcome and propensity models are fitted (Section 5.6.2). Here in the context of survival analysis, note that the censoring indicator $\delta = I(T >= C)$ is analogous to the missingness indicator R in the conventional missing data problems. Thus, fitting a model of δ on the auxiliary variables is executed by fitting a survival analysis model for the time to censoring.

2. Create two risk scores, $\hat{RS}_f = X\hat{\beta}_f$ and $\hat{RS}_c = X\hat{\beta}_c$, where $\hat{\beta}_f$ and $\hat{\beta}_c$ are the coefficient estimates from the two working Cox models of the event time and censoring time, respectively. Standardize and center them across the individuals, and denote them as $\hat{RS}^*_f$ and $\hat{RS}^*_c$, respectively.

3. Define a distance measure between subjects i and j as

$$d(i,j) = \sqrt{w_f\{\hat{RS}^*_f(i) - \hat{RS}^*_f(j)\}^2 + w_c\{\hat{RS}^*_c(i) - \hat{RS}^*_c(j)\}^2}, \qquad (8.6)$$

where w_f and w_c are nonnegative weights that sum to 1. For each censored subject i, this distance is used to define a set of nearest neighbors. This neighborhood, $R(i^+, NN)$, consists of NN subjects who survive longer than the censoring time of subject i and has a small distance from the censored subject i. Note that this distance is based on predicted means from two working models. Similar to the number of donors in PMM, the number of nearest neighbors concerns the trade-off between the bias and variance of estimates.

4. Once the risk set, $R(i^+, NN)$, is formed, the donors can be randomly selected with equal probability from the risk set, as is done typically in PMM.

5. Although Step 4 appears straightforward, it often happens that the risk set can include a sizable proportion of the censored cases. Some special adjustments are implemented to prevent selecting censored cases from risk sets to replace the censored subject i, which is not desirable. On the other hand, simply drawing from only observed event times in the risk set with equal probability would bias the imputation inference. Specifically, the adjustment draws an event time from a KM estimator of the distribution of failure times based on the imputing risk set, using both censored and event time data. Thus, the procedure imputes only the observed event times unless the longest observed time in the risk set is censored, in which case some imputed event times may include this censoring time. Suppose, among the donors, the observed event times are t^*_j, $j = 1, \ldots J$; then by the KM estimator, the failure probability at these time points are $Pr(T <= t^*_j) = \hat{F}(t^*_j) = 1 - \hat{S}(t^*_j)$. Then we select the donor among t^*_j's using the probability, $F(t^*_1)$, $F(t^*_2) - F(t^*_1)$, ..., $1 - F(t^*_{J-1})$, if the longest time point

in the risk set is an event time. Otherwise, we sample the censoring time point (the last one) using probability $1 - F(t^*_{J-1})$.

6. In order for the multiple imputation to be proper, the above steps are executed using nonparametric bootstrap samples of the data repeatedly M times.

In the absence of auxiliary variables or covariates (i.e. the risk set for a censored observation consists of all of the subjects still at risk), Taylor et al. (2002) provided a connection between standard survival analysis methods and multiple imputation and showed that the above imputation procedures lead to asymptotically the same results from applying the KM estimator to censored data. The fact that the KM estimator can be reproduced asymptotically using multiple imputation of future event times provides a basis for the use of imputation to handle unobserved event times due to censoring in survival analysis. With auxiliary variables, Hsu et al. (2006) and Hsu and Taylor (2009) showed that using the above semiparametric imputation method reduces the bias caused by dependent censoring (i.e., CAR) and increases the efficiency of estimates estimating or comparing the marginal survival functions.

Example 8.4. *A simulation study for the semiparametric imputation of event times*

We present a simulation study demonstrating the performance of the semiparametric imputation procedure. Following Hsu et al. (2006), we consider five X covariates, $X_1, \ldots X_5$, independently generated from $Uniform(0,1)$. The event time distribution is generated by setting the hazard function as $h(t) = t^4 exp(-2X_1 + \frac{1}{2}X_2 - 2X_3 + 2X_4 + 2X_5)$. The censoring time distribution is generated by setting $h_c(t) = t^3 exp(-3X_1 + \frac{1}{2}X_2 - 2X_3 + \frac{3}{2}X_4 + 2X_5)$. Since both the event and censoring time distributions depend on X_1 through X_5, this creates a dependent censoring mechanism, or CAR but not CCAR. The censoring rate is around 25%. The simulation is based on sample size 200 with 500 replicates.

To set up the two working models (i.e., Cox model) for the imputation, we consider four scenarios:

I. Both working models are correctly specified and include all five covariates.

II. The working model for event time is correctly specified, and the working model for censoring time is misspecified by including only X_1 to X_3.

III. The working model for censoring time is correctly specified, and the working model for event time is misspecified by including only X_1 to X_3.

IV. Both working models are misspecified, and they include only X_1-X_3.

The size of risk set NN is fixed at 10. NNMI is conducted for $M = 10$ times. We fix w_f at 0.8 and w_c at 0.2. Similar to Example 8.2, the estimand in the evaluation is the probability of the survival time greater than the population median, that is, $Pr(T >= T_{50}) = 0.5$, where T_{50} is estimated as the average of the median of the survival time (before censoring is applied) across 500 replicates.

Table 8.5 (see Table III from Hsu et al. 2006) shows the simulation results. Similar to Example 8.2, when the censoring mechanism is related to fully observed covariates, directly using the KM estimates (i.e., the PO method) would result in biases and low coverage rates in all four scenarios. NNMI can effectively correct the bias. This is evident when the working model for the failure time includes all five covariates in Scenarios I and II, in which the biases are minimal, and the coverage rates are around the nominal level. When the working model for the failure time is misspecified and that for the censoring time is correctly specified in Scenario III, the bias increases yet is still acceptable. Again, this pattern demonstrates the advantage of the dual modeling strategy, which is related to the notion of "double robustness." That is, by correctly specifying the working model for the censoring time if the working model for the event time is misspecified, some protection of the bias is expected. Interestingly, in Scenario IV where both working models are misspecified, the bias and coverage rates from NNMI are still acceptable.

TABLE 8.5
Example 8.4. Simulation results

Method	RBIAS (%)	SD	$\overline{SE}$	COV (%)
		Scenario I		
FO	0	0.037	0.035	94.4
PO	12.4	0.040	0.040	62.0
NNMI	0.4	0.041	0.040	94.8
		Scenario II		
FO	0	0.036	0.035	94.0
PO	13.0	0.040	0.040	61.2
NNMI	1.4	0.042	0.040	93.2
		Scenario III		
FO	0	0.036	0.035	93.6
PO	13.8	0.040	0.039	58.4
NNMI	4.2	0.042	0.041	91.0
		Scenario IV		
FO	0	0.037	0.035	93.2
PO	13.0	0.041	0.040	63.0
NNMI	3.2	0.042	0.041	90.8

This semiparametric imputation strategy includes several advantages. First, instead of directly imputing censored event times based on parametric models, the imputations are drawn from nonparametric estimates of the survival distribution using donors in the risk set. Even if the models for constructing the risk scores might be somewhat misspecified, the donors contained in the risk set might still be similar to the censored subject with respect to auxiliary variables. For example, Hsu et al. (2015) demonstrated this by using the risk scores estimated from Cox models to impute event times for censored observations when the underlying models for event and censoring times are AFT models: NNMI still yields satisfactory results despite the apparent model misspecification. Second, the construction of risk sets follows the dual modeling strategy, involving using information from both the event time and censoring time models. The imputed values exhibit some protection against biases caused by misspecification of either model. Third, the setup of w_f and w_c allows the practitioners to conduct sensitivity analysis, varying the relative weight of the event time model to the censoring time model as well as assessing its impact on the imputation estimates.

Example 8.5. *Multiple imputation analysis for a prostate cancer dataset*

We demonstrate the semiparametric imputation approach using a prostate cancer dataset, which consists of 503 patients with localized prostate cancer treated with external-beam radiation therapy at the University of Michigan and affiliated institutions between July 1987 and February 2000. This dataset had been used to develop a weighted KM estimation approach to adjust for dependent censoring using linear combinations of prognostic variables where the linear combination is categorized to define risk groups, and the final KM estimate is the weighted average of the KM estimates from all of the risk groups (Hsu and Taylor 2010).

There are several variables collected at baseline, including age, Gleason score, PSA (prostate specific antigen) score, tumor (T) stage, and total radiation dose that are established prognostic variables of prostate cancer. In addition, age and total radiation dose are expected to be predictive of the patient's survival or censoring time. Here, we treat these five variables as the auxiliary variables for estimating the distribution of recurrence/prostate cancer-free survival. To assess the proportional hazards (PH) assumption, time-dependent variables consisting of an interaction between the auxiliary variables and log(time) are included. Violations of the PH assumption are detected for age and Gleason score with a p-value of 0.04 and 0.02, respectively. When the PH assumption is violated, one alternative is to use an AFT model. However in NNMI, the two working models are only fitted to derive the two risk scores. In addition, based on the relationship between Cox and AFT models: the relative importance of the covariates derived from the Cox model remains unchanged approximately when the true model is an AFT model (Struthers and Kalbfleisch 1986; Hutton and Solomon 1997). Therefore we simply use two working Cox models to derive the two risk scores.

Table 8.6 shows the results from fitting the two working Cox models: Gleason score, T stage, and total radiation dose are significantly associated with event time, and all of the five auxiliary variables are significantly associated with censoring time. The risk scores derived from the two working Cox models are used to calculate the distance between subjects and then to form the imputing risk set for each censored observation. As expected, the two derived risk scores are highly correlated with a Spearman correlation coefficient of −0.77.

TABLE 8.6
Example 8.5. Prostate cancer data: estimation of two working Cox models

	Event time model			Censoring time model		
Predictor	EST	SE	p-value	EST	SE	p-value
Age	−0.024	0.017	0.15	0.031	0.008	<0.01
Log(PSA)	0.173	0.127	0.17	−0.115	0.054	0.03
Gleason	0.405	0.104	<0.001	0.092	0.049	0.06
T stage	1.355	0.218	<0.001	−0.679	0.085	<0.001
Total dose	−0.111	0.030	<0.001	0.176	0.014	<0.001

Table 8.7 shows the estimates of the recurrence-free probability from various methods. As before, the partially observed (PO) analysis is based on the KM estimation. PMI is a parametric imputation, where a Weibull model is fitted to the observed data to impute event time for each censored observation (Example 8.3). Both PMI and NNMI are conducted $M = 10$ times. They produce slightly higher estimated survival at both 5 and 10 years and slightly lower associated estimated standard errors than PO at 5 years. Figure 8.3 displays the estimated survival curves for all of the aforementioned methods. PMI and NNMI consistently produce slightly higher estimated survival compared with PO until the very late stage of the follow-up. The plateau of the survival probability from NNMI in the late stage of the follow-up is similar to that of PO because the former method largely uses observed event times as the imputation. In contrast, the survival probability of PMI does not exhibit such a pattern because the imputations are random numbers generated from the assumed Weibull distribution. When the CCAR assumption is violated (likely in this case), both imputation methods have the potential to reduce bias caused by dependent censoring (i.e., CAR) compared with PO.

8.2.4 Merits

We have introduced several strategies of using multiple imputation for censored survival data. There is a growing trend of applying this strategy in the literature (e.g., Carpenter and Kenward 2013, Chapter 8). One might wonder about the advantage of doing this, given the fact that many standard, nonimputation methods have been developed to account for censoring (e.g., Kalbfleisch and Prentice 2002 and references therein). Based on our limited knowledge, there might exist several advantages.

TABLE 8.7
Example 8.5. Estimation of recurrence-free probability at 5 years and 10 years.

	$t = 5$ years		$t = 10$ years	
Method	$\hat{S}(t)$	SE	$\hat{S}(t)$	SE
PO	0.852	0.018	0.742	0.029
PMI	0.864	0.016	0.769	0.024
NNMI	0.863	0.017	0.766	0.030

Note: $S(t) = Pr(T >= t)$, the survival probability.

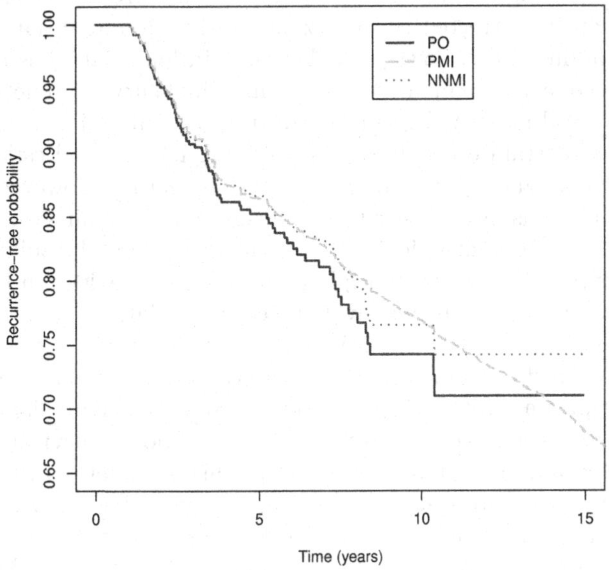

FIGURE 8.3
Example 8.5. Prostate cancer study: recurrence-free survival curves derived from various methods. PO (solid line): KM estimates are derived from the observed censored data; PMI (dashed line): a Weibull model is fitted to impute event times. NNMI (dotted line): two Cox models are fitted to define imputing risk sets and imputation.

First, through the coarsened data system established by Heitjan and Rubin (1991) and Heitjan (1993; 1994), we can formally recognize that censored data is a special type of coarsened data, which is a generalization of missing data. Therefore many notions for the missing data methodology such as the missingness mechanism also apply to censored data. For example, CAR (including CCAR as the special case) vs. CNAR can be connected to the notion of noninformative censoring vs. informative censoring.

Second, some direct methods, such as the KM estimator, can incur bias for estimating the marginal survival function if the censoring mechanism is not independent of the distribution of the event time given fully observed covariates (i.e, CAR but not CCAR). By imputing event times for the censored observations, we can use the information embedded in the auxiliary variables to correct such bias using multiply imputed data. The assumption is that the auxiliary variables are predictive of the censoring mechanism. This is similar to the scenario with a missing outcome variable Y and its missingness being MAR (not MCAR) through fully observed X-covariates, applying a complete-case analysis for estimating the marginal mean of Y would produce biased results. Yet the bias can be corrected by analyzing multiply imputed data based on the model including the X-covariates. The imputation strategy is also useful if the analysis concerns comparing the two marginal survival functions using the commonly used log-rank test or Wilcoxon test (Hsu and Taylor 2009).

Note that in certain cases, biases of the marginal survival estimates can be corrected by weighting (e.g., weighted KM estimator). However, multiple imputation analysis is often easier to implement and disseminate in practice. Overall, because of the connection between censored data and traditional missing data, we expect that more multiple imputation strategies can be smartly applied to solve certain problems involving censored data.

With these advantages, if the estimand of interest lies in the relationship between the survival outcome and fully observed auxiliary variables, we surmise that multiple imputation analysis using the same set of auxiliary variables would provide little gain compared with the conventional survival regression strategies if the same model is used in both. This is similar to the fact that imputing the missing outcome using the same set of auxiliary variables would provide essentially the same results as directly regressing the incomplete outcome on the auxiliary variables under the same model, assuming MAR.

8.3 Survival Analysis with Missing Covariates

8.3.1 Overview

Missing covariate problems frequently occur in survival analysis. Following the theme of this book, it is desirable to apply multiple imputation to handle missing covariate problems in general including for survival analysis. In Chapters 4 and 5 we provided an overview of using multiple imputation to handle missing covariates in the GLM-type regression models. The basic principle also applies here. Most importantly, for both survival analysis and conventional regression analysis, it is essential to include the outcome variable in the imputation model since imputation is drawn from the predictive distribution of the missing covariates conditional on the observed data, which includes the outcome variable (Little 1992).

However, a general challenge of imputing missing covariates in survival analysis is that popular survival models often have a semiparametric or nonlinear structure (survival function can be a nonlinear function of time even if parametric survival models are used). Such semiparametric or nonlinear structure often makes it difficult to obtain the exact posterior predictive distribution of missing values for imputation. Sometimes an approximation strategy might be needed.

In addition, the outcome variable in survival analysis is subject to censoring (i.e. partially observed) unlike that of conventional regression analysis. This fact adds another level of complexity to the posterior predictive distribution of missing covariate values. The occurrence of the censored outcome can even make the missingness mechanism of covariates more sophisticated. As we have shown before (e.g., Section 4.5), complete-case analysis (CC) in regression analysis with missing covariates can produce biased regression coefficient estimates under MAR. However, this is not always the case in survival analysis. For survival data, MAR can be further classified into two scenarios: failure-ignorable MAR (i.e. missingness does not depend on failure/event time) and censoring-ignorable MAR (i.e. missingness does not depend on censoring time but may depend on failure time) (Rathouz, 2007). When missingness is censoring-ignorable MAR, CC may produce biased regression coefficient estimates. However, when missingness is failure-ignorable MAR, CC can still produce valid regression coefficient estimates. Of course this subtle distinction might not affect the performance of the multiple imputation, which is expected to work well under general MAR settings with a correctly specified imputation model. However, this indicates that handling survival analysis with missing covariates is not as straightforward as one would expect.

In this section, we discuss a few strategies for imputing missing covariates in survival analysis. First, we assume that censoring is at random (CAR) or noninformative censoring conditional on observed covariates (i.e. dependent censoring). Second, we assume that the missing covariates are MAR conditional on observed survival outcome and other fully observed covariates. Using the conditional independence notation, (Dawid 2006; see also Section 1.4), these assumptions can be summarized by: $T \perp C|X_{obs}$, where X denotes the observed component of covariates, and $R \perp X_{mis}|Y, \delta, X_{obs}$, where R is the response indicator of X, $Y = min(T, C)$ and δ is the censoring indicator. The multiple imputation strategies largely fall into three categories: (a) JM; (b) FCS; (c) semiparametric methods.

8.3.2 Joint Modeling

In principle, once the survival regression analysis model is specified, we only have to specify the covariate model to impute the missing values using JM. This follows the framework listed in Section 6.6. The survival regression model can be parametric (e.g., AFT models) or semiparametric (e.g., Cox models). The nonlinear or semiparametric structure of survival analysis models can make the posterior predictive distribution of the missing covariate values rather complicated and often without closed forms. Applying JM would have to involve advanced Bayesian sampling algorithms. As illustrated before, it might be more effective by using WinBUGS or related packages to implement such DA algorithms. For example, Hemming and Hutton (2012) modeled missing covariates for a log-normal AFT model using WinBUGS. Their program code can be conveniently used or modified to generate multiple imputations for missing covariates. However, this general strategy also depends on the capabilities of WinBUGS for running different types of Bayesian survival analysis models. Practitioners are recommended to check the software package for possible options.

Example 8.6. *A Bayesian multiple imputation approach to missing covariates in Cox models*

Arguably, the mostly used survival analysis model is the Cox model (8.5). The missing covariate problem in Cox models often occurs in practice and has to be handled. To apply JM, we need to set up the analysis model (i.e., the Cox model) and the covariate model. The latter component is relatively simple to build as we can apply, for example, the models introduced in Chapter 6 with multiple covariates. The former component is somewhat sophisticated because of the semiparametric structure of the Cox model. However, there exists some literature that approach the Cox model using the Bayesian estimation method so that the posterior distribution of the regression coefficients can be obtained. For example, Clayton (1994) formulated the Cox model using the counting process notation introduced by Anderson and Gill (1982) and discussed estimation of the baseline hazard and regression parameters using MCMC methods. This approach is implemented in WinBUGS. Technical discussions of these and related methods are beyond the scope of this book, and some of them can be found in Ibrahim et al. (2001).

Here we consider a simple example that includes two covariates X_1 and X_2. Suppose the Cox model is specified as $h(t|X_1, X_2) = h_0(t)exp(X_1\beta_1 + X_2\beta_2)$, X_1 and X_2 are from a bivariate normal distribution (i.e., the covariate model), and X_1 is incomplete while X_2 is fully observed. Suppose the missingess of X_1 is MAR; then the WinBUGS code for imputing X_1 is included as follows:

```
model
{
# Set up data
for(i in 1:N) {
for(j in 1:T) {
```

```
# risk set = 1 if obs.t >= t
Y[i,j]<-step(observetime[i]-failtime[j]+eps)
# counting process jump = 1 if obs.t in (t[j], t[j+1])
# i.e. if t[j]<=obs.t<t[j+1]
dN[i,j]<-Y[i,j]*step(failtime[j+1]-observetime[i]-eps)
*case[i]
}
}
# The Cox Model
for(j in 1:T) {
#   beta0[j] ~ dnorm(0, 0.001);
for(i in 1:N) {
# Likelihood
dN[i, j]~dpois(Idt[i, j])
# x1 is the missing covariate
# x2 is the fully observed covariate
# Intensity
Idt[i,j]<-Y[i,j]*exp(betaM*x1[i]+betaC*x2[i])*dL0[j]
}
dL0[j]~dgamma(mu[j],c)
# prior mean hazard
mu[j]<-dL0.star[j]*c
          }
# covariate model
for (i in 1:N)
{
x1[i]~dnorm(mu_zm[i], tau);
mu_zm[i]<-alpha0+alpha1*x2[i];
}
c <- 0.001
r <- 0.1
for (j in 1 : T)
{dL0.star[j]<-r*(failtime[j+1]-failtime[j])
}
betaM~dnorm(0.0,0.001)
betaC~dnorm(0.0,0.001)
alpha0~dnorm(0.0,0.001);
alpha1~dnorm(0.0,0.001);
tau~dgamma(0.001,0.001);
sigma<-sqrt(1/tau);
}
```

Detailed information about the sample code can be found in the Leukemia example (Example 18, WinBUGS Example Volume 1 from http://www.mrc-bsu.cam.ac.uk/wp-content/uploads/WinBUGS_Vol1.pdf)), which explains how the counting process format of the survival data is set up and how the Cox model is parameterized in the Bayesian form. To impute the missing values in X_1, we just need to augment the original code by implementing the conditional distribution of X_1 given X_2 (under the comment "covariate model"). WinBUGS can therefore automatically provide the posterior samples of all model parameters as well as posterior draws (imputations) of missing X_1 cases.

We conduct a simple simulation study to assess the performance of the JM strategy. The complete data assume a Weibull proportional hazards model with the baseline hazard setting as $h_0(t) = t$. Two standard normal covariates (X_1 incomplete and X_2 complete) with correlation = 0.5 are generated, and $\beta_1 = \beta_2 = 2$. The censoring time follows a standard exponential distribution. We generate missing values in X_1 under an MAR mechanism: the missingness of X_1 is related to both X_2 and the survival outcome (observed time and the censoring indicator) and results in around 50% of missing cases. The complete sample size n is set as 100 and the number of simulations is also set as 100. The number of imputations M is set as 50. Table 8.8 shows the simulation results: CC yields biased results for both β_1 and β_2 and low coverage rates, and JM yields estimates with little bias and satisfactory coverage rates.

TABLE 8.8
Example 8.6. Simulation results

Parameter	Method	RBIAS (%)	SD	$\overline{\text{SE}}$	COV (%)	CI
$\beta_1 = 2$	BD	0	0.329	0.292	93	1.146
	CC	−12.4	0.364	0.331	80	1.299
	JM	0.2	0.351	0.334	95	1.331
$\beta_2 = 2$	BD	0	0.306	0.287	95	1.124
	CC	−11.3	0.380	0.329	81	1.290
	JM	0	0.353	0.319	95	1.271

Although JM can be implemented using WinBUGS, the computational burden is not trivial based on our limited experience. This might be due to the counting process format of the survival data that is needed for the Bayesian estimation. This is more apparent when the number of distinct observed event times (or the number of parameters) increases. We use R nimble to conduct the above simulation.

8.3.3 Fully Conditional Specification

Unlike JM, FCS attempts to avoid deriving or obtaining the exact posterior predictive distribution of missing covariates, X_{mis}. Instead the solution is to specify a good conditional imputation model for X_{mis} given other variables, which have to include the information from Y and δ, the observed event/censoring time and censoring indicator. For pragmatic reasons, such specification is not expected to be complicated. We provide some discussions here.

First we consider missing covariate problems in parametric survival models and use AFT models to illustrate the ideas. In Example 7.6, we showed that an effective conditional specification in FCS can be obtained by approximating the posterior predictive distribution of missing data using simple regression models. This strategy was used in Qi et al. (2018), which proposed an FCS strategy for missing covariates in AFT models. In brief, we consider an AFT model (8.2) with a log-normal error distribution. Suppose $X = (X_1, X_2)$, where X_1 is continuous and X_2 is binary. The analysis model (or the event time model) is

$$log(T) = \beta_0 + \beta_1 X_1 + \beta_2 X_2 + \tau W, \tag{8.7}$$

where $W \sim N(0,1)$.

One choice to model X is to use the general location model (GLOM) framework as

$$\begin{aligned} X_2|p &\sim Bernoulli(p), \\ X_1|X_2 = l, \mu_l, \sigma^2 &\sim N(\mu_l, \sigma^2), \end{aligned} \tag{8.8}$$

where $l = 0, 1$.

Following the Bayes rule, it can be derived from Models (8.7) and (8.8) that for X_1, we have

$$\begin{aligned} log(f(X_1|Y_o, \delta, X_2)) = -\frac{\delta}{2\sigma^2}[X_1 - \mu_0 - (\mu_1 - \mu_0)X_2]^2 \\ -\frac{\delta}{2\tau^2}(Y_f - \beta_0 - \beta_1 X_1 - \beta_2 X_2)^2 - \frac{1-\delta}{2\sigma^2}[X_1 - \mu_0 - (\mu_1 - \mu_0)X_2]^2 \\ +(1-\delta)log(\Phi(\frac{\beta_0 + \beta_1 X_1 + \beta_2 X_2 - Y_c}{\tau})) + A_1, \end{aligned} \tag{8.9}$$

where $Y_o = min\{Y_f = log(T), Y_c = log(C)\}$, C is the censoring time, and A_1 is some constant free of Y_o, X_1, δ, and X_2.

Similarly for X_2, we have

$$\begin{aligned} log(f(X_2 = 1|Y_o, \delta, X_1)) - log(f(X_2 = 0|Y_o, \delta, X_1)) = \\ -\frac{\delta}{2\sigma^2}(\mu_1 - \mu_0)(\mu_1 + \mu_0 - 2X_1) + \frac{\delta}{2\tau^2}\beta_2[\beta_2 + 2(\beta_1 X_1 + \beta_0 - Y_f)] \\ -\frac{1-\delta}{2\sigma^2}(\mu_1 - \mu_0)(\mu_1 + \mu_0 - 2X_1) \\ +(1-\delta)log(\Phi(\frac{\beta_0 + \beta_2 + \beta_1 X_1 - Y_c}{\tau})/\Phi(\frac{\beta_0 + \beta_1 X_1 - Y_c}{\tau})) + A_2, \end{aligned} \tag{8.10}$$

where A_2 is some constant free of Y_o, X_1, δ, and X_2.

For Eq. (8.9), the first two terms, $-\frac{\delta}{2\sigma^2}[X_1 - \mu_0 - (\mu_1 - \mu_0)X_2]^2$ and $-\frac{\delta}{2\tau^2}(Y_f - \beta_0 - \beta_1 X_1 - \beta_2 X_2)^2$, are quadratic functions of X_1, suggesting a normal linear regression model for X_1 conditioning on Y_f and X_2 for the observations with observed failure times (i.e., $\delta = 1$).

The fourth term, $(1-\delta)log(\Phi(\beta_0 + \beta_1 X_1 + \beta_2 X_2 - Y_c)/\tau))$, is complicated and not a quadratic function of X_1 in general. Yet it might be reasonably approximated by a quadratic function of X_1 in some scenarios. Together with the third term $-\frac{1-\delta}{2\sigma^2}[X_1 - \mu_0 - (\mu_1 - \mu_0)X_2]^2$, also a quadratic function of X_1, this implies a normal linear regression model for X_1 conditioning on Y_c and X_2 for the censored observations (i.e., $\delta = 0$) can be used to approximate the posterior predictive distribution of missing X_1-values.

Above reasoning suggests that a reasonable conditional specification for X_1 might be: fitting a linear regression model for X_1 using Y_f (logarithm of failure time) and X_2 as predictors for observations with failure time available; fitting a separate linear regression model for X_1 using Y_c (logarithm of censoring time) and X_2 as predictors for censored observations. Note that there exists no apparent evidence that these two lines are identical or parallel.

Similarly for Eq. (8.10), the first two terms, $-\frac{\delta}{2\sigma^2}(\mu_1 - \mu_0)(\mu_1 + \mu_0 - 2X_1)$ and $\frac{\delta}{2\tau^2}\beta_2(\beta_2 + 2(\beta_1 X_1 + \beta_0 - Y_f))$, are linear functions of X_1 and Y_f. This suggests that for the observations with observed failure time (i.e., $\delta = 1$), the appropriate conditional model specification is a logistic regression of X_2 on Y_f and X_1. The fourth term, $(1 - \delta)log(\Phi(\frac{\beta_0+\beta_2+\beta_1 X_1 - Y_c}{\tau})/\Phi(\frac{\beta_0+\beta_1 X_1 - Y_c}{\tau}))$, is complicated and nonlinear. It might be well approximated by a quadratic function of X_1 and Y_c. However, some approximation adequacy might be sacrificed for simpler implementation by using a linear function. Together with the third term $-\frac{1-\delta}{2\sigma^2}(\mu_1 - \mu_0)(\mu_1 + \mu_0 - 2X_1)$, also a linear function of X_1, this suggests that a logistic regression for X_2 conditional on X_1 and Y_c for the censored observations (i.e., $\delta = 0$) might be used to approximate the posterior predictive distribution of missing X_2-values.

The above reasoning suggests that a possible conditional model specification for X_2 might include two pieces: fitting a logistic regression model for X_2 using Y_f and X_1 as predictors for cases with $\delta = 1$; fitting a separate logistic regression for X_2 on Y_c and X_1 as predictors for cases with $\delta = 0$.

After considering other types of covariate models for X, Qi et al. (2018) proposed an FCS strategy as a generalization of the preceding idea:

(a) For an incomplete continuous variable X: fitting a normal linear regression model for X using Y_f and other X-covariates as predictors for cases with $\delta = 1$; fitting a separate linear regression model for X using Y_c and other X-covariates as predictors for cases with $\delta = 0$.

(b) For an incomplete categorical (e.g., binary) variable X: fitting a GLM for X using Y_f and other X-covariates as predictors for cases with $\delta = 1$; fitting a separate GLM for X using Y_c and other X-covariates as predictors for cases with $\delta = 0$.

In this FCS strategy, a log-transformation of the observed failure or censoring time is used as a predictor for the missing X-variable. More importantly, the conditional models are separated by the censoring indicator δ. This separation essentially interacts δ with other predictors in the imputation for the whole sample. The proposed FCS strategy is straightforward to implement in practice.

Example 8.7. *A simulation study for the FCS imputation of missing covariates in AFT Models*

We conduct a simulation study to demonstrate the performance of the FCS imputation strategy. For complete-data generation, we consider two continuous covariates (X_1, X_2) and one binary covariate X_3 taking values 0 and 1. The log-normal AFT model for the log-transformed failure time Y and the complete covariates is

$$Y = \beta_0 + \beta_1 X_1 + \beta_2 X_2 + \beta_3 X_3 + \epsilon, \tag{8.11}$$

where $(\beta_0, \beta_1, \beta_2, \beta_3) = (0, 1.5, 3, 1.5)$ and $\epsilon \sim N(0, 1)$. For the distribution of covariates, we generate data from a GLOM:

$$\begin{aligned} X_3|\pi &\sim Bernoulli(p) \\ (X_1, X_2)|X_3 = l, \mu_l, \Sigma &\sim N(\mu_l, \Sigma), \end{aligned} \tag{8.12}$$

where $l = 0, 1$, $p = 0.5$, $\mu_0 = (0, 0)$, $\mu_1 = (1, 1)$, and $\Sigma = \begin{pmatrix} 1 & 0.5 \\ 0.5 & 1 \end{pmatrix}$.

Around 25% or 50% of the failure times are censored, generated from a uniform distribution. We let X_2 be fully observed and assign some missing data to both X_1 and X_3. The missingness probability of covariates is set to be MAR by relating it to the survival outcome (both the observed failure or censoring time and censoring indicator) and observed covariates. As a result, around 20% or 40% of subjects are with either X_1 or X_3 unobserved, respectively. The simulation consists of 1000 replicates and the complete dataset has sample size $n = 500$ per simulation.

We use the R mice function to implement the FCS strategy. For imputing the continuous X_1, we consider both the normal linear imputation and its PMM version. We also include a simplified FCS strategy that does not impute the missing covariates separately by δ. We collect the imputed data after running FCS for 20 iterations. Each incomplete dataset is imputed $M = 20$ times. For each imputed dataset, the maximum likelihood method, adopted in the survreg function in R for the log-normal AFT models, is used for estimating the parameter β.

Table 8.9 (see Table 2 from Qi et al. 2018) shows the simulation results. The results from CC are biased due to the MAR setup for missing covariates. The norm-log(t)-SEP method, which fits separate imputation models for the observed and the censored failure times, yields reasonable results. Its PMM version, the PMM-log(t)-SEP method, produces little bias, yet it can have a low coverage rate in some cases (e.g. 77.0% when $\beta_2 = 3$). The simple FCS specifications (i.e., norm-log(t) and PMM-log(t)) without separating between the censoring and the observed failure times can result in sizable biases and low coverage rates. Apparently, the FCS option that separates the imputation by the censoring indicator is consistently better than the option that does not. Qi et al. (2018) performed comprehensive simulation assessments and demonstrated that the proposed FCS strategy works well overall for a wide variety of AFT models and covariate distributions.

TABLE 8.9
Example 8.7. Simulation results

Parameter	Method	RBIAS (%)	SD	$\overline{\text{SE}}$	COV (%)	CI
$\beta_1 = 1.5$	BD	−0.1	0.070	0.067	95.2	0.26
	CC	−10.3	0.087	0.084	54.2	0.33
	norm-log(t)	−10.8	0.084	0.102	68.1	0.40
	norm-log(t)-SEP	0	0.081	0.079	94.7	0.31
	PMM-log(t)	−6.0	0.116	0.108	85.6	0.43
	PMM-log(t)-SEP	1.0	0.100	0.086	90.9	0.34
$\beta_2 = 3$	BD	0	0.075	0.073	94.5	0.29
	CC	−10.2	0.102	0.105	17.1	0.41
	norm-log(t)	3.1	0.115	0.131	92.9	0.53
	norm-log(t)-SEP	0.5	0.107	0.106	95.1	0.42
	PMM-log(t)	6.0	0.150	0.115	63.4	0.46
	PMM-log(t)-SEP	3.9	0.113	0.098	77.0	0.39
$\beta_3 = 1.5$	BD	−0.1	0.130	0.125	94.2	0.49
	CC	−12.2	0.131	0.130	68.7	0.51
	norm-log(t)	6.2	0.151	0.179	95.5	0.71
	norm-log(t)-SEP	−1.5	0.142	0.143	94.7	0.56
	PMM-log(t)	5.7	0.162	0.178	95.0	0.70
	PMM-log(t)-SEP	2.7	0.152	0.150	94.9	0.59

How do we carry out FCS for missing covariates in Cox models? In one of the earliest papers on FCS, van Buuren et al. (1999) proposed to impute missing X's based on a GLM framework, including the censoring indicator, δ, the observed event or censoring time, Y, and the logarithm of Y as predictors. For illustration, suppose X_1 is missing; then the conditional model can be specified as

$$E(g(X_1)) = \beta_0 + \beta_2 X_2 + \ldots + \beta_p X_p + \gamma_0 \delta + \gamma_1 Y + \gamma_2 log(Y), \qquad (8.13)$$

where g is the link function for X_1. If more than one of the X-variables are missing, then similar conditional imputation models can be used for them to set up an FCS approach.

This strategy seems to work well in several empirical applications (e.g., Clark and Altman 2003; Brazi and Woodward 2004; Giorgi et al. 2008). In addition, Qi et al. (2010) considered and assessed an FCS strategy that omitted $log(Y)$ from Model (8.13), and the corresponding results are not as good.

White and Royston (2009) showed that the predictive distribution of a missing covariate in a Cox model conditional on the observed data is a function of the cumulative baseline hazard function, censoring indicator, and other fully observed covariates. Here the cumulative baseline hazard function is defined as $H_0(t) = \int h_0(t)dt$. Note that the cumulative baseline hazard function is not observed and is required to be estimated as well. In general the predictive distribution of the missing covariate does not have a closed form. However, in certain cases it might be well approximated by commonly used regression models. Supposing the incomplete X_1 is either continuous or binary, they proposed two conditional models for an FCS strategy:

$$E(X_1) = \beta_0 + \beta_2 X_2 + \ldots + \beta_p X_p + \gamma_0 \delta + \gamma_1 \hat{H}_0(t), \quad (8.14)$$
$$logit(f(X_1 = 1)) = \beta_0 + \beta_2 X_2 + \ldots + \beta_p X_p + \gamma_0 \delta + \gamma_1 \hat{H}_0(t), \quad (8.15)$$

where Models (8.14) and (8.15) suggest that if X_1 is continuous (binary), then an imputation model can be a linear regression (logistic regression) model regressing X_1 on δ, the cumulative baseline hazard function estimate $\hat{H}_0(t)$, and other covariates. The semiparametric structure of the Cox model, therefore, is accounted for by including $\hat{H}_0(t)$ as a predictor, which can be estimated using the Nelson-Aalen method from complete cases. With more than one missing covariate, this specification can be used for each of the missing covariates. The proposal from White and Royston (2009) has been implemented in R mice and is very popular in practice.

To gain some insights into why $\hat{H}_0(t)$ is needed in the conditional model specification, Fig. 8.4 shows the scatter plots between a missing covariate X_1 and various forms of survival function (i.e., t, $log(t)$, and $\hat{H}_0(t)$) from a simulation sample. The LOWESS fit (solid line) and linear fit (dashed line) from simulated samples are also shown. The complete data assume a Weibull proportional hazards model with the baseline hazard being set as $h_0(t) = 2t$, two standard normal covariates (X_1 incomplete and X_2 complete) with correlation $= 0.5$, and $\beta_1 = \beta_2 = 1$. The censoring time follows a standard exponential distribution. The sample size is set to be 1000. The relationships between X_1 and different types of survival functions are displayed. However among them, the scatter plot between X_1 and $\hat{H}_0(t)$ appears to be closer to one implied by a normal linear regression model because the LOWESS fit matches well with the linear fit.

Example 8.8. *A simulation study for the FCS imputation of missing covariates in Cox models*

We use the simulation design of Example 8.7 to demonstrate the performance of the FCS imputation strategy for missing covariates in Cox models. We change the log-normal error in Model (8.2) to an extreme-value error distribution (with location $= 0$, scale $= 1$, and shape $= 0$) so it becomes a Weibull AFT model, for which the PH assumption also holds. We consider β's

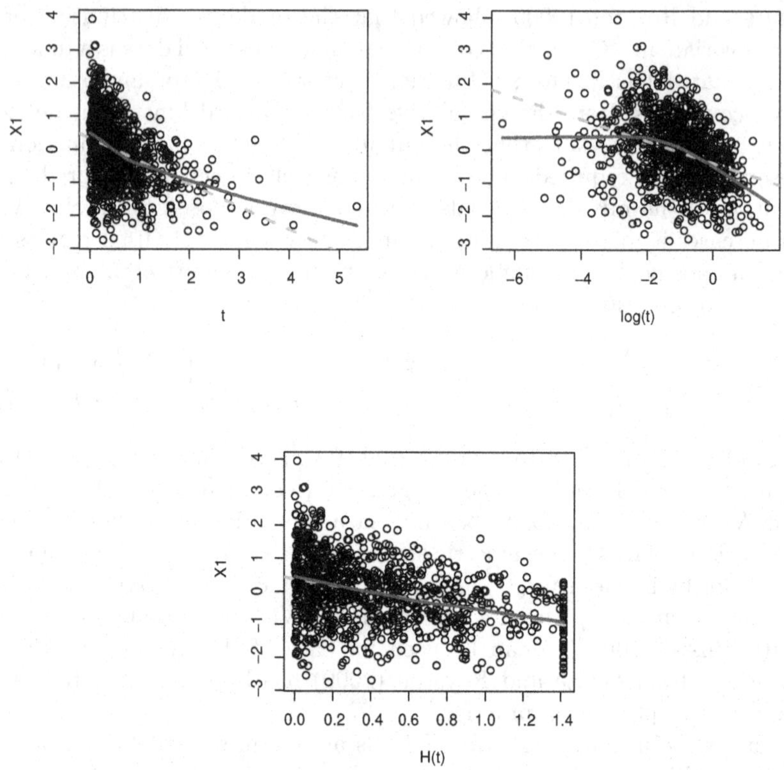

FIGURE 8.4
Scatter plots between a covariate X_1 and different functions of survival time under a Weibull proportional hazards model based on simulation. Top left: X_1 vs. t; top right: X_1 vs. $log(t)$; bottom: X_1 vs. $\hat{H}_0(t)$. Solid line: LOWESS fit; dashed line: linear fit.

with different magnitudes including $(\beta_0, \beta_1, \beta_2, \beta_3) = (0,\ 0.1,\ 0.15,\ 0.2)$ and $(\beta_0, \beta_1, \beta_2, \beta_3) = (0, 1.5, 3, 1.5)$. The other simulation setups are similar.

To implement the FCS strategy proposed by White and Royston (2009), we consider both the linear regression imputation and its PMM version for continuous X_1. We also include a more general FCS strategy, that is, impute the missing covariates separately by δ, motivated by the FCS proposal for the AFT models in Example 8.7. Table 8.10 (see Tables 6 and S6 from Qi et al. 2018) shows the results. When β's are small, all imputation methods work well. However, with large β's, there is some degradation of the performance for estimating β_1 and β_2. Using PMM or imputing the missing covariates separately by δ does not seem to help much. These results are consistent with White and Royston (2009), which commented that the proposed FCS strategy would not work well in situations with strong covariate effects and large cumulative incidences. Additional research is warranted to address such issues.

8.3.4 Semiparametric Methods

One possible option for robustifying the imputation is to use the dual modeling strategy (Section 5.6.2). Hsu et al. (2014) applied this strategy to the scenarios in which the missing variable is a covariate of a typical regression analysis. Hsu and Yu (2019) applied the dual modeling strategy to the setting of a Cox regression analysis with a missing covariate. See also Section 8.2 for its application to censored event times.

Specifically, suppose X_1 is the missing covariate. Two working models are constructed. The outcome model, which predicts X_1, follows Eqs (8.14) or (8.15). The propensity model, which predicts the response propensity of X_1, is constructed as a logistic regression model with Y, δ, and other fully observed covariates as predictors. Once the two predictive scores are derived, the imputing set is defined by a distance function based on the weighted combination of the two predictive scores and then used to randomly select values for each missing observation. Simulation studies from Hsu and Yu (2019) demonstrated that the performance of this approach can be robust against some misspecifications of either working model. This imputation strategy is implemented by R NNMIS .

Example 8.9. *Semiparametric imputation for a breast cancer dataset*

We demonstrate the approach in Hsu and Yu (2019) on a dataset which consists of 7050 women diagnosed with stage IV breast cancer between 2005 and 2011 in California. This dataset was extracted from the breast cancer registries under the Surveillance, Epidemiology and End Results (SEER) Program. Of the 7050 patients, besides survival outcome (i.e. survival status/death status and event/censoring time) after being diagnosed with breast cancer, for each patient there are several variables collected at diagnosis such as age (years), race (White/Black/Other), HER2 (positive/negative), radiation status (yes/no) and surgery status (yes/no). These variables are summarized in Table 8.11 (Table 1 from Hsu and Yu 2019). The missing variable is HER2, which is a member of the human epidermal growth factor receptor family and has been shown to be strongly associated with increased disease recurrence and a poor prognosis for breast cancer patients. Among the 7050 patients, 1293 (18.3%) had missing HER2 values.

Table 8.12 (Table 2 from Hsu and Yu 2019) identifies variables predictive of HER2 value and its missingness probability. Specifically, based on complete cases, age, race, surgery, and the cumulative baseline hazard estimate are predictive of HER2 value from a logistic regression. The results indicate that younger patients, patients who did not have surgery, and patients who

TABLE 8.10
Example 8.8. Simulation results

Parameter	Method	RBIAS (%)	SD	$\overline{SE}$	COV (%)	CI
$\beta_1 = 0.1$	BD	0.3	0.069	0.067	95.0	0.26
	CC	8.1	0.074	0.069	92.9	0.27
	norm-H(t)	−0.2	0.078	0.075	93.7	0.29
	norm-H(t)-SEP	1.3	0.078	0.076	94.6	0.30
	PMM-H(t)	0.6	0.078	0.075	94.0	0.29
	PMM-H(t)-SEP	1.6	0.079	0.076	94.5	0.30
$\beta_2 = .15$	BD	0.6	0.070	0.067	94.7	0.26
	CC	15.8	0.077	0.070	91.9	0.28
	norm-H(t)	−0.6	0.074	0.070	94.1	0.28
	norm-H(t)-SEP	0.2	0.075	0.071	94.4	0.28
	PMM-H(t)	−0.6	0.074	0.070	94.6	0.28
	PMM-H(t)-SEP	0.5	0.075	0.070	93.6	0.28
$\beta_3 = .2$	BD	−5.3	0.137	0.133	94.6	0.52
	CC	−21.6	0.140	0.134	92.6	0.52
	norm-H(t)	1.2	0.164	0.158	94.2	0.62
	norm-H(t)-SEP	−4.1	0.167	0.162	94.3	0.64
	PMM-H(t)	0.5	0.165	0.158	94.2	0.62
	PMM-H(t)-SEP	−3.9	0.169	0.164	93.8	0.65
$\beta_1 = 1.5$	BD	0.1	0.071	0.069	94.0	0.27
	CC	−9.4	0.085	0.081	58.5	0.32
	norm-H(t)	−11.9	0.179	0.157	74.9	0.63
	norm-H(t)-SEP	−4.2	0.171	0.128	92.4	0.51
	PMM-H(t)	−5.9	0.103	0.115	89.7	0.46
	PMM-H(t)-SEP	−4.4	0.096	0.103	88.3	0.41
$\beta_2 = 3$	BD	0.1	0.078	0.075	93.6	0.30
	CC	−9.4	0.100	0.102	21.9	0.40
	norm-H(t)	7.5	0.371	0.208	53.0	0.84
	norm-H(t)-SEP	13.1	0.202	0.148	10.2	0.59
	PMM-H(t)	5.4	0.124	0.141	79.1	0.57
	PMM-H(t)-SEP	13.3	0.130	0.126	10.1	0.50
$\beta_3 = 1.5$	BD	−0.6	0.132	0.129	94.0	0.50
	CC	−11.7	0.132	0.125	69.0	0.49
	norm-H(t)	−0.5	0.222	0.213	96.7	0.85
	norm-H(t)-SEP	−1.4	0.188	0.192	96.4	0.76
	PMM-H(t)	−4.9	0.152	0.178	94.0	0.70
	PMM-H(t)-SEP	−1.6	0.150	0.174	95.4	0.68

had a higher hazard rate are more likely to have a positive HER2 value. In addition, age, surgery status, radiation status, survival status, and the cumulative baseline hazard estimate are predictive of the missingness probability. The results indicate that older patients, patients who did not have surgery, patients who did not have radiation, and patients who had a lower hazard rate are more likely to have a missing HER2 value. These predictive covariates are then used to derive the predictive scores. For the outcome model, a working logistic regression model for HER2 positive indicator with age, race, surgery, survival status, and the cumulative baseline hazard estimate as covariates is fitted. For the propensity score model, a working logistic regression model for the HER2 missingness indicator with age, radiation, surgery, survival status, and the cumulative baseline hazard estimate as covariates is fitted. To conduct the semiparametric imputation (NNMI), the two predictive scores derived are then used to calculate the distance between patients and then an imputing set for each patient with missing HER2 is selected. We also use the outcome model, that is, the predictive score for the missing HER2, to conduct a PMM imputation. The number of imputations M is set at 50.

The post-imputation analysis includes a Cox regression analysis with age, race (White as the reference group), HER2, radiation status, and surgery status as the covariates. Table 8.13 (Table 3 from Hsu and Yu 2019) displays the hazard ratio (HR) estimate of each covariate along with the associated 95% confidence interval (CI) and p-value. CC results indicate that age, black, and surgery status are significantly associated with survival after diagnosis

TABLE 8.11
Example 8.9. Description statistics of the stage IV breast cancer dataset

Predictor	Mean/Frequency	SD/Percentage
Age	60.91	14.41
Race		
White	5585	79.2%
Black	721	10.2%
Others	744	10.6%
HER2		
Negative	4180	59.3%
Positive	1577	22.4%
Missing	1293	18.3%
Surgery		
No	3916	55.6%
Yes	3134	44.4%
Radiation		
No	4484	63.6%
Yes	2566	36.4%

TABLE 8.12
Example 8.9. Identification of predictors associated with missing value and missingness probability of HER2

	Missing value			Missingness probability		
Predictor	OR	95% CI	p-value	OR	95% CI	p-value
Age	0.987	(0.983, 0.991)	<0.0001	1.031	(1.026, 1.035)	<0.0001
Black	1.187	(0.983, 1.433)	0.08	1.007	(0.825,1.229)	0.94
Other	1.360	(1.135, 1.629)	<0.001	0.908	(0.742, 1.112)	0.35
No Radiation	0.913	(0.811,1.028)	0.13	1.580	(1.385, 1.804)	<0.0001
No Surgery	0.884	(0.787,0.993)	0.04	2.146	(1.885, 2.443)	<0.0001
Dead	0.997	(0.888, 1.120)	0.96	2.205	(1.937, 2.510)	<0.0001
$\hat{H}_0(t)$	1.416	(1.235, 1.624)	<0.0001	0.641	(0.549, 0.747)	<0.0001

Note: OR: odds ratio estimate.

TABLE 8.13
Example 8.9. Missing data analysis results from the Cox regression

	CC		PMM		NNMI	
Predictor	HR (95% CI)	p-value	HR (95% CI)	p-value	HR (95% CI)	p-value
Age	1.015 (1.012, 1.017)	<0.01	1.018 (1.015, 1.020)	<0.01	1.018 (1.015, 1.020)	<0.01
Black	1.437 (1.286, 1.605)	<0.01	1.443 (1.308, 1.591)	<0.01	1.442 (1.307, 1.591)	<0.01
Others	0.887 (0.781, 1.007)	0.06	0.879 (0.786, 0.984)	0.03	0.879 (0.785, 0.983)	0.02
No radiation	1.056 (0.978, 1.140)	0.16	1.044 (0.976, 1.118)	0.21	1.044 (0.976, 1.118)	0.21
No surgery	1.893 (1.755, 2.042)	<0.01	1.885 (1.762, 2.016)	<0.01	1.896 (1.773, 2.028)	<0.01
HER2	0.940 (0.867, 1.020)	0.14	0.932 (0.860, 1.010)	0.09	0.939 (0.864, 1.021)	0.14

Note: HR: hazard ratio estimate.

with stage IV breast cancer. Specifically, older patients tend to have a higher hazard rate than younger patients, black patients tend to have a higher hazard rate than white patients, and patients without surgery tend to have a higher hazard rate than patients with surgery. Patients in the other race group have a slightly lower hazard rate than white patients but not significant at the level of 5%. Radiation status and HER2 are not significantly associated with survival after diagnosis with stage IV breast cancer. PMM and NNMI produce similar results as the CC method, except for the predictor of the other race group. The results of PMM and NNMI methods indicate that patients with other race might have a significantly lower hazard rate than white patients. In addition, PMM and NNMI produce tighter 95% CI than CC, improving the precision of the estimates compared with the latter method.

8.4 Summary

In this chapter, we focus on right-censored data to illustrate the idea of using multiple imputation in survival analysis. Besides right censoring, survival data can also be subject to left or interval censoring. Multiple imputation has been used to impute event times for interval-censored survival data to simplify the estimation procedures. Some examples can be found in Pan (2000) and Hsu et al. (2007). Multiple imputation has also been used to impute left-censored survival data (Karvanen et al. 2010). Furthermore, it is expected that the idea of multiple imputation can be applied to more complicated survival data (e.g., subject to competing risks by Ruan and Gray 2008).

9

Multiple Imputation for Longitudinal Data

9.1 Introduction

In longitudinal studies, information from the same set of subjects is measured repeatedly over time. The aims of longitudinal studies often are to estimate/predict the mean or individual response at certain time points, to relate time-invariant or time-dependent covariates to repeatedly measured response variables, or to relate the response variables with each other. Longitudinal design is often used in clinical trials, intervention studies, and panel surveys. Missing data frequently occur in longitudinal studies because subjects may miss some visits during the study, some variables may not be measured at particular visits, or subjects may drop out prematurely in the middle of the study (i.e., loss to follow-up). The absence of complete data is a serious impediment to longitudinal data analysis.

There exists a large body of literature analyzing longitudinal data (e.g., Hedeker and Gibbons 2006; Fitzmaurice et al. 2009) as well as handling related missing data problems (e.g., Daniels and Hogan 2008; Molenberghs and Kenward 2007). Many of the methods are nonimputation approaches (e.g., likelihood-based and estimation equation methods). In this chapter we present some ideas using the multiple imputation strategy. Section 9.2 briefly introduces commonly used mixed effects models for longitudinal data. Section 9.3 presents some imputation strategies based on mixed effects models. Section 9.4 discusses some alternative imputation strategies. Section 9.5 briefly discusses the topic of applying multiple imputation to multilevel data since longitudinal data can be classified as a type of multilevel data. Section 9.6 provides a summary.

9.2 Mixed Models for Longitudinal Data

A main modeling approach to longitudinal data is based on mixed effects models (or mixed models) (e.g., Laird and Ware 1982). To illustrate the main idea, suppose the outcome y_{it} from subject i ($i = 1, \ldots, n$) is measured at time t_j ($j = 1, \ldots T$). For complete data, a univariate linear mixed model for

DOI: 10.1201/9780429156397-9

y_{it}'s can be expressed as

$$Y_i = X_i\beta + Z_i b_i + \epsilon_i, \tag{9.1}$$

where $Y_i = (y_{it_1}, y_{it_2}, \ldots, y_{it_T})$ is a $T \times 1$ ($T \geq 2$) vector containing y_{it}'s, X_i is a known $T \times p$ design matrix in subject i associated with the common $p \times 1$ fixed effects vector β, and Z_i is a known $T \times q$ design matrix in subject i associated with the $q \times 1$ random effects vector b_i. Subjects can also be viewed as clusters of longitudinal observations. The term "mixed" usually means a mixture of both fixed effects and random effects here. Sometimes mixed models are also referred to as random effects models in the literature.

Unlike the fixed effects, β, some distributional assumptions are made for the random effects b_i's in Model (9.1). It is often assumed that they are independently and interchangeably normally distributed, such as $b_i \sim N(0, \Omega)$. The number of random effects (q) is typically smaller than the number of fixed effects (p). In many cases if we set $q = 1$ and $Z_i = 1$, then Model (9.1) is also called a random intercepts model. Symbol ϵ_i denotes the $T \times 1$ vector of error terms, which are independently normally distributed as $\epsilon_i \sim N(0, R_i)$. For simplicity, it is often assumed that $R_i = \sigma^2 I_T$ so that the error variance is equal for all subjects measured at all time points, although more complex variance structure can be applied. In addition, b_i's and ϵ_i's are assumed to be independent of each other. Typically, parameters Ω and σ^2 are termed the variance components of mixed models.

There are two keys to the modeling of longitudinal data. One is to capture the temporal trend exhibited by the data. To achieve this, we can include functions of time t in both X- and Z-covariate matrices. For example, if the X-matrix includes $\{1, t_j\}$, then the model implies linear curves (after adjusting for other baseline covariates) for both the overall (mean) trend and individual trajectories. If the Z-matrix also includes $\{1, t_j\}$, then this model assumes both random (i.e. different) intercepts and slopes for each subject. It is often termed a random coefficients model. If we drop the $\{t_j\}$ from Z_i, then this model is reduced to a random intercepts model, assuming different intercepts yet identical slopes for all subjects. In theory, we can specify X-covariates and/or Z-covariates to characterize the overall or individual temporal trend as polynomial functions of time such as t^2, t^3, etc. Apparently data collected from more time points are needed to precisely estimate higher-order functions of time. In practice, it is often the case that random effects are used only up to linear slopes to avoid model complexity, although the adequacy of simple models needs to be assessed and defended. Besides the function of time, other covariates in X can include baseline/time-invariant variables such as demographics. Time-variant covariates can also be included, yet the model might become more complicated.

The other key is to account for the correlation among the repeated observations clustered within the same subject. In Model (9.1), the random effects b_i's are used to induce such correlation. It can be shown that $Var(y_{it_1}, y_{jt_2}) = 0$ if $i \neq j$ and $t_1, t_2 \in \{1, \ldots T\}$ (i.e., observations from different subjects are uncorrelated), $Var(y_{it_1}) = Z_{it_1}\Omega Z_{it_2}^t + \sigma^2$, and $Cov(y_{it_1}, y_{it_2}) = Z_{it_1}\Omega Z_{it_2}^t$ if $t_1 \neq t_2$, where Z_{it_1} and Z_{it_2} are the t_1-th and t_2-th row of Z_i, respectively (i.e., observations from the same subject at different time points are correlated).

For longitudinal data that are not continuous, a generalized linear mixed model can be specified in the similar spirit of generalized linear models (e.g., McCulloch et al. 2008). To sketch the idea, suppose that y_{it} is generated from an exponential dispersion family of the form:

$$f(Y_{it} = y_{it}; \theta_{it}, \phi) = exp\{\frac{\theta_{it}y_{it} - k(\theta_{it})}{a(\phi)} + c(y_{it}; \phi)\},$$

where θ_{it} and $k(\cdot)$ are known functions that determine the distribution of y_{it}, and $c(y_{it}; \phi)$ is a known function indexed by a scale parameter ϕ. The expectation of y_{it}, denoted as μ_{it}, is related to both x_i and z_i through the link function g, that is,

$$\mu_{it} = E(y_{it}|X, Z) = g^{-1}(x_{it}\beta + z_{it}b_i),$$

where x_{it} and z_{it} are the t-th row of the covariate matrix X_i and Z_i, respectively. Coefficients β and b_i's are the fixed and random effects, respectively. A distributional assumption for random effects, such as $b_i \sim N(0, \Omega)$, can be specified. For example, a logistic link function for $g(\cdot)$ can be used to model binary data in y_{it}'s.

Parameter estimation in mixed models has been an important and complicated topic area in statistics. Estimation procedures are available in various software packages (e.g., SAS PROC MIXED, SAS PROC GLIMMIX, and R lme4). Suppose some of the y_{it}'s are missing and the missingness is assumed to be MAR. We emphasize that here MAR implies that the missingness, after conditioning on observed data, is unrelated to not only the missing values of Y, X, and Z, but also unrelated to the unknown random effects b_i's. If we further suppose that covariates X and Z are fully observed, that is, only longitudinal outcomes are missing, then valid estimates for parameters of mixed models (e.g., β, Ω, and σ^2 in Model (9.1)) can be obtained by directly fitting the model using complete cases without the need of imputing missing y_{it}'s. Similar to the setting of cross-sectional data (Section 4.5), this is because the missing responses in y_{it}'s do not contribute any extra information for estimating their relationship with covariates in mixed models. In fact, when the number of imputations is moderate, the results from the multiple imputation analysis might be slightly less efficient if the estimands of interest are the mixed model parameters. However, as discussed before, imputing the missing responses in longitudinal data can be helpful if the analysis of interest is not limited to the parameters of mixed models (e.g., regressing the later measurements on earlier measurements after adjusting for covariates), or if covariates are also subject to missing data.

9.3 Imputation Based on Mixed Models

9.3.1 Why Use Mixed Models?

For drop-outs in longitudinal data, a simple and popular imputation procedure is known as "last observation carried forward" (LOCF), where for each subject the most recent observed value is used to replace all subsequent missing values. Although there may be situations where the mean structure for the missing values is reasonably well captured by the implicit model underlying the LOCF procedure, such an assumption can be violated in many cases where trajectories of repeated measures can trend upward or downward. In these scenarios, LOCF might introduce substantial biases (e.g., Little and Yau 1996; Tang et al. 2005; National Research Council 2010). As discussed in Section 9.2, however, temporal profiles in longitudinal data can be readily modeled using mixed models such as Model (9.1).

Another important issue is to incorporate the subject (cluster) effects and the correlation of repeated observations from the same subject. This concerns the validity of the multiple imputation inference. To illustrate the rationale, we consider the following simple example.

Example 9.1. *A univariate missing data problem in a one-way random effects model*

In Example 3.1, we discussed a univariate missing data problem where the data rise from an independent normal model with an unknown mean and variance. Here we consider its extension to repeated measurements. Specifically, the complete-data model can be expressed as

$$\begin{aligned} y_{ij} &= \mu + \alpha_i + \epsilon_{ij}, \\ \alpha_i &\sim N(0, \tau^2), \\ \epsilon_{ij} &\sim N(0, \sigma^2), \end{aligned} \tag{9.2}$$

for $i = 1, \ldots, n$, $j = 1, \ldots, T$, where y_{ij} is the random variable, μ is the population mean, α_i's are between-subject random effects, and ϵ_{ij}'s are the error terms. Model (9.2) is often referred to as a balanced, one-way normal random effects model. It can be viewed as a special case of Model (9.1) assuming flat lines for subjects with random intercepts.

For ease of notation, we assume that y_{ij}'s are observed for $i = 1, \ldots, n$, $j = 1, \ldots, r_i$ and missing otherwise. Suppose that missing data follow MCAR so that $E(r_i) = r$. For simplicity, let $r_i = r$, and the rate of missingness is $\frac{T-r}{T}$. Under these assumptions and supposing that the estimand of interest is the population mean μ, it can be verified that the grand mean of the observed data $\hat{\mu}_{obs} = \overline{y}_{\cdot\cdot,obs} = \frac{\sum_{i=1}^{n}\sum_{j=1}^{r} y_{ij}}{nr}$ is unbiased: $E(\hat{\mu}_{obs}) = \mu$, and its variance, $Var(\overline{y}_{\cdot\cdot,obs}|\tau^2, \sigma^2) = \frac{\tau^2}{n} + \frac{\sigma^2}{nr}$.

There are three possible multiple imputation strategies for missing y_{ij}'s in this case:

1. IGN: Impute y_{ij}'s ignoring the subject/cluster effects. That is, treat y_{ij}'s as if they come from an independent normal distribution with an unknown mean and variance (Example 3.1).

2. FIX: Impute y_{ij}'s treating α_i as fixed effects. That is, instead of assuming $\alpha_i \sim N(0, \tau^2)$, assume that $\sum_i \alpha_i = 0$. The imputation is then based on a normal linear regression model in which the explanatory variable is the subject indicator (Section 4.2).

3. RAN: Impute y_{ij}'s based on Model (9.2). The general imputation algorithm will be introduced in Section 9.3.2.

All imputation methods yield unbiased point estimates for μ. However, multiple imputation variance estimates of different methods are different. Details of the derivations can be found in Andridge (2011) and He et al. (2016). A summary of the variance estimators can be found in the following. As $n \to \infty$,

1. IGN: $Var(\hat{\mu}_{MI,IGN}) \to \frac{\tau^2}{n} + \frac{\sigma^2}{nr}$, $E(\hat{Var}(\hat{\mu}_{MI,IGN})) \to \frac{T^2+r^3-r^2}{T^2 r}\frac{\tau^2}{n} + \frac{\sigma^2}{nr} < Var(\hat{\mu}_{MI,IGN})$.

2. FIX: $Var(\hat{\mu}_{MI,FIX}) \to \frac{\tau^2}{n} + \frac{\sigma^2}{nr}$, $E(\hat{Var}(\hat{\mu}_{MI,FIX})) \to \frac{\tau^2}{n} + \frac{3T-2r}{T}\frac{\sigma^2}{nr} > Var(\hat{\mu}_{MI,FIX})$.

3. RAN: $Var(\hat{\mu}_{MI,RAN}) \to \frac{\tau^2}{n} + \frac{\sigma^2}{nr}$, $E(\hat{Var}(\hat{\mu}_{MI,RAN})) \to Var(\hat{\mu}_{MI,FIX})$.

The imputation ignoring the clustering (IGN) would lead to a variance estimator that (asymptotically) underestimates the true variance (i.e., $E(\hat{Var}(\hat{\mu}_{MI,IGN})) < Var(\hat{\mu}_{MI,IGN})$). On the other hand, the imputation treating clusters as fixed effects (FIX) would lead to a variance estimator that overestimates the true variance: although the subject effects are included, the within-subject correlation is not considered in this method. Not surprisingly, the imputation based on Model (9.2) (RAN) would yield an unbiased variance estimator.

Andridge (2011) used simulation studies to assess the performance of the variance estimators of the marginal mean of Y when Model (9.2) contains fully observed covariates. The general pattern is similar to that with no covariate. Imputation ignoring the clustering at all would lead to a underestimated variance, yet treating cluster effects as fixed effects would lead to an overestimated variance. The former would lead to undercoverage and the latter would lead to overcoverage for drawing the inference about the mean of Y. The biases of the variance estimates can be related to several major factors including the missingness rate of Y, the correlation between Y and covariates, and the intraclass correlation coefficient, $\rho = \tau^2/(\tau^2 + \sigma^2)$. In addition, the fixed effects imputation strategy would become more problematic if the number of clusters gets large, implying a large number of covariates in the model. The problem of fixed effects imputation is also demonstrated by Van Buuren (2011), in which the method would either fail to converge or yield suboptimal coverage rates for the estimates for regression parameters when the number of clusters increases or the number of observations per cluster reduces.

In summary, it is important to incorporate the subject or clustering effects in longitudinal data, and how these effects are incorporated also matters for the multiple imputation inference. Mixed model-based imputation is a good strategy.

9.3.2 General Imputation Algorithm

Suppose incomplete data Y has a longitudinal structure. As reasoned before, a principled strategy is to model Y using Model (9.1) and adopt a Bayesian, model-based imputation strategy for missing values. This follows the JM strategy for multivariate missing data (Chapter 6). The key component in the imputation process is to derive and implement the DA algorithm for mixed models.

Example 9.2. *Multiple imputation for missing data in multivariate linear mixed models*

Schafer and Yucel (2002) pioneered computational strategies for multivariate mixed effects models with missing values. They considered a model as follows:

$$Y_i = X_i\beta + Z_ib_i + \epsilon_i, \tag{9.3}$$

where Y_i denote an $n_i \times r$ matrix of multivariate responses for subject i, X_i ($n_i \times p$) and Z_i ($n_i \times q$) are known covariate matrices, β ($p \times r$) is a matrix of regression coefficients common to all units, and b_i ($q \times r$) is a matrix of coefficients specific to unit i. Further, n_i rows of ϵ_i are independently distributed as $N(0, \Sigma)$, and the random effects are distributed as $vec(b_i) \sim N(0, \Phi)$ independently for $i = 1, \dots m$ (the "vec" operator vectorizes a matrix by stacking its columns). Model (9.1) can be viewed as a special case by setting $n_i = T$ and $r = 1$ in Model (9.3).

To complete the Bayesian model specification, Schafer and Yucel (2002) assumed a flat prior for β and proper prior for variance components. Specifically, they are inverse Wishart prior $\Sigma^{-1} \sim Wishart(\nu_1, \eta_1)$ and $\Phi^{-1} \sim Wishart(\nu_2, \eta_2)$. The DA algorithm for drawing model parameters and missing values can be sketched as follows:

1. $b_i^* \sim f(b_i|Y_{obs}, Y_{mis}, \theta)$. Specifically, $vec(b_i) \sim N(vec(\tilde{b}_i), U_i)$, where $vec(\tilde{b}_i) = U_i(\Sigma^{-1} \otimes Z_i^T)vec(y_i - X_i\beta)$, and $U_i = (\Phi^{-1} + (\Sigma^{-1} \otimes Z_i^T Z_i))^{-1}$, where $\otimes$ is the Kronecker product of two matrices.

2. $\theta^* \sim f(\theta|Y_{obs}, Y_{mis}, B)$, where B is the collection of b_i's. Specifically, first draw Φ^{-1} from a Wishart distribution with degrees of freedom $\nu_2' = \nu_2 + m$ and scale $\eta_2' = (\eta_2^{-1} + B^T B)^{-1}$. Second, calculate $\hat{\beta} = (\sum_{i=1}^m X_i^T X_i)^{-1}(\sum_{i=1}^m X_i^T(y_i - Z_ib_i))$ and residuals $\hat{\epsilon}_i = y_i - X_i\hat{\beta} - Z_ib_i$,

and draw Σ^{-1} from a Wishart distribution with degrees of freedom $\nu_1' = \nu_1 - p + \sum_{i=1}^{m} n_i$ and scale $\eta_1' = (\eta_1^{-1} + \sum_{i=1}^{m} \hat{\epsilon}_i^T \hat{\epsilon}_i)^{-1}$. Finally, draw β from a multivariate normal distribution centered as $\hat{\beta}$ with covariance matrix $\Sigma \otimes V$, where $V = (\sum_{i=1}^{m} X_i^T X_i)^{-1}$.

3. $y_{i,mis}^* \sim f(y_{i,mis}|Y_{obs}, B, \theta)$. Specifically, note that the rows of $\epsilon_i = y_i - X_i\beta - Z_i b_i$ are independent and normally distributed with mean zero and covariance matrix Σ. Therefore, in any row of ϵ_i, the missing elements have an intercept-free multivariate normal regression on the observed elements; the slopes and residual covariances for this regression can be calculated by inverting the square submatrix of Σ corresponding to observed variables. Drawing the missing elements in ϵ_i from the regression and adding them to the corresponding elements of $X_i\beta + Z_i b_i$ completes the simulations of $y_{i,mis}$.

4. Repeat the above steps independently M times to obtain multiple imputations.

The above algorithm has been implemented in R pan. Yucel (2008; 2011) provided additional details and studied extensions of Model (9.3). Note that Model (9.3) can be used more generally for multilevel data that are not limited to longitudinal data. Imputation algorithms for different types (i.e., continuous, categorical, and mixed) of multilevel data have also been developed, some of which will be briefly discussed in Section 9.5.

9.3.3 Examples

In this section we present a few examples of using mixed model-based imputation for longitudinal data, where missing values can occur for outcomes and/or covariates.

Example 9.3. *Imputation for missing weights of rats*

The original data of this example are taken from Section 6 of Gelfand et al. (1990), and include 30 young rats whose weights were measured weekly for five weeks. The data are shown in Table 9.1. This dataset is also used to illustrate the coding of a hierarchical Bayesian model in WinBUGS (example 1 of Volume 1 and example 8 of Volume 2, from http://www.mrc-bsu.cam.ac.uk/wp-content/uploads/WinBUGS_Vol1.pdf)). We follow the identical procedure as in the WinBUGS reference manual by deleting the last observation of cases 6-10, the last two from 11-20, the last 3 from 21-25, and the last 4 from 26-30. This results in 60 missing values from 150 observations. Fig. 9.1 shows the plot of observed cases, which appear to follow linear trends over time.

We consider a linear mixed model as follows:

$$y_{ij} = \beta_{0i} + \beta_{1i} t_j + \epsilon_{ij}, \tag{9.4}$$

where y_{ij} records the weight of rats for $i = 1, \ldots, 30$, $j = 1, \ldots, 5$, $t_j =$

TABLE 9.1
Example 9.3. Weights of 30 rats in the wide format

Rat	Day 8	Day 15	Day 22	Day 29	Day 36
1	151	199	246	283	320
2	145	199	249	293	354
3	147	214	263	312	328
4	155	200	237	272	297
5	135	188	230	280	323
6	159	210	252	298	331*
7	141	189	231	275	305*
8	159	201	248	297	338*
9	177	236	285	350	376*
10	134	182	220	260	296*
11	160	208	261	313*	352*
12	143	188	220	273*	314*
13	154	200	244	289*	325*
14	171	221	270	326*	358*
15	163	216	242	281*	312*
16	160	207	248	288*	324*
17	142	187	234	280*	316*
18	156	203	243	283*	317*
19	157	212	259	307*	336*
20	152	203	246	286*	321*
21	154	205	253*	298*	334*
22	139	190	225*	267*	302*
23	146	191	229*	272*	302*
24	157	211	250*	285*	323*
25	132	185	237*	286*	311*
26	160	207*	257*	303*	345*
27	169	216*	261*	295*	333*
28	157	205*	248*	289*	316*
29	137	180*	219*	258*	291*
30	153	200*	244*	286*	324*

Note: * denotes values that are set as missing.

$(8, 15, 22, 29, 36)$, $\beta_i = (\beta_{0i}, \beta_{1i})^t \sim N_2(\mu_\beta, \Omega_\beta)$, and $\epsilon_{ij} \sim N(0, \sigma^2)$. Here $\mu_\beta = (\beta_0, \beta_1)$ is a 2×1 mean vector for case-level intercepts and slopes, and Ω_β is an unstructured 2×2 covariance matrix. To complete the Bayesian model specification, we apply a diffuse prior for μ_β as $\mu_\beta \sim N(0, 10^6 I_2)$, $\sigma^2 \sim Inverse - Gamma(0.001, 0.001)$, and $\Omega_\beta^{-1} \sim Wishart(R, 2)$, where $R = \begin{pmatrix} 200, 0 \\ 0, 0.2 \end{pmatrix}$.

Data in Table 9.1 are presented in the so-called wide format (i.e., one record line for each subject). To fit longitudinal data using mixed models, we typically need to arrange the data into the so-called long format (Table 9.2).

WinBUGS is used to implement the imputation based on Model (9.4), and some sample code is included:

```
model
{
for(i in 1:N) {
for(j in 1:T) {
Y[i,j] ~ dnorm(mu[i,j], tauC)
mu[i,j] <- beta[i,1] + beta[i,2] * t[j]
}
beta[i,1:2] ~ dmnorm(mu.beta[], R[,])
}
mu.beta[1:2] ~ dmnorm(mean[],prec[,])
R[1:2,1:2] ~ dwish(Omega[,], 2)
tauC ~ dgamma(0.001, 0.001)
}
```

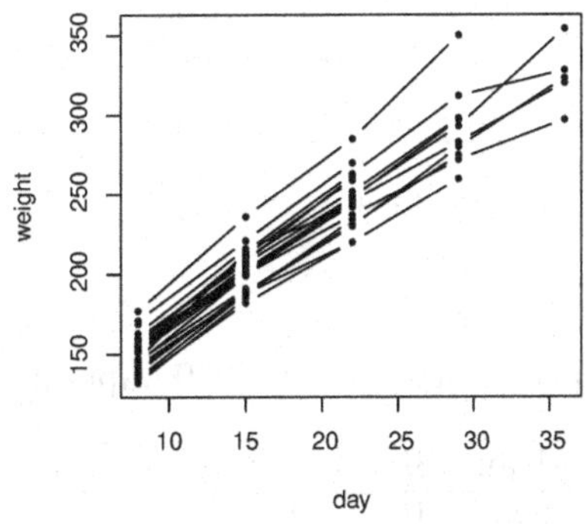

FIGURE 9.1
Example 9.3. The trajectory plot of rats' observed weights.

Note that there exist several imputation software programs (e.g., R pan) that can be used to execute the linear mixed model-based imputation. We use WinBUGS here to show how the mixed model is coded, which is not very obvious in some of the alternative packages. After running a data augmentation chain for 10000 iterations with a burn-in of 5000 iterations, we take draws for every 100 iterations to obtain 50 sets of imputations. Since we know true values of the missing cases in this example, the quality of imputations can be assessed by comparing the true and imputed values. For illustration, we choose the 26th case and present the summary statistics (mean, standard deviation, and percentiles) for the 50 imputations of the four missing weights in Table 9.3. The true weights all fall within the 25-75%-tile of the imputed values, and some of them are close to the center of the distribution of imputed values.

TABLE 9.2
Example 9.3. Weights of 30 rats in the long format

Rat	Weight	Day
1	151	8
1	199	15
1	246	22
1	283	29
1	320	36
2	145	8
2	199	15
...	...	...
11	160	8
11	208	15
11	261	22
11	M	29
11	M	36
...	...	...

Note: "M" symbolizes missing values.

TABLE 9.3
Example 9.3. Summary posterior statistics of 50 imputations for the 26th case

True weight	Mean	SD	25%-tile	50%-tile	75%-tile
207	205	10	198	206	213
257	252	14	244	250	256
303	298	16	286	298	307
345	345	19	335	344	354

An appealing feature of using WinBUGS is that the posterior estimates of model parameters can also be obtained more easily because WinBUGS is devised as a Bayesian analysis software. In this example, the posterior means are: $\mu_\beta = (102.1, 6.6)^t$, $\Omega_\beta = \begin{pmatrix} 85.04, 0.028 \\ 0.028, 0.30 \end{pmatrix}$, and $\sigma^2 = 36.8$. In addition, WinBUGS allows more flexible specifications of the prior distribution of model parameters.

Example 9.4. *Imputation for nonresponses in a longitudinal survey using cubic smoothing splines*

Sometimes functional forms of the temporal trend in longitudinal responses are not obvious. If the number of time points is sufficiently large, then smoothing techniques introduced in Section 5.3 might be applied. He et al. (2011) considered a cubic smoothing spline-based imputation to incomplete longitudinal data. The method was motivated by the problem of handling nonresponses in the Panel Study of Income Dynamics (PSID). In brief, PSID is a longitudinal survey of a nationally representative sample of U.S. families. Originated in 1968, PSID interviewed and re-interviewed individuals from sampled families every year until 1996. Since 1997, the interviewing has taken place every other year. The panel is based on a complex survey design for households. It follows adults through the full life course and also includes families when children of the original sample grow up and establish separate households or when marriage goes separate ways. PSID emphasizes the dynamic aspects of economic and demographic behavior, but it contains a wide range of measures, including health, sociological, and psychological ones. In 1997, PSID supplemented its main data collection with additional data on 0- to 12-year-old children, their parents, and other care givers. The Child Development Supplement (CDS) cohort of PSID studies a broad array of developmental outcomes such as physical health, emotional well-being, cognitive abilities, achievements, and social relationships with family and peers. More information about PSID and CDS can be found from (http://psidonline.isr.umich.edu).

He et al. (2011) considered a subsample of 2172 families included in the PSID cohort from 1979 to 1997, and augmented it with the data from their children collected in CDS. For illustration, we use their dataset and brief their work here. For simplicity, we only include the data from the eldest child in each family and ignore the sampling design in our illustration. The longitudinal response variables are annual records of family income-to-needs ratio from 1979 to 1997, which is defined as the ratio of household income to the poverty threshold for the corresponding family size. The average missingness rate for the income data is approximately 15.1% per year, with a declining trend at the end. The missingness pattern is general. The subset also included a few fully observed covariates: the child's gender (GENDER: 1=Male, 51.8%; 0=Female, 48.2%); age (AGE: mean=6.8 years, std=3.7 years), health status (HEALTH: 1=Excellent/very good, 82.4%; 0=Good/fair/poor, 17.6%), the household head's race (RACE: 1=White, 51.8%; 0=Non-white, 48.2%), and education level (EDU: 1=More than high school, 43.1%; 0=High school or less, 56.9%).

A log-transformation to the income-to-needs ratio (LOGINR=log(Income-to-needs+1)) was applied to accommodate the skewness of income data and carry out the imputation on the transformed scale. Fig. 9.2 includes the histogram and QQ plots of LOGINR, which shows a reasonable approximation of normality. Fig. 9.3 plots the LOGINR over a 19-year period from the dataset. The mean trajectory (thick line) is increasing over time as expected, and the trend appears to be linear. Yet the individual series (thin lines) from 25 randomly chosen families exhibit more noisy patterns, and their temporal patterns are not easy to be determined by visual inspections (unlike in Example 9.3).

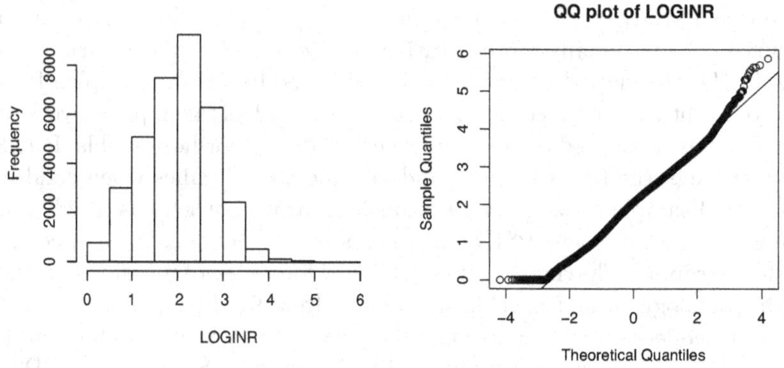

FIGURE 9.2
Example 9.4. Left: the histogram of observed LOGINR; right: the QQ plot.

To allow more flexible assumptions of the temporal trends, He et al. (2011) modeled longitudinal response profiles as smooth curves of time. These curves were estimated using cubic smoothing splines. Briefly speaking, let $\mathbf{t} = (t_1, \ldots, t_T)^t$, y_{t_j} denote the value of y measured at t_j for $j = 1, \ldots T$, a cubic smoothing spline is a curve estimate $\hat{f}$ that minimizes the penalized sum of squares,

$$\sum_{t=t_1}^{t_T} (y_{t_j} - f(t_j))^2 + \lambda \int_{t_1}^{t_T} \{f''(x)\}^2 dx,$$

where λ is a positive smoothing parameter. Wahba (1978) showed that the estimate of a cubic smoothing spline can also be obtained through the posterior estimate of a Gaussian stochastic process (i.e., integrated Weiner process). That is,

$$f(t) = B_1 + B_2 t + \lambda^{-1/2} \int_0^t W(s)ds, \tag{9.5}$$

where B_1 and B_2 have diffuse priors and are treated as fixed in the terminology of linear mixed models (i.e., $(B_1, B_2)^t \sim N_2(0, \tau I_2)$ with $\tau \to \infty$), $W(s)$'s is a Weiner process, and λ is the smoothing parameters controlling the balance between smoothness and "goodness-of-fit" for $f(t)$. Therefore, $\lambda^{-1/2} \int_0^t W(s)ds$ model the nonlinear curvature of $f(t)$.

As shown by Green (1987) and Zhang et al. (1998), the estimation of $f(t)$ in Eq. (9.5) at the design points $\mathbf{t}$ can be obtained through the following linear mixed model:

$$f(\mathbf{t}) = \mathbf{T}\boldsymbol{\zeta} + \mathbf{Ba}, \tag{9.6}$$

where $\mathbf{T} = \{\mathbf{1}, \mathbf{t}\}$ is a $T \times 2$ matrix, $\boldsymbol{\zeta} = (B_1, B_2)^t$ is a 2×1 vector, $\mathbf{B}$ is a $T \times (T-2)$ matrix which satisfies $\mathbf{BB}^T = \mathbf{R}^1$, where $\mathbf{R}^1$ is the integrated Weiner covariance matrix evaluated at $\mathbf{t}$, and the random effect $\mathbf{a} \sim N(0, \frac{1}{\lambda} I_{T-2})$. In mixed model (9.6), $\boldsymbol{\zeta}$ models the fixed intercept and slope for function $f(t)$, $\mathbf{a}$ is a $(T-2) \times 1$ vector of random effects so that $\mathbf{Ba}$ models the departure of $f(t)$ from the straight line $\mathbf{T}\boldsymbol{\zeta}$. The inverse of smoothing parameters, $1/\lambda$, is a variance component of the mixed model (9.6). When $\lambda \to \infty$, the estimated function $f(t)$ is close to a straight line. On the other hand, when $\lambda \to 0$, the estimated function tends to interpolate all the data points. The advantage of using the mixed model format of the smoothing spline is that the smoothing parameter λ is automatically estimated as a variance component following a data-driven principle.

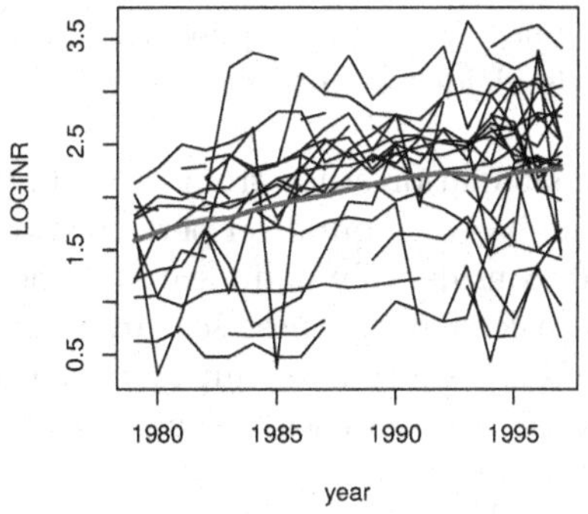

FIGURE 9.3
Example 9.4. The trajectory plot of LOGINR. Thick line: the mean profile from all families; thin lines: individual plots from 25 randomly selected families.

In the context of PSID data, He et al. (2011) modeled the longitudinal data via a functional mixed model (Guo 2002):

$$y_{ij} = X_{ij}\beta(t_{ij}) + Z_{ij}\alpha_i(t_{ij}) + \epsilon_{ij}, \epsilon_{ij} \sim N(0, \sigma^2), \tag{9.7}$$

where $X_{ij} = (x_{ij1}, \ldots, x_{ijp})(1 \times p)$ and $Z_{ij} = (z_{ij1}, \ldots, z_{ijq})(1 \times q)$ are design matrices, $\beta(t) = (\beta_1(t), \ldots, \beta_p(t))^T$ is a $p \times 1$ vector of fixed functions, $\alpha_i(t) = (\alpha_{i1}(t), \ldots, \alpha_{iq}(t))^T$ is a $q \times 1$ vector of random functions that are modeled as realizations of Gaussian processes $A(t) = (a_1(t), \ldots, a_q(t))^T$(collection of processes) with zero means; and ϵ_{ij} is the measurement error. Model (9.7) can be viewed as a variation of Model (9.1) where both the fixed and random effects are characterized as smooth functions of time.

Estimating both $\beta(t)$ and $\alpha_i(t)$'s using cubic smoothing splines and expressing them using the linear mixed model format as in Eq. (9.6), Model (9.7) can be expressed as a linear mixed model as in Model (9.1). Imputation of missing values and estimation of model parameters can therefore be conducted using the DA algorithm. Details can be found in He et al. (2011).

Fig. 9.4 plots the observed values, multiply imputed values, as well as the estimated LOGINR from three randomly selected families. The family-specific curve estimates are posterior means of the smooth functions from Model (9.7), and they are similar across independent Gibbs chains from the DA algorithm. There exists some curvature for the estimated trends, yet the deviation from linearity is not strong. The imputed values for each family are scattered around the estimated trend, and they are different across multiple Gibbs chains, incorporating the uncertainty of imputation.

In the meanwhile, the smoothing parameter estimates for both the fixed and random functions are large (on the scale of hundreds), indicating that both $\beta(t)$ and $\alpha_i(t)$'s are little different from linear curves. The imputation method assuming linear functions for $\beta(t)$ and $\alpha_i(t)$'s (i.e, models assuming random intercept and slopes) was also applied to the dataset using R pan. Both imputation methods yielded similar results for several post-imputation analyses.

On the other hand, a simulation study from He et al. (2011) demonstrated that the smoothing spline-based imputation performs well for scenarios with temporal trends that are considerably nonlinear, yet the imputation method assuming linear trends breaks down in these scenarios. Therefore, making correct assumptions for the temporal trend is important for imputing incomplete longitudinal data, especially for data with a large number of measurements over time.

In terms of practical implementation, many of the smoothing spline models (for both the continuous or noncontinuous outcomes) can be expressed as linear mixed models or generalized linear mixed models. They can also be conveniently coded and estimated using Bayesian software such as WinBUGS (e.g., Crainiceanu et al. 2005). Imputation based on these models using WinBUGS is a possible option.

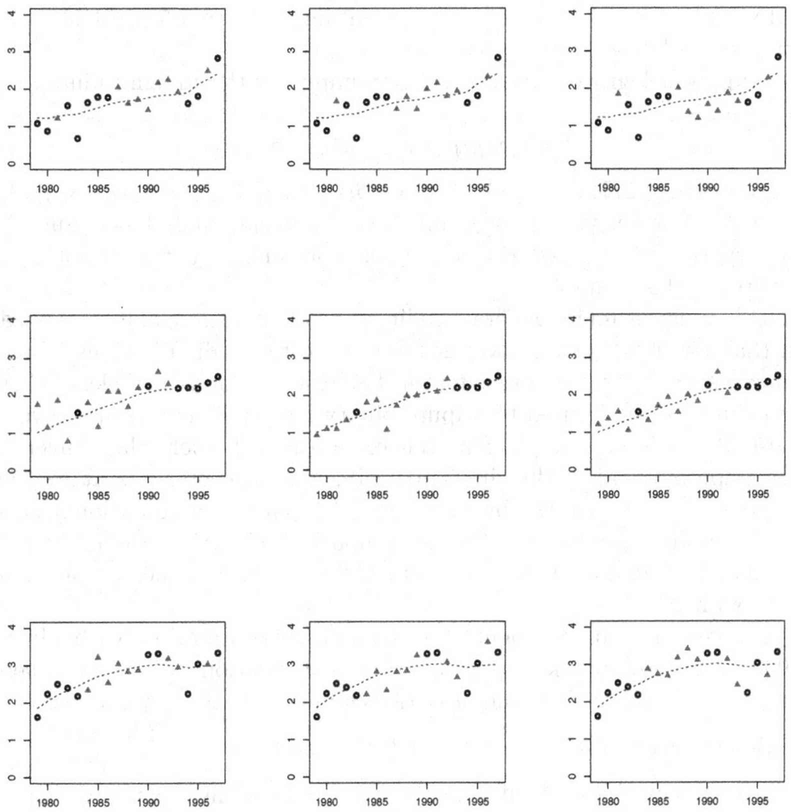

FIGURE 9.4
Example 9.4. Trajectory plots of the observed and imputed values of LOGINR; row (from top to bottom): 3 selected subjects; column (from left to right): 3 imputations per subject. Dashed line: family-specific curve estimates; circle: observed values; filled triangle: imputed values.

Example 9.5. *Imputation for missing binary data using logistic mixed models*

In Section 9.2 we briefly introduced the idea of generalized linear mixed models, which can be used to impute categorical longitudinal data. We continue using the PSID data from Example 9.4 to illustrate the method. In the dataset, the values of LOGINR range from 0 to 6, and the median and mean are both around 2. We create a "dichotomized" dataset by setting $Y = 1$ if LOGINR >= 2 and $Y = 0$ otherwise, resulting in around 50% of 1's and 0's for the whole dataset.

We apply a logistic mixed model for imputing the missing binary data. The model is

$$logit(f(y_{ij} = 1)) = \beta_{0i} + \beta_{1i}t_j, \tag{9.8}$$

where $i = 1, \ldots, 2172$, $t_j = (1, \ldots, 19)$, $\beta_i = (\beta_{0i}, \beta_{1i})^t \sim N_2(\mu_\beta, \Omega_\beta)$. Here μ_β is a 2×1 mean vector for family-level intercepts and slopes, and Ω_β is an unstructured 2×2 covariance matrix. For simplicity, we do not include covariates in this example.

The DA algorithm for generalized linear mixed models is more complicated than that for linear mixed models because posterior distributions of model parameters do not have closed forms. For example, it is not clear whether R pan can be directly used to impute binary longitudinal data. However, we can use WinBUGS to implement the imputation. To help the convergence of the imputation algorithm in WinBUGS, we can center t_j's at its mean. We also reparameterize Ω_β by modeling $f(\beta_{1i}|\beta_{0i})$ as a conditional normal distribution. To complete the Bayesian model specification, we apply diffuse prior distributions for model parameters. For simplicity, the sample code is not shown here.

After we run a data augmentation chain for 10000 iterations with a burn-in of 5000 iterations, we take draws for every 100 iterations to obtain 50 sets of imputations. Here we show the posterior estimates of model parameters. The posterior means are: $\mu_\beta = (-2.401, 0.264)^t$, and $\Omega_\beta = \begin{pmatrix} 13.65, -0.4142 \\ -0.4142, 0.0833 \end{pmatrix}$, for which the correlation coefficient (between random intercepts and slopes) is -0.4142. This makes sense because a higher baseline (intercept) is very likely to be associated with a less steep slope (in the logit scale) if the data are bounded between 0 and 1. This can be seen from the trajectory plots (i.e., $\beta_{0i} + \beta_{1i}t$) from 25 randomly chosen families in Fig. 9.5. Fig. 9.5 also shows the histograms of posterior means of random intercepts and slopes, as well as the scatter plot between them. Interestingly, the marginal distributions of posterior means of intercepts and slopes somewhat deviate from normality. The scatter plot also clearly shows that there is a negative correlation between the intercepts and slopes.

In general, however, fitting logistic mixed models (or generalized linear mixed models) for imputation can be more difficult than linear mixed models because categorical data contain less information than continuous data for the same number of parameters. This is more obvious if the number of time points

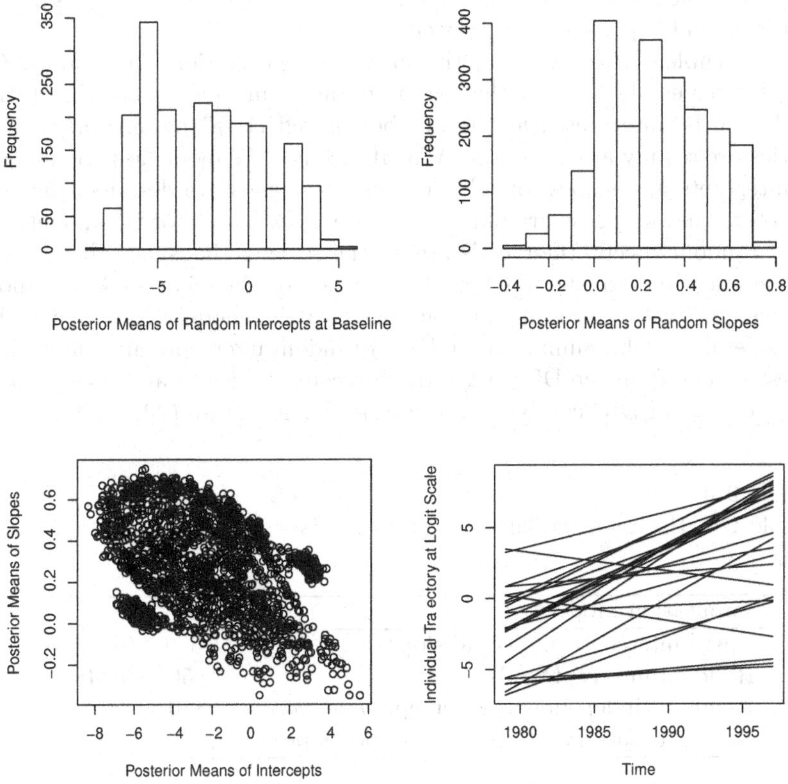

FIGURE 9.5
Example 9.5. Top left: the histogram of posterior means of random intercepts; top right: the histogram of posterior means of random slopes; bottom left: the scatter plot of posterior means of random intercepts and slopes; bottom right: individual trajectory on the logit scale for 25 randomly chosen families.

and covariates is not sufficiently large. Therefore we suggest always starting from the simplest model, for example, assuming that $logit(f(y_{ij} = 1))$ follows a model with identical linear trends (i.e., without random intercepts nor slopes) and then increasing the model complexity step by step (i.e., assuming random intercepts only, then assuming both random intercepts and slopes which are independent, and finally assuming correlated random intercepts and slopes). The estimates from the simpler model can be used as initial values for more complex models. The use of WinBUGS can accommodate this model exploration and expansion process well.

For example, Table 9.4 lists the deviance information criterion (DIC) (Spiegelhalter et al. 2002) of the four imputation models, where the statistics pD can be interpreted as the number of "effective" parameters. These statistics are readily available from WinBUGS. For the model assuming common intercepts and slopes for all subjects, the trends are described by two parameters, and its pD is around 1.5, which makes sense. For the models assuming random effects, their pD's are on the scale of thousands that match with the magnitude of the number of subjects. For Bayesian model diagnostics, lower DIC values generally indicate better-fitting models. Apparently, the most general model assuming (correlated) random intercepts and slopes has the best fit based on the DIC criterion. A recent discussion about the utility and limitations of DIC can be found in Spiegelhalter et al. (2014).

TABLE 9.4
Example 9.5. Bayesian model diagnostics statistics for different imputation models

Imputation Model	pD	DIC
Equal intercepts and equal slopes	1.502	46224
Random intercepts only	1850	26448
Random (independent) intercepts and slopes	2810	22980
Random (correlated) intercepts and slopes	2795	22914

In addition to missing repeated measurements in y_{ij}'s, some of the covariates (X's or Z's) used in mixed models might also be subject to missing data. As outlined in Section 6.5, a general JM imputation strategy is to specify an analysis model (i.e., the mixed model) and a covariate model that models the distribution of missing covariates.

Liu et al. (2000) presented a multiple imputation analysis for longitudinal data with both incomplete responses and covariates. The analysis model is a multivariate version of Model (9.1) (e.g., Model (9.3)). For continuous, time-invariant covariates X, Liu et al. (2000) assumed a conditional normal model for $f(X_{mis}|X_{obs})$. This specification follows the conditional modeling framework (Section 6.6.2). The full complete-data model consists of the product of the mixed model and the conditional normal model, plus prior distributions for the corresponding model parameters. A DA algorithm was applied to impute missing responses and covariates.

Example 9.6. *Imputation for missing LOGINR and baseline covariates in the PSID dataset*

We continue using the PSID dataset for illustration. In Examples 9.4 and 9.5, we did not include covariates in the imputation models. Here we include three baseline covariates: RACE, EDU, and HEALTH. These covariates were fully observed in the working dataset, yet we randomly delete 10% of each variable to create some missing covariates for illustration.

The analysis model is a linear mixed model:

$$y_{ij} = \beta_{0i} + \beta_{RACE} RACE_i + \beta_{EDU} EDU_i + \beta_{HEALTH} HEALTH_i + \beta_{1i} t_j + \epsilon_{ij}, \tag{9.9}$$

where $i = 1, \ldots, 2172$, $t_j = (1, \ldots, 19)$, $\beta_i = (\beta_{0i}, \beta_{1i})^t \sim N_2(\mu_\beta, \Omega_\beta)$, and $\epsilon_{ij} \sim N(0, \sigma^2)$. Here $\mu_\beta = (\beta_0, \beta_1)^t$ is a 2×1 mean vector for case-level intercepts and slopes, and Ω_β is an unstructured 2×2 covariance matrix.

The covariate model is specified using the conditional modeling framework as a series of logistic regressions :

$$\begin{aligned}
logit(f(RACE_i) = 1)) &= \mu_{race}, \\
logit(f(EDU_i = 1)) &= \gamma_{0,EDU.RACE} + \gamma_{1,EDU.RACE} RACE_i, \\
logit(f(HEALTH_i = 1)) &= \gamma_{0,HEALTH.EDU.RACE} + \gamma_{1,HEALTH.EDU} EDU_i + \\
&\quad \gamma_{1,HEALTH.RACE} RACE_i + \\
&\quad \gamma_{1,HEALTH.EDU.RACE} EDU_i \times RACE_i. \qquad (9.10)
\end{aligned}$$

Note that there exist multiple ways of specifying the covariate model. For example, a log-linear model for RACE, EDU, and HEALTH can also be specified for these variables.

Diffuse prior distributions can be applied to all model parameters. WinBUGS is used to implement the model-based imputation. Some sample code is included:

```
model
{
for( i in 1 : N ) {
# analysis model;
for( j in 1 : T ) {
Y[i,j] ~ dnorm(Y_mu[i,j], tau.error)
Y_mu[i,j] <- alpha[i] + alpha.c + alpha.health*health[i]
+alpha.edu*edu[i]+alpha.race*race[i]
+ beta[i] * (x[j] - xbar) + beta.c*(x[j]-xbar)
}
# covariate model;
```

```
race[i] ~ dbern(p_race[i])
edu[i] ~ dbern(p_edu[i])
health[i] ~ dbern(p_health[i])
logit(p_race[i]) <- mu_race
logit(p_edu[i]) <- alpha.edu.race+beta.edu.race*race[i]
logit(p_health[i]) <-
alpha.health.edu.race+beta.health.edu*edu[i]
+beta.health.race*race[i]+beta.health.edu.race*edu[i]*
race[i]
# random effects
alpha[i] ~ dnorm(0,tau.alpha)
beta[i] ~ dnorm(mu[i], tau.beta.alpha)
mu[i] <- slope*alpha[i]
}
tau.error <- 1/(sigma.error*sigma.error)
tau.alpha <- 1/(sigma.alpha*sigma.alpha)
tau.beta.alpha <- 1/(sigma.beta.alpha*sigma.beta.alpha)
# Prior
alpha.c ~ dnorm(0.0,1.0E-6)
alpha.health ~ dnorm(0.0, 1.0E-6)
alpha.edu ~ dnorm(0.0, 1.0E-6)
alpha.race ~ dnorm(0.0, 1.0E-6)
beta.c ~ dnorm(0.0,1.0E-6)
slope ~ dnorm(0.0, 1.0E-6)
mu_race ~ dnorm(0.0, 1.0E-6)
alpha.edu.race ~ dnorm(0.0, 1.0E-6)
beta.edu.race ~ dnorm(0.0, 1.0E-6)
alpha.health.edu.race ~ dnorm(0.0, 1.0E-6)
beta.health.edu ~ dnorm(0.0, 1.0E-6)
beta.health.race ~ dnorm(0.0, 1.0E-6)
beta.health.edu.race ~ dnorm(0.0, 1.0E-6)
sigma.error ~ dunif(0,100)
sigma.alpha~ dunif(0,100)
sigma.beta.alpha ~ dunif(0,100)
}
```

Based on our limited experience, choices of initial values of model parameters and random effects might impact running the WinBUGS program. Again we suggest starting with simpler imputation models to get some ideas about possible ranges of parameter values and then using them as initial values for more complex models.

Table 9.5 shows the posterior mean and standard deviation, as well as the 95% credible intervals of some parameters. These estimates are based on 5000 iterations after discarding a burn-in of 5000 iterations. The direction of these coefficients makes intuitive sense. For example, the baseline level of LOGINR is higher for white, higher education, and better health groups, judged by estimates for β_{RACE}, β_{EDU}, and β_{HEALTH}, respectively. In addition, in terms of the distribution of covariates, there exists a positive association between a higher education level and being white ($\gamma_{1,EDU.RACE} = 0.760$), a positive association between having better health and being white ($\gamma_{1,HEALTH.RACE} = 0.686$), and a positive association between having better health and having a higher education level ($\gamma_{1,HEALTH.EDU} = 0.333$). Imputed values are expected to retain such associations from observed data.

TABLE 9.5
Example 9.6. Posterior estimates of regression coefficients from the imputation model

Parameter	Mean	SD	2.5%-tile	97.5%-tile
β_0	1.332	0.0082	1.315	1.347
β_1	0.037	0.000965	0.035	0.039
β_{RACE}	0.569	0.0065	0.556	0.582
β_{EDU}	0.466	0.0065	0.453	0.479
β_{HEALTH}	0.208	0.0087	0.191	0.225
μ_{RACE}	0.071	0.042	−0.011	0.154
$\gamma_{1,EDU.RACE}$	0.760	0.088	0.581	0.932
$\gamma_{1,HEALTH.EDU}$	0.333	0.162	0.0226	0.648
$\gamma_{1,HEALTH.RACE}$	0.686	0.152	0.392	0.994
$\gamma_{1,HEALTH.EDU.RACE}$	0.127	0.246	−0.352	0.602

Note: Posterior estimates are based on 5000 iterations after discarding a burn-in of 5000 iterations.

Example 9.7. *Imputation of missing time-invariant covariates in nonlinear mixed models*

Wu and Wu (2001) tackled a missing time-invariant covariates problem in nonlinear mixed models as follows:

$$y_{ij} = f(t_{ij}, \beta_i) + e_{ij}, \quad (9.11)$$
$$\beta_i = X_i\beta + u_i, \quad (9.12)$$

where $f(\cdot)$ can be a nonlinear function, t_{ij} indicates the time (dose) of the measurement y_{ij}, $E(e_{ij}) = 0$, β_i is the regression covariate for individual i, X_i is the corresponding design matrix containing time-invariant covariates, and u_i is the random effects with $u_i \sim N(0, D)$. A simple example of a nonlinear function in Eq. (9.11) is $y_{ij} = \beta_{0i} + \beta_{2i}e^{-\beta_{1i}t_{ij}} + e_{ij}$. More discussion about nonlinear mixed models can be found, for example, in Wu (2010).

When some of the X_i's are missing, deriving the exact posterior distribution of $f(X_{mis}|Y, X_{obs})$, or even producing posterior simulations, might be difficult because of the nonlinear feature of Models (9.11) and (9.12). Wu and Wu (2001) proposed to set up a regression-based imputation model as

$f(X_{i,mis}|X_{i,obs}, \hat{\beta}_i)$, where $\hat{\beta}_i$ are the estimates of the individual-level regression coefficient (i.e., the random effects). This can be justified by the belief that β_i summarizes the association between Y_i and X_i, which is necessary to be included for imputing missing covariates (Little 1992). With incomplete X_i's, the initial values of $\hat{\beta}_i$ can be obtained by fitting Models (9.11) and (9.12) without using the incomplete covariates, and then they can be updated each time when plugging in the multiply imputed missing covariates and using Rubin's combining rules. The performance of this method was shown to be satisfactory by simulation studies.

A more "exact" solution to this problem might be implemented using WinBUGS, which can be used to fit nonlinear mixed models automatically (e.g., Gurrin et al. 2003). A similar strategy as in Example 9.6 can be applied as long as a covariate model for missing X's is formed.

9.4 Wide Format Imputation

In previous sections we covered the Bayesian mixed model-based imputation for longitudinal data. Mixed models are widely used to model longitudinal data and have appealing features. However, the computational sophistication can be burdensome for many practical problems. For example, the computational algorithm can be complicated if the mixed model includes a large number of random effects. On the one hand, practitioners can seek help from specific imputation software packages (e.g., R pan, R mice, R jomo, and WinBUGS). On the other hand, it is important to have some alternative options that might be easier to apply.

In a balanced longitudinal data setting as in Examples 9.3, an effective alternate strategy is to treat the single missing variable measured at different time points as separate variables based on the wide format. This will transform the original problem into a multivariate missing data problem (e.g., in $Y_1, \ldots, Y_T$) in a cross-sectional setting. In theory, a univariate linear mixed model with normal random effects and error distributions (e.g., Model (9.1)) can be transformed into a multivariate normal model for the repeated measurements with structured mean and variance components (Tang et al. 2005). Therefore, imputing the missing repeated measurements using a more general multivariate normal model with unstructured mean and variance components is a feasible alternative strategy.

Example 9.8. *Multivariate normal model imputation vs. linear mixed model imputation for rats weight data*

In Example 9.3 we used a random coefficients model (Eq. (9.4)) for the rats weight data. We can also treat data in Table 9.1 consisting of five variables ($Y_1 \ldots Y_5$), which measure the weights at $t = 8, 15, 22, 29, 36$, respectively. Under Model (9.4), it can be shown that Y_1 through Y_5 follow a five-variate normal distribution. Specifically, $E(Y_j) = \beta_0 + \beta_1 t_j$, $Var(Y_j) = (1, t_j)\Omega_\beta(1, t_j)^t + \sigma^2$, $j = 1, \ldots 5$; and $Cov(Y_{j_1}, Y_{j_2}) = (1, t_{j_1})\Omega_\beta(1, t_{j_2})^t$ for $j_1 \neq j_2$. For the purpose of imputation, however, we can specify a more general multivariate normal model for Y_j's by assuming $(Y_1 \ldots, Y_5) \sim N_5(\mu, \Sigma)$, where μ is a 5×1 vector and Σ is a 5×5 unstructured covariance matrix. This specification is less parsimonious than the mean and covariance matrix implied by Model (9.4). In this example, the multivariate normal model-based imputation is implemented by R mice.

We compare the two imputation methods for the missing rat weight data. We estimate the means of rats' weights on waves 2 to 5. Table 9.6 shows the results based on $M = 50$ imputations. Compared with CC, both imputation methods improve the precision because the former method completely discards information from incomplete observations. Results from the two imputation methods are largely similar. From waves 2 to 4, the point estimates from both imputation methods are rather close. In wave 5, the mean estimate based on the mixed model imputation appears to be relatively higher than the normal model imputation (i.e., 336 vs. 324). This might be due to the fact that mixed model imputation assumes a linear trend for the temporal pattern, even for wave 5 when there are only 20% observed data to support such an assumption. Judging from Fig. 9.1, however, the observed data might show a slight drop from the linear trend at wave 5. In addition, the standard errors from the mixed model imputation are somewhat and consistently larger than those from the normal model imputation. This might also be due to the fact that given more drop-outs at later waves, a mixed model assuming linear trends might need to account for more uncertainty, which is reflected in the increased standard errors.

TABLE 9.6
Example 9.8. Mean estimates of rats' weights on wave 2 to wave 5

Mean	CC	Mixed model imputation	Normal model imputation
Wave 2	201.8 (2.563)	201.7 (2.443)	202.2 (2.295)
Wave 3	246.4 (3.597)	245.4 (3.191)	245.2 (2.883)
Wave 4	292.0 (8.003)	291.7 (4.129)	289.1 (4.031)
Wave 5	324.4 (9.114)	336.0 (4.976)	324.4 (4.547)

Note: Standard errors are included in parentheses.

The normal model-based imputation does not explicitly model the temporal trend as in the mixed model, yet implicitly accounts for that through the relationship among different variables. On the other hand, the former approach might offer some robustness if the temporal trend is misspecified in mixed models. The correlation among repeated measurements within each subject is also accounted for in the former approach through the covariance matrix of the multivariate normal model, albeit in a less refined manner.

In general, Raghunathan et al. (2018, Page 145) recommended using the wide format imputation for balanced longitudinal data because of its flexibility and generality. For example, a wide class of multivariate joint models can be applied for $Y_1, \ldots, Y_T$ and other covariates. When repeated measurements are categorical variables, imputing the data using generalized linear mixed models can be especially computationally demanding, yet imputing them using established multivariate categorical data models (e.g., log-linear models, latent variable models, etc.) is relatively straightforward. The wide format imputation can be conducted using either a JM or FCS strategy. As discussed in Chapter 7, it is more convenient to use FCS to account for complex structures of data. In terms of practice, the wide format imputation strategy is perhaps more appealing if the main purpose is to construct a completed database to accommodate a wide variety of analyses using longitudinal data.

9.5 Multilevel Data

Longitudinal data are a special case of multilevel data. Multilevel data structures arise when observations of individual units cannot be considered as independent, but instead are correlated because they are nested within groups or clusters of various kinds. Typical examples include observations repeatedly collected over time in longitudinal studies, patient observations nested within practices/providers in health outcomes and services contexts, students nested in classes within schools in educational settings, or survey samples collected following the cluster sampling design. There exists extensive literature for statistical analysis of multilevel data in various settings. For example, see Goldstein (2011) and references therein.

Because of the generality of multilevel data, as well as the complexity and variety of multilevel data analysis in various fields (e.g., biostatistics, social and behavioral statistics, and survey statistics), multiple imputation of multilevel data possesses many challenges. For example, for some type of multilevel data (e.g., missing test scores from unequal numbers of students nested within schools), the wide format imputation (Section 9.4) makes less sense, and it is not immediately clear that this method can be directly applied. The mainstream imputation strategy is to use the long format of the multilevel data and use mixed (or multilevel) models. There exists much ongoing research in this topic area, and this field is moving rapidly. Some detailed discussion can be found in Carpenter and Kenward (2013, Chapter 9) and Van Buuren (2018, Chapter 7).

As for regular data, JM and FCS are the two main approaches to imputation of multilevel data. The key to the JM strategy is to specify a complete-data model for multilevel data including both responses and covariates, which can both be subject to missing values. There are many ways of specifying multilevel models (or hierarchical Bayesian models from a Bayesian perspective). Here we use a model proposed by Goldstein et al. (2009) for illustration. The type of models is termed multilevel models for multivariate mixed response types by the relevant literature. Let $i = 1, \ldots, n$ index the level-2 units (i.e., clusters or groups), and $j = 1, \ldots n_i$ index level-1 units, nested within the level-2 units. A two-level model for outcome y_{ij} can be specified as:

$$y_{ij}^{(1)} = X_{1ij}\beta^{(1)} + Z_{1ij}b_i^{(1)} + \epsilon_{ij}^{(1)}, \tag{9.13}$$

$$y_i^{(2)} = X_{2i}\beta^{(2)} + Z_{2i}b_i^{(2)}, \tag{9.14}$$

$$\epsilon_{ij}^{(1)} \sim N(0, \Omega_1), \tag{9.15}$$

$$b_i = (b_i^{(1)}, b_i^{(2)})^t \sim N(0, \Omega_2). \tag{9.16}$$

The superscripts denote the level at which the variable is measured or defined. Thus, $y_{ij}^{(1)}(1 \times p_1)$ contains p_1 (latent or actual) normal responses that are defined at level 1, and $y_i^{(2)}(1 \times p_2)$ contains p_2 responses that are defined at level 2. Let $X_{1ij}(1 \times f_1)$ contain the level-1 predictor variables and $\beta^{(1)}(f_1 \times p_1)$ be the matrix containing the fixed coefficients for these predictors. Similarly, $Z_{1ij}(1 \times q_1)$ is the vector that contains the predictor variables for the q_1 level-1 random effects denoted by $b_i^{(1)}(q_1 \times p_1)$ for the level 1 responses. The level-1 residuals $\epsilon_{ij}^{(1)}$ are calculated by subtracting the current estimate of the linear component of the model from each of the level-1 responses, with their distribution assumed by Eq. (9.15). At level 2, $X_{2i}(1 \times f_2)$ is the vector that contains predictor variables for higher level unit i and $\beta^{(2)}(f_2 \times p_2)$ contains the fixed coefficients. The vector $Z_{2i}(1 \times q_2)$ contains the predictor for level-2 random effects $b_i^{(2)}(q_2 \times p_2)$ for the level-2 responses. These random effects can be correlated with the level-2 residuals for the level-1 responses $b_i^{(1)}$ by assuming Eq. (9.16). Prior distributions are also imposed for fixed effects β's and variance components Ω's to complete the Bayesian model specification.

Needless to say, the multilevel model specified by Eqs. (9.13)-(9.16) is very general and sophisticated. It models the responses at different levels via Eq. (9.13) and (9.14), as well as connects the two by Eq. (9.16). The mixed models discussed in longitudinal data (e.g., Model (9.1) and (9.3)) can be viewed as a special case of the multilevel model by dropping Eq. (9.14) and only considering the level-1 responses. The multilevel model can be further extended or modified to accommodate different types of data (e.g., ordinal or nominal for y's) by modeling them using the actual or latent scale. In addition, since the multilevel model defines an analysis model, models for covariates X and Z (at both levels) can be specified to define a covariate model. Both models are needed for imputation if missing data at least occur for covariates. A series of related work can be found, for example, in Carpenter et al. (2011), Carpenter and Kenward (2013), and Goldstein et al. (2014). The relevant imputation models are implemented via R jomo.

On the other hand, given the flexibility of FCS in handling multiple incomplete data with complicated data structure such as skip patterns and bounds (Chapter 7), there exists vibrant research on applying this strategy to multilevel data. To sketch the general idea, we use the formulation in Yucel et al. (2018). For variable $y_{ij}^{(1)}$ (using the same superscript as in Models (9.13)-(9.16)), a generalized linear mixed model can be specified as:

$$g(E(y_{ij}^{(1)}|x_{ij}, z_i, \theta)) = \eta_{ij}^{(1)}, \tag{9.17}$$

where $g(\cdot)$ denotes a function linking the expected value of response y_{ij} to the linear predictor,

$$\eta_{ij}^{(1)} = x_{ij}\beta^{(1)} + z_i b_i^{(1)}, \tag{9.18}$$

where x_{ij} and z_i are vectors of covariates corresponding to individual-level (i.e., level 1) and cluster-specific characteristics (i.e., level 2), $\beta^{(1)}$ and $b_i^{(1)}$ are the corresponding coefficients (population-average fixed effects and cluster-specific random effects, respectively), and θ denotes the unknown parameters such as regression coefficients ($\beta^{(1)}$) and variance components of the random effects governing the underlying model.

Note that in the context of FCS, covariates x_{ij} and z_i should be generally defined as variables in the data excluding y_{ij}. They are either completely observed or imputed in the previous cycles of the FCS imputation iterations. The imputation based on Models (9.17) and (9.18) is cycled through all the variables in the data.

Although the idea of FCS in multilevel data seems straightforward, many sophisticated issues can arise. For example, specifications in Models (9.17) and (9.18) are directly applicable to responses measured at level 1. What might be a good specification if a level-2 variable also has missing values? Other issues include derived variables that involve both level-1 and level-2 variables, unbalanced data and possible heterogeneity across clusters, cross-level interactions, and the selection of donors if PMM is to be used. There might not exist a unique solution to some of the questions, and certain strategies might be found by simulation studies and empirical evaluations. However, these complex issues, while adding to the existing compatibility problem in FCS, make the relevant research a very interesting and practically useful topic. Van Buuren (2018, Chapter 7) provided a comprehensive review of the methods, findings, and practical guidelines for using the FCS imputation for multilevel data. Many of the methods recommended there have been implemented by R mice, R miceadds, R micemd, and R mitml. On the other hand, it remains an interesting question to see whether WinBUGS and its variants can be used to conduct the FCS imputation for multilevel data as we explored its relevant utility in Chapter 7.

9.6 Summary

In this chapter we discussed some methods for imputing missing longitudinal data which can occur for responses and/or covariates. Since mixed models are the mainstream approach to modeling longitudinal data, multiple imputation based on mixed models is a natural strategy. In some cases, the wide format imputation can also be applied due to its simplicity and flexibility. In practice, the computational complexity associated with mixed models, or more generally, multilevel models, can be accommodated by using statistical software. We mainly used WinBUGS for illustration. However, specific imputation programs such as R pan, R jomo, and R mice provide many more options and frequent updates with new techniques. Practitioners are encouraged to explore and use these programs when necessary.

A key assumption behind methods discussed in this chapter is that missing longitudinal data are MAR. This assumption might not hold in some cases, especially in the scenarios of drop-outs. Drop-out or attrition refer to a specific situation where subjects are observed without interruption from the beginning of the study until a certain time point prior to the scheduled end of the study. Subjects who drop out in the middle do not return to the study. In clinical trials, possible reasons behind drop-outs include adverse events from the treatment/intervention, ineffective (or effective) medication in the middle of the study, or protocol violation. If the probability of dropping out from a study is related to the subject's response, then the missingness mechanism is not MAR and is nonignorable. Nonignorable missing data problems are generally very challenging, and we offer some discussions in Chapter 13. Details on the topic of analyzing longitudinal drop-outs can be found in, for example, Molenberghs and Kenward (2007) and Daniels and Hogan (2008), which covered many nonimputation missing data methods.

10

Multiple Imputation Analysis for Complex Survey Data

10.1 Introduction

The original motivation behind multiple imputation was to handle the nonresponse problem in public-use databases (e.g., large-scale population surveys) administered by organizations (Rubin 1978). This was first briefly mentioned in Section 3.1. Additional justifications are provided here, which summarize main messages from several key references such as Rubin and Schenker (1991), Rubin (1996), Barnard and Meng (1999), and Schafer (2003).

Public-use/shared databases are analyzed by many data users with varying degrees of statistical expertise and computing power, and with different scientific questions and objectives. Typically such users are able to conduct standard complete-data analyses (e.g., linear regression, logistic regression, survival analysis, mixed models, etc.) using various software packages such as SAS and R. In addition, program routines are readily available for complete-data management such as merging files, subsetting data, deleting cases and variables, and various resampling (e.g., bootstrap) and simulation programs.

Essentially all public-use datasets have missing values. In general, data users who do not have sophisticated statistical/programming expertise may have neither the knowledge nor the tools to address missing data problems satisfactorily. Applying ad hoc missing data methods (e.g., Section 2.5), which seems straightforward, can lead to invalid or suboptimal statistical inferences. In addition, data users are supposed to focus overall on their substantive scientific analyses, and missing data generally can be treated as a nuisance problem.

An effective solution is multiple imputation conducted by the organizations under carefully chosen models, and data users use the repeated-imputation inference (i.e., multiple imputation combining rules) for analysis to adequately incorporate imputation uncertainty. This is because multiple imputation is substantially easier for the data users than any other current methods that can satisfy the dual objectives of reliance only on complete-data methods and general validity of inference. This process is entirely general because it allows the consideration of any multiple imputation method and any complete-data analysis.

DOI: 10.1201/9780429156397-10

For the sake of argument, even if some data users do have adequate resources for modeling and computation and can conduct their own multiple imputation, imputers from the organization typically know more about reasons behind survey nonresponse and have better access to detailed information that can be useful for modeling missing data. Some of this information is confidential and might not be released for public use (e.g., exact addresses and neighborhood relationships). For instance, in some of the examples of this chapter, variables measuring distributions of demographics at geographic unit level (i.e., geocode-summary variables such as U.S. counties, census blocks, etc.) were used in imputation models. Yet these geocode-summary variables might not be always available on public-use data due to data confidentiality reasons.

On the other hand, it also has to be realized that, in general, imputers may not foresee all types of analyses from data users and may lack subjective knowledge behind some of the analyses. This asymmetry between imputer and data user in terms of their information about nonresponse and analysis corresponds to the issue of imputation uncogeniality (Meng 1994), which will be discussed in Section 14.2. However, this challenge does not imply that multiple imputation should not be used; it rather suggests that the imputation model should be sufficiently general to accommodate a wide variety of possible analyses, following the inclusive imputation strategy.

Despite the advantage of multiple imputation in principle, applying this strategy to complex probability survey data, which are the major data sources for some organizations, involves many complicated issues. From the theoretical perspective, a main challenge is to have imputation models conform to survey sampling design features of survey data. This is based on the fact that the design-based approach is the main inferential tool for probability survey data. From the applied perspective, in many real surveys the imputation often has to deal with a large number of variables and complex data structure (e.g. skip patterns, restrictions, bounds, and multiple survey forms). As discussed in Chapter 7, a promising solution is to use the FCS imputation strategy. In addition, from the organization's perspective, if multiply imputed data are to be released for external use, there exist additional issues in data processing and production.

In spite of these challenges, multiple imputation continues to be a useful tool to address survey nonresponse problems, as can be witnessed by multiple successful real-life applications. Its success benefits from the rapid development/expansion of the underlying theory, effective modeling strategies, and software implementation over the years. In this chapter we discuss several major issues for applying multiple imputation to survey data. Section 10.2 briefly introduces the design-based estimation for probability survey data. Section 10.3 presents several imputation modeling strategies targeted to survey data. Section 10.4 lists a few successful practical examples. Section 10.5 discusses the process of managing and releasing multiply imputed data from the organization's perspective. Section 10.6 provides a summary.

10.2 Design-Based Inference for Survey Data

A finite population sample survey consists of selecting a sample from a well defined list (i.e., the sampling frame). The selection process is based on a probability sampling process or mechanism so that the sample can be viewed as "representative" of the target population. A probability sample requires that every individual in the sampling frame has a positive probability of being included in the sample. Survey data are collected from the selected sample using a questionnaire or other means.

In practice, to capture important features of the target population, the sampling mechanism uses stratification of the population (based on the information in the sampling frame) and unequal probabilities of selection to achieve specific goals. To reduce the cost of administration (for example, interview traveling cost for face-to-face surveys), the sample is often selected in clusters and then the clusters might be further subsampled for individual units. That is, the sample selection can consist of multiple stages. All these considerations introduce weighting, stratification, and clustering as standard features in many sample surveys. These features need to be incorporated in the design-based inference for survey data, which will be briefly reviewed in the following.

As discussed in Chapter 2, traditional statistical inferences are based on statistical models built on the likelihood/probability function of data Y. The fundamental assumption is that Y can be treated as some random quantities generated from an infinite population following some distribution functions. However, for data collected from probability-based survey samples, the classic inferential approach has a different perspective and is often referred to as "design-based". Here "design" means the sampling mechanisms that are used to draw samples from a finite population. Examples of sampling mechanisms include simple random sample (SRS), stratified sample, cluster sample, and their combinations in multiple stages.

In the design-based inference for sample data, the estimands of interest are usually some finite population quantities. We denote y_i, $i = 1, \ldots, N$ as the observation for the finite population of interest. For unit i, we can define the sample indicator function $I_i = 1$ if unit i is included in the sample, and $I_i = 0$ otherwise. In a probability sampling scheme, each element y_i has a well defined probability to be selected (included), and the selection process follows the sampling mechanism. In Rubin (1987, Chapter 2), this process is characterized as a distribution function for I given data Y, the collection of y_i's, and design information Z as $\pi_i = f(I_i = 1|Y, Z)$, also termed as "the sample selection probability/function." The design information Z is typically from the sampling frame, and the selection process is known for the sampler.

A probability-based sample s would only include y_i's for which $I_i = 1$. Without loss of generality, denote these sampled cases as y_i, $i = 1, \ldots, n$

and assume all the sampled cases respond. Note that s typically takes only a small fraction of the population, that is $n < N$. Suppose the estimand of interest is the sum of y_i's as $T = \sum_{i=1}^{N} y_i$ (i.e., the population total), a finite-population quantity. The inferential problem for us is to estimate T based on the observations in sample s. In the design-based inferential framework, data Y are treated as fixed, yet the sampling indicator I is treated as random with respect to the sampling mechanism. Therefore, the vehicle for statistical inference is on the sampling indicator I.

More specifically, the design-based statistical inference hinges on the sampling weights $w_i = 1/\pi_i$ $(i = 1, \ldots, n)$, which is the inverse of the selection probability. An intuitive way of understanding the role of weights is that a unit selected from a target population with probability π_i is representing w_i units in the population and hence should be given the weight w_i in the estimate of a population quantity. For example, the population total T can be estimated as $\hat{T}_{HT} = \sum_{i=1}^{n} y_i w_i$, the well known Horvitz-Thompson (HT) estimator (Horvitz and Thompson 1952). For other estimands of interest such as means, ratios, and regression coefficients, sampling weights are also included in the estimators.

Variance estimation for weighted estimates are often complicated. This is because it involves the double inclusion probabilities of any two cases (e.g., i and j), $\pi_{ij} = f(I_{ij} = 1|Y, Z)$, which many times have no closed forms for the usual sampling mechanisms. Approximation methods such as Taylor linearizations or jackknife repeated replication techniques have been developed for variance estimation for commonly used sampling designs. More details about the design-based approach to sample data can be found in several classic textbooks such as Cochran (1977) and Sarndal et al. (1992). In addition, there also exist model-based or model-assisted analysis approaches to survey sample data, for example, reviewed by Little (2004).

Finally, we briefly introduce an important statistic in survey data analysis, the design effect originally proposed by Kish (1965). Generally speaking, design effect (Deff), is the ratio of the variance estimator under a complex design to the variance that would have been obtained from a SRS of the same number of units. Symbolically,

$$Deff(\hat{\theta}) = \frac{V(\hat{\theta})}{V_{SRS}(\hat{\theta})}, \tag{10.1}$$

where $\hat{\theta}$ is an estimator of some population quantities, V denotes variance under the current sampling design used, and V_{SRS} is the variance of the same quantity assuming SRS. The sample size for V_{SRS} is the same as the sample size of units used in the numerator estimate.

Design effect can be used for sampling design and planning (Valliant et al. 2013). Sometimes the design effect can be used for calculating important statistics such as confidence intervals for proportion estimates (e.g., Korn and Graubard 1999). In practical surveys, design effects are often greater than

1, reflecting the increase of variance (compared with SRS) due to unequal sampling weights and/or clustering in the design. The increase of variance can also be described conveniently using the notion of effective sample size n_{eff}, which is defined as $n_{eff} = n/Deff$ where n is the sample size of units in the survey. When $Deff > 1$, $n_{eff} < n$, showing the inflation of the variance or reduction of the precision due to the sampling design used.

As a matter of fact, calculation of design effect can be complicated for different estimators under various sampling designs, and some of them are implemented in certain software packages. A simple case is for a proportion estimator (i.e, $\theta = p$), and its design effect can be expressed as $Deff(\hat{p}) = \frac{V(\hat{p})}{\frac{\hat{p}(1-\hat{p})}{n}}$, where $\hat{p}$ and $V(\hat{p})$ are the proportion estimate and its variance obtained under the current sampling design, and n is the sample size in the survey. Additional discussion on design effect and its estimation can be found, for example, in Valliant et al. (2013).

Example 10.1. *Estimation of the proportion of the health insurance coverage from a web survey*

In Example 2.1 and 2.2, we used a variable measuring subjects' health insurance coverage to illustrate some basic concepts in statistical inference. This variable was collected from a web survey (RANDS I) based on a probability panel. For illustration, there we ignored the sampling design such as survey weights. Here we calculate the weighted estimate, $\frac{\sum_i w_i y_i}{\sum_i w_i}$, where y_i is 1 or 0 (yes or no) and w_i is the survey weight, $i = 1, \ldots, 2304$. The weighted estimate is 0.934, the standard error estimate is 0.008, and the corresponding 95% confidence interval is (0.918, 0.950). Compared with results from Example 2.1, here the confidence interval is wider, exhibiting the increased variability due to unequal sampling probabilities. The corresponding design effect estimate is $0.008^2/(0.934 * (1 - 0.934)/2304) = 2.392$, and the effective sample size is $2304/2.392 \approx 963$.

Fig. 10.1 plots the histogram of the weights. In this case, the survey weights are standardized so that they add up to the sample size 2304. Such standardization does not affect estimates for means and ratios. The ratio between the maximum and minimum weight is 41, suggesting a large variation of the "representativeness" among respondents.

Ideally, the sampling weights are expected to add up to the size of the target population, that is, $\sum_{i=1}^{n} w_i = N$. As a matter of fact, in Example 10.1 and also for many real-life probability surveys, the development of survey weights can be rather complicated. Typically, the original sampling weights (i.e., the inverse of the selection probability) have to be adjusted so that their totals in certain groups (e.g., classified by age, sex, and race) match with the known totals from the population. This adjustment process is usually called post-stratification, which is a special case of the general weight calibration approach (Sarndal et al. 1992).

In addition, the weighting strategy has been the main approach to handling survey unit nonresponse (Table 1.4). Unit nonresponses refer to subjects who are contacted yet refused to participate in the survey and hence provide little or no information for the survey questions. Nonresponse weighting can be justified by treating the unit nonresponse process as a second-stage selection in the sampling process. That is, the first-stage selection is from the population to the intended sample, and the second-stage selection is from the intended sample to actual respondents. Let R be the response indicator; we can use $f(R = 1|I = 1, Y, Z)$ to describe the nonresponse mechanism. However, a major challenge is that unlike the first-stage sample selection process, $f(I = 1|Y, Z)$, the exact response mechanism in the second stage is often unknown. Therefore, we often have to estimate $f(R = 1|I = 1, Y, Z)$ based on some assumptions.

For simplicity, suppose r of the n sampled cases respond. Let $\hat{\pi}_R = f(R = 1|I = 1, Y, Z)$ be the estimated response probability from the n sampled cases, and we can set up the response weights $w_R = 1/\hat{\pi}_R$ for the r respondents. In the two-stage sampling scheme and using practical terms, we have $f(\text{sample selection and response}) = f(\text{sample selection}) \times f(\text{response}|\text{sample selection}) = \pi \times \hat{\pi}_R$. The final weights for the r respondents are therefore the product of the sampling weights and response weights $w_{i,f} = w_i \times w_{R,i}$ for respondent i. Similar to the implication from sampling weights, an intuitive reasoning behind the nonresponse weighting is that the sum of the final weights from the respondents (i.e., $w_{i,f}$'s) shall add up to the total of population. That is, one respondent with an attached weight $w_{i,f}$ can be viewed as representing $w_{i,f}$ cases in the population after we conduct the weighting adjustment. Suppose the inferential problem is to estimate the

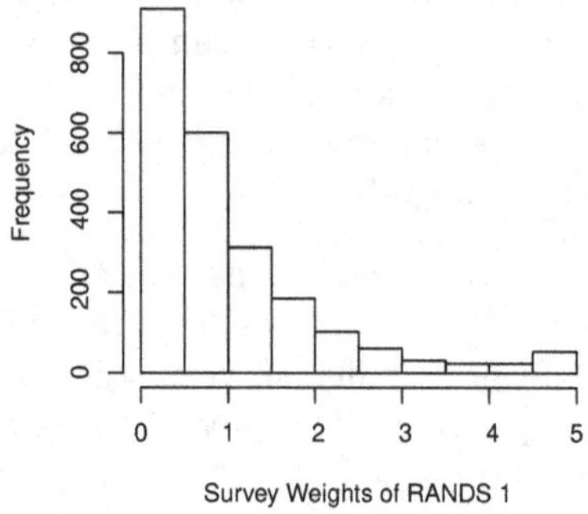

FIGURE 10.1
Example 10.1. The histogram of survey weights from RANDS I.

population quantity, such as T, from the r respondents. Again using the HT estimator, the estimate from the r respondents would be $\hat{T}_{HT,R} = \sum_{i=1}^{r} y_i w_{i,f}$. Variance estimation for nonresponse weighted estimates can also be constructed but is generally complicated.

In summary, nonresponse weighting is a traditional missing data approach to reducing unit nonresponse bias when the probability of being selected in the survey differs between respondents and nonrespondents. This strategy is widely used by organizations that produce and release large surveys. The method is relatively simple since only one set of weights is needed for all incomplete variables from respondents. Yet this simplicity might cast a doubt about whether this single set of weights can really capture the nonresponse behavior for all the variables in the data. More technical details can be found, for example, in Sarndal and Lundstrom (2005) and Raghunathan (2016, Chap. 2).

On the other hand, for item nonrespondents with general missingness patterns, the process of developing weights might involve discarding data by listwise deletion. In general it is not straightforward to handle item nonresponse, especially multivariate missing data problems, using weighting methods. In addition, the variance estimation of nonresponse weighting can get very complicated if they are built upon complex survey designs. Multiple imputation is often regarded as a more viable approach to survey item nonresponse problems.

10.3 Imputation Strategies for Complex Survey Data

10.3.1 General Principles

10.3.1.1 Incorporating the Survey Sampling Design

Before the systematic use of multiple imputation analysis to survey data, imputation methods based on explicit regression models had been applied as a competitive, sometimes more promising, alternative to the hot-deck imputation system used by organizations (e.g., David et al. 1986). Of course, variance estimation had been an issue for the single imputation approaches including hot-deck imputation, and Rubin's proposal of proper multiple imputation has provided an adequate answer for this problem.

Some of the early applications of multiple imputation to survey data focused on the modeling and combining rules. For example, Heitjian and Little (1991) applied a PMM method to impute the blood alcohol content variables in the fatal accident reporting system administered by the National Highway Traffic Safety Administration. Clogg et al. (1991) and Schenker et al. (1993) used multiple imputation to harmonize the coding for the occupation and industry code between the 1970 and 1980 U.S. censuses. A main contribution of these studies was to demonstrate the use of multiple imputation as a reasonable approach to real-world survey nonresponse problems.

In principle, when applying multiple imputation to survey data, it is necessary to incorporate survey design into the imputation model. Here we provide a heuristic justification. Recall that for nonsurvey types of data Y, the idea of multiple imputation is to draw Y_{mis} from $f(Y_{mis}|Y_{obs}, R)$ independently multiple times, where R is the response indicator. Under the ignorable missingness assumption, multiple imputations of Y_{mis} are drawn from $f(Y_{mis}|Y_{obs})$ since R can be "ignored". Suppose we now consider data Y as a sample from a finite population, with the sampling indicator I following a sampling mechanism $f(I|Y, Z)$, where Z contains all the design information that are needed to select the sample from the population. According to Rubin (1987, Chapter 3), the principled multiple imputation strategy for Y is to draw Y_{mis} from $f(Y_{mis}|Y_{obs}, I, R, Z)$. Suppose that both the sampling and response mechanisms are ignorable, that is, I, R, and Y_{mis} are mutually independent conditional on Y_{obs} and Z; then this leads to drawing imputations from $f(Y_{mis}|Y_{obs}, Z)$, where Z is conditioned (included) in the imputation model.

Example 10.2. *A simulation study for a multiple imputation of an incomplete sample selected using unequal probabilities*

We conduct a simple simulation study to gauge the impact of inclusion (or exclusion) of the sampling design on the imputation estimates. We set up a finite population with $N = 100000$. In this population, $Z \sim N(1, 1)$ and $Y = -2+Z+\epsilon$, where $\epsilon \sim N(0, 1)$. Note that marginally $Y \sim N(-1, 2)$. We sample (with replacement) from this population using the inclusion probability $f(I = 1|Z) \propto \frac{exp(-1.4+Z)}{1+exp(-1.4+Z)}$ with sample size $n = 1000$. This sampling mechanism creates unequal probabilities as when a unit's Z-value increases, its inclusion probability also increases. This can be viewed as an approximation to the probability-proportional-to-size sampling often used in practice. Each sample consists of Y, Z, sampling selection probability $\pi = P = f(I = 1|Z)$, and sampling weight $W = 1/P$. For the drawn sample, we then randomly delete Y-values for around 30% of the cases to create missing data in Y under MCAR. Values for Z, P, and W are fully observed for sampled cases.

Suppose the estimand of interest is $\overline{Y}$ in the finite population. We implement multiple missing data methods as follows:

1. CC: complete-case analysis using response weights, which is the sampling weight, W, multiplied by n/n_{obs}, where n_{obs} is the number of complete cases for Y.

2. UW: unweighted CC analysis, that is, treating Y as a simple random sample of the finite population.

3. MI: perform multiple imputation of Y ignoring the sampling design, that is, by assuming $Y \sim N(\mu, \sigma^2)$ such as in Example 3.1.

4. MIWE: use the same idea as in MI although the sufficient statistics in the imputation algorithm are adjusted by the weights. Specifically, both $\overline{Y}_{obs}$ and s^2_{obs} are replaced by the weighted versions, and n_{obs} is replaced by the effective sample size. The latter is n_{obs} multiplied by the design effect (Kish 1965), which can also be estimated using Y and the weights as in Eq. (10.1).

5. MI-Z: impute Y using the correct normal linear model for Y given Z.

6. MI-W: impute Y assuming a normal linear model for Y given W.

7. MI-LOGW: impute Y assuming a normal linear model for Y given $log(W)$.

8. MI-P: impute Y assuming a normal linear model for Y given P.

9. MI-ZW: impute Y assuming a normal linear model for Y given Z and W.

For all multiple imputation methods, the estimate of the population mean is the weighted mean by the sampling weight, $\frac{\sum_s w_i y_i}{\sum_s w_i}$, and the variance estimate is based on the Taylor linearization approximation. They are obtained by R survey. The simulation consists of 1000 replicates, and the imputation analysis is based on $M = 30$ imputations.

Table 10.1 shows the simulation results. Because the missingness of Y is MCAR, CC analysis using the correct response weights provides unbiased estimates. However, UW is badly biased because it completely ignores the unequal sampling weights. In MI, the imputation model for Y assumes a correct marginal normal distribution. However, ignoring the unequal selection process leads to a considerable bias (RBIAS=-14.9%) and a very low coverage rate (39.5%). All other multiple imputation methods attempt to incorporate the sampling design in one way or another. In MIWE, the sufficient statistics in the univariate normal imputation model are replaced by the weighted versions. The bias is little, although the multiple imputation variance seems to somewhat underestimate the true variance. In MI-Z, MI-W, MI-LOGW, MI-P, and MI-ZW, the imputation models explicitly include the design variable Z (or its variants such as weights and selection probabilities) as a predictor. Estimates from all these methods have little bias and reasonable coverage rates. Among them, MI-Z, MI-P, and MI-ZW provide similar estimates as the imputation model either exactly matches with or is close to the true-data generating model. Note that MI-ZW uses information from both Z and W, which are certainly overlapped, and yet it does not seem to have negative impacts on the imputation inference. Compared with these methods, the estimates from MI-W seem to be more variable because the distribution of Y given W deviates somewhat from a normal linear model. After applying a log transformation to sampling weights, the estimates from MI-LOGW have less variability than those of MI-W.

TABLE 10.1
Example 10.2. Simulation results

Method	RBIAS (%)	SD	$\overline{\text{SE}}$	MSE	CI	COV (%)
BD	−0.1	0.062	0.062	0.00383	0.241	94.6
CC	−0.3	0.075	0.073	0.00557	0.287	94.4
UW	−48.6	0.052	0.051	0.2391	0.2	0
MI	−14.9	0.067	0.068	0.0265	0.268	39.5
MIWE	−0.3	0.075	0.069	0.00556	0.271	93.1
MI-Z	−0.2	0.066	0.067	0.00429	0.264	94.7
MI-W	0.5	0.093	0.088	0.00866	0.341	94.9
MI-LOGW	1.4	0.069	0.071	0.00491	0.279	95.6
MI-P	−0.7	0.065	0.065	0.00421	0.254	94.0
MI-ZW	−0.3	0.067	0.067	0.00447	0.264	94.2

Fig. 10.2 shows the distribution of (observed and imputed) Y over Z and W for 250 randomly chosen observations from a simulation replicate. These scatter plots show the relationship between the variable of interest and sampling design. They demonstrate that the imputed values from MI somewhat distort the original relationship between Y and Z in the population, while those from MI-Z, MI-P, and MI-W more or less retain such relationship. Because Y and Z are related in the original population and the sampling selection probability is also related to Z, ignoring Z in the process of imputing Y as in MI would break off this connection and lead to invalid inference even if the missingness of Y is unrelated to Z. This cannot be sufficiently rescued by merely analyzing the corresponding imputed data using the sampling design.

In this simple simulation, we purposefully do not allow the missingness of Y (i.e. R) to depend on Z to illustrate that bias can arise from failing to account for the survey design even when the ignorable missing data mechanism does not depend on the sampling design. Of course, if R is also related to Z, then it further justifies the importance of including design in the imputation model. In addition, if R is also related to other fully observed variables X, then X has also to be included in the imputation model as we discussed in previous chapters of the book.

Example 10.2 shows the importance of including the design variable Z in imputation, although there exist several alternative ways of including such information. Beyond just sampling weights, Reiter et al. (2006) provided a detailed assessment of the impact of excluding typical survey design information (i.e., sampling strata and clusters) on multiple imputation inferences. The estimands of interest include both the marginal mean of the nonresponse variable and the coefficients of regressing the incomplete variable on fully observed covariates. The results demonstrated that the bias can arise and can be

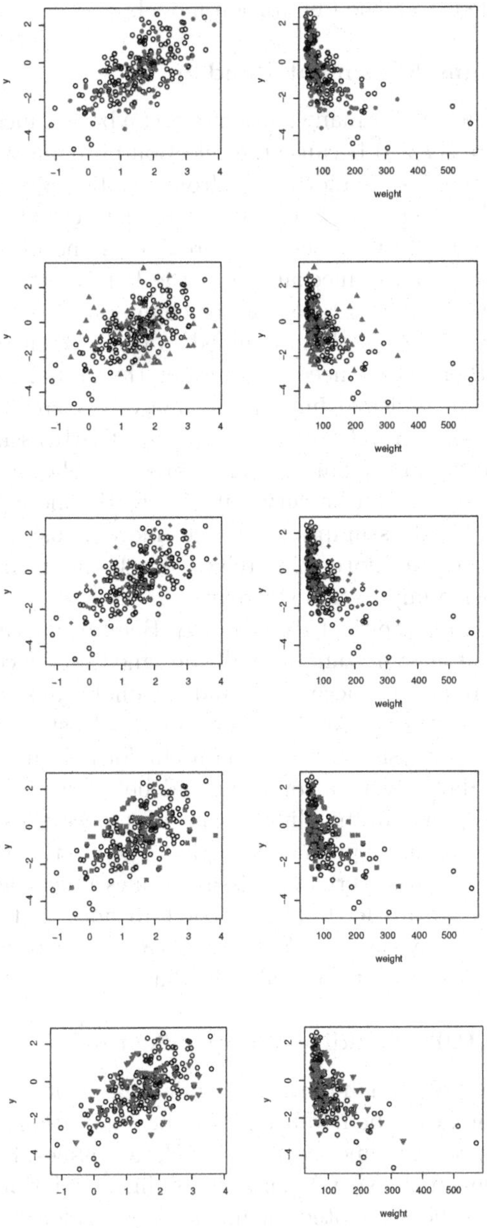

FIGURE 10.2
Example 10.2. Left: scatter plots of Y and Z; right: scatter plots of Y and W. Circle: observed values; filled circle: deleted values (the 1st row); filled triangle: imputations from MI (the 2nd row); filled diamond: imputations from MI-Z (the 3rd row); filled square: imputations from MI-P (the 4th row); filled reverse triangle: imputations from MI-W (the 5th row).

severe when strata and clusters are related to the survey variables of interest and are yet excluded from the imputation model.

10.3.1.2 Assuming Missing at Random

The sampling mechanism for many real surveys can be assumed to be ignorable (i.e., $f(I|Y, R, Z) = f(I|Z)$) because the selection probability is controlled by the sampler using the design information from Z. Missingness mechanisms of incomplete variables Y can be more difficult to characterize due to the complexity of survey data. In Chapter 1 we provided some illustrative analyses on the possible nonresponse mechanism of the family total income variable in NHIS 2016, which appears to be rather complicated. However, the missingness of survey variables is often assumed to be MAR in real applications. Assuming MAR, there is no need to consider the actual process of nonresponse mechanism in multiple imputation analysis. It makes the modeling and computation more straightforward than the MNAR assumption because the latter often needs additional and subjective speculations and the corresponding analysis results can be very sensitive to the modeling assumptions (Chapter 13). The MAR assumption becomes more plausible as the model is enriched to include more information related to the nonresponse mechanism following the inclusive imputation strategy.

Results from many empirical studies (e.g., Belin et al. 1993; Rubin et al. 1995) suggest that at least in some carefully designed survey contexts, imputation under MAR can provide acceptable and reasonable answers. Therefore, it is recommended to adopt the MAR assumption (at least as a starting point) for the imputation analysis of survey data. In formulating the imputation models for survey data, every attempt should be made to incorporate variables related to response probabilities. Note that once these variables have been included, it is no longer possible to verify or refute the MAR assumption by examining rates and patterns of missing values unless additional unverifiable assumptions are made. It is important to note that MAR is not an inherent property of any dataset. Rather, MAR is an assumption made for the nonresponse model used to describe the data.

10.3.1.3 Using Fully Conditional Specification

As discussed in Chapter 7, handling missing data problems in complex survey data is a strong motivation behind using FCS for imputation (Raghunathan et al. 2001). Here we provide some recap of the relevant issues. First, correlations among the variables in the survey, such as nonlinear relationships as well as interactions, need to be accounted for in the imputation. Second, large-scale surveys usually incorporate multiple incomplete variables of various types and distributional shape, and it is necessary to include all of these variables in the imputation. Third, questionnaires are to a large extent hierarchical in nature, with skip patterns defining the set of variables to be imputed on an individual level. Fourth, either by nature or by the hierarchy of questions, certain ranges or limits apply to possible values of a variable. Fifth, the sampling design information needs to be incorporated into the imputation. Finally, as we discussed in Chapter 7, conditional models specified in FCS are deemed as an approximation of the true joint model.

There were some real-life applications of multiple imputation using the JM strategy before the full development of FCS (Example 10.3). Nowadays it is fair to say that FCS is the mainstream approach to imputing survey data in practical applications. The merits of FCS are also enhanced by established software packages such as R mice, SAS PROC MI (FCS option), and IVEware.

10.3.2 Modeling Options

We now discuss some modeling options for including survey design information in imputation. Although the general principle sounds straightforward, actual implementation can be complicated. In Example 10.2 we use a single variable Z to characterize the design variable, and this is an oversimplified illustration. In many complex surveys with multi-stage sampling, the information embedded in the design is not straightforward to handle. For example, the strata and clusters in NHIS consist of geographic units (e.g., U.S. census blocks) that are not simple to list and operate in data form. The selection probability, although controlled by the sampler, is complicated by the actual data collection operations. In addition, oversampling of certain demographic groups, post-stratification, and nonresponse adjustments for the final survey weights further complicate the process.

Despite the complexity of exact design information, it is often reasonable to assume that this information can be approximated by the final survey weights (after post-stratification and nonresponse adjustment) W, sampling strata D, and sampling cluster E, which are nested within each sampling stratum. For simplicity, we still use Z to denote variables that are used in designing the sampling and can be included in the imputation analysis. The general principle is to include W, D, E, and Z in the imputation modeling. Information from them might have some overlap. Yet we believe it is of less concern following the guideline of the inclusive imputation strategy.

Under the FCS framework, we start with a single incomplete variable Y and let X denote fully observed (or completed) other variables excluding the design information. In principle, the specification for imputing Y should include X, W, D, E, and Z. As a starting point, a prototype of the imputation model might be sketched as:

$$g(E(Y|X, W, D, E, Z)) = X\beta_X + D\beta_D + E\beta_E + W\beta_W + Z\beta_Z, \qquad (10.2)$$

where $g(\cdot)$ denotes a function linking the expected value of response Y to predictors in the GLM framework, and β's are the regression coefficients.

Based on our limited understanding, there seem to exist different proposals in the literature on how to complete the specification of Model (10.2), perhaps driven by somewhat different contexts. Due to the complexity of survey data, a one-size-fits-all modeling strategy might not exist. In addition, practical considerations need to come into play because sophisticated models might not work well or can be hard to implement for real survey data. We summarize a few ideas that have been used in the literature. Yet practitioners need to make the best decision in their own contexts.

1. Impose a multilevel structure for Model (10.2). Reiter et al. (2006) proposed two strategies, both of which treat the stratum indicator (D) as dummy variables (i.e., fixed effects). One strategy treats the cluster effects (E) as fixed effects, and the other treats the cluster effects as random effects. The fixed effects model is relatively easy to implement, yet might not be the optimal approach if samples from the same clusters are correlated (Section 9.3.1). The random effects model might be theoretically more preferred but can be computationally more complex. Apparently, when the number of strata or clusters increases, implementing either model would encounter some difficulty. In some surveys the design may be so complicated that it is impractical to include dummy variables for every cluster. Reiter et al. (2006) suggested that in these cases, imputers can simplify the model, for example, by collapsing cluster categories or including proxy variables (e.g., cluster size) that are related to Y as a predictor.

2. Consider transformed weights and the interactions of weights with other variables. Including W as a predictor is intuitive yet might not be sufficient from a theoretical perspective for some completed-data analyses (e.g., a weighted regression of Y on X). The fundamental reason behind this is that Rubin's variance (i.e., combining rule) estimator would not be consistent if the analysis model is uncongenial with the imputation model (e.g., Kim et al. 2006). Some general discussion related to this point will be deferred to Section 14.2.

   A better strategy might be, according to Seaman et al. (2012), Carpenter and Kenward (2013, Section 11.3), and Quartagno et al. (2020), is to include W and its interactions with X in the imputation model if a weighted analysis of regressing Y on X is to be conducted. Practically, this might be implemented by including their interactions as predictors in Model (10.2), by conducting separate imputations within strata formed by weights, or by conducting a multilevel model-based imputation using weight-based strata as level 2 of the model and allowing the covariance matrix of Y and X to vary across level-2 units (Section 9.5). In addition, Zhou et al. (2016a, 2016b, 2016c) suggested that using the log-transformed weights and having them interacting with major predictors perform well.

   Based on the aforementioned literature, our recommendation is that survey weights W (perhaps in the log-transformed form) should be at least included in the imputation model as a main-effect predictor. Note that in Example 10.2 we included different forms of weights (i.e., W, $P = 1/W$,

$log(W)$) as an imputation predictor. An expanded model including weights interacting with other major predictors can be tested to see if there exist any practical differences.

3. Include external information. Broadly speaking, X can be expanded to include contextual variables that are external to original variables in the survey. For example, to impute incomplete variables in U.S. national surveys, demographical information at geographical units level (e.g., percentages of racial-ethnic groups at U.S. census blocks level) is often included as predictors. On the one hand, they can be associated with missing variables of interest. On the other hand, they might be used to compensate for the information in the survey design that is not easily incorporated (e.g., when sampling cluster effects are treated as random effects yet the model is difficult to execute).

Although the aforementioned options aim to incorporate the survey design information in the imputation model, there still exist a few challenges. First, not all design information is typically available (e.g., D and external variables in X) in public-use data due to disclosure risk concerns. This can create some problems if data users would like to conduct their own multiple imputations of survey data. Second, as pointed out before, the modeling task in Model (10.2) may be complicated by attempting to include all interactions of weights (or weight-related design variables) with other covariates in the model or considering a multilevel version of the model.

In a series of papers, Zhou et al. (2016a, 2016b, 2016c) adopted a Bayesian analysis approach to survey data. They proposed a two-step multiple imputation procedure. In the first step, a weighted finite population Bayesian bootstrap (the weighting is to validly impute/synthesize the original population including item nonresponses) from the observed data. Item-level missing data then become a part of these "uncomplexed" synthetic populations. Thus in the second step, having generated posterior predictive distributions of the entire population, standard imputation models (without incorporating the survey design) can be applied. Their suggested procedure allows the parametric imputation model to no longer need to model interactions between weights and covariates in the imputation regression model to account for model misspecifications. The performance of the proposed procedures is vindicated by simulation studies. However, this approach is not designed as an alternative strategy to release multiply imputed datasets for complex sample design data, but rather as an analytic strategy. The combining rules from multiply imputed datasets are different from Rubin's combining rules. In addition, the computation might be a burden for practitioners because it involves a bootstrap process, which can easily go towards hundreds or thousands of simulations. This method is implemented in IVEware, and a practical guideline can be found in Raghunathan et al. (2018, Chapter 9).

10.4 Some Examples from the Literature

In this section we briefly review a few examples of applying multiple imputation analysis to complex survey data from past literature. We focus on applications that explicitly include survey design and data structure in the imputation. The main goals for some of these imputation projects are to release imputed datasets for external users.

Example 10.3. *Applying multiple imputation to NHANES III data*

The National Health and Nutrition Examination Survey (NHANES) collects important nutritional and health-related data on the civilian noninstitutionalized U.S. population and important subgroups. General information about NHANES can be found in (https://www.cdc.gov/nchs/nhanes/index.htm). It is administered by NCHS. NHANES III is the seventh in a series of NHANES. Data were collected from 1988 till 1994 with a total sample size of approximately 40000. NHANES III is a complex, multi-stage area sample with oversampling of young children (under 5), the elderly (60+) Mexican Americans and African Americans. Details of the sampling design can be found in Ezzati-Rice et al. (1992).

One of the earliest applications of constructing public-use databases is the multiple imputation project for NHANES III. It provided a first-hand experience for this objective, as it involved collaborative efforts of both academia and government. The data product and documentation can be found at (https://www.cdc.gov/nchs/nhanes/nhanes3/data_files.htm#augmented at *7A, Multiple Imputation*). Five sets of multiple imputations were created.

In summary, of the 39695 individuals selected in the NHANES III sample, household interviews were obtained for 33994 (86%). Among these interviewed persons, 30818 (91%) were subsequently examined in a Mobile Examination Center (MEC) and 493 (1.4%) received limited physical examinations at home. Rates of response varied across demographic subgroups decided by age, racial-ethnic groups, and size of the household. A set of 67 key variables was designated for imputation, including body measurements, key variables from bone densitometry, fundus photography, blood pressure, and laboratory results from the analysis of blood and urine samples. These variables spanned over both the questionnaires from household interviews and physical examinations from MEC.

A JM strategy was used in the imputation. The interviewed persons in NHANES III were divided among nine age classes, and a multivariate linear mixed model (Section 9.3.2) was applied to each class. Since sampled persons in NHANES III came from 89 survey locations, these locations were treated as clusters in the multivariate linear mixed model to reflect the correlation of the responses within each survey location. Specifically, random intercepts across clusters were assumed. However, location indicators are viewed as confidential and therefore are not released to the public.

For continuous missing variables, transformation was applied to preserve the approximate normality when necessary. Categorical variables were treated as continuous, and rounding was applied to the fractional imputed values after the imputation. Note that rounding is not recommended nowadays as an optimal imputation strategy for categorical data (Section 4.3.2.3). Covariates in mixed models include typical demographic variables and the logarithm of household size. This is because household size, along with the race/ethnicity and age, affected the probability that an individual was selected into the NHANES III sample. Household size is also strongly related to rates of nonresponse. Finally, additional items from the household interview questionnaires were used to form model covariates. These predictors were fully observed and not themselves imputed. These variables were chosen in consultation with subject matter experts either because (a) they might be related for obvious medical or physiological reasons to the missing variables, or because (b) they are likely to appear in a variety of analyses by users of NHANES III data.

Note that the NHANES III sampling weights were not included in the imputation model because it was then believed including variables determining the survey design was sufficient to maintain design consistency for the estimates using multiply imputed datasets. Simulation studies were conducted to assess the performance of the imputation strategy, and the results showed that the imputation model was successful in creating valid design-based repeated-sampling inferences for a wide range of estimands (e.g. population and subdomain means). In addition, the project emphasized that the post-imputation analysis should still use design-based analysis for survey data.

Example 10.4. *Multiple imputation for family total income in NHIS*

We began this book (Example 1.1) with the family income nonresponse problem in NHIS, which is a cross-sectional survey conducted annually since 1957. In terms of the sampling design, the NHIS uses complex geographically based cluster sampling methods that result in a probability sample of households. The sampling procedure of NHIS is redesigned following every decennial U.S. census. The most recent sample design was implemented in 2016 to account for changes in the distribution of the U.S. population since 2006, when the previous sample design was implemented (National Center for Health Statistics, 2017). Additional details of the sampling and weighting methods used for NHIS (e.g., from 2006 to 2015) can be found in Parsons et al. (2014).

Through 2018, all families within a selected household were included in the survey. Within a family, one adult and one child (if any) were randomly selected and face-to-face interviews were conducted with that sample adult and with an adult respondent for the sample child. This probability design permits representative sampling of the civilian, noninstitutionalized U.S. population. Since 2019, the NHIS questionnaire was redesigned and removed the family component of the survey.

The imputation project (Example 1.2) started in 1997 and has been executed on an annual basis. The major modeling strategy had been consistent until 2018 (before the questionnaire redesign in 2019), despite some sporadic changes of variables used in the model. Since the questionnaire redesign in 2019 led to either reduction or modification of many original variables, a new imputation modeling strategy is being developed for and applied to NHIS starting from 2019. For example, the imputed files and some of the related documentation for NHIS 2019 can be found in (https://www.cdc.gov/nchs/nhis/2019nhis.htm).

Discussion in this example (and throughout this book) is only pertaining to the imputation modeling from 1997 to 2018, for which the detailed scholarly work was presented in Schenker et al. (2006). For NHIS 2016, for example, the documentation can be found in (https://www.cdc.gov/nchs/nhis/nhis_2016_data_release.html). Unlike Example 10.3, where more than 60 variables were selected for imputation in the NHANES III imputation project, the main target variable here is the family total income. A cubic-root transformation was applied to the family income variable in the imputation (Example 5.2). In addition, personal earnings, which is highly associated with family income, was also imputed. Because personal earnings were collected only for employed adults, employment status was also imputed for adults for whom it was unknown. Finally, the ratio of family income to the applicable federal poverty thresholds was derived for families with missing incomes, based on the imputed values.

In the NHIS imputation project, a large number of predictors (around 60) were included in the imputation. These predictors included variables describing demographic characteristics, family structure, geography, education, employment status, hours worked per week, sources of income, limitations of activities, health conditions that caused limitations, overall health, use of health care, and health insurance. The rationale behind choosing these predictors is similar to that of the NHANES III imputation project, both of which follow the inclusive imputation strategy. That is, when multiple imputations are being created, it is beneficial to include a large number of predictors in the imputation model, especially variables that will be used in subsequent analyses of the multiply imputed data.

The sampling design of NHIS was also incorporated in the imputation. Somewhat different from the NHANES III imputation project, no multilevel structure was imposed. Instead indicators for the distinct combinations of stratum and primary sampling units were used as predictors. In addition, summary of family income (i.e., the (log) mean and (log) standard deviation of reported family income) at secondary sampling unit level (i.e., small clusters of housing units) was used as contextual predictors. Survey weights were also used as a predictor. All these predictors are treated as fixed effects in conditional regression models for imputation.

A unique challenge in the NHIS income imputation project is that variables in the imputation are reported at different levels of the survey and are hierarchical in nature. For example, family income was reported at the family level and personal earning (for NHIS 1997-NHIS 2018) was reported at the person level. To account for this, the imputation procedure was devised to include a series of steps:

(a) Impute missing values of person-level covariates for adults.

(b) Create family-level covariates.

(c) Impute missing values of family income and family earnings, and any missing values of family-level covariates due to missing person-level covariates for children.

(d) Impute the proportion of family earnings to be allocated to each employed adult with missing personal earnings, and calculate the resulting personal earnings.

Note that in Step (c), family earnings becomes an intermediate variable before the imputation of the personal earnings. In Step (d), imputing the proportion ensures that the family earnings are always greater than or equal to personal earnings. This strategy is used to impute compositional data (e.g., Van Buuren 2018, Section 6.4.4). Another major constraint that needs to be accounted for is that the imputed family income value has to fall within the range of the bracketed income questions from respondents (Example 1.1).

An FCS strategy was used in the imputation, which is a natural choice to tackle the complexity from both the modeling and hierarchical data structure. Several evaluations of the multiply imputed data showed that they often yielded lower standard error estimates than those from analyses without imputation (e.g., complete-case analysis). On the other hand, multiple imputation estimates correctly have higher standard errors than using just a single imputation because the latter does not appropriately account for the imputation uncertainty.

Example 10.5. *Multiple imputation of DXA scan variables in NHANES 1999-2006*

Schenker et al. (2011) presented a multiple imputation project for NHANES dual-energy x-ray absorptiometry (DXA) scan data (from 1999 to 2006). The datasets and documentation can be found in (https://wwwn.cdc.gov/Nchs/Nhanes/Dxa/Dxa.aspx). In 1999, DXA scans were added to the NHANES to provide information on soft tissue composition and bone mineral content. However, in the specified time periods, DXA data (more than 30 DXA-related measurements) were missing in whole or in part for about 20% of the NHANES participants eligible for the DXA examination. More specifically, DXA measurements include fat mass, body fat percentage, fat-free mass (includes bone mineral content), lean soft tissue mass (excludes bone mineral content), bone area, bone mineral content, and bone mineral density for the total body and several body subregions depending on the variable being considered.

An FCS strategy was used in the imputation project. The imputation was conducted within predefined gender and age groups. The imputation predictors included demographic, socioeconomic, and geographic variables, body measurements, health indicators, dietary and medication use variables, and blood test results. In addition, a lower-bound was applied to the imputed DXA values so that the range of the imputations is reasonable.

To reflect the NHANES sample design in the DXA imputation model, the MEC survey weights were included, as were several variables related to the selection of primary sampling units: metropolitan-area status, data release cycle, self-reported race/ethnicity, family income, and education. Moreover, several strong person-level predictors for the DXA variables, such as other DXA variables, age, race/ethnicity, gender, body mass index, waist circumference, triceps and subscapular skinfolds, report of previous fracture, and family history of osteoporosis were included.

Since DXA variables were not normally distributed, an iterative Box-Cox transformation was applied to these continuous variables during the FCS iteration process. The iterations continued until the set of estimated transformations stabilized, that is, did not change from one iteration to the next. The exponents for the final estimated power transformations of the DXA variables varied. Note that this is different from the NHIS income imputation project (Example 10.5), in which the transformation was fixed at the cubic root.

Example 10.6. *Multiple imputation for survey data of cancer care*

As briefly mentioned in Example 1.4, the Cancer Care Outcomes Research and Surveillance (CanCORS) Consortium was a multi-site, multi-mode, and multi-wave study of the quality and patterns of care delivered to population-based cohorts of diagnosed patients with lung and colorectal cancer from 2003 to 2005 (Ayanian et al. 2004). It consisted of seven data collection and research sites (academic medical centers and health care organizations) and a statistical coordinating center. Each site identified appropriate samples to obtain combined cohorts of approximately 5000 patients diagnosed with each cancer. The coordinating center assisted all the study sites in the collection of standardized data across the individual research sites and served as the central repository for the pooled data.

CanCORS collected data from multiple sources including patient surveys, medical records, and surveys of physicians who treated these patients. Data from these primary sources were supplemented with cancer registry data and other publicly available datasets such as Medicare claims. As is typical in observational population studies, missing data were a serious concern for CanCORS, following complicated patterns that impose severe challenges to the consortium investigators. Multiple imputation was applied to both patient and providers surveys to construct a centralized completed database that was used by investigators from multiple sites. Although the datasets were only shared among CanCORS investigators, the idea of constructing centralized databases shared by multiple users still follows the original motivation of applying multiple imputation by Rubin (1978).

The technical details were presented in He et al. (2010), which focused on imputation in the patient surveys. These surveys obtained information from participants regarding their cancer diagnosis and treatment, quality of life, experiences of care, care preferences, health habits, other medical conditions, and demographic information. The survey items were organized into 12 to 14 topic-related sections, most of which were identical for lung and colorectal cancer patients. Depending on the status of the patients, different versions of surveys were conducted. The baseline survey uses four forms, including the full survey for the patient, a brief version for patients who cannot complete the full interview, a survey for surrogates when the patients are alive, but unable to complete the interview, and a survey for surrogates of dead patients. The surrogate surveys contained parts of the full survey and a few additional items that pertain specifically to the surrogate's experiences of the patient's cancer care. The follow-up survey was attempted for all participants who were alive at the time of the baseline survey, but not for those who had already died before initial contact.

Unlike previous examples, one major challenge of this project was that there were no specific target variables. That is, ideally all missing variables in surveys were to be imputed. There were hundreds of variables included in each survey form. Although the survey variables were organized into topic-related sections in the questionnaire, imputing sections (or other divisions) of the survey separately may ignore important associations among variables, as the best predictors for a variable can be in both the same and different sections. Imputing the survey as a whole also makes the MAR assumption more plausible. Another major challenge was to handle the feature of multiple survey formats. Note that although different survey forms had different purposes and targeted patient subgroups, there existed a large number of overlapping items and hence it would be inefficient to impute them separately.

An FCS strategy was used for multiple imputation. It concatenated all of the surveys to create a combined rectangular data set with around 800 variables and imputed all included items simultaneously. Based on input from clinical investigators, concatenation and imputation were carried out separately for patients with lung and colorectal cancer to avoid complex interactions, since the two groups have different disease etiology and care patterns. However, the concatenation procedure causes block missingness in the combined data set for the variables that were not used in all survey forms. Some of these missing data may not be used in any meaningful analysis, such as follow-up survey items for patients who died before attempted contact. Others may be of potential use, depending on the context of the analysis. For example, questions about patients' income were omitted from the brief survey, but the missing values are meaningful for the brief survey participants and could be imputed and used if the sample for an analysis involving the income variables includes that group. These block missing values, as well as most of the skip patterns, were treated as imputable variables and were imputed. After imputation, missingness for the block missing data that were not eligible for analyses or not used by investigators, as well as skip patterns were restored. The justification behind this strategy was based on practicality rather than methodological rigor. However, some evaluation results showed that this strategy provided reasonable results in the actual application. See Section 7.4.2 for more discussion about handling skip patterns in FCS. Some diagnostics were also run to ensure the selected imputation models provide reasonable results. For example, it was shown that the multiple imputation-based inferences gain precision advantages over the nonresponse weighting method which treated the block missingness as unit nonresponses.

Specifications of the conditional models in the FCS imputation follow the commonly used options as discussed in Chapter 7, depending on the type of the variables. Due to the large number of variables included in the dataset, for any missing variable, it was virtually impossible to include all of the rest of variables as predictors in the FCS imputation. The imputation used the automatic model selection feature of IVEware (the increment of marginal R^2 for the fitting of each conditional regression model and the maximum number of predictors) to finish the imputation in a reasonable time.

Technically, the survey was an observational study and not a strict probability-based survey. Therefore there existed no such notion of design-related variables or survey weights. However, since all the survey variables were included in the imputation, the selection bias due to possible deviation of MAR might be minimal.

This project also documented difficulties when imputing categorical data with rather unbalanced distributions (e.g., a binary variable with most of the values concentrated on either 1 or 0). This is due to data separation in the logistic regression imputation for categorical data (Section 4.3.2.4). As we discussed there, nowadays there exist more satisfactory imputation methods to account for data separation.

In addition, multiple imputation was also carried out for some targeted analyses of CanCORS data when necessary. For example, He et al. (2010) illustrated a multiple imputation project study assessing the effect of cardiovascular diseases on hospice discussion for late stage lung cancer patients, where many of the variables including the response and covariates suffer from missing values. Example 11.9 will show an example using multiple imputation to correct measurement error in CanCORS data.

Example 10.7. *Single vs. multiple imputation for highly clustered surveys*

The National Ambulatory Medical Care Survey (NAMCS) has been administered by NCHS since 1973 (https://www.cdc.gov/nchs/ahcd/index.htm). While aspects of the sample design and survey instrument have evolved over the past years, its objective has always been to collect and disseminate nationally representative data on office-based physician care. The ultimate sample unit is a doctor-patient encounter, drawn systematically from the terminus of a multi-stage, clustered sampling design. For example, sampling design and weighting related documentation for NAMCS 2008 can be found in National Center for Health Statistics (2009).

Patient race, an important demographic variable, has been subject to a steadily increasing item nonresponse rate. In 1999, race was missing for 17% of cases; by 2008, the nonresponse rate had risen to 33%. Over this period, single hot-deck imputation had been the main adjustment method. Lewis et al. (2014) evaluated the possibility of applying model-based multiple imputation to the missing race variable in NAMCS 2008. An FCS imputation strategy was used, using the logistic regression imputation for the missing race. The imputation predictors included patient age, sex, urban/rural indicator based on metropolitan statistical area, physician specialty group, reason for visit, logarithm of time spent with physician, and an indicator of who entered data into the survey.

To incorporate the survey design, the imputation model included sampling stratum and primary-sampling unit (PSU) indicators, as well as sample weights, but encountered convergence issues for the logistic regression parameters that did not cease until the PSU indicators were omitted. In NAMCS data, the PSUs (clusters) are physician practices, for which the numbers were on the scale of thousands. In this case, treating them either as fixed effects or random effects would lead to model convergence issues. As a compromise, contextual variables from local geographic regions were used as predictors. Specifically, two variables based on U.S. Census 2000 estimated proportions of non-Hispanic whites and non-Hispanic blacks at the patient/provider ZIP code tabulation area level were used.

However, that imputation project identified many estimates' ratios of multiple imputation to single imputation standard errors close to 1. That is, there is little difference between multiple and single imputation variance estimates for the means of the missing race variable. Due to this reason and other logistic factors, multiple imputation was not used to impute the missing NAMCS race variable.

He et al. (2016) further investigated this phenomenon. Note that the ratio of multiple imputation variance (T_M) to single imputation variance (U_M) can be approximately expressed as $\frac{T_M}{U_M} = \frac{U_M+(1+\frac{1}{M})B_M}{U_M} = \frac{1}{1-FMI}$, where $FMI = \frac{(1+1/M)B_M}{(1+1/M)B_M+U_M}$ is the fraction of missing information. Consider a simple one-way random effects model as follows

$$\begin{aligned} y_{ij} &= \mu + \alpha_i + \epsilon_{ij}, \\ \alpha_i &\sim N(0, \tau^2), \\ \epsilon_{ij} &\sim N(0, \sigma^2), \end{aligned} \tag{10.3}$$

for $i = 1, \ldots, m$, $j = 1, \ldots, n$, where y_{ij} is the random variable, μ is the population mean, α_i's are between-cluster random effects, and ϵ_{ij}'s are the error terms. This model was also considered in Section 9.3.1.

Model (10.3) and its variants are frequently used as a data-generating model in the analysis of clustered surveys (Valliant et al. 2000, Chapter 8). From the perspective of design-based inference, it can be shown that the design effect for complete data is $Deff_{com} = \frac{\tau^2/m+\sigma^2/mn}{(\tau^2+\sigma^2)/mn} = 1 + (n-1)\rho$

where $\rho = \tau^2/(\tau^2 + \sigma^2)$ is the intraclass correlation coefficient. Suppose that missing data occur in the original sample. For ease of notation, suppose that y_{ij}'s are observed for $i = 1, \ldots, m$, $j = 1, \ldots, r$ and missing otherwise, that is, an equal missingness rate across clusters with MCAR.

A Baysian imputation algorithm can be applied to Model (10.3) for the missing y_{ij}'s. The variance of both the multiple imputation and single imputation estimator for μ can be derived (He et al. 2016). It can be shown that the population FMI has a relationship with the design effect, which can be expressed as:

$$FMI \approx \frac{p_{mis}}{(1 - p_{mis})Deff_{com} + p_{mis}}, \tag{10.4}$$

where $p_{mis} = (n - r)/n$ quantifies the rate of missingness.

According to Eq. (10.4), a multiple imputation mean estimator for a dataset with a larger design effect tends to have a smaller FMI. In the case of the NAMCS imputation project, the design effects of the mean estimates from the working dataset range between 10 and 40 due to its highly clustered nature, and this would lead to a FMI quite close to 0, implying that the ratio between multiple imputation and single imputation variance is close to 1. For example, suppose the design effect of the complete data is 10, for a 30% missingness rate, the FMI is about 4% based on Eq. (10.4). This corresponds to $100/96 \approx 1.04$ as the ratio of the multiple imputation vs. single imputation variance estimates. The inverse relationship between the design effect and FMI provided an explanation of the phenomenon identified in Lewis et al. (2014). However, the general pattern for other types of multiple imputation estimates (e.g., regression coefficients) with highly clustered samples was not studied.

Example 10.8. *Multiple imputation for missing data in linked files*

Record linkage is a valuable and efficient tool for connecting information from different data sources. Survey data are often linked with administrative databases (often across organizations) to enhance the utility of the information in these data sources. For example, NCHS has linked its population-based health surveys with various administrative data. However, the linked files are often subject to missing data: first, not all survey participants agree to record linkage, and second, nonresponses exist in the original files as usual. Multiple imputation can be an effective approach to handling the missing data problems in linked files. In this context, the extra variables from one source can be helpful predictors for imputing missing variables in another data source, following the principle of the inclusive imputation strategy.

For example, Zhang et al. (2016) examined the usefulness of multiple imputation for handling missing data in linked NHIS–Medicare files administered by NCHS. They considered a study of mammography status from 1999 to 2004 among women aged 65 years and older enrolled in the Fee-for-Service (FFS) program. In their example, mammography screening status and FFS/Medicare Advantage (MA) plan type are missing for the NHIS survey participants who were not linkage eligible. Mammography status is also missing for linked participants in an MA plan because Medicare claims data are only consistently available for beneficiaries enrolled in the FFS program, not in MA plans. In addition, the incomplete variables were longitudinal because the linked file in the application dataset spanned multiple years. Through linkage, variables in NHIS were used as effective predictors in the imputation. In addition, since NHIS is a probability survey aimed to represent the general U.S. population, the target population of the imputation inference is well defined. Zhang et al. (2016) explored alternative imputation approaches: (i) imputing screening status first, (ii) imputing FFS/MA plan type first, (iii) and imputing the two variables from multiple years simultaneously. All imputation approaches were based on the FCS strategy. Their simulation studies showed that method (ii) and (iii) were better than method (i). Their assessment also showed that multiple imputation methods yielded better performances than ad hoc missing data methods such as complete-case analysis.

10.5 Database Construction and Release

Examples in the preceding section mainly focus on methodological issues such as the setup of the models, variables included, and estimation process. In general, public-data release in organizations can be a comprehensive and daunting task that goes beyond only statistical and modeling considerations. The main purpose of this section is to provide some related discussions so that practitioners can have a better idea of the general process of releasing multiply imputed datasets to the public.

10.5.1 Data Editing

As an indispensable step of data publishing, survey data editing is the inspection and alteration of collected data, prior to or after statistical analysis. The general goal of editing is to verify that the data have properties intended in the original measurement design and make common sense. Editing is accomplished through different kinds of checks such as:

1. Range edits (e.g., recorded age should lie between 1 month and 120 years).

2. Comparisons to historical data (e.g., the diabetes prevalence in general should not change dramatically from year to year).

3. Balance edits (e.g., the percentages of income from various sources such as salary, tips, stock earnings should add up to the total household income).

4. Checks of highest and lowest values in the dataset or other detection of implausible outliers.

5. Consistency edits (e.g., when recorded age is less than 12, then marital status should be recorded as "never married").

6. Checks of survey weights (e.g, whether the sum of weights overall or in some subdomains is consistent with some known population control totals or historical records).

Another important function of data editing is to preserve the data confidentiality and minimize disclosure risk. For some survey respondents, their data values from some variables in the public-use version are changed to minimize the probabilities that these respondents can be identified based on the survey data as well as other publicly available information. Practical procedures for preserving data confidentiality include top-coding and random swapping. In addition, certain geographic units of the sample survey are also blocked from external data users.

In an important contribution, Fellegi and Holt (1976) invented a system of editing that integrates editing and imputation and has a fixed set of steps, replicable given a set of editing rules. Note that in their context, the term "imputation" means changing of implausible values by these rules. A detailed discussion of editing is beyond the scope of this book. Additional references can be found, for example, in Groves et al. (2009, Section 10.3).

The presence of survey nonresponse apparently adds another level of complexity to data editing as the existing procedures following Fellegi and Holt (1976) are targeted to complete datasets. In some cases, it is possible to conduct data editing and handling survey nonresponse simultaneously. For example, Little and Smith (1987) proposed a three-stage strategy for cleaning survey data (continuous variables) with missing and outlying values. See also Ghosh-Dastidar and Schafer (2003) for a more complex model for conducting both multiple imputation and correcting measurement error. However, these models are often complicated and might not be feasible for large-scale datasets with multiple incomplete variables. This fact reflects the complexity of imputation tasks for real-world surveys, as pointed out early by Sande (1982).

In many cases before the model-based multiple imputation, a logic-based (or deterministic) imputation can be conducted. This step uses other observed variables to determine the missing value. For example, if a person's sex is missing, the person's name might be used to determine the sex. This step can also be included in the aforementioned data editing step.

Therefore a practical procedure in the context of multiple imputation analysis can include:

1. Run consistency checks and edits and conduct some logic-based imputation before data are multiply imputed.

2. In the process of multiple imputation, ensure imputed values also follow the consistency and logic requirements as much as possible. This is better implemented using the FCS imputation (e.g., restricting imputed values,

maintaining skip patterns, etc.) Some editing rules might need to be directly incorporated in the imputation algorithm.

3. Run consistency checks and edits for multiply imputed data. This might involve modifying the imputed values that are random numbers generated from the imputation model. However, it is expected that few modifications would be necessary for imputations based on a reasonably good model. Therefore, the post-imputation editing can also be viewed as a quality check or diagnostics of the imputation model (Chapter 12). In addition, modify the imputed data to maintain the requirement of data confidentiality.

Example 10.9. *Data editing and top-coding in the NHIS income imputation project*

In the brief summary of the NHIS family income imputation project (Example 10.4), family earnings were included as an important predictor for total income and were imputed as well. Some procedures were developed to ensure that the family earnings were always greater than or equal to personal earnings in the imputed data. More specifically, for any family that had only one employed adult with missing personal earnings, once the family earnings were imputed, the person's missing earnings could be determined by subtracting the observed personal earnings for members of the family from the imputed family earnings. For families that had more than one employed adult with missing personal earnings, in the imputation of the proportion of family earnings to be allocated to each employed adult with missing personal earnings, the logit (log-odds) transformation was applied to the proportions, and a normal linear regression model was used for the logit. After the logits were imputed, they were transformed back to proportions. Then, within each family, the proportions for the employed adults with missing personal earnings were rescaled so that they would sum to the total proportion of family earnings not accounted for by persons whose earnings had been observed. Imputed personal earnings were calculated from an imputed proportion via multiplication of the proportion by the family earnings.

In terms of releasing the public-use data, the family income is top coded. In NHIS 2016 public-use data, for example, the family income was top coded at the 95th percentile ($206,000) from that year, and the top five percent of values are set to this top-coded value. Note that data modifications for confidentiality (e.g., top-coding) would need to be conducted after imputation rather than before imputation because these procedures are expected to reduce or distort the information from the original data to some degree.

10.5.2 Documentation and Release

Documentation and data release of multiply imputed datasets can also be complicated. Inside the organization, this typically involves a joint effort between different teams. Of course, there is no uniform guideline on how to do it as different organizations might have different structure and customs. We continue using the NHIS family income imputation project to illustrate some of the issues.

Example 10.10. *NHIS income imputation project documentation*

The NHIS income imputation project covers the annual surveys starting from 1997. For NHIS 2016, for example, the documentation can be found in (https://www.cdc.gov/nchs/nhis/nhis_2016_data_release.html). Although the main methodology had been published in Schenker et al. (2006) and remained largely consistent from 1997 to 2018, a separate document with more specifics is publicly available for each year's income data (e.g., by going to the above URL and clicking the link "Technical documentation: methods and examples"). This is necessary because, like all other surveys, NHIS has gone through changes (e.g., sampling and questionnaire designs) over the years. It is important to record the corresponding changes (e.g., variables used in the model) in the imputation modeling in the public documentation.

As opposed to the methodological paper by Schenker et al. (2006), the imputation documentation provides more details for data users. Here we comment on two major features. One is that the document lists all the variables used in the imputation modeling. Information about these variables can be helpful for data users to understand the general idea and quality of imputations. The other is how to use multiply imputed datasets for analysis. The document not only presents multiple imputation combining formulas (Section 3.3), but also includes some sample code for accessing the datasets and running the analysis using commonly used software packages (e.g., SAS). Note that the original imputation code is not included in the document. This is mainly because the actual code involves many steps of data processing and modeling.

The multiply imputed data files for NHIS 1997 to 2018 are separate from the original (unimputed) NHIS data. They only contain actual imputed values, imputation flags, and necessary cross-walk variables. All of this important information, as well as a description of how to access the imputed data and merge them with the original data, can be obtained by going to the above URL and clicking the link "data set documentation" and "Sample SAS input program."

Finally, a basic yet challenging question for releasing multiply imputed datasets is, how many imputations (M) should be released to the public? In the examples covered in this chapter, M was usually set as five, following the original recommendation by Rubin (1987). More recent research shows that the number of imputations might need to be increased so that the multiple imputation efficiency is sufficiently high for a wide variety of estimands (Section 3.3.3). In addition, computational burden and data storage for a large number of imputations are of less concern nowadays. More considerations might be taken for this statistical custom from the organization's perspective.

10.6 Summary

In this chapter we have discussed several major methodological issues for applying multiple imputation to survey data, as well as constructing multiply imputed databases for public use. Multiple examples were used for illustration, which constitute only a small fraction of past and ongoing real-life applications. Here we reiterate some of the key points in developing effective imputation and data-release strategies for survey data.

1. Use the FCS strategy that provides flexible options of imputation modeling and handling complex data features.
2. Use the inclusive imputation strategy by including a large number of variables in the imputation model. This makes the MAR assumption more plausible. The selection of variables also needs input from subject-matter perspectives. On the other hand, it is important to find a balance between model complexity and practicality.
3. Incorporate survey designs in the imputation model. There might not exist a unique rule, as appropriate and feasible options need to be considered in different contexts.
4. Run data edits and checks on imputations.
5. Prepare clear analytical documentations and guidelines for imputation data users.

In addition, all surveys are subject to error, both systematic and random (Groves 1989). This chapter focuses on the use of multiple imputation for correcting item nonresponse error of surveys. In Chapter 11, we will illustrate the use of multiple imputation for correcting measurement error including those for survey data.

11

Multiple Imputation for Data Subject to Measurement Error

11.1 Introduction

Data collected in many studies are subject to measurement error. Measurement error for categorical data is also termed as misclassification. Measurement error can take a variety of forms. For example, in survey data, response errors such as recall bias might happen from some survey participants. When one survey is administered at different times, the coding for some variables can change and lead to measurement error problems. Such inconsistency can also happen between different data sources or instruments that are targeted to measure the same constructs. Similar to missing data problems, if left unaddressed or approached incorrectly, measurement error problems could lead to invalid or suboptimal statistical inferences.

There exists a large body of literature on statistical analysis with mismeasured data. Classic references can be found, for example, in Carroll et al. (2006), Buonaccorsi (2010), Gustafson (2004), and Fuller (2006). Many of the established methods are not explicitly based on imputation. This chapter provides a discussion on using the multiple imputation approach to measurement error problems. As pointed out in Reiter and Raghunathan (2007), the multiple imputation framework is well suited for some measurement error problems: the imputation approach replaces values with stochastic measurement errors with draws from probability distributions designed to correct the errors, creating "corrected" datasets. Then the error-corrected datasets can be analyzed using standard complete-data methods rather than some specific measurement error techniques, because adjustments for measurement error are automatically included in the corrected datasets.

In general, for any measurement error method (including multiple imputation), it would be crucially helpful to include information from validation data source(s) that includes both the true values and mismeasured ones for the variable(s) of interest. The relation between the true values and mismeasured values, which is quantified by measurement error models, can be inferred using the validation data. Such a relation can then be used to correct the measurement error in the original data. However, there exist some scenarios where

DOI: 10.1201/9780429156397-11

apparent validation data are not available, and measurement error models used there can have rather strong assumptions.

The rest of Chapter 11 is organized as follows. Section 11.2 lays out the main rationale behind the multiple imputation approach to measurement error problems. Section 11.3 presents alternative imputation modeling strategies. Section 11.4 discusses the problem where two different coding/measurement systems are attempted for the same construct. Section 11.5 provides a general discussion of using multiple imputation to combine information from multiple data sources. Section 11.6 discusses the imputation of a missing composite variable. Section 11.7 provides a summary.

11.2 Rationale

There are alternative ways of describing and formulating measurement error problems and related models. In this chapter we treat variables with measurement error as being incomplete because we do not know their true values. We also treat their measured values and other variables (i.e., covariates or auxiliary variables) as being observed. The idea is to multiply impute true values of the mismeasured variables using the information from observed components of the data based on some reasonable models. Multiple imputation analysis can be carried out using the imputed "true values" of the mismeasured variables.

This rationale then closely resembles that of multiple imputation analysis for typical missing data problems (Section 3.2.1). Let Y_T denote the variable(s) of interest (here T symbolizes "true"), Y_O denote the measurement (here O symbolizes "observed" so that it can connect to the typical notations in missing data analysis), X denote the collection of other covariates without measurement error. Let θ denote the parameters governing analysis models of interest. For example, θ might parameterize a regression model relating Y_T with X.

When validation data are available where both Y_T and Y_O are observed, the measurement error problem becomes a missing data problem (Carroll et al. 2006, Page 30). See also Little and Rubin (2020, Section 11.7). That is, Y_T is incomplete: their values are missing in the main data yet observed in the validation data. From a Bayesian perspective, the analytical aim is to infer about θ from its posterior distribution, $f(\theta|Y_O, X)$. To integrate out unobserved values in Y_T, we have,

$$f(\theta|Y_O, X) = \int f(\theta, Y_T|Y_O, X)dY_T \quad (11.1)$$
$$= \int f(\theta|Y_T, Y_O, X)f(Y_T|Y_O, X)dY_T \quad (11.2)$$
$$= \int f(\theta|Y_T, X)f(Y_T|Y_O, X)dY_T \quad (11.3)$$
$$\approx \frac{1}{M}\sum_{m=1}^{M} f(\theta|Y_T^{(m)}, X), \quad (11.4)$$

where $M > 1$ and $Y_T^{(m)}$ is drawn from $f(Y_T|Y_O, X)$.

Note that $f(\theta|Y_T, Y_O, X) = f(\theta|Y_T, X)$ in general because the mismeasured version of Y_T (i.e., Y_O) would not contribute any extra information to the analysis when Y_T is fully available. This is the so-called nondifferential measurement error assumption (Carroll et al. 2006). Eqs. (11.1)-(11.4) are analogous to Eqs. (3.1)-(3.4) in Chapter 3.

The key is to draw, or multiply impute, Y_T from its posterior predictive distribution, $f(Y_T|Y_O, X)$. In addition, an important assumption behind Eqs. (11.1)-(11.4) is that the missingness of Y_T is MAR, which will be further discussed in Section 11.3.1.

Example 11.1. *A simulation study for a continuous variable subject to measurement error*

We continue using the simulation setting in Example 2.5 to illustrate the performance of multiple imputation in handling measurement error problems. The complete-data model is: $X \sim N(1, 1)$, and $Y_T = -2 + X + \epsilon$, where $\epsilon \sim N(0, 1)$. However, here Y_T is subject to some measurement error in the actual data. For the measurement error model, we assume that the error-prone measurement Y_O has the distributional form: $Y_O = c_0 + c_1 Y_T + \epsilon_O$, where $\epsilon_O \sim N(0, 1)$. We also assume that 20% of the original data is used as a validation subset in which both Y_T and Y_O are available. The selection of the validation data is purely random so that the missingness of Y_T is MCAR. The structure of the observe data is illustrated in Table 11.1, where Y_T is missing for around 80% of the data and is observed in the validation subset, and both Y_O and X are fully observed for the whole data.

This measurement error problem can thus be regarded as a missing data problem as shown in Table 11.1. A solution is to multiply impute missing Y_T-values in the data and conduct a multiple imputation analysis. Suppose the estimands of interest include the mean of Y_T, the slope coefficient from regressing Y_T on X, and the slope coefficient from regressing X on Y_T. In addition to before-deletion (BD) analysis that uses the complete Y_T-values, we consider three measurement error methods:

1. Naive: use the mismeasured variable, Y_O, for the whole data.

2. Validation: restrict the analysis only to the validation subset. This corresponds to the complete-case (CC) analysis from the perspective of missing data methods.

3. MI: multiple imputation analysis for missing Y_T. In this case, the imputation model is a normal linear regression model for Y_T given Y_O and X. The number of multiple imputation M is set as 80.

The simulation study considers two measurement error processes. In Scenario I, $c_0 = 0$ and $c_1 = 1$ so that Y_O is unbiased for Y_T yet with additional noises for ϵ_O. In Scenario II, $c_0 = -0.2$ and $c_1 = 1.2$ so that Y_O incurs both biases and noises with respect to Y_T. The complete-data sample size is set as 1000, and the simulation includes 1000 replicates.

Table 11.2 shows the simulation results for Scenario I. When Y_O is set to be unbiased for Y_T yet with additional noises, the naive method (i.e., replacing Y_T with Y_O) is still unbiased for the mean estimate of Y_T as well as the coefficient from regressing Y_T on X. The corresponding coverage rates are also close to the nominal level. This pattern can also be verified by theoretical derivations. Both the validation and MI methods yield results with little bias and nominal coverage rates. The latter gains more efficiency than the former for both estimands. The advantage of MI is not surprising for estimating the mean of Y_T. Yet for estimating the regression coefficient for Y_T on X, the imputation of Y_T here uses information from both X and Y_O and is expected to be better than just using the information from X as in the validation method. The former approach follows the inclusive imputation strategy (e.g., similar patterns shown in Example 5.8).

However, when Y_T is used as a regressor in the regression, the naive method shows a considerable bias (RBIAS= -33.3%) for the coefficient estimate. This phenomenon is the well documented "attenuation" when covariates are with measurement error in regression settings (Carroll et al. 2006, Chapter 3). In

TABLE 11.1
Example 11.1. An illustrative dataset from simulation

Y_T	Y_O	X
M	1.86	.55
M	−1.49	1.04
M	1.21	1.64
...	...	...
−3.43	−3.66	−0.21
−2.36	−4.45	−0.37
...	...	...

Note: "M" symbolizes unobserved values for Y_T. The validation dataset has observed Y_T-values.

this setting, the attenuation factor, also called the reliability ratio (Fuller 2006), can be calculated as $\frac{Var(Y_T)}{Var(Y_T)+Var(Y_T|Y_O)} = 2/(2+1) = 2/3$, which corresponds to a relative bias as $-1/3$. On the other hand, MI estimates correct for such attenuation bias and retain the association. It also gains extra efficiency compared with the validation method. Fig. 11.1 plots a random subset (sample size 250) from a simulation replicate using scatter plots and fitted lines. Compared with Y_T (top left panel), the scatter plot of X vs. Y_O (top right panel) shows more variability and a weaker association. The bottom panel of Fig. 11.1 shows the scatter plot of X vs. completed Y_T from MI, which retains such association.

Table 11.3 shows the simulation results for Scenario II. In this case, since the measurement error includes a mean shift, the naive method is also biased for estimating the mean of Y_T and the regression coefficient for Y_T on X. The MI method still performs well for all three estimands, consistently showing improved precision over the validation method.

Example 11.2. *A simulation study for a binary variable subject to misclassification*

In the simulation, the complete-data model is: $X \sim N(1, 1/4)$, and $logit(f(Y_T = 1)) = -2 + 2X$. However, Y_T is subject to some misclassification in the actual data. For the measurement error model, we assume that the error-prone measurement Y_O has the distributional form: $f(Y_O = 1|Y_T = 1) = 0.9$ (sensitivity) and $f(Y_O = 0|Y_T = 0) = 0.8$ (specificity). We also assume that

TABLE 11.2
Example 11.1. Simulation results in Scenario I

Method	RBIAS (%)	SD	$\overline{\text{SE}}$	MSE	CI	COV (%)
			Mean of Y_T			
BD	0	0.043	0.045	0.00188	0.175	96.1
Naive	0	0.053	0.055	0.00276	0.214	95.1
Validation	0.5	0.100	0.100	0.0101	0.392	94.9
MI	0.1	0.062	0.063	0.00384	0.248	96.0
			Regression coefficient for Y_T on X			
BD	0	0.032	0.032	0.000995	0.124	95.0
Naive	0	0.045	0.045	0.00202	0.175	95.2
Validation	−0.1	0.068	0.071	0.00469	0.279	95.9
MI	0.5	0.054	0.055	0.00296	0.216	94.8
			Regression coefficient for X on Y_T			
BD	0	0.016	0.016	0.00025	0.062	95.0
Naive	−33.3	0.015	0.015	0.02799	0.058	0
Validation	−0.1	0.037	0.036	0.00133	0.139	94.5
MI	−0.3	0.025	0.026	0.00062	0.102	96.4

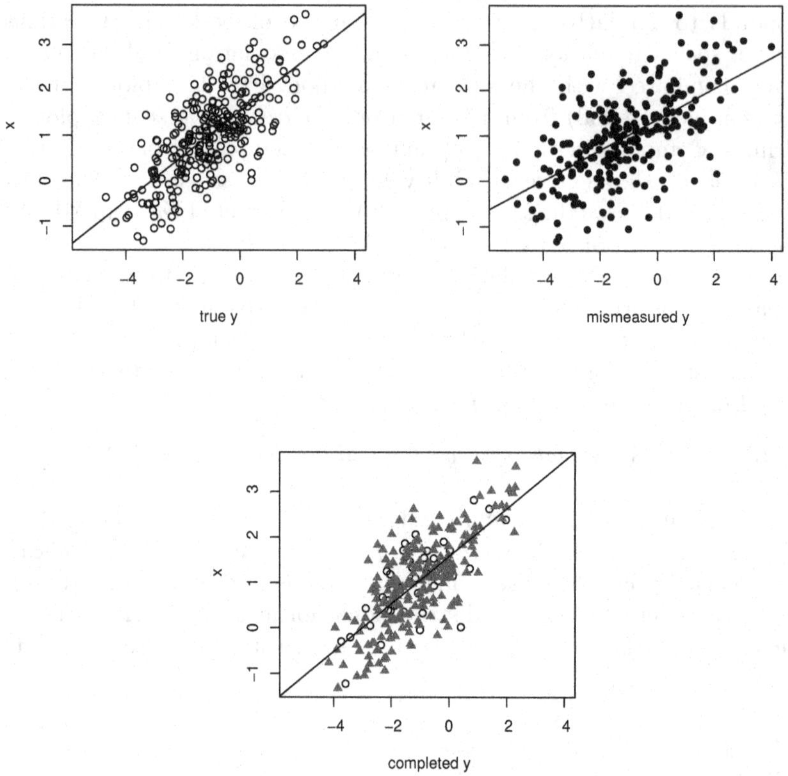

FIGURE 11.1
Example 11.1. Scatter plots and fitted lines from a subset of a simulation replicate. Top left: between X and Y_T; top right: between X and Y_O; bottom: between X and completed Y_T. Circle: observed Y_T; filled circle: Y_O; filled triangle: imputed Y_T.

20% of the original data is used as a validation subset in which both Y_T and Y_O are available. The selection of the validation data is purely random (i.e., MCAR). The structure of the observed data is illustrated in Table 11.4, where Y_T is missing for around 80% of the data and is observed in the validation subset, and both Y_O and X are fully observed for the whole data. The simulation consists of 1000 replications, and the complete-data sample size is 1000.

Similar to Example 11.1, four methods are applied to the simulated data (i.e., BD, naive, validation, and MI). In MI, the imputation model for Y_T is a logistic regression imputation using both Y_O and X as predictors, and it is

TABLE 11.3
Example 11.1. Simulation results in Scenario II

Method	RBIAS (%)	SD	$\overline{\text{SE}}$	MSE	CI	COV (%)
			Mean of Y_T			
BD	0	0.043	0.045	0.00188	0.175	96.1
Naive	40.0	0.060	0.062	0.1632	0.244	0
Validation	0.5	0.100	0.100	0.0101	0.392	94.9
MI	0.1	0.059	0.060	0.00348	0.236	95.9
		Regression coefficient for Y_T on X				
BD	0	0.032	0.032	0.000995	0.124	95.0
Naive	20.0	0.050	0.049	0.04241	0.194	1.7
Validation	−0.1	0.068	0.071	0.00469	0.279	95.9
MI	0.4	0.051	0.051	0.00262	0.203	94.4
		Regression coefficient for X on Y_T				
BD	0	0.016	0.016	0.00025	0.062	95.0
Naive	−38.1	0.013	0.013	0.03655	0.050	0
Validation	−0.1	0.037	0.036	0.00133	0.139	94.5
MI	−0.1	0.024	0.025	0.00057	0.098	96.1

TABLE 11.4
Example 11.2 An illustrative dataset from simulation

Y_T	Y_O	X
M	1	0.55
M	0	1.04
M	1	1.64
...	...	...
1	0	−0.21
1	1	−0.37
0	1	0.32
0	0	−1.37
...	...	...

Note: "M" symbolizes unobserved values for Y_T. The validation dataset has observed Y_T-values.

conducted $M = 80$ times. Table 11.5 summarizes the simulation results for estimating the mean of Y_T as well as the slope coefficient for running a logistic regression of Y_T on X. In this simulation setup, the naive method is biased for both the mean of Y_T and logistic regression coefficient. For the former, it can be verified that the naive method would be unbiased if $\frac{E(Y_T)}{1-E(Y_T)} = \frac{\text{1−specificity}}{\text{1−sensitivity}}$ in the misclassification process, which it not satisfied in this simulation setup. For the latter, the simulation results are consistent with the pattern that a misclassified binary response variable would lead to biased logistic regression coefficient estimates (Neuhaus 1999). Similar to the pattern in Example 11.1, MI corrects the biases for both estimates from the naive method and has coverage rates close to the nominal level. It also improves the precision over the validation method. Note that the correction for bias is associated with an increase of the variance estimate, which reflects the phenomenon of bias-variance trade-off. As a result, the confidence interval from MI is considerably wider than that of the naive method, especially for the logistic regression coefficient.

TABLE 11.5
Example 11.2. Simulation results

Method	RBIAS (%)	SD	$\overline{\text{SE}}$	MSE	CI	COV (%)
			Mean of Y_T			
BD	0	0.015	0.016	0.000239	0.062	95.8
Naive	10.0	0.016	0.016	0.00274	0.062	11.5
Validation	0.1	0.035	0.035	0.00113	0.139	95.3
MI	0	0.025	0.026	0.000643	0.103	95.5
			Logistic regression coefficient for Y_T on X			
BD	0	0.163	0.167	0.027	0.163	95.6
Naive	−36.0	0.147	0.145	0.544	0.194	0.4
Validation	1.4	0.388	0.381	0.151	1.494	95.5
MI	−0.3	0.331	0.317	0.110	1.26	95.5

11.3 Imputation Strategies

Examples 11.1 and 11.2 can be used as a starting point to gauge the utility of multiple imputation analysis for mismeasured data. In the following we provide some details of how to impute Y_T from $f(Y_T|Y_O, X)$, depending on specific study designs and analytical goals.

11.3.1 True Values Partially Observed

11.3.1.1 Basic Setup

As illustrated in Examples 11.1 and 11.2, in many practical cases, the error-prone measurement, Y_O, is available for the full sample, and the true value, Y_T, is only observed in a validation sample. For brevity, suppose the full data S consists of two subsamples, S_1 and S_2, where S_1 is the main study sample and S_2 is the validation sample. Assume that Y_T is not available in S_1 yet available in S_2, and both Y_O and X are available in S_1 and S_2. We also assume that X is without measurement error, and there exist no extra missing values for Y_T, Y_O, and X, which will be relaxed in Section 11.3.1.4. Table 11.6 sketches the basic data structure.

TABLE 11.6
A sketch of the data structure when Y_T is partially observed

Sample	Y_T	Y_O	X
S_1	M	O	O
S_2	O	O	O

Note: S_1: the main study sample; S_2: the validation dataset. Y_T: true values of Y; Y_O: error-prone measurements of Y; X: covariates. "M" symbolizes missing values; "O" symbolizes observed values.

Clearly, the measurement error problem in Table 11.6 resembles the problem of one incomplete variable Y_T with fully observed variables including Y_O and X. This topic had been discussed in Chapters 4 and 5. The general solution is to multiply impute missing Y_T in S_1 using information from Y_O and X based on some models.

The missingness at random assumption (MAR) for Y_T can be expressed as:

$$f(Y_{T,S_1}|Y_{O,S_1}, X_{S_1}) = f(Y_{T,S_2}|Y_{O,S_2}, X_{S_2}), \qquad (11.5)$$

where Y_{T,S_1} denotes the values of Y_T in S_1 (and similar notations apply for other variables). Eq. (11.5) can be also understood as the "transportable" assumption in measurement error methods literature (Carroll et al. 2006). That is, both the model, $f(Y_T|Y_O, X)$, as well as the parameters governing that model have to be identical between S_1 and S_2. It also (implicitly) assumes that corresponding variables from S_1 and S_2 are measuring the same quantities. This assumption is very important because the validation data S_2 can be an internal subset of the original data S, or S_2 can be from an external and independent data source from S_1. In the former situation, the transportability assumption seems to more natural as long as the selection of the validation sample is not dependent on Y_T after conditioning on Y_O and X. In Examples 11.1 and 11.2, the selection of S_2 is completely at random so that the transportability (or MAR) assumption clearly holds. In the latter situation where

S_2 and S_1 come from two different studies or data sources, the transportability assumption would be generally more difficult to assess.

If S_2 is an internal subset of S, the multiple imputation analysis can be applied to the whole sample S. Typically, analyzing S_2 only (e.g., the validation method or complete-case analysis in Examples 11.1 and 11.2) is usually not the optimal approach because the sample size is relatively smaller compared with S_1. This can be seen from the comparison between the validation and MI method in Examples 11.1 and 11.2. In addition, if the selection of S_2 within S is not completely at random, then estimates for some estimands (e.g., the mean of Y_T) can be biased by analyzing only S_2.

11.3.1.2 Direct Imputation

We can use methods discussed in Chapters 4 and 5 to multiply impute Y_T in S_1 using the information from Y_O and X based on a complete-data model $f(Y_T|Y_O, X)$. In Example 11.1, the imputation model is a simple normal linear regression model; in Example 11.2, the imputation model is a logistic regression model. From the perspective of making predictions, Raghunathan and Siscovick (1998) argued that the observed Y_O provide means of forming a prior distribution of the actual Y_T, and the model, $f(Y_T|Y_O, X)$, uses the input of this prior information to impute Y_T. This strategy can be viewed as a direction imputation approach. However, the specification of imputation models can be more complicated in real data with different study designs and analytical objectives.

Example 11.3. *Improving on analyses of self-reported data in a large-scale health survey by using information from an examination-based survey*

The response bias associated with self-reports in surveys is well documented (Groves 1989). Schenker et al. (2010) presented a multiple imputation analysis aiming to correct for the error associated with self-reports of health status (e.g., hypertension, diabetes, and obesity) from a large survey (e.g., NHIS), using information from more accurate clinical measurements in a small survey (e.g., NHANES). Their descriptive analyses showed that estimates based on clinical data were larger (i.e., worse health conditions) than the estimates based on self-reported data using the two surveys. For example, the estimates of diabetes prevalence based on clinical data was higher than self-reports of doctor diagnosis of diabetes. This can be expected because of cases with diabetes but no diagnosis. Yet such a pattern suggests that collecting self-reported data on disease diagnosis can lead to underestimation of the disease prevalence without some corrections.

The proposed statistical strategy in Schenker et al. (2010) can be well explained by the modeling framework outlined here: Y_T includes clinical measurements of disease status or health outcomes (e.g., hypertension, diabetes, obesity), and Y_O is the less accurate, self-reported version of Y_T. In their studies, S_1 is the NHIS study sample in which only the self-reported survey responses are available (i.e., Y_T is lacking), and S_2 is the NHANES study sample in which both the clinical measurement (Y_T) and self-reports (Y_O) from survey respondents are available.

To summarize the idea from Schenker et al. (2010), we use the hypertension status (a binary variable) for illustration. A person was classified as having hypertension based on self-reported data (i.e., Y_O) if a doctor or other health professional had told the person two or more times that he/she had hypertension or high blood pressure. For clinical data, the classification of hypertension (i.e. Y_T) was based on having systolic blood pressure greater than 140 mmHg, having diastolic blood pressure greater than 90 mmHg, or taking medication to control blood pressure. A set of common variables including basic demographics and general health variables (e.g., have health insurance, tobacco use, self-reported health status) as well as their interactions were used as the covariates X.

One unique challenge in their application is that the two surveys (i.e, NHIS and NHANES) are separate and not from one integrated study. Therefore some statistical techniques were applied to make the transportability assumption more appropriate. Specifically, let R be the sample inclusion indicator (i.e. the survey membership): $R = 1$ if a subject is in S_2 and $R = 0$ if a subject is in S_1. First a propensity score model, $logit(f(R = 1)) = X\alpha$, was fitted to create K subgroups of S. Within each subgroup, the distributions of covariates were made to be similar across two surveys by the propensity score analysis. Note that Y_O should not be included as a predictor of this propensity model because it is the outcome of interest. Second, within each subgroup, another propensity score model $logit(f(Y_O = 1)) = X\beta$ was fitted. Let $\hat{g}_1(X)$ and $\hat{g}_2(X)$ be the corresponding estimated propensity score from the two models.

Within each subgroup, the imputation model $f(Y_T|Y_O, X)$ is a logistic regression model since Y_T is binary. The predictors of the imputation model for Y_T include Y_O, $\hat{g}_1(X)$, $\hat{g}_2(X)$, and their two-way interactions. The utility of the propensity scores has two folds. One was to make the transportability assumption more plausible by conducting the imputations within each subgroup formed by the propensity score estimates for the survey membership. Second was to use the propensity score estimates, $\hat{g}_1(X)$ and $\hat{g}_2(X)$, as the summary of the information embedded in Y_O and X (Section 5.6.2). This could yield some robustness against possible misspecifications of the model, $f(Y_T|Y_O, X)$, and is also simpler to operate if some higher-order interactions among X needed to be included as predictors.

The multiple imputation analysis strategy appeared to be effective in correcting the biases associated with self-reports. In the evaluation of Schenker et al. (2010), estimates using multiply imputed clinical measurements in NHIS were much closer to those from NHAHES than the self-reports from NHIS. In addition, multiply imputing clinical values for NHIS resulted in more precise estimates than those based on the NHANES. This is consistent with the pattern observed in Example 11.2.

Schenker et al. (2010) also commented upon several possible limitations. One is that, apart from the information contained in X-variables, the survey design information from two surveys (e.g., survey weights) were not explicitly accounted for (Chapter 10). In addition, in their setting, the analysis of multiply imputed data was confined to NHIS (i.e., S_1 only), yet not the combination of two surveys (i.e., $S = S_1 \cup S_2$). This makes practical sense because here the validation data is "external" to the data that have error-prone variables. Yet such a situation invokes the issue of imputation uncongeniality (Meng 1994) since in theory analyzing the whole sample S is expected to be more efficient than just a subset. That is, the post-imputation analysis does not use some of the information from S_2 that is used in the imputation model. As a result, Rubin's variance formula in combining estimates from multiply imputed datasets tends to overestimate the actual variance. However, when the relative sample size of S_1 over S_2 is large (around 18 in their application), the bias of variance estimation is relatively small. A general solution to this issue can be found in Reiter (2008). See also Little and Rubin (2020, Page 281).

In Example 11.3, the idea of using models fitted to one set of data (S_2), with information about an item under two types of coding or reporting systems (both Y_T and Y_O), to impute values for Y_T in a separate set of data (S_1), with information under only one of the systems (Y_O), has been used in other applications and will be illustrated again throughout this chapter. For example, Raghunathan (2006) used a similar strategy as in Schenker et al. (2010) to correct the self-reported disease status in NHIS for assessing health disparities.

11.3.1.3 Accommodating a Specific Analysis

The direct imputation strategy is straightforward if the primary purpose is to construct multiply completed databases for some general statistics. However, sometimes the scientific interpretation of the model, $f(Y_T|Y_O, X)$, is not straightforward. This issue is more salient if a specific post-imputation analysis is to be applied to Y_T and X. In general, a majority of studies dealing with measurement error or misclassification of variables in observational or epidemiological investigations have well defined analyses for targeted scientific goals.

Suppose Y_T is treated as a response variable in an analysis and X includes the regressors in this analysis. We consider a decomposition of $f(Y_T|Y_O, X)$ as follows:

$$\begin{aligned} f(Y_T|Y_O, X) &\propto f(Y_T, Y_O, X) \\ &= f(Y_T|X) f(Y_O|Y_T, X). \end{aligned} \tag{11.6}$$

The first component of Eq. (11.6), $f(Y_T|X)$, characterizes the scientific relationship between Y_T and X, and its formulation matches with the analysis of interest. The second component, $f(Y_O|Y_T, X)$, characterizes the pattern of measurement given the true values and other covariates, and is a formulation of the possible measurement error model. Since in the measurement error model, Y_T is being conditioned on and unobserved in S_1, the missing data problem can be perceived as a problem of imputing a missing covariate as discussed in Section 4.4.

Example 11.4. *Imputation of underreported adjuvant treatment status in cancer registry data*

Yucel and Zaslavsky (2005) proposed an imputation approach to an underreporting problem of adjuvant chemotherapy for stage II/III colorectal cancer patients in data from a cancer registry. To sketch their idea, Y_T is the true status (Yes/No) of adjuvant chemotherapy, Y_O is what gets reported in the cancer registry; S is the cancer registry sample; S_2 is a validation study applied to a subsample of the registry that collects medical records data about the treatment status (Y_T); X includes clinical and demographic covariates at both patient- and aggregate(hospital)-level.

To set up an appropriate model for $f(Y_T|Y_O, X)$, The underreporting assumes that $Y_O = 0$ if $Y_T = 0$ (deterministic); otherwise Y_O can be 1 or 0 if $Y_T = 1$ (stochastic). That is, if $Y_O = 1$, then Y_T has to be imputed as 1. The stochastic imputation for Y_T only comes if $Y_O = 0$. Note that the underreporting problem here is different from the setup in Example 11.2, where the misclassification of Y_T can go in either direction. Table 11.7 shows a schematic structure of the data.

TABLE 11.7
Example 11.4. Schematic data structure

Y_T	Y_O	X	Sample
M(= 1)	1	0.55	S_1
M= (?)	0	1.04	S_1
M(= 1)	1	1.64	S_1
...	...	...	...
1	1	−0.21	S_2
0	0	−0.37	S_2
1	0	−1.36	S_2
...	...	...	

Note: "M" symbolizes unobserved values for Y_T. When $Y_O = 1$, then Y_T is imputed as 1 under the assumption of underreporting. The stochastic imputation for Y_T only happens when $Y_O = 0$. S_1: the subsample where Y_T is unobserved; S_2: the validation sample.

In Yucel and Zaslavsky (2005), $f(Y_T|X)$ and $f(Y_O|Y_T, X)$ were termed as the "outcome model" and "reporting model", respectively. In their context,

Eq. (11.6) can be further elaborated as follows:

$$\begin{aligned} f(Y_{T,S_1}|Y_{T,S_2},Y_O,X) &= \int f(Y_{T,S_1},\theta|Y_{T,S_2},Y_O,X)d\theta \\ &\propto \int f(Y_{T,S_1},Y_{T,S_2},Y_O,X,\theta)d\theta \\ &= \int f(Y_{T,S_1},Y_{T,S_2}|X,\theta_O)f(Y_O|Y_{T,S_1},Y_{T,S_2},X,\theta_R)d\theta \\ &= \int f(Y_{T,S_2}|X_{S_2},\theta_O)f(Y_O|Y_{T,S_1},Y_{T,S_2},X,\theta_R)d\theta, \end{aligned} \tag{11.7}$$

where θ can be decomposed into two distinct sets of parameters, θ_O and θ_R. The parameter θ_O governs the outcome model, $f(Y_T|X,\theta_O)$, and the parameter θ_R governs the reporting model (measurement error model), $f(Y_O|Y_T,X,\theta_R)$. Since both Y_T and Y_O are binary, logistic regression models are used to model both the outcome and reporting models. In the context of actual data, θ_O consists of logistic regression coefficients that relate patients' demographic and clinical covariates with their probability of receiving the adjuvant therapy; θ_R consists of logistic regression coefficients that relate those variables with the probability of having the treatment status underreported in the registry data. In addition, to account for similar patterns for both the outcome and reporting processes within the same providers (e.g., hospitals), random effects at hospital level were assumed in both models.

Note that in Eq. (11.7), $f(Y_{T,S_1},Y_{T,S_2}|X,\theta_O) = f(Y_{T,S_2}|X_{S_2},\theta_O)$ because Y_{T,S_1} (missing) does not contribute any information as the response variable. However, $f(Y_O|Y_{T,S_1},Y_{T,S_2},X,\theta_R) \neq f(Y_{O,S_2}|Y_{T,S_2},X_{S_2},\theta_R)$ because the missing Y_{T,S_1} is a covariate in the implied logistic regression here. Therefore the imputation procedure cannot be accomplished in one step, and has to be implemented in an iterative fashion. However, if the missingness of Y_T (i.e. the relative sample size of S_1 to S_2) is relatively large as in the usual case, the iterative process tends to be less stable. This is because the imputation of Y_T relies on information from two parts: one is based on $f(Y_{T,S_2}|X_{S_2},\theta_O)$, and the other is based on $f(Y_O|Y_{T,S_1},Y_{T,S_2},X,\theta_R)$. The iterative imputation process needs to find a tuned balance between the two parts.

Although the coefficient estimates for θ_O and θ_R become secondary if the primary purpose is to multiply impute Y_T, whether these estimates make practical sense can help assess the quality of imputations. The direction of estimates in the application from Yucel and Zaslavsky (2005) matched well with findings from the literature on the quality of cancer care. In their applications, coefficient estimates from the outcome model indicated that chemotherapy was used more often among younger, married, higher-income, and stage III rectal cancer patients and less often among those with stage II rectal cancer or comorbidities. Coefficient estimates from the reporting model indicated that high-volume hospitals reported chemotherapy more completely than others, as did teaching hospitals. Treatment of older, low-income, and married patients was less likely to be reported. Reporting was better for patients with rectal cancer (for whom additional treatment information was often received from radiation facilities) and those who were diagnosed and treated in the same hospital. In terms of using multiply imputed data in S_1, again, their estimates were more precise than just using data from the validation sample S_2.

He et al. (2008) provided a simple illustration of Yucel and Zaslavsky (2005) for the lay audience. He and Zaslavsky (2009) extended the modeling strategy to simultaneously handle multiple variables in Y_T, which include both adjuvant chemotherapy and radiation therapy for stage II/III colorectal cancer patients.

The mismeasured Y_T can also be used as a covariate in an analysis. Let $X = (X_1, Z)$, where X_1 is the response variable of the targeted analysis, and Z includes confounders or other covariates that are to be adjusted in the analysis. An example would be regressing X_1 on Y_T (the main predictor if fully observed) adjusted by Z. When Y_T is categorical, a classic example is the misclassification of the exposure variables (regressors) in a logistic regression model characterizing the exposure-outcome relationship.

The principle of the imputation, as suggested by Eqs. (11.1)-(11.4), still applies in this setup. That is, the goal is to impute Y_T from $f(Y_T|Y_O, X) = f(Y_T|Y_O, X_1, Z)$. A key note here is that X, the covariate in the sense of imputation, should include the outcome variable X_1 of the targeted analysis. This follows the general rule of including the outcome variable for imputing a missing covariate in a targeted analysis (Section 4.4).

There exist various options of forming the model $f(Y_T|Y_O, X_1, Z)$, depending on the study design, actual data, targeted analysis, and modeling assumptions. One strategy is to consider a JM strategy similar to the imputation of missing covariates (Section 6.6) by considering the following decomposition:

$$\begin{aligned} f(Y_T|Y_O, X_1, Z) &\propto f(Y_T, Y_O, X_1, Z) \\ &= f(X_1|Y_T, Y_O, Z) f(Y_O|Y_T, Z) f(Y_T|Z) \\ &= f(X_1|Y_T, Z) f(Y_O|Y_T, Z) f(Y_T|Z). \end{aligned} \tag{11.8}$$

The first component, $f(X_1|Y_T, Z)$, formulates the analysis model. For example, in the studies of the exposure-disease relationship, $f(X_1|Y_T, Z)$ can be modeled using a logistic regression. The second component, $f(Y_O|Y_T, Z)$, can be used to formulate the measurement error model. The third component, $f(Y_T|Z)$, is the model characterizing the distribution of the true values of Y_T given other covariates (i.e., the covariate model). This can be viewed as an extension of the JM modeling strategy for imputing missing covariates, where both the analysis model and covariate model need to be specified. In addition for imputing covariates subject to measurement error, in Eq. (11.8) we need to specify an additional model for the measurement error, $f(Y_O|Y_T, Z)$.

Oftentimes using the decomposition in Eq. (11.8) and plugging in usual model specifications for $f(X_1|Y_T, Z)$, $f(Y_O|Y_T, Z)$, and $f(Y_T|Z)$, the posterior predictive distribution of the mismeasured covariate, $f(Y_T|Y_O, X_1, Z)$, does not have a closed form (Messer and Natarajan 2008; Guo and Little 2011). Advanced Bayesian computation techniques might need to be used. As we have illustrated in this book, practically the multiple imputation algorithm can be conveniently coded using WinBUGS in many cases. In addition, based on Eq. (11.8), the parameter estimates pertaining to the analysis of interest, $f(X_1|Y_T, Z)$, can be directly obtained from the posterior draws.

Note that Eq. (11.8) has several assumptions. First, $f(X_1|Y_T, Y_O, Z) = f(X_1|Y_T, Z)$, indicating that having Y_O adds no extra information to the analysis if Y_T and Z are fully available. This is the so-called "nondifferential" measurement error assumption (Carroll et al. 2006). In addition, modeling measurement error via $f(Y_O|Y_T, Z)$ (the conditional distribution of the error-prone measurement given the true values and other covariates is used for examples throughout this chapter) falls in the framework of classical measurement error. Alternatively, we can also try the decomposition $f(Y_T, Y_O, Z) \propto f(Y_T|Y_O, Z)f(Y_O|Z)$ in Eq. (11.8), where $f(Y_T|Y_O, Z)$ is the conditional distribution of the true values given the error-prone measurement and other covariates, following the Berkson error assumption (Berkson 1950). The distinction between two common mechanisms of measurement error was discussed in Carroll et al. (2006, Chapter 1).

The imputation modeling via Eq. (11.8) and the subtleties can be related well with the large body of measurement error models in the literature. Various statistical techniques (largely nonimputation methods) have been developed to address measurement error problems in regressors of analyses. Extensive discussions and references can be found, for example, in Carroll et al. (2006). Note that a different notation system was adopted in Carroll et al. (2006), where X is usually used to denote the true values of the variable, W is used to denote the measurement of X, and Y is used to denote the outcome in the analysis.

Besides the JM strategy, a more practical alternative is the direct imputation of Y_T by specifying a model for $f(Y_T|Y_O, X_1, Z)$ based on standard practices and convenience, as well as some exploratory analysis, ignoring the decomposition. It was discussed in Section 11.3.1.2 and can also be viewed as an approximation if the exact specification of each component in Eq. (11.8) is difficult. More generally, this option can often be extended to the FCS strategy when other variables (e.g., X_1 or Z) also have missing values and/or various components of Eq. (11.8) are complicated to specify, which will be discussed in Section 11.3.1.4. For both the JM and direct imputation strategies, again, the key is to include X_1 (the outcome of the targeted analysis) in the imputation of the error-prone analysis covariate Y_T.

Example 11.5. *Imputation of mismeasured covariates in a survival analysis*

Cole et al. (2006) was perhaps one of the earliest applications of using multiple imputation for the covariate measurement-error problems. In their study, Y_T is a binary exposure variable and X_1 is a survival outcome, and the analysis model $f(X_1|Y_T, Z)$ involves estimating the hazard ratio under a

Weibull survival model. The imputation modeled Y_T as a logistic regression using predictors including Y_O, the censoring indicator, and logarithm of the event/censoring time. This modeling strategy was discussed in Section 8.3. The estimates of this regression model are obtained from the validation study S_2 and then used to impute the missing Y_T in the main study sample S_1. Once Y_T is multiply imputed, the survival estimate can be obtained using multiple imputation combining rules. Their simulation study results showed that the multiple imputation analysis was effective.

Note that a model behind the direct imputation strategy might not be consistent with that derived from the JM approach based on the decomposition. This issue was discussed in Chapter 7 when comparing FCS with JM for imputing general missing data problems. See also Example 4.11. In the context of measurement error problems, this issue was also raised in both Cole et al. (2006) and Messer and Natarajan (2008), and their study results suggested that a convenient specification of $f(Y_T|Y_O, X_1, Z)$ in the direct imputation strategy might work well even if it is incoherent with what can be derived based on the decomposition. Therefore, it is important to understand the existence of the two imputation options and their distinctions. A variety of imputation models including alternative specifications of $f(Y_T|Y_O, X_1, Z)$ can be applied for a comparison analysis.

Example 11.6. *A simulation study comparing the JM and direct imputation methods for a covariate subject to measurement error*

We conduct a simple simulation study to assess and compare the performance between the JM and direct imputation method. Suppose X_1 is binary and Y_T and Z are both continuous. The analysis model is $logit(f(X_1 = 1)) = -1 + 0.5Y_T + 0.5Z$, the covariate model is $(Y_T, Z) \sim N_2((0,0)^t, \begin{pmatrix} 1, 0.5 \\ 0.5, 1 \end{pmatrix})$, and the measurement error model is $Y_O = -0.2 + 1.2Y_T + \epsilon_O$, where $\epsilon_O \sim N(0,1)$. Similar to Examples 11.1 and 11.2, we also assume that 20% of the original data is used as a validation subset in which both Y_T and Y_O are available. The selection of the validation data is purely random (i.e., MCAR). The complete-data sample size is 1000 and the simulation study consists of 1000 replicates.

The estimands of interest include the coefficients by running a logistic regression for X_1 on Y_T and Z. We consider two multiple imputation methods: JM and direct imputation. For the former, we use WinBUGS to implement the imputation, for which some sample code is included as follows. For the latter, we assume a normal linear imputation model of Y_T conditional on X_1, Y_O, and Z. For both methods, the number of multiple imputation M is set as 80.

```
model
{
for( i in 1 : M ){
# analysis mode
```

```
x1[i] ~ dbern(p[i])
logit(p[i]) <-beta0+beta1*y_t[i]+beta2*z[i]
# covariate model
z[i] ~ dnorm(mu, tau)
y_t[i] ~ dnorm(theta_12[i], psi_12)
theta_12[i] <- alpha0_12+alpha1_12*z[i]
# measurement error model
y_o[i] ~ dnorm(theta_m1[i], psi_m1)
theta_m1[i] <- alpha0_m1+alpha1_m1*y_t[i]
}
# prior for model parameters
beta0 ~ dnorm(0.0, 1.0E-3)
beta1 ~ dnorm(0.0, 1.0E-3)
beta2 ~ dnorm(0.0, 1.0E-3)
mu ~ dnorm(0.0, 1.0E-3)
alpha0_12 ~ dnorm(0.0, 1.0E-3)
alpha1_12 ~ dnorm(0.0, 1.0E-3)
alpha0_m1 ~ dnorm(0.0, 1.0E-3)
alpha1_m1~ dnorm(0.0, 1.0E-3)
tau ~ dgamma(1.0E-3, 1.0E-3)
psi_12 ~ dgamma(1.0E-3, 1.0E-3)
psi_m1 ~ dgamma(1.0E-3, 1.0E-3)
}
```

Table 11.8 shows the simulation results. Similar to Examples 11.1 and 11.2, the naive method, which replaces Y_T by Y_O in the logistic regression analysis, yields a large bias for the coefficient of Y_T. It is also biased for the coefficient of Z (RBIAS=21.6%) and invokes a low coverage rate (76.9%), although Z is not subject to any measurement error. Both JM and the direct imputation method correct the biases for the logistic regression coefficients and produce good coverage rates. They are also more efficient than the validation method. Between the two imputation methods, JM seems to be more efficient than the direct method as the former has smaller standard deviation and shorter confidence intervals. This is expected because the decomposition in JM follows the true data-generating model and the direction imputation only approximates it.

11.3.1.4 Using Fully Conditional Specification

In many practical settings, missing values can happen for Y_T in the validation sample, as well as for Y_O and X for the full data. The reasons behind these missing values are similar to those discussed for general missing data problems. The data structure can be sketched as in Table 11.9, which can be viewed as adding more holes of missingness to Table 11.6. One strategy used in some examples discussed in the preceding section is to first impute Y_O and X and make them completed and then focus more on the modeling of

$f(Y_T|Y_O, X)$. This might not be an optimal approach if the rate of missingness is nontrivial because Y_T, Y_O, and X are expected to be related to each other. In addition, the relationship between Y_T, Y_O, and X can be complicated to model, especially when multiple incomplete variables are involved or the data have complex features (e.g., survey data as discussed in Chapter 10).

Therefore, a more practical option is to use FCS (Chapter 7). That is, to impute all the missing variables in Y_O, Y_T, and X altogether simultaneously.

TABLE 11.8
Example 11.6. Simulation results

Method	RBIAS (%)	SD	$\overline{\text{SE}}$	MSE	CI	COV (%)
Logistic regression coefficient of Y_T						
BD	0	0.092	0.089	0.00847	0.350	94.3
Naive	−57.1	0.052	0.052	0.086	0.204	0
Validation	2.8	0.210	0.204	0.0442	0.801	94.7
JM	2.8	0.122	0.121	0.015	0.478	94.5
Direct	2.5	0.154	0.148	0.024	0.587	95.1
Logistic regression coefficient of Z						
BD	0	0.088	0.089	0.00776	0.350	95.1
Naive	21.6	0.084	0.085	0.019	0.333	76.9
Validation	1.4	0.214	0.204	0.0458	0.801	94.2
JM	−1.3	0.095	0.097	0.00909	0.379	94.9
Direct	−1.0	0.105	0.105	0.011	0.414	95.7

TABLE 11.9
A sketch of data structure when Y_T is collected in a validation sample S_2 and missing data occur for all variables

Sample	Y_T	Y_O	X
S_1	M	O	O
S_1	M	M	O
S_1	M	O	M
S_1	...	...	...
S_2	O	O	O
S_2	M	O	O
S_2	O	M	M
S_2	...	...	...

Note: "M" symbolizes missing values; "O" symbolizes observed values. S_1: the main study sample; S_2: the validation dataset. Y_T: true value of Y; Y_O: error-prone measurements of Y; X: covariates.

This strategy puts less focus on the interpretation of various components as in Eq. (11.8) and focuses more on drawing plausible values of unobserved Y_T. The FCS strategy imputes Y_T using a model $f(Y_T|Y_O, X)$, imputes missing Y_O-values using a model $f(Y_O|Y_T, X)$, and imputes missing values in X using other specified conditional models (e.g., suppose $X = (X_1, X_2)$; then the models are $f(X_1|Y_T, Y_O, X_2)$ and $f(X_2|Y_T, Y_O, X_1)$).

Example 11.7. *Multiple imputation for mismeasured variables in linked data*

Self-reports of some variables in surveys can be subject to response bias. In Example 11.3, a multiple imputation strategy was used to correct such bias in one large survey by combining it with more accurate clinical measurements from a small survey. Such a combining process can be viewed as vertically concatenating or stacking two datasets. On the other hand, if two data sources are linked for the same set of subjects (i.e., combined in a horizontal way), then the error-prone variable in one data source can be replaced by more accurate measurements from another data source. However, missing data issues in linked files (e.g., Example 10.8) can add more complexity to this strategy.

Rammon et al. (2019) presented a multiple imputation analysis-based solution. This can be viewed as an extension of Example 10.8 by handling measurement error problems in linked files. In their context, the main variable of interest Y_T is the true Medicaid/Children's Health Insurance Program (CHIP) enrollment status of NHANES participants. This is due to the fact that the self-reported version of the insurance enrollment status in NHANES (Y_O) is often underreported, that is, the so-called phenomenon of "Medicaid undercount in health surveys" (Davern et al. 2009). On the other hand, the true insurance enrollment status (Y_T) was recorded in U.S. Centers for Medicare and Medicaid Services' Medicaid Analytic eXtract (CMS MAX) files. If a linked dataset between NHANES and CMS MAX files were produced, the linked cases would be enrolled and unlinked cases would be un-enrolled, generating correct enrollment status for Y_T in the survey.

However, a complexity of the linked data is that not all survey participants can be linked to administrative files. For example, NHANES participants who do not provide sufficient information from personal identifiers such as their social security number or their health insurance claim number are ineligible for linkage. For subjects who were linkage ineligible, their Y_T-values, as well as other information from the CMS-MAX files, are missing. The data structure can be sketched in Table 11.10. In summary, S_1 is the linkage ineligible subsample of NHANES, and S_2 is the NHANES subsample that is linkage eligible and falls into two groups: participants who can be linked to CMS-MAX ($Y_T = 1$) and those who cannot be linked ($Y_T = 0$). Of course, other types of missing values also occur for all these variables.

An FCS strategy was developed to impute all incomplete variables in the example dataset. For Y_T (the administrative Medicaid status), the conditional model is a logistic regression model including Y_O (the error-prone survey Medicaid status) and other predictors including both demographic and socioeconomical variables, NHANES survey design variables, as well as health characteristics that are related to Medicaid/CHIP status. Note that all these X-variables come from both NHANES and CMS-MAX data. Since a post-imputation analysis is to relate children's serum cotinine levels (collected in NHANES) with their Medicaid/CHIP status, the former outcome was also included in the imputation for the latter. Note that the cotinine variable itself had around 20% of missing cases, which was also imputed by FCS simultaneously. Some continuous variables were transformed (e.g., a log-transformation is applied to the cotinine measure) in the imputation.

This example also illustrates why it is generally suboptimal to only use the information in S_2 for analysis, despite that both Y_T and Y_O are available. One way to analyze incompletely linked data is to limit analyses to the linkage eligible individuals. However, survey respondents with sufficient personal identification for linkage are self-selected. If the linkage eligible subset differs systematically from those who are not eligible, then eliminating the linkage ineligibles (S_1) without adjustments could lead to biased estimates. In the notion of missing data analysis, the self-selection of the linkage eligible subjects could violate the MCAR assumption of S_2. Multiple imputation analysis can be an effective approach to adjusting for such selection bias after including sufficient information from covariates, as we have illustrated throughout the book.

11.3.1.5 Predictors under Detection Limits

Another type of measurement error problem pertains to values under some detection limits. For example, environmental and biomedical researchers often encounter laboratory data constrained by a lower limit of detection (LOD). Because values less than LOD cannot be determined precisely, data are missing at the lower end of the distribution. They can also be viewed as left-censored (sometimes interval-censored if the measurements have also to be positive for biochemical assays). In Chapter 8 we discussed methods for imputing right-

TABLE 11.10
Example 11.7. A sketch of data structure when Y_T is collected from a linked file in the presence of linkage ineligibles

Sample	Y_T	Y_O	X
S_1 (NHANES subsample ineligible for linkage)	M	...	...
S_2 (NHANES subsample linked to CMS-MAX)	1	...	...
S_2 (NHANES subsample not linked to CMS-MAX)	0	...	...

Note: "M" symbolizes missing values. Y_T: true value of Medicaid status from the administrative data; Y_O: underreported Medicaid status from NHANES; X: covariates that are included in both data sources.

censored data (e.g., event times) that are used as a survival outcome variable. Here the left-censored variable is often used as a predictor in a targeted analysis.

Following the preceding notations, suppose Y_T is the variable subject to the detection limit and Y_O is the observed version in the data, X_1 is the outcome of the analysis, and Z denotes other covariates in the analysis. In general, the complete-data model requires the specification of the likelihood function as:

$$\prod_{Y_O > \text{LOD}} f(X_1|Y_T, Z, \beta) \prod_{Y_O = \text{LOD}} \int_{-\infty}^{\text{LOD}} f(X_1|Y_T, Z, \beta) f(Y_T|Z, \alpha) dY_T, \quad (11.9)$$

where $f(X_1|Y_T, Z, \beta)$ is the distribution function of the analysis model relating X_1 to Y_T and Z, and $f(Y_T|Z, \alpha)$ characterizes the distribution of the Y_T given other covariates.

A corresponding imputation model can be sketched as:

$$g(E(X_1|Y_T, Z)) = \beta_0 + \beta_1 Y_T + \beta_2 Z, \quad (11.10)$$

$$Y_O = Y_T I(Y_T > \text{LOD}) + \text{LOD} I(Y_T <= \text{LOD}), \quad (11.11)$$

$$Y_T = \alpha_0 + \alpha_1 Z + \epsilon, \quad (11.12)$$

where $I(\cdot)$ is the identity function and $\epsilon \sim N(0, \sigma_\alpha^2)$. The analysis model in Eq. (11.10) used a GLM-type model to relate the expected value of response X_1 to predictors by a link function $g(\cdot)$ including β's as the regression coefficients. The measurement error model in Eq. (11.11) specifies that when Y_T is above the detection limit, its actual value is recorded and used; yet when Y_T is lower than the detection limit, only the detection limit is recorded. The covariate model in Eq. (11.12) relates Y_T with Z using a normal linear regression model as an example. These specifications follow the general modeling principle laid out in Eq. (11.8).

The multiple imputation approach would draw unobserved Y_T from Models (11.10)-(11.12) under the detection limit multiple times and then apply combining rules to analyze the datasets. Because of the censoring due to the detection limit, the imputation algorithm for Y_T lower than LOD often consists of drawing from a truncated distribution or drawing under the constraint that these values should be lower than LOD.

Some of the applications can be found in Chen et al. (2011; 2013) and Arunajadai and Rauh (2012). These studies showed that using the multiple imputation approach yields a better performance than some of the ad hoc methods such as replacing the unobserved Y_T with LOD/2 or LOD. However, we emphasize again that a complete imputation model should include the analysis model, $f(X_1|Y_T, Z, \beta)$, which is specified in Eq. (11.10). Imputation using the information only from the measurement error model and covariate model (i.e., Eqs. (11.11) and (11.12)) would still bias the regression coefficient β_1 towards the null value. This is similar to the situation of imputing missing covariates in a targeted analysis (Section 4.4).

11.3.2 True Values Fully Unobserved

In some cases, Y_T is completely unobserved as there is no apparent validation data such as S_2 in Table 11.6. This is illustrated in Example 1.4. In this scenario, imputing missing Y_T would purely rely on the modeling assumption behind $f(Y_T|Y_O, X)$. There exist several modeling strategies in the literature.

Example 11.8. *Mixture models for mismeasured variables*
To sketch the idea, suppose Y_T is continuous and we can assume that

$$\begin{aligned} Y_O &= Y_T + \epsilon, \\ \epsilon &= 0 \text{ if } K = 1, \\ \epsilon &\sim N(0, \sigma^2) \text{ if } K = 0, \\ K &\sim Bernoulli(p). \end{aligned} \tag{11.13}$$

Therefore Y_T is correctly measured when $K = 1$ and has additional noises when $K = 0$. However, the indicator K for the correctness is unknown and stochastic, and needs to be estimated in the imputation process. The data sample can therefore be viewed as a mixture of groups with different measurement properties. For example, Ghosh-Dastidar and Schafer (2003) and Zhang et al. (2014) used normal mixture models to characterize the relationship between Y_O and Y_T.

Sometimes there might exist multiple data sources with alternative measurements for Y_T, denoted by $Y_{O,1}$, $Y_{O,2}$, etc. Models can be posited to relate unobserved Y_T to Y_O's from multiple data sources for imputation. Note that in general, since Y_T is not observed at all, some constraints might need to be imposed to the imputation model to ensure the model is identifiable especially if Y_T is categorical. These constraints need to be well specified and justified.

Example 11.9. *Combining information from two data sources with misreporting and missing data to impute hospice use among cancer patients*
In the CanCORS study (Examples 1.6 and 10.6), multiple data sources (e.g., surveys, cancer registry, medical records, Medicare claims) were used to collect information from study participants. He et al. (2014) used information from medical records and Medicare claims data to impute true hospice use status of late-stage cancer patients. This was motivated by the hypothesis that underreporting of hospice use could have happened in both data sources when the data were collected. In addition, data from both sources suffered from missing values. The data structure can be sketched in Table 11.11. Similar to Example 11.4, due to the assumption of underreporting, Y_T is imputed as 1 deterministically if either $Y_{O,1}$ or $Y_{O,2}$ is 1. Otherwise Y_T is stochastically imputed as 1 or 0.

Table 11.12 (see Table I from He et al. 2014) presents some of the actual numbers laid out in a 3×3 table (Yes/No/Missing). This problem can be considered as a variant of the dual-system estimation (DSE) (Sekar and Deming 1949) of population counts and other capture-recapture problems. The classic

DSE approach conceptualizes that each person in a population is either in or not in the two lists. Analogously in this setting, a patient's hospice-use status is either reported or not in the two data sources, medical records and Medicare claims data. For illustration, first ignore the cells containing missing values. Note that the cell with "No" from both sources (617) can be viewed as the count of cases excluded from both lists in the DSE framework. The reporting completeness (inclusion probability or capturing rate) for data source 1 (or 2) can be simply calculated as $E(Y_{O,1}|Y_{O,2} = 1)$ (or $E(Y_{O,2}|Y_{O,1} = 1)$). For example in Table 11.12, note that the 449(= 395 + 54) cases who were reported as "Yes" from the medical records data are true positives. Yet only 395 of them were reported as "Yes" from the Medicare claims data. This suggests that the reporting completeness rate for the Medicare claims data is $395/(395 + 54) \approx 88\%$. Similarly, the capturing rate for medical records is estimated as $395/(395 + 445) \approx 44\%$. In this simple illustration, the reporting accuracy from Medicare claims data is much better than that from medical records data. In addition, the total number of "Yes" on hospice use in the sample is estimated as $(395 + 54)(395 + 445)/395 \approx 955$. That is, among the 617 cases with "No" reports from both sources, $445 \times 54/395 \approx 61$ cases would be imputed as "Yes" in hospice use.

However, the classic DSE approach does not handle missing values in either source, and it also assumes that the reporting processes from two lists are independent. In He et al. (2014), a multiple imputation approach was developed to impute the true hospice status for these cases (i.e., unobserved $Y_{O,1}$ and/or $Y_{O,2}$), as well as for the cases whose hospice status were reported to be "No" in both cases. In addition, information from covariate X (i.e., patient- and aggregate-level demographic and clinical characteristics) was included in

TABLE 11.11
Example 11.9. A sketch of the data structure

Y_T	$Y_{O,1}$	$Y_{O,2}$	X
M(= 1)	1	1	...
M(= 1)	1	0	...
M(=?)	0	M	...
M(=?)	0	0	...
M(= 1)	0	1	...
M(= 1)	M	1	...
M(=?)	M	0	...
M	...	...	...

Note: "M" symbolizes missing values. Y_T: true hospice use (1=Yes and 0=No); $Y_{O,1}$: reports from medical records; $Y_{O,2}$: reports from the Medicare claims; X: covariates. Y_T is imputed as 1 when either $Y_{O,1}$ or $Y_{O,2}$ is 1.

the model, and the reporting processes from the two data sources was assumed to be independent conditional on X.

More specifically, He et al. (2014) considered the following decomposition of the complete-data model,

$$f(Y_{O_1}, Y_{O_2}, Y_T|X, \theta) = f(Y_T|X, \theta_O) f(Y_{O_1}, Y_{O_2}|Y_T, X, \theta_R). \quad (11.14)$$

Similar to Example 11.4, the first component in Eq. (11.14), $f(Y_T|X, \theta_O)$, is the outcome model. It relates the hospice use to covariate X, with regression parameters θ_O which might be of subject-matter interest. The second component is the reporting model (or measurement error model), $f(Y_{O,1}, Y_{O,2}|Y_T, X, \theta_R)$, which characterizes reporting in the two sources given true hospice-use status, covariates, and parameters θ_R. The reporting model rests on two assumptions:

Assumption 1: Reporting in the two sources is independent conditional on true status and observed covariates:

$$f(Y_{O,1}, Y_{O,2}|Y_T, X, \theta_R) = f(Y_{O,1}|Y_T, X, \theta_{R,1}) f(Y_{O,2}|Y_T, X, \theta_{R,2}).$$

Assumption 2: Both sources may be subject to underreporting but not overreporting: $f(Y_{O,l} = 1|Y_T = 0, X, \theta_{R,l}) = 0$ for $l = 1, 2$.

In the application dataset, both the outcome and reporting models were modeled using probit regression models with random effects from the CanCORS study sites. Results from the outcome model suggested that hospice was used less often by colorectal cancer patients than lung cancer patients, more often in the older groups (65-69 years old or $>$ 80 years old), more often among late (stage 3 or 4) or missing stage patients, and more often among patients with depression symptoms. Patients who lived longer after diagnosis were more likely to use hospice services. Results from reporting models suggested that medical record data were less complete for hospice use among patients aged between 65 and 69 years old, and much less complete among patients with stage missing information; Medicare claims data were less complete

TABLE 11.12
Example 11.9. Hospice-use reports from medical records and Medicare claims, a subsample of CanCORS data

		Medicare Claims		
		Yes	No	Missing
	Yes	395	54	260
Medical	No	445	617	646
Records	Missing	136	116	358

Note: The subsample consists of 3027 CanCORS patients who died within 15 months of diagnosis.

for hospice use among Hispanic patients, and more complete among patients with heart failure or diabetes, or those who lived longer. These associations made intuitive sense.

The multiple imputation analysis results were compared with alternative methods, notably using data from either source alone. Since the alternative methods did not appropriately account for measurement errors in the response variable, the resulting coefficients were susceptible to bias (Neuhaus 1999) which might lead to misleading scientific conclusions. For example, using medical records only, which contain considerable measurement errors in the application dataset, might be less capable of detecting true association effects. On the other hand, using Medicare claims only, which might be more accurate yet were limited to the sample over 65 years old, might produce association estimates that cannot be generalizable to younger patients. By using the information from both sources, the model-based multiple imputation analysis overcame the weakness of either data source alone.

The problems of misreporting of the treatment status are related to the problems of response misclassification in logistic regression models (Gustafson 2004; Carroll et al. 2006). In Examples 11.4 and 11.9, underreporting was a reasonable assumption because the report of treatment status could be incomplete prior to the data collection in the study, depending on the time line. Overreporting is theoretically possible yet largely unlikely in real practice. Assuming underreporting in imputation models also makes the model less prone to identifiability issues. On the other hand, allowing the misclassification of Y_T for both directions can make the model less identifiable, especially in the scenario with no validation sample.

11.4 Data Harmonization Using Bridge Studies

In preceding sections the discussion focused on measurement error problems with well defined "true values" (Y_T) and "error-prone" measurements (Y_O). In some cases, the measurement issues concern two different instruments/coding systems of the same construct from two datasets. For example, in large and ongoing surveys, it often happens that a newly developed coding or measurement system is used to replace old ones periodically to reflect updated knowledge and usage of the construct. In this case, if there is an analytical need to compare the values of the variable of interest from this survey in different time periods, then it is necessary to convert one measurement system to another. This situation is briefly mentioned in Table 1.6.

We denote the two measurements as $Y_{T,1}$ and $Y_{T,2}$, collected in data S_1 and S_2, respectively. The data structure can be sketched in Table 11.13. Note that the setup here is different from that in Example 11.9, where the two measurements ($Y_{O,1}$ and $Y_{O,2}$) are collected for the same subject. Suppose the analysis of interest is to assess the change of $Y_{T,2}$ between S_2 and S_1. A solution is to impute the missing $Y_{T,2}$ in S_1. Oftentimes there exists a bridge study sample B that collects both $Y_{T,1}$ and $Y_{T,2}$ for a comparison between the two instruments. We can impute $Y_{T,2}$ in S_1 from $f(Y_{T,2}|Y_{T,1}, X)$ derived from data B. Similarly we can also impute $Y_{T,1}$ in S_2 from $f(Y_{T,1}|Y_{T,2}, X)$ derived from data B.

TABLE 11.13
A sketch of the data structure in a bridge study

$Y_{T,1}$	$Y_{T,2}$	X	S
...	...	...	...
O	M	...	S_1
O	M	...	S_1
O	O	...	B
O	O	...	B
M	O	...	S_2
M	O	...	S_2
...	...	...	...

Note: "O" symbolizes observed values; "M" symbolizes missing values. $Y_{T,1}$: the instrument used in S_1; $Y_{T,2}$: the instrument used in S_2; B: the bridge study.

A classic example came from a multiple imputation project aiming to harmonize the coding for the occupation and industry code between the 1970 and 1980 censuses (e.g., Clogg et al. 1991, Schenker et al. 1993). In that project, $Y_{T,2}$ is the (later) coding system following the rule of the 1980 census, $Y_{T,1}$ is the (earlier) coding system following the rule of the 1970 census, S_1 is the analytic subsample in the 1970 census, S_2 is the analytic subsample in the 1980 census, and B is a subsample of the 1980 census in which both codings are available. The imputation models considered were logistic regression models.

In another example (Sternberg 2017), a similar multiple imputation strategy was used to adjust an old assay/measurement of Vitamin D to a new measurement in NHANES data. In this setting, it is important to evaluate the potential impact of a change so that any observed fluctuations in concentrations of biomarkers over time are not confounded by changes in the assay. Specifically, B was the bridge (cross-over) study sample containing both the old and new assay from a selected sample of NHANES participants. This cross-over sample was obtained from the lab which is in charge of measuring nutritional biomarkers for NHANES. Since the imputed variable is continuous, the imputation model was a normal linear regression model. The simulation study demonstrated advantages of the multiple imputation approach over alternative methods including the adjustment equation method (i.e, the regression prediction imputation) and stochastic conditional regression method (i.e, the single imputation) for a variety of estimands including the percentile distributions of the measurements.

Of course, when there exists no bridge study, the imputation of $Y_{T,1}$ or $Y_{T,2}$ becomes more difficult. This is because no data are available to estimate the association between the two variables conditional on X, that is, $f(Y_{T,1}, Y_{T,2}|X)$. This situation, where $Y_{T,1}$ and $Y_{T,2}$ are never jointly observed, is also referred to as statistical file matching: one dataset (S_1) has $Y_{T,1}$ and X recorded, another dataset (S_2) has $Y_{T,2}$ and X recorded, and the desire is to have the concatenated file having $Y_{T,1}$, $Y_{T,2}$, and X completed for all units (Rubin 1986). Some assumptions have to be made for $f(Y_{T,1}, Y_{T,2}|X)$ (e.g., Moriarity and Scheuren 2001) in order to solve this problem.

11.5 Combining Information from Multiple Data Sources

In preceding sections, we listed multiple examples of using observed Y_O, presumably a mismeasured version of Y_T, and covariate X to build a model to impute partially or fully unobserved Y_T. Imputation can also be conducted to harmonize different versions of Y_T's such as in bridge studies. The information behind the imputation models often rely on validation data, which can be based on internal or external samples, bridge samples, or through the data linkage. The connection between the original data and extra data is schematically described using notations S_1 and S_2, respectively. As illustrated by these examples, multiple imputation has been used as an effective analytic approach to combine information from multiple data sources (studies) to enhance the utility of the original data. This can lead to completed data, which account for missing data and/or measurement error problems in the original data. Multiple imputation analysis applied to completed data can lead to improved statistical inference. It has to be emphasized that when data from multiple sources are involved, the transportability assumption (Carroll et al. 2006) behind the models and parameters is important. In some cases, it is equivalent to the MAR assumption in typical missing data problems. Related to this point, the target population for the imputation inference, especially when S_1 and S_2 come from different data sources, needs to considered carefully and defined clearly.

Schenker and Raghunathan (2007) provided an overview of the strategy of combining information from multiple surveys. They argued that this strategy can be an effective way of improving estimates. More specifically, survey estimates are often affected by both sampling and nonsampling errors. Though a particular survey may have been planned to achieve a certain level of sampling error, it is often difficult to assess nonsampling errors. Sources of such nonsampling errors include: missing data, which can occur due to survey nonresponse; coverage error, which occurs when the population from which the sample is drawn differs from the population of interest; and measurement or response error, which occurs, for example, when survey responses are not completely accurate. Combining information from more than one survey can take advantage of the strengths of the different surveys and can use one survey to supply information that is lacking in another, thereby adjusting for nonsampling errors. Furthermore, the resulting combined estimates may have lower levels of sampling error. Thus, combining information from multiple surveys may provide enhanced estimates of quantities of interest.

As we have shown, multiple imputation is an effective strategy for combining information from multiple data sources especially if information at the individual (micro)-level is available. However, depending on the analysis of interest, study designs, and data structure, the combining strategies are not necessarily limited to multiple imputation. A wide variety of statistical approaches, ranging from the traditional calibration estimators in survey statistics to more sophisticated hierarchical Bayesian models, can be used. Additional references in this topic area, for example, can be found in Elliott et al. (2018) and Lohr and Raghunathan (2017). Detailed discussion of these methods is beyond the scope of this book. In the following we provide an example that is not using multiple imputation per se, yet can still be perceived in a missing data framework.

Example 11.10. *Combining information from two interview surveys to enhance small-area estimation*

Large, multi-purpose complex surveys based on probability sampling are often designed to produce reliable estimates for the target population or subgroups of the population known as planned domains. However, unplanned small domains (i.e., small areas) often have insufficient sample sizes that are associated with large sampling errors, which limit the capability of obtaining reliable direct survey design-based estimates for these areas. Examples of unplanned small domains include subpopulations formed by cross-classifications of different demographic variables such as age, gender, and racial/ethnic groups, or domains on small geographic scales such as at the state or county level from national surveys. In survey statistics, the methods aiming to produce estimates for the small areas are termed "small area estimation." Over the last few decades, various small area estimation techniques have been developed and applied. For the history and reviews of small area estimation, see Rao (2003).

As expected, features of individual surveys would have large impacts on the quality of their small area estimates. Raghunathan et al. (2007) presented a Bayesian model to combine information from two surveys to enhance the small area estimates. Table 11.14 sketches the basic data structure. Similar to Table 11.6, here $S = S_1 \bigcup S_2$ includes small areas from a large survey and S_2 includes small areas from a small survey. These small areas are U.S. counties in Raghunathan et al. (2007). Based on the sampling design, unbiased design-based small-area level estimates $\hat{Y}_T$ can be obtained from S_2 (the small survey), yet small-area level estimates $\hat{Y}_O$ from S (the large survey) are subject to some bias. The analytic goal is to obtain correct estimates for all small areas in S. Note that here the missing elements are small-area level estimates and are not individual records as in previous examples.

TABLE 11.14
Example 11.10. A sketch of the data structure for combining information from two surveys to enhance small area estimation

Sample	$\hat{Y}_T$	$\hat{Y}_O$	θ	X
S_1	M	O	?	...
S_2	O	O	?	...

Note: "O" symbolizes observed/available values; "M" symbolizes missing values. $\hat{Y}_T$: unbiased direct small-area estimates; $\hat{Y}_O$: possibly biased direct small-area estimates; θ: model-based small-area estimates; X: area level covariates.

We can sketch the following framework for a complete-data model; let i index the small areas of S:

$$\hat{Y}_{i,T} \sim N(\theta_{i,1}, \hat{s}^2_{i,1}), \quad (11.15)$$

$$\hat{Y}_{i,O} \sim N(\theta_{i,2}, \hat{s}^2_{i,2}), \quad (11.16)$$

$$(\theta_{i,1}, \theta_{i,2})^t \sim N_2(X_i\beta, \Omega). \quad (11.17)$$

Eqs. (11.15)-(11.17) are a multivariate version of typical Fay-Herriot models (Fay and Herriot 1979) for small area estimation. At the area level (Eqs. (11.15) and (11.16)), the design-based estimates, $\hat{Y}_{i,T}$ and $\hat{Y}_{i,O}$, are assumed to be unbiased for the respective area level means, $\theta_{i,1}$ and $\theta_{i,2}$. Their variance estimates, $\hat{s}^2_{i,1}$ and $\hat{s}^2_{i,2}$, can be obtained using design-based procedures. Across the areas, the area level means, $\theta_{i,1}$ and $\theta_{i,2}$, are assumed to have a bivariate normal random distribution after including the area level covariate X. Note that $\theta_{i,2}$'s, which can be estimated directly from $\hat{Y}_{i,O}$'s in the large survey, are presumably biased. The estimands of interest include $\theta_{i,1}$'s for all small areas in S, despite that they cannot be directly estimated (i.e., missing) for small areas in S_1. However, the model predicts $\theta_{i,1}$'s based on Eq. (11.17), which relates $\theta_{i,1}$ with $\theta_{i,2}$ for small areas in S_2.

In Raghunathan et al. (2007), S is BRFSS (briefly introduced in Example 4.7) and S_2 is NHIS. The analytic interests include U.S. county-level estimates of the prevalence of cancer risk factors and cancer screening. More specifically, BRFSS is a large, state-based survey that is conducted by telephone (e.g., the largest telephone survey in the world). Almost all of the counties in the U.S. are included in its sample, so it obtains data that can be used to calculate direct estimates at small areas such as counties. Because BRFSS is a telephone survey, however, it does not include households without telephones in its sample, and it also tends to have relatively high nonresponse rates, as is the case for telephone surveys in general. In summary, the coverage error and nonresponse error might be a concern for BRFSS small-area estimates. On the other hand, NHIS is a face-to-face survey that includes both telephone and nontelephone households in its sample, and it has generally lower nonresponse rates than does BRFSS. NHIS is small relative to the BRFSS, and it includes only about 25% of the counties in the U.S. in its sample. However, NHIS was designed for calculating direct estimates at national and some subnational levels (e.g. for census regions and some states) but not to produce reliable estimates at the county level. For producing better county-level estimates, the project in Raghunathan et al. (2007) aimed to reduce sampling error by including information from both surveys and to reduce coverage error as well as possibly errors due to missing data by supplementing the information available from BRFSS with that available from NHIS.

Raghunathan et al. (2007) used a more sophisticated model than the one outlined in Eqs. (11.15)-(11.17). More specifically, they further separated NHIS to telephone and nontelephone households, considered an arc-sin-square-root transformation to the proportion estimates, and used the relative biases from the BRFSS to NHIS (telephone households) (i.e., $\frac{\theta_{i,2}-\theta_{i,1}}{\theta_{i,1}}$) to model the relationship between the two sets of small area estimates. Their empirical evaluations suggested that the combined estimation procedure helps to address noncoverage and nonresponse issues in BRFSS as well as to expand beyond areas that are covered by NHIS. More references can be found in Davis et al. (2010) and Liu et al. (2019). Based on this project, U.S. county-level estimates for several cancer risk factors and screening behaviors based on the model can be found in (https://sae.cancer.gov/nhis-brfss/).

This project is now entering a new phase to use more recent data and account for changes of study designs. More specifically, the U.S. Preventive Services Task Force cancer screening guidelines have changed for several screening procedures. There is an interest in updating the results for the years 2011-2016, and BRFSS has added a cell phone frame since 2011. This new phase of the project is expected to report estimates for cancer screening tests/risk factors (e.g., PSA testing, former smoking, hysterectomy prevalence) which were not considered during the prior phase.

More generally, especially in the era of big data when a large amount of data from different data sources is available in many forms, combining information is not only limited to survey data and can be applied to both survey and nonsurvey data (e.g., administrative data, social media data, data from specific research studies such as clinical trials) with different aims and purposes. Citro (2014), for example, emphasized the need to rely on multiple data sources for producing official statistics.

11.6 Imputation for a Composite Variable

We switch gears somewhat in this section. In many studies, there exist multiple incomplete variables in the study data, yet the analysis of interest only focuses on a summary variable constructed from the multiple variables. In this book we refer to the summary variable as a composite variable and refer to the original variables as source variables. The composite variable can be derived from applying a simple arithmetic operation to source variables. For example, a scale variable is often a simple or weighted average (or sum) of individual items in survey data. A composite variable can also be constructed using a rather complex algorithm that takes information piece by piece from each of the source variables (Example 11.11). We refer to the process of deriving the composite variable as a "combining process or algorithm". Clearly, if any of the source variables is incomplete, the composite variable is also likely to suffer from missing values.

In principle the composite variable is subject to error because some information from source variables is lost during the combining process. This section provides some insights on how to impute a composite variable. A natural question is, "Do we impute the composite variable directly without considering source variables, or do we impute source variables first and then combine them to derive the composite variable?" The former strategy can be termed "imputation after combining" and appears to be simpler as it targets in one incomplete variable, yet it might fail to capture all the information from source variables. The latter approach can be termed "combining after imputation" and appears to be more rigorous, yet modeling multiple variables in the imputation can be a complex task.

More formally, we consider source variables $Y_1, Y_2, \ldots Y_p$. Suppose a composite variable Y is defined as:

$$Y = C(Y_1, Y_2, \ldots Y_p), \tag{11.18}$$

where C indicates a deterministic combining algorithm. We also assume that some covariate X is available. If some of Y_1-Y_p's have missing values, depending on the specification of C, Y can have missing values correspondingly. However, we note that sometimes the combining process C might implicitly impute the missing values of Y in a deterministic way. For example, say C is a simple average of source variables. If a case does not have all of them observed, then a possible algorithm C can be specified as taking Y as the average of observed values for this subject. This process can often guarantee that Y does not have missing values as long as at least one of the source variables is observed for each subject. This is essentially an ad hoc, available-case analysis method that is usually not preferred. On the other hand, the focus of our discussion is not to assess the performance of the combining process (and its implicit imputation process). Instead we focus on the two imputation strategies given that the combining process C is fixed and accepted.

In general we can consider three imputation approaches:

1. Active imputation (imputation after combining): directly impute Y using X as predictors by considering a complete-data model $f(Y|X)$; this method completely ignores the roles of the source variables (Y_1-Y_p) in the imputation process.

2. Just another variable imputation (JAV) (imputation after combining): directly impute Y using Y_1-Y_p and X as predictors by considering a joint model for them; although the missing values in the source variables are also imputed in this method, we do not use completed source variables to derive Y (as opposed to the passive imputation Strategy); we instead directly use the imputed Y.

3. Passive imputation (combining after imputation): first we impute Y_1-Y_p using X as the covariates by considering a joint model for Y_1-Y_p and X, and we then derive the imputed Y by applying the combining algorithm C to completed source variables.

Note that "active imputation", "just another variable", and "passive imputation" are established terms in the literature (e.g., Van Buuren 2018). To handle multiple incomplete variables encountered, we can either use the JM or FCS strategy to set up models although we think FCS is often easier to implement. In JAV, note that theoretically all the information in Y can be traced back to Y_1-Y_p in a deterministic manner. Therefore to form a stochastic statistical model between the Y and Y_1-Y_p can be theoretically invalid. However practically such models can still be fitted in FCS. In passive imputation, one question is whether we need to take into account the information in the combining algorithm in specifying the model. We provide some discussions of these issues in the following example.

Example 11.11. *Imputation of HIV transmission categories*

Pan et al. (2020) compared the three imputation approaches using data from the CDC National HIV Prevention Program Monitoring and Evaluation (NHME), which collects information from HIV-positive men who have a valid risk profile and who were identified in a nonhealth care setting, which provides neither diagnostic nor treatment services. In the NHME system, a hierarchical combining algorithm is applied to the three risk factors (i.e. source variables) to assign the appropriate HIV transmission category for each individual. Thus, the HIV transmission category (XMODE) is the composite variable that summarizes multiple risk factors to identify the factors most likely to have resulted in HIV transmission. However, a considerable number of missing data are observed for transmission category in the NHME system.

The algorithm used to define the transmission category for a subset of data from the 2014 NHME dataset can be described as follows. See also Table 11.15 (Table 1 from Pan et al. 2020).

1. If an HIV positive male indicated ever having sex with a male (MSM) and having injected drugs (IDU), then he is assigned the HIV transmission category of MSM+IDU.

2. If he indicated having sex with a male but not having injected drugs, then his assigned HIV transmission category is MSM.

3. If he indicated not having sex with a male but having injected drugs, then he is assigned the category of IDU. Up to this point, other risk factors are not considered. Note that Steps 1-3 attempt to define the transmission category only based on the information from the first two source variables. That is, the category "MSM+IDU", "MSM", and "IDU" are defined regardless of the values of the third source variable.

4. If he indicated not having sex with a male and not having injected drugs, but indicated contact with female known to have, or to be at high risk for HIV infection then his HIV transmission category was "Heterosexual" (or HET).

5. If an HIV positive male indicated not having sex with a male, not having injected drugs and not having sex with a female, then he is assigned the HIV transmission category of "Other". The "Other" category includes hemophilia, blood transfusion, and perinatal exposure.

6. In any other scenario, transmission category is assigned to "Missing".

TABLE 11.15
Example 11.11. Determination of transmission categories (the composite variable) based on the hierarchy of three binary risk factors (source variables) in the NHME System.

Are you MSM? R_1	Are you IDU? R_2	Are you HET? R_3	Transmission Category XMODE
Yes ($R_1 = 1$)	Yes ($R_2 = 1$)		MSM and IDU ($R_1 = 1$ and $R_2 = 1$)
Yes ($R_1 = 1$)	No ($R_2 = 0$)	Yes/No/Unknown	MSM ($R_1 = 1$ and $R_2 = 0$)
No ($R_1 = 0$)	Yes ($R_2 = 1$)		IDU ($R_1 = 0$ and $R_2 = 1$)
		Yes ($R_3 = 1$)	HET ($R_1 = R_2 = 0$ and $R_3 = 1$)
No ($R_1 = 0$)	No ($R_2 = 0$)	No ($R_3 = 0$)	Other ($R_1 = R_2 = 0$ and $R_3 = 0$)
		Unknown ($R_3 = M$)	Other ($R_1 = R_2 = 0$ and $R_3 = M$)
Yes/No	Unknown ($R_2 = M$)		
Unknown ($R_1 = M$)	Yes/No	Yes/No/Unknown	Unknown ($R_1 = M$ or $R_2 = M$)
Unknown ($R_1 = M$)	Unknown ($R_2 = M$)		

Note: "M" symbolizes missing values. Steps 1-3 of the combining algorithm attempt to define the transmission category only based on the information from of R_1 and R_2. That is, the category "MSM+IDU", "MSM", and "IDU" are defined regardless of the values of R_3.

Pan et al. (2020) proposed and compared five imputation methods targeted to the missing composite variable XMODE from three binary source variables R_1, R_2, and R_3.

(A) Active imputation: impute XMODE directly ignoring source variables (R_1-R_3) but including X as predictors.

(B) JAV: impute XMODE directly including both source variables and X as predictors.

(C) Passive imputation with information from the algorithm (Passive I): impute source variables R_1, R_2, and R_3 through a two-stage procedure compatible with the variable-combining algorithm. Specifically, first impute source variables R_1 and R_2, and then impute R_3 for a subset of the sample in which R_1 and R_2 are both 0, based on the combining algorithm. After source variables R_1,R_2, and R_3 are imputed, they are combined using the combining algorithm to produce the composite variable XMODE.

(D) Passive imputation without interactions (Passive II): impute the source variables and X without considering the variable-combining algorithm as in (C), and only main effects are included in the imputation model.

(E) Passive imputation with interactions (Passive III): extending from Passive II, main effect, two-way and possible three-way interactions among all variables are included in the imputation model. This method follows the inclusive imputation strategy to build up a general imputation model.

Both the active imputation and JAV methods are "imputation after combining". For them, the discriminant analysis imputation method (Section 4.3.3.2) was used to impute XMODE due to its many categories. All passive imputation methods are "combining after imputation". For them, logistic regression imputation methods were used for imputing R_1-R_3 before they are combined to yield completed XMODE. These methods were implemented by SAS PROC MI FCS. .

A simulation study was conducted to compare the performance of the above five imputation methods. To mimic the distribution of real data in NHME, the simulation datasets were generated by considering correlated R_1, R_2, R_3 and a binary covariate X. The simulation includes 1000 replicates with around 20% of missingness for each of the source variables under MAR. The complete-data sample size is 1000. In the evaluation, the mean estimate of XMODE is the target estimand. Some results are presented in Table 11.16 (see also Table 3 from Pan et al. 2020). Neither the active nor JAV method performs well for categories with fewer observations (e.g., category "Other".) For example, the relative bias is −18.7% and coverage probability is as low as 75% for the "Other" category in the active imputation. The relative bias is −11.3% and coverage probability is 84.4% for the "Other" category in JAV, showing some improvement over the former. All passive imputation methods generally work better. Among them, the passive III method performed the best with minimal relative biases in major categories, and their coverage probabilities were all close to the nominal level.

TABLE 11.16
Example 11.11. Simulation results

Method	XMODE	True	RBIAS (%)	SD	$\overline{\text{SE}}$	COV (%)
Active	MSM+IDU	0.09	0	0.012	0.012	95.1
	MSM	0.73	1.9	0.018	0.018	87.5
	IDU	0.05	4.3	0.009	0.009	94.6
	HET	0.08	−8.2	0.010	0.010	86.8
	Other	0.05	−18.7	0.008	0.008	74.5
JAV	MSM+IDU	0.09	2.4	0.012	0.014	95.6
	MSM	0.73	0.4	0.017	0.019	96.6
	IDU	0.05	−5.4	0.009	0.009	91.9
	HET	0.08	4.3	0.011	0.011	95.7
	Other	0.05	−11.3	0.009	0.008	84.4
Passive I	MSM+IDU	0.09	0	0.012	0.012	95.1
	MSM	0.73	−0.1	0.017	0.017	95.9
	IDU	0.05	0.1	0.008	0.008	95.1
	HET	0.08	−7.9	0.010	0.010	89.2
	Other	0.05	13.8	0.008	0.009	93.8
Passive II	MSM+IDU	0.09	−0.5	0.012	0.012	94.0
	MSM	0.73	0.1	0.017	0.017	95.9
	IDU	0.05	2.0	0.008	0.008	94.5
	HET	.08	−6.1	0.010	0.010	90.8
	Other	0.05	7.5	0.008	0.009	94.5
Passive III	MSM+IDU	0.09	1.2	0.012	0.012	94.6
	MSM	0.73	−0.2	0.017	0.017	95.0
	IDU	0.05	1.3	0.008	0.009	95.7
	HET	0.08	−0.4	0.011	0.011	95.5
	Other	0.05	−0.4	0.009	0.009	94.8

The simulation study also included other designs. Overall the results showed that combining after imputation is better than imputation after combining for a composite variable. This conclusion is consistent with evidence from past literature (e.g., Gottschall et al. 2012; Eekhout et al. 2014). In addition, the passive III method yields the best performance in terms of estimating the marginal distribution of the composite variable. This is consistent with the principle of the inclusive imputation strategy. The evaluation results indicated that we should prefer using an inclusive, passive imputation strategy when dealing with a missing composite variable. In addition, there is no need to incorporate information in the combining algorithm in the imputation model as attempted in the passive I method. This is in fact a blessing because many of the combining algorithms can be very complicated to digest and understand. Of course, how to specify adequate FCS models for the source variable and covariates is usually not a simple task and warrants further research especially when the number of source variables is large (e.g., Eekhout et al. 2018).

11.7 Summary

In this chapter we mainly presented some multiple imputation approaches to measurement error problems, especially with validation data including values from both Y_T and Y_O. Sometimes measurement error problems are coupled with existing missing data problems in the data. Since analysis of data with measurement error is a very general area that includes many developed nonimputation methods, there has been a growing interest in comparing the performance of the multiple imputation with some of the conventional measurement error methods (e.g., Freedman et al. 2008; Messer and Natarajan 2008; Guo and Little 2011). In general, there exists no apparent disadvantage of the multiple imputation approach in the evaluations. The two strategies can also be used jointly. For example, when Y_O has some missing values, Gorfine et al. (2007) suggested multiply imputing Y_O first, and then applying the classic measurement error methods to the multiply imputed Y_O.

Overall, there is an increasing use of the multiple imputation strategy for measurement error problems when they can be framed as missing data problems. In many such applications, using this strategy also incorporates information from different data sources. A practical advantage of using the multiple imputation approach is that multiple imputation software programs are readily available for implementing flexible measurement error models. On the other hand, sometimes the study designs for measurement error problems are not so easily framed as the typical missing data problems. For example, the validation study can have repeated measurements of Y_O's instead of having the true values of Y_T collected. The utilities and limitations of multiple imputation in various study designs for correcting measurement error (e.g., Keogh and White 2014) remain to be investigated.

12

Multiple Imputation Diagnostics

12.1 Overview

Like all other statistical modeling techniques, researchers and practitioners might often wonder whether there exist any sound and convenient strategies for assessing and diagnosing imputed values and/or imputation models. That is, how do we know the imputation-based analysis results are of good quality? Multiple imputation diagnostics is also a natural need given that there exist many imputation models and strategies developed for many types of missing data problems, some of which have been discussed throughout this book. So which method do we use for the problem at hand? Another key motivation behind conducting imputation diagnostics stems from the increasing use of imputation software programs. Obviously these programs have greatly improved the capability of users for conducting multiple imputation analysis on real problems. However, available methods in these programs are often fixed by the software designer, and sometimes the imputation software package might be like a black box for some users. There is no guarantee that results based on some modeling options in the imputation programs are optimal for every problem at hand. It is therefore important for users to assess/check the analysis results via imputation diagnostics instead of mechanically using output from the imputation programs.

In general, we believe that imputation diagnostics is a comprehensive process, which might not be limited to steps taken after imputed values are drawn (i.e., post-imputation). Obviously, if we start with a reasonably good imputation model, then it is likely that fewer problems would be identified by running diagnostics and checks. In addition, it is not clear whether there exists a perfect imputation method or the single best imputation model per se in many real scenarios, and therefore it would be unrealistic to pursue these goals. In many cases, analysis results from several alternative imputation models might be similar to each other. Therefore, it is also important to conduct comparisons among multiple candidate models in imputation diagnostics.

Based on our understanding of related literature and our own experiences, we suggest paying attention to the following (related) questions in the process of diagnostics:

DOI: 10.1201/9780429156397-12

1. How are the variables/predictors selected when forming the imputation model?

2. Are the relationships among variables well retained by the imputation?

3. Are the distributional features of observed data well retained by the imputation?

4. Are the analysis results sensitive to different candidate imputation methods?

Answers to these questions can be pursued at different stages of the imputation analysis. This chapter discusses some of the ideas and issues involved. The rest of Chapter 12 is organized as follows. Section 12.2 reiterates some general principles in the stage of developing imputation models. Sections 12.3-12.5 focus on diagnostic analysis after data are imputed: Section 12.3 discusses several methods for comparing observed with imputed values; Section 12.4 discusses a strategy of comparing completed data with their replicates generated from the same model; Section 12.5 discusses a strategy that monitors and compares the fraction of missing information (FMI) of estimates from different imputation models. Section 12.6 briefly discusses about the use of prediction accuracy in multiple imputation analysis. Section 12.7 stresses the importance of comparing among alternative imputation models and other missing data methods. Section 12.8 provides a summary.

12.2 Imputation Model Development

12.2.1 Inclusion of Variables

As we pointed out in previous chapters, a general rule for forming good imputation models is to use every piece of available information (i.e., the inclusive imputation strategy) to yield multiple imputations that would reduce the bias and/or increase the efficiency of the estimates based on imputed data. Applying this strategy also tends to make the MAR assumption more plausible. In practice, for large datasets that contain many variables (e.g., hundreds or more), it is usually not feasible to include all of these variables due to multicollinearity and computational burden. In many cases, it might not be necessary either. Some empirical research suggests that the increase in explained variance in linear regression is typically negligible after at most, say, 10-15 variables have been included (Van Buuren 2011). For developing the model, therefore, a suitable balance needs to be sought between the principle of being inclusive and the practice of being manageable.

We list a few key strategies in selecting variables in the imputation model, which are largely based on recommendations from Van Buuren et al. (1999):

1. Include all variables that will be used in post-imputation analysis. Failure to do so may bias the completed data analysis, especially if the completed-data analysis contains strong predictive relations. For example, we pointed out before (Section 4.4) that if the missing variable is used as a predictor in a regression analysis, then not including the regression response variable in the imputation will tend to bias the association estimates towards zero.

2. Include variables that are related to the nonresponse. Factors that are known to have influenced the occurrence of missing data (e.g., reasons for nonresponse) are to be included on substantive grounds. Other variables of interest are those for which the distributions differ between the response and nonresponse groups. These can be found by running a logistic regression for the response indicator on a pool of candidate variables (i.e., the propensity score model) and then identifying the significant predictors.

3. Include variables that are correlated with or predictive of imputation variables even if they are not directly included in the post-imputation analysis (i.e., auxiliary variables in Example 5.8). These variables usually explain a considerable amount of variation of the missing variable. Having such variables in the model helps reduce the uncertainty of the imputations.

4. In the context of complex survey data, include sample design-related variables (e.g., stratification, clustering, survey weights, and contextual variables) (Chapter 10). Sometimes this would require obtaining variables from external data sources.

5. In the context of measurement error problems, include the error-prone measurement for imputing true values of the variable of interest.

6. Always seek subject-matter input if possible.

Note that the aforementioned strategies only serve as some general guidelines. There exist no unique statistical criteria for the selection of variables. For example, if we decide to run a correlation analysis between possible predictors with the imputation variable of interest, then the cut-off point for the correlation coefficient (or the corresponding p-value) has to be decided within the context of the problem at hand.

12.2.2 Specifying Imputation Models

After key variables for the imputation are identified and included, then the extent to which their correct relations are retained depends on how they are modeled statistically. In previous chapters we discussed various strategies for forming imputation models depending on the type of variables, data structure, study design, and analysis of interest. To provide some quick recap here that is related to imputation diagnostics, an imputation model in general is expected to be suitable with the type of missing variables (e.g., continuous, binary, censored data, etc.) For example, it is usually not advised to impute binary data by rounding off fractional imputations. When multiple incomplete variables are present, it is often easier to use the FCS strategy to accommodate different types of variables in the model.

In addition, modeling the relationship among variables is important, and this can be helped with some exploratory analysis. For example, imputations for missing variables that are largely linearly related would be appropriate if the imputation model captures well such linear relationships (e.g., via the normal linear or multivariate normal models). We can plot missing variables and selected predictors and then decide whether normal linear models are sufficient to capture their relationship for the imputation (e.g., residual plots for regression models). On the contrary, the imputation model that assumes a linear relationship between two nonlinearly related variables may only capture a partial relationship. For instance, in Example 5.5, we identified that the relationship between the birth weight and gestational age deviated from a linear one. Further in Example 5.10, we used the regression R^2 to assess the fit of several alternative models for imputing the gestational age. Sometimes using PMM or other adjustments can guard against certain model misspecifications.

Implementing the aforementioned strategies carefully in the process of formulating imputation models would likely render good imputations. Of course, however, exploratory analyses in the model developing stage also suffer from incompleteness because they only use observed data and many times only use complete cases. The corresponding results and pattern are informative yet still bear with the uncertainty caused by missing data. Therefore we still need to check the quality of imputed data after the imputations are drawn, partly addressing some of the uncertainty encountered in the model formulation stage. More importantly, we need to relate imputation diagnostics with completed-data analysis results that are ultimate goals of projects. Some of the ideas will be discussed in subsequent sections.

12.3 Comparison between Observed and Imputed Values

12.3.1 Comparison on Marginal Distributions

As discussed in Section 10.5.1, running data editing checks on imputed data can help identify implausible imputations (e.g., the imputed age is 10 for a mother) and thus flag the model when this happens. Beyond this value-specific check, a natural question in imputation diagnostics would be, "Are imputed values similar to observed data?" Intuitively, dramatic differences between the observed and imputed values might suggest a problem in the imputation model that needs to be further checked.

One quick way to evaluate the quality of an imputation procedure is to compare imputed and observed values for each variable to see if they are similar in terms of their marginal distributions. When the number of missing cases is relatively small, the comparison can be done more conveniently between completed and observed values. Some imputation programs (e.g., IVEware) output simple statistics (e.g., means and variances) from observed and imputed values of missing variables separately to facilitate this comparison. Other techniques can also be used. Examples include plotting the marginal distribution of observed and imputed values separately for a visual comparison (this procedure has been used multiple times in previous examples), and possibly using the Kolmogorov–Smirnov test (KS-test) for the difference between the two marginal distributions. See applications of these diagnostics in Abayomi et al. (2008), Van Buuren (2011), and Nguyen et al. (2013).

Results from comparisons on marginal distributions are informative yet should be interpreted with caution. Discrepancies between the marginal distributions of imputed and observed values do not necessarily reveal a shortcoming of the imputation method, because subjects with missing values may systematically differ from those with observed values in a variety of ways. For example, suppose that the probability of nonresponse is higher for elderly persons than for nonelderly ones; if an imputation method is working properly, then the imputed values should more closely resemble those of elderly persons than the observed sample. More generally, many types of systematic differences between observed and imputed values are allowed by a plausible imputation model under the MAR assumption.

To illustrate this point in a more technical way, we assume that Y is the incomplete variable. The diagnostics are pertaining to the comparison between $f(Y_{obs})$ and $f(Y_{imp})$, where Y_{imp} denote imputed values under some working models, and $f(\cdot)$ is the distribution function. As reasoned before, using this strategy correctly has a hidden assumption, that is, the missing data in Y are likely MCAR. To see that, we consider a simple complete-data model:

$$Y = \beta_0 + \beta_1 X + \epsilon, \tag{12.1}$$

where X is a fully observed covariate and $\epsilon \sim N(0, \sigma^2)$. Let R be the response indicator for Y. Let $X_{R=1}$ ($X_{R=0}$) be the cases for which the corresponding Y-value is observed (missing). If the imputation model is correctly specified as in Eq. (12.1), then the marginal distribution of Y_{obs} and Y_{imp} are not necessarily similar unless the missingness is MCAR. On the other hand, if the missingness is MAR but not MCAR, then the distribution of Y_{obs} can be different from that of Y_{imp}, and the difference is related to how strong the MAR mechanism is and how Y is related to X (i.e., the magnitude of β's). Therefore in general, considerable or significant differences between $f(Y_{obs})$ and $f(Y_{imp})$ do not necessarily suggest that the imputation model is inadequate unless MCAR holds well.

Example 12.1. *The issue of comparing marginal distributions in imputation diagnostics*

We use the simulation study setup in Example 4.1 to illustrate our points. The complete-data model is $Y = -2 + X + \epsilon$, where $X \sim N(1, 1)$ and $\epsilon \sim N(0, 1)$. The complete data sample size $n = 1000$. Two types of missingness mechanisms are considered: MCAR and MAR, and for MAR the nonresponse mechanism can be described as $logit(f(R = 0|X)) = -1.2+0.4X$. In both cases, the missingness rate is around 40%. We apply the correct imputation model in both cases. For simplicity, we only use one simulation replicate and compare Y_{obs} with Y_{imp} from one imputation. Table 12.1 lists summary statistics including the mean and variance of the observed and imputed values, as well as the KS-test results for comparing the two distributions. Under MCAR, the means of the observed and imputed values are close, and the KS-test result does not suggest a significant difference. However, under MAR, the means of the observed and imputed values are quite apart, and the KS-test result is highly significant. In both cases, the variance gets inflated around 25% (i.e., from around 2 to 2.5). Fig. 12.1 plots the density histograms of these data. Under MCAR, both observed and imputed values are symmetric and normally distributed. Yet under MAR, the distribution of observed values is slightly left-skewed, and that of imputed values is slightly right-skewed.

TABLE 12.1
Example 12.1. Mean and variance (in parentheses) of observed and imputed values under MCAR and MAR

Missingness Mechanism	Y_{obs}	Y_{imp}	KS-test
MCAR	−1.024 (2.178)	−1.204 (2.553)	$p = 0.2$
MAR	−1.33 (1.933)	−0.522 (2.468)	$p = 2.0 \times 10^{-9}$

In this simple example, we show that comparing the marginal distribution between observed and imputed values and concluding the model/imputations are problematic if large differences are detected can be misleading when the missing data are MAR. Nguyen et al. (2013) drew a similar conclusion, as they found out that significance results can be obtained after applying the KS test to compare the marginal distributions of Y_{obs} and Y_{imp} under the correct imputation model, which suggested otherwise. In addition, relationships between variables (e.g., Y and X) cannot be directly checked only using the marginal distributions.

Nevertheless, comparisons on marginal distributions between observed and imputed values are still useful for detecting gross problems and should be conducted whenever possible. For example, imputed values are usually not expected to lie outside the range of physical plausibility displayed by observed data. If the systematic differences between respondents and nonrespondents are not unusually strong, say weakly MAR judged from their respective distributions on X-covariates, then the distributions of observed and imputed values should be more or less similar in location, scale, and shape.

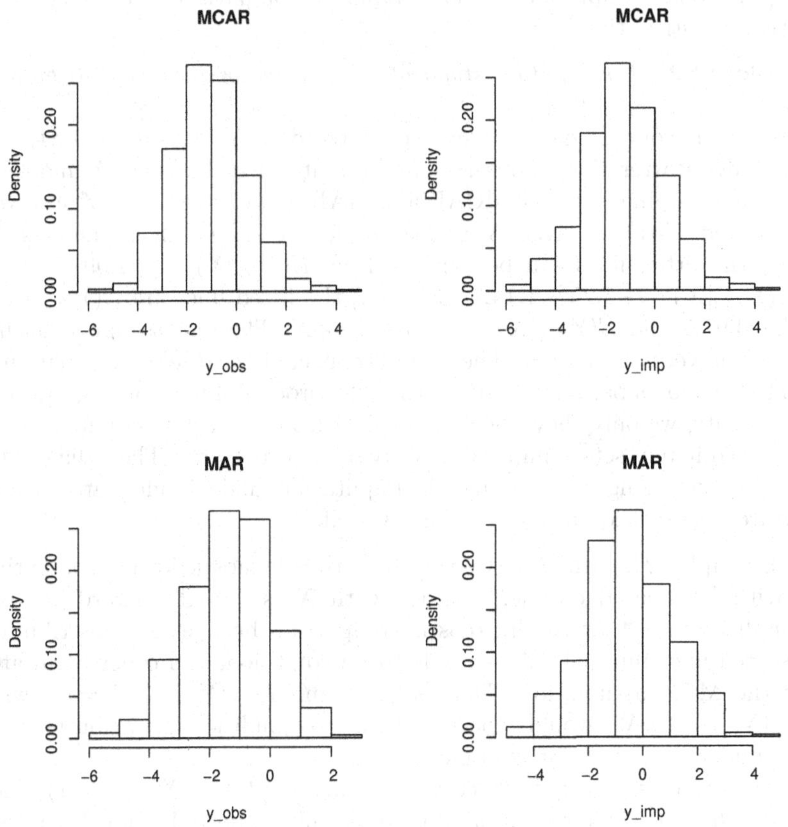

FIGURE 12.1
Example 12.1. Top left: the histogram of observed values under MCAR; top right: the histogram of imputed values under MCAR; bottom left: the histogram of observed values under MAR; bottom right: the histogram of imputed values under MAR. Displayed data are from one simulation replicate with one imputation.

12.3.2 Comparison on Conditional Distributions

12.3.2.1 Basic Idea

What would be the principled way of comparing observed and imputed values? In Example 12.1, whether it is in MAR or MCAR, the complete-data model is a normal linear regression model for Y conditional on X. Therefore, we expect that $f(Y_{obs}|X, R = 1)$ would be similar to $f(Y_{imp}|X, R = 0)$ if the imputations are appropriate. This implies a comparison of the respective conditional distributions.

Example 12.2. *A simulation study of comparing conditional distributions for imputation diagnsotics*

We continue our diagnostics using simulated data from Example 12.1. Fig. 12.2 includes scatter plots of observed and imputed Y-values over X and their respective fitted lines. In both MCAR and MAR, the distributions of observed and imputed Y-values given X appear to follow similar linear models. In MCAR, the fitted lines can be expressed as: $E(Y_{obs}|X) = -2.09 + 1.03X$ and $E(Y_{imp}|X) = -2.25 + 1.14X$; in MAR, the fitted lines are $E(Y_{obs}|X) = -1.97 + 0.97X$ and $E(Y_{imp}|X) = -2.09 + 1.08X$. The regression coefficients are close and yet not identical. The latter is expected because some variation is added to the model parameter estimates in the process of multiple imputation. For simplicity, we only show the scatter plots from one set of imputations as those from different sets of imputations have similar patterns. These diagnostic results provide strong evidence that the imputation model is adequate because it captures the relation between Y and X well.

In Example 12.2 with only two variables, the diagnostic idea of comparing Y_{imp} with Y_{obs} based on their relation with X is straightforward. A general implication is that the diagnostic comparison between the distribution of observed and imputed values needs to be conditional on other covariates under the MAR assumption. That is, we compare $f(Y_{obs}|X, R = 1)$ with $f(Y_{imp}|X, R = 0)$. We would expect that the two conditional distributions are similar under a good imputation model.

However, it is not always easy to compare $f(Y_{obs}|X, R = 1)$ with $f(Y_{imp}|X, R = 0)$ in many situations, especially if both Y and X are expanded to contain multiple variables. Schafer (1997) suggested that a representative selection of bivariate scatter plots of observed and imputed values can be used to assess if important relationships between the two variables are retained. For each pair of continuous variables (e.g. Y and X in Example 12.2) in question, scatter plots displaying all subjects for which both variables are observed, as well as those displaying subjects for which one or both variables are imputed, can be shown. For the latter, plots from multiple sets of imputations can also be shown to assess if the relationship is stably retained. If X consists of categorical variables and the dimension of X is not high, Abayomi et al. (2008) compared the distributions between Y_{obs} and Y_{imp} within the subgroups defined by X.

12.3.2.2 Using Propensity Score

Suppose incomplete data are limited to one variable Y, one commonly used strategy for handling the multi-dimensionality of X in missing data analysis is to use the propensity score, denoted by $g(X)$, to summarize the information included in X (Section 5.6.2). Let $X = (X_1, \ldots, X_p)$ and $g(X) = f(R = 1|X)$, which is the propensity score of the response. Bondarenko and Raghunathan

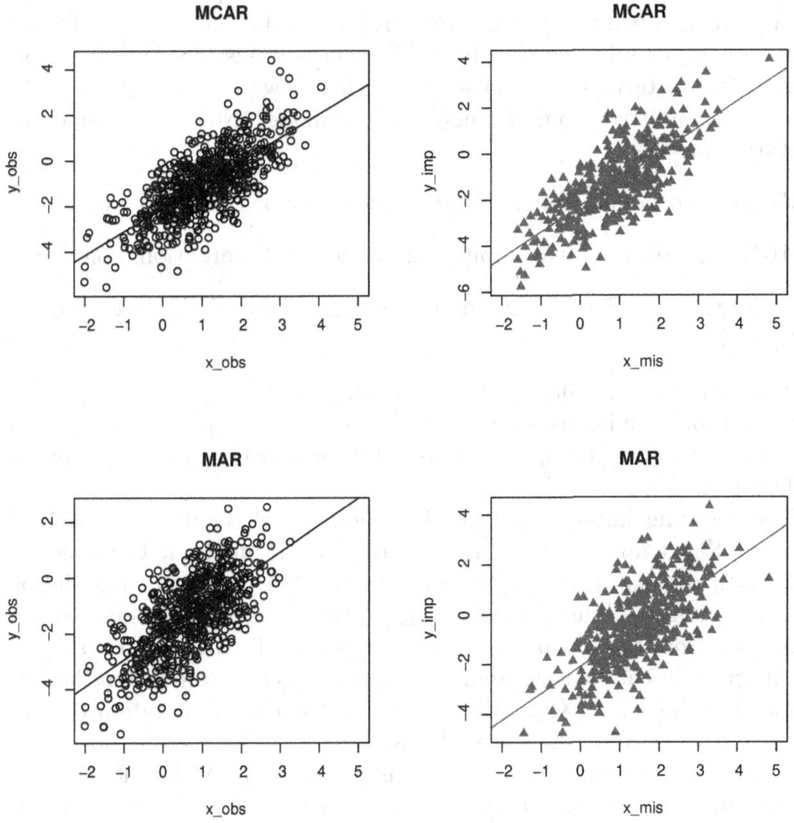

FIGURE 12.2
Example 12.2. Top left: the scatter plot of Y_{obs} and X under MCAR; top right: the scatter plot of Y_{imp} and X under MCAR; bottom left: the scatter plot of Y_{obs} and X under MAR; bottom right: the scatter plot of Y_{imp} and X under MAR. Imputed values are from one set of imputations. Circle: observed values; filled triangle: imputed values. Fitted lines are added to the scatter plots.

(2016) proposed to compare $f(Y_{obs}|g(X), R = 1)$ with $f(Y_{imp}|g(X), R = 0)$ using graphics for imputation diagnostics. The justification is that, under MAR and conditional on $g(X)$, Y_{mis} can be viewed as a random sample from the data prior to nonresponse. Large differences between the two distributions would suggest problematic imputations.

Example 12.3. *A simulation study to compare Y_{obs} with Y_{imp} conditional on $g(X)$*

We consider a complete-data model, $Y = -2 + X_1 + X_2 + \epsilon$, where X_1 and X_2 are independently generated from $N(1,1)$ and $\epsilon \sim N(0,1)$. The complete-data sample size $n = 1000$. We generate missing data for Y based on $logit(f(R=0|X_1)) = -1.4 + X_1$ which follows MAR. Note that X_2 is intentionally omitted from the nonresponse model. Now we consider three imputation models:

1. MI-X1 : a normal linear imputation model for Y conditional on X_1;
2. MI-X2 : a normal linear imputation model for Y conditional on X_2;
3. MI-X1X2 : a normal linear imputation model for Y conditional on both X_1 and X_2.

Before conducting the diagnostics, we know in advance that the MI-X1 and MI-X2 methods are inadequate because both of them ignore a predictor. The MI-X1X2 method is the optimal imputation model because it matches with the data-generating model.

Based on one imputed dataset from using each model, Fig. 12.3 plots Y_{imp} and Y_{obs} against X_1 or X_2. For the MI-X1 method, the conditional distribution of Y_{imp} given X_1 appears to be well preserved. On the other hand, the conditional distribution of Y_{imp} given X_2 is not well retained, which can be seen from the much attenuated slope when fitting Y_{imp} over X_2. The opposite pattern can be seen from the imputations generated from the MI-X2 method. For the MI-X1X2 method, however, the conditional distributions of Y_{imp} given either X_1 or X_2 are well preserved.

In this artificial example, since $g(X)$ is equivalent to X_1 by the missingness mechanism, we can see that only checking the conditional distribution of Y_{imp} given $g(X)$ (essentially X_1) is not sufficient for imputation diagnostics. If we only do this and do not check $f(Y_{imp}|X_2)$, then the problem of the MI-X1 method cannot be revealed. More generally, having $f(Y_{obs}|g(X), R=1) = f(Y_{imp}|g(X), R=0)$ does not necessarily guarantee that $f(Y_{obs}|X, R=1) = f(Y_{imp}|X, R=0)$ because $g(X)$ only contains partial information from X.

How can we fetch the information not included in $g(X)$ in the comparison? Note that $f(Y|X) = \frac{f(Y,X)}{f(X)} = \frac{f(Y,g(X),X)}{f(X,g(X))} = \frac{f(Y|g(X))f(X|Y,g(X))}{f(X|g(X))}$. Under MAR, $g(X)$ also balances the distribution of X, that is, $f(X|g(X), R=1) = f(X|g(X), R=0)$. Thus $f(Y_{obs}|X, R=1) = f(Y_{imp}|X, R=0)$ is equivalent to two conditions:

(a) $f(Y_{obs}|g(X), R=1) = f(Y_{imp}|g(X), R=0)$.

(b) $f(X|Y_{obs}, g(X), R=1) = f(X|Y_{imp}, g(X), R=0)$. Let $K(X)$ be any scalar function of X, then condition (b) also implies $f(K(X)|Y_{obs}, g(X), R=1) = f(K(X)|Y_{imp}, g(X), R=0)$.

The above reasoning suggests that in principle we need to compare two components in comparing the conditional distributions. One is to compare

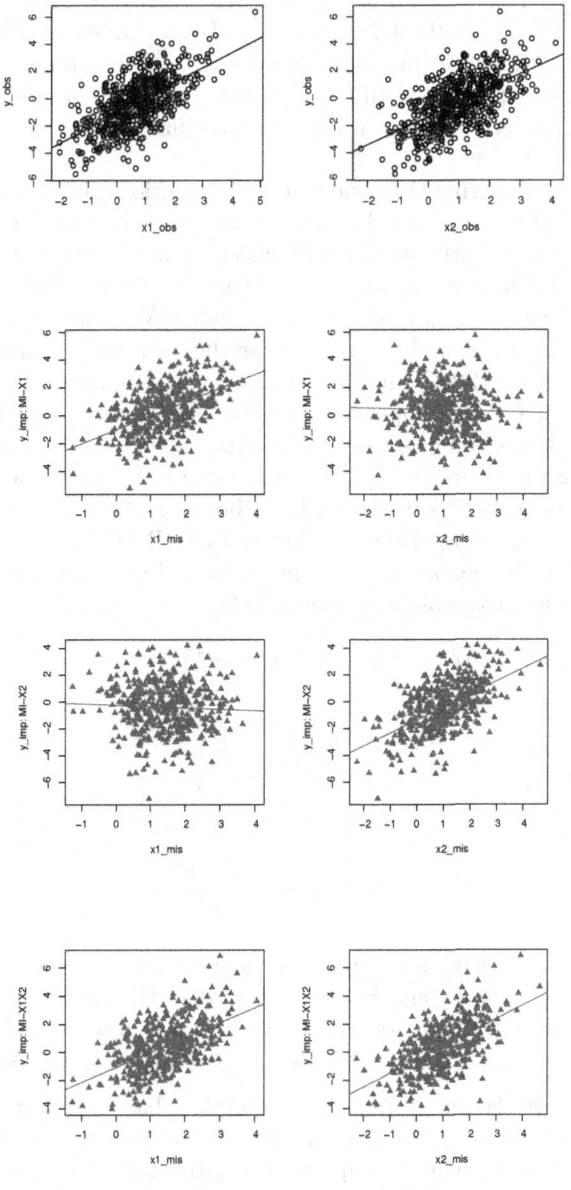

FIGURE 12.3
Example 12.3. 1st row left: the scatter plot of Y_{obs} and X_1; right: the scatter plot of Y_{obs} and X_2. 2nd row (MI-X1) left: the scatter plot of Y_{imp} and X_1; right: the scatter plot of Y_{imp} and X_2. 3rd row (MI-X2) left: the scatter plot of Y_{imp} and X_1; right: the scatter plot of Y_{imp} and X_2. 4th row (MI-X1X2) left: the scatter plot of Y_{imp} and X_1; right: the scatter plot of Y_{imp} and X_2. Circle: observed values; filled triangle: imputed values. Imputed values are from one set of imputations. Fitted lines are added to the scatter plots.

$f(Y_{obs}|g(X), R=1)$ with $f(Y_{imp}|g(X), R=0)$ as in (a), and the other is to compare $f(X|Y_{obs}, g(X), R=1)$ with $f(X|Y_{imp}, g(X), R=0)$ as in (b). Note that in (b), X on the left-hand side of the conditional distribution is p-dimensional. However in practice, we do not need to compare the full joint conditional distribution of X. Instead we can focus on comparing some scalar functions of X, $K(X)$, which can be the individual X-variable or some interactions between X-variables.

The idea of comparing the conditional distributions in both (a) and (b) is similar to that of checking the balance of covariates by estimated propensity score in observational studies. That is, checking if $f(X|\hat{g}(X), R=1)$ is similar to $f(X|\hat{g}(X), R=0)$, where $\hat{g}(X)$ is the estimated propensity score for the treatment indicator in observational studies. We can take advantage of these well developed procedures and adapt them for imputation diagnostics. For example, Austin (2009) and Stuart (2010) provided reviews on balancing diagnostics for matched samples in observational studies, and recommended using several diagnostic measures. For brevity, we focus on two simple numerical metrics that measure the location and scale of the data: one is the standardized difference (STD) of the variable between the two matched groups, and the other is the ratio of their variances (VARRATIO).

More specifically, denote $\hat{g}(X)$ as the estimated response propensity score; we can use the two diagnostic measures as follows:

(a) $f(Y_{imp}|\hat{g}(X), R=0)$ vs. $f(Y_{obs}|\hat{g}(X), R=1)$:

$$STD_{Y^*} = \frac{\overline{Y}_{imp} - \overline{Y}_{obs,match}}{\sqrt{\frac{s^2_{Y_{imp}} + s^2_{Y_{obs,match}}}{2}}}, \quad (12.2)$$

and

$$VARRATIO_{Y^*} = \frac{s^2_{Y_{imp}}}{s^2_{Y_{obs,match}}}, \quad (12.3)$$

where $\overline{Y}_{imp}$ denotes the sample average of the imputed values from a single set of imputations, $\overline{Y}_{obs,match}$ denotes the sample average of the observed values from the one-to-one matched sample by $\hat{g}(X)$, and $s^2_{Y_{imp}}$ and $s^2_{Y_{obs,match}}$ denote the corresponding sample variances, respectively.

Note that here the one-to-one match sample is based on a propensity score matching between the cases with missing and observed Y. We treat the former as a "treatment" group and the latter as a "control" group as in observational studies for estimating some treatment effects. In many real cases, the rate of missingness is less than 50%, and thus we can construct a one-to-one match to the "treatment" group from the "control" group by $\hat{g}(X)$. The propensity matching can be conducted once and for all prior to the imputation.

(b) $f(K(X)|\hat{g}(X), Y_{imp}, R=0)$ vs. $f(K(X)|\hat{g}(X), Y_{obs}, R=1)$:

$$STD_X = \frac{\overline{K}_{X|R=0} - \overline{K}_{X|R=1,match}}{\sqrt{\frac{s^2_{K_{X|R=0}} + s^2_{K_{X|R=1,match}}}{2}}}, \tag{12.4}$$

and

$$VARRATIO_X = \frac{s^2_{K_{X|R=0}}}{s^2_{K_{X|R=1,match}}}, \tag{12.5}$$

where $\overline{K}_{X|R=0}$ and $s^2_{K_{X|R=0}}$ are the average and sample variance of $K(X)$ for $X|R = 0$ (i.e., the samples for which Y is imputed), respectively, and $\overline{K}_{X|R=1,match}$ and $s^2_{K_{X|R=1,match}}$ are their counterparts from the one-to-one matched sample in $X|R = 1$ (i.e. from observed cases).

Note that here the matched sample is constructed using the information from both completed Y^* and $\hat{g}(X)$ through a bivariate matching. For example, we can construct a one-to-one matched sample using Mahalanobis distance matching. The balance of scalar statistics $K(X)$ can then be assessed on the matched sample. The bivariate matching needs to be carried out separately for each of the M completed datasets.

For the cut-off points associated with STD and VARRATIO, there are some suggestions from past literature. In observational studies, for example, if the STD for a covariate is higher than $c_{STD} = 0.1(10\%)$, then it might suggest some meaningful imbalance (Austin et al. 2009). Other cut-off points have also been used in practice, say, 0.25 (25%) in Stuart (2010). In addition, we can set $\{c_{VARRATIO_L}, c_{VARRATIO_U}\}$ as $F-$ distribution critical values for which the degrees of freedom are equal to the size of the matched sample minus 1.

A straightforward application of these assessment rules to imputation diagnostics is to calculate the simple average of the balance diagnostics over M completed datasets and compare the results with the thresholds. That is, we assess, referring back to Eqs. (12.2) and (12.3),

$$I(|STD^{MI}_{Y^*}| > c_{STD}), \tag{12.6}$$

where $|STD^{MI}_{Y^*}| = \frac{\sum_{m=1}^{M} |STD^{(m)}_{Y^*}|}{M}$, and

$$I(VARRATIO^{MI}_{Y^*} \notin (c_{VARRATIO_L}, c_{VARRATIO_U})), \tag{12.7}$$

where $VARRATIO^{MI}_{Y^*} = \frac{\sum_{m=1}^{M} VARRATIO^{(m)}_{Y^*}}{M}$. In Eqs. (12.6) and (12.7), $I(\cdot)$ is the identity function, $STD^{(m)}_{Y^*}$ and $VARRATIO^{(m)}_{Y^*}$ are STD_{Y^*} (in Eq. (12.2)) and $VARRATIO_{Y^*}$ (in Eq. (12.3)) evaluated for the m-th imputed dataset, respectively, and c_{STD} and $\{c_{VARRATIO_L}, c_{VARRATIO_U}\}$ are some predetermined thresholds for STD and VARRATIO, respectively.

Moreover, we can account for the uncertainty of the diagnostic estimates using a Bayesian-like strategy. Similar to the idea of posterior predictive checking (Gelman et al. 1996), which will be further discussed in Section 12.4, we estimate the probability that the diagnostic statistic exceeds the threshold over multiple imputations. For example, in addition to Eqs. (12.6) and (12.7), we can use

$$p_{STD_{Y*}} = \frac{\sum_{m=1}^{M} I(|STD_{Y*}^{(m)}| > c_{STD})}{M}, \tag{12.8}$$

and

$$p_{VARRATIO_{Y*}} = \frac{\sum_{m=1}^{M} I(VARRATIO_{Y*}^{(m)} \notin (c_{VARRATIO_L}, c_{VARRATIO_U}))}{M} \tag{12.9}$$

for assessment: if $p_{STD_{Y*}}$ or $p_{VARRATIO_{Y*}}$ is relatively large, say over 95%, then the results might suggest considerable imbalance between the imputed and observed values.

In addition to evaluating the diagnostics from the matched sample, it is important to further evaluate them in subclassified groups from the matched sample. This is to assess the local balance/similarity between the conditional distributions of the imputed and observed values. Without this step, the diagnostic results can be misleading because STD and $VARRATIO$ evaluated on the entire matched sample only quantifies the overall/marginal difference for the mean and variance. For example, in the case of MCAR, an inadequate imputation model which ignores the correlation between Y and X can still preserve well the marginal mean and variance of Y. Thus the comparison in (a) would reveal little imbalance unless such a comparison is conducted by subgroups.

The subgroups for comparing $f(Y_{obs}|\hat{g}(X), R = 1)$ with $f(Y_{imp}|\hat{g}(X), R = 0)$ can be easily formed by $\hat{g}(X)$ using stratification. For comparing $f(K(X)|Y_{imp}, \hat{g}(X), R = 1)$ with $f(K(X)|Y_{obs}, \hat{g}(X), R = 0)$, since the matched sample is created via a bivariate matching, we can create two sets of subgroups, one by $\hat{g}(X)$ and the other by Y^* (Y_{imp} or matched Y_{obs}). With a relatively smaller sample size, the number of subgroups might need to be reduced to ensure enough data for comparison in each subgroup. For large datasets, we might consider a higher number of subgroups to assess the balancing at a finer level.

Example 12.4. *Comparing and diagnosing alternative imputation models for gestational age using propensity score*

As in Examples 4.2, 5.5, and 5.9, we continue using the dataset with missing gestational age to illustrate the comparison strategy based on the propensity score. In Example 5.9, we fitted a logistic propensity model for the nonresponse indicator of DGESTAT, using the selected predictors as main effects and including some of their interactions. We then conduct some simple diagnostics (Austin 2009; Stuart 2010) for the estimated propensity score. The results (not shown) appear to suggest that the estimated propensity score reasonably balances the marginal distributions of the predictors, as their STDs are less than 4%, well below 10%.

We implement four imputation models for DGESTAT, and the differences of these models focus on the effect of DBIRWT in predicting DGESTAT:

1. Model I: a linear regression imputation model for DGESTAT, for which the predictors include all the X-covariates except for DBIRWT.

2. Model II: Model I plus including DBIRWT as a linear predictor (the NMI method in Example 5.5).

3. Model III: Model II plus including the quadratic term of DBIRWT (to capture the nonlinear relationship between DGESTAT and DBIRWT) (the NMI-QUAD method in Example 5.5).

4. Model IV: a PMM version of Model III so that the imputed values are all integers and fall in the range of observed values (the PMM-QUAD method in Example 5.5).

Model I is apparently inappropriate because it ignores DBIRWT, an important predictor for DGESTAT. In Example 5.5, we used plots to show that Model III and IV are better than Model II in terms of preserving the nonlinear relationship between DBIRWT and DGESTAT. Here we examine the diagnostic statistics based on conditional distributions. We conduct $M = 100$ imputations under each model. For simplicity, we only present the diagnostic results from comparing $f(DGESTAT^*|\hat{g}(X))$ and $f(DBIRWT|\hat{g}(X), DGESTAT^*)$ between imputed and observed values, omitting the diagnostics for other predictors. In this application, we set $c_{STD} = 10\%$. Since there are 4142 incomplete cases, for the overall comparison we set the critical values as $c_{VARRATIO_L} = F(4141, 4141, .025) = 0.94$ and $c_{VARRATIO_U} = F(4141, 4141, .975) = 1.06$. We also conduct the comparison on five subgroups and set critical values as $c_{VARRATIO_L} = F(4142/5-1, 4142/5-1, .025) = 0.87$ and $c_{VARRATIO_U} = F(4142/5-1, 4142/5-1, .975)$=1.15.

Table 12.2 shows the diagnostic results from all four imputation models. Diagnostics exceeding the thresholds are highlighted. Although Model I excludes the rather important predictor, DBIRWT, the overall STD from the matched sample is well below 10%. However, the STDs in the subgroups suggest an obvious imbalance. Again, this indicates the necessity of checking the balance on subgroups from the matched sample. In addition, the variance of imputed values is affected, as can be seen from the strong evidence of unequal variance between matched samples. For Model II, the STDs are all within the acceptable range when comparing $f(DGESTAT^*|\hat{g}(X))$. However, there exists a large imbalance in the fifth stratum (grouped by $DGESTAT^*$)

when comparing $f(DBIRWT|\hat{g}(X), DGESTAT^*)$. Further graphical investigations (not shown) reveal that this region exhibits a strong nonlinear relationship between DBIRWT and DGESTAT, indicating that using DBIRWT as a linear predictor can be suboptimal. This is consistent with findings from Example 5.5.

When the quadratic term of DBIRWT is included in Model III, all STDs (for both the overall and subgroup comparisons) fall within the acceptable range, suggesting a significantly better fit of the model. Yet we still see two flags indicating unequal variances. A slight improvement in preserving the variance can be seen using Model IV which has only one flag. We examine in more depth the issue of the unequal variance of DGESTAT (i.e., variance ratio=0.78 in the third stratum of the matched sample by $\hat{g}(X)$) from Model IV. By plotting the data (not shown), we find that this is largely caused by the fact that DBIRWT is not balanced very well in this subgroup by using the estimated propensity score $\hat{g}(X)$, although DBIRWT is well balanced over the full sample judged by the overall STD. As expected, this example demonstrates the fact that the diagnostic methods based on the propensity score can be sensitive to the accuracy of the estimated propensity score.

Assisted by the propensity score, we can compare the conditional distribution of observed and imputed values given other variables more conveniently to detect any problematic imputations. Future research will need to extend the idea to handle multiple incomplete variables.

12.4 Checking Completed Data

Once missing values are imputed, the data become completed, for which complete-data procedures (analyses) can be applied. Since the ultimate goal of a multiple imputation analysis is to obtain statistical results from completed-data analysis, it would be necessary to relate imputation diagnostics with completed-data statistics and analysis. In this section we discuss diagnostic strategies that aim to check completed data.

For example, graphic displays of completed data can be used to check for unusual patterns. The quality of imputation models can also be checked based on some numerical summaries. Instead of directly comparing Y_{obs} and Y_{imp} (Section 12.3), we can use the same model to generate Y_{obs}, denoted as Y_{obs}^{rep} (i.e., the re-imputations/replicates of Y_{obs} under the complete-data model) and compare Y_{obs} with Y_{obs}^{rep}. For a reasonable imputation model, Y_{obs} and Y_{obs}^{rep} are expected to be similar. On the contrary, a large discrepancy of the two might indicate some inadequacy of the imputation model. Some of the early applications of this idea can be found in Clogg et al. (1991) and Gelman et al. (1998). Here we first briefly introduce a popular Bayesian diagnostics

Example 12.4. Diagnostic results

Conditional distribution	Diagnostic statistics	Overall sample	Stratified by	Strata 1	2	3	4	5
Model I: DBIRWT excluded								
$DGESTAT^*\|\hat{g}(X)$	STD ×100	6.0 (0)	$\hat{g}(X)$	**22.5 (100)**	9.6 (44)	5.3 (6)	-4.6 (5)	-10.9 (63)
	VARRATIO	**0.82 (100)**		**0.48 (100)**	**0.76 (99)**	0.99 (1)	**1.37 (100)**	**1.5 (100)**
$DBIRWT\|\hat{g}(X)$,	STD ×100	-1.2 (0)	$DGESTAT^*$	**51.6 (100)**	-4.2 (10)	**-20.7 (100)**	**-37.2 (100)**	**-51.0 (100)**
$DGESTAT^*$	VARRATIO	**1.22 (100)**		**1.51 (100)**	**1.80 (100)**	**1.71 (100)**	**1.67 (100)**	**1.52 (100)**
$DBIRWT\|\hat{g}(X)$,	STD ×100		$\hat{g}(X)$	**-20.1 (100)**	-4.3 (3)	1.0 (1)	5.1 (7)	**16.6 (96)**
$DGESTAT^*$	VARRATIO			**1.73(100)**	1.24 (94)	0.94 (5)	0.95 (4)	0.92 (16)
Model II: DBIRWT included as a linear predictor								
$DGESTAT^*\|\hat{g}(X)$	STD ×100	1.5 (0)	$\hat{g}(X)$	2.7 (0)	0.3 (0)	2.6 (0)	-1.0 (0)	2.6 (1)
	VARRATIO	**0.86 (100)**		**0.59 (100)**	**0.81 (96)**	0.94 (9)	1.25 (94)	**1.40 (100)**
$DBIRWT\|\hat{g}(X)$,	STD ×100	1.0 (0)	$DGESTAT^*$	0.2 (0)	-14.6 (84)	-9.3 (41)	1.2 (4)	**31.2 (100)**
$DGESTAT^*$	VARRATIO	**1.13 (100)**		**1.27 (100)**	0.97 (7)	0.96 (6)	0.97 (5)	1.09 (12)
$DBIRWT\|\hat{g}(X)$,	STD ×100		$\hat{g}(X)$	-6.9 (7)	0.4 (0)	2.5 (1)	1.9 (0)	9.5 (46)
$DGESTAT^*$	VARRATIO			**1.46 (100)**	1.12 (25)	0.94 (7)	0.96 (2)	0.98 (2)
Model III: both the linear and quadratic terms of DBIRWT included								
$DGESTAT^*\|\hat{g}(X)$	STD ×100	0.2 (0)	$\hat{g}(X)$	-3.7 (9)	-1.4 (0)	5.1 (1)	2.5 (0)	2.0 (1)
	VARRATIO	0.94 (48)		0.94 (1)	0.93 (4)	**0.80 (97)**	1.00 (0)	1.02 (1)
$DBIRWT\|\hat{g}(X)$,	STD ×100	-0.4 (0)	$DGESTAT^*$	-1.2 (0)	-3.9 (10)	-0.1 (2)	1.7 (4)	3.4 (7)
$DGESTAT^*$	VARRATIO	1.04 (11)		1.14 (58)	1.13 (35)	1.03 (3)	0.90 (28)	**0.75 (99)**
$DBIRWT\|\hat{g}(X)$,	STD ×100		$\hat{g}(X)$	-3.0 (0)	0.1 (0)	-2.0 (0)	-1.9 (0)	5.7 (6)
$DGESTAT^*$	VARRATIO			1.09 (2)	1.04 (0)	0.97 (1)	1.01 (0)	1.04 (2)
Model IV: the PMM version of Model III								
$DGESTAT^*\|\hat{g}(X)$	STD ×100	0.6 (0)	$\hat{g}(X)$	-3.4 (0)	-0.8 (0)	5.4 (3)	3.4 (0)	1.9 (0)
	VARRATIO	0.96 (16)		1.02 (0)	0.96 (2)	**0.78 (98)**	0.95 (10)	0.94 (11)
$DBIRWT\|\hat{g}(X)$,	STD ×100	-0.7 (0)	$DGESTAT^*$	-0.5 (0)	1.1 (1)	0.7 (1)	-0.8 (3)	-7.6 (35)
$DGESTAT^*$	VARRATIO	1.03 (3)		1.05 (6)	1.14 (36)	1.06 (7)	1.00 (2)	0.91 (29)
$DBIRWT\|\hat{g}(X)$,	STD ×100		$\hat{g}(X)$	-3.0 (0)	0.9 (0)	-3.1 (1)	-2.4 (1)	5.0 (9)
$DGESTAT^*$	VARRATIO			1.05 (0)	1.04 (2)	.95 (4)	1.04 (1)	1.03 (3)

Note: The column labeled "Conditional distribution" includes the conditional distribution to be compared between imputed and observed values; the column labeled "Diagnostic statistics" includes STD and VARRATIO for these comparisons. The column labeled "Overall sample" includes the statistics calculated from one-to-one matched samples; the columns labeled "1"-"5" under "Strata" present statistics calculated from subgroups of matched samples; the grouping is based on either $\hat{g}(X)$ or Y^* under the column labeled "Stratified by". In each cell, the numbers outside the parenthesis are the averages of diagnostic statistics across 100 imputations as in Eqs. (12.6) and (12.7), and the numbers inside the parenthesis are the corresponding exceeding probabilities (%) as in Eqs. (12.8) and (12.9). Both values are highlighted if the exceeding probability is greater than 95%, flagging the possible imbalance of the imputations.

tool, posterior predictive checking and then discuss how this tool can be used in the context of multiple imputation analysis to check completed data.

12.4.1 Posterior Predictive Checking

The essence of comparing observed data with their replicates is to check the goodness-of-fit of the imputation model applied to the observed data. This idea is consistent with a popular model checking method, posterior predictive checking (PPC) (Gelman et al. 1996). Now suppose data Y are complete; PPC compares Y to draws of replicates Y^{rep} from their posterior predictive distribution (under the assumed model $f(Y|\theta)$) using a discrepancy function Q, a scalar function of the data and model parameters θ. One commonly used diagnostic summary in PPC is the posterior predictive p-value (Gelman et al. 1996), defined as the probability that the replicated data would be more extreme than the observed data, as measured by Q,

$$\begin{aligned} p_B &= Pr(Q(Y^{rep}, \theta) \geq Q(Y, \theta)|Y) \\ &= \int I(Q(Y^{rep}, \theta) \geq Q(Y, \theta)) f(Y^{rep}, \theta|Y) \, d\theta \, dY^{rep} \\ &= \int I(Q(Y^{rep}, \theta) \geq Q(Y, \theta)) f(Y^{rep}|\theta) f(\theta|Y) \, d\theta \, dY^{rep}, \quad (12.10) \end{aligned}$$

where $I(\cdot)$ is the indicator function.

The use of p_B accounts for uncertainty about θ in evaluating the similarity between Y and Y^{rep}. An extreme p_B-value (close to 0 or 1) suggests that the discrepancy between Y and Y^{rep} may not be easily explained by chance and hence casts suspicion on the model fit (Gelman et al. 2013, Chapter 6).

PPC has been widely used for model checking in applied Bayesian analysis. To test the general fit of the model for complete data, the choices of discrepancy functions may include (a) common descriptive statistics for the data, such as means, variances, quantiles, and correlations; (b) summaries of model fit, such as a χ^2-discrepancy; (c) graphs of residuals measuring discrepancies between the model and data; and (d) features of the data not directly addressed by the probability model.

We can calculate p_B by simulation. For each of L draws $\{\theta^l\}$ from $f(\theta|Y)$ ($l = 1, \ldots, L$), we draw one $Y^{rep,l}$ from $f(Y^{rep}|\theta = \theta^l)$, hence obtaining L draws from $f(Y^{rep}, \theta|Y)$. PPC compares the realized test quantities $Q(Y, \theta^l)$ to the predictive distribution of test quantities $Q(Y^{rep,l}, \theta^l)$; the estimated p_B-value is the proportion of these L simulations for which $Q(Y^{rep,l}, \theta^l) \geq Q(Y, \theta^l)$.

12.4.2 Comparing Completed Data with Their Replicates

When data Y are incomplete and we can describe completed data as $Y_{com} = (Y_{obs}, Y_{mis})$, the definition of the replicates of the observed data Y_{obs} becomes a little tricky. Here we use Y_{com} to emphasize the replication process for completed data in the imputation. Because Y_{obs} is a function of Y_{com} and R, the response indicator, their replicates should be applied to both the completed data Y_{com} as well as R to set up a correct reference distribution for the idea of PPC being applied to Y_{obs}. However, with a general missingness pattern from multivariate incomplete variables, it is hard to describe both Y_{obs} and R.

In many real situations, when we assume that the missingness is ignorable, the model for R is unnecessary for the purpose of analyzing completed data. Gelman et al. (2005) extended PPC to handle incomplete data, keeping R fixed in the replication process. They also argued that the diagnostics applied to Y_{obs} only are in many cases loosely related to the ultimate interest of analysts, who are often interested in estimands or information from the completed data. For example, in the normal linear model (12.1), the analytic interest is often on $\overline{Y}_{com}$ (mean of completed data), not $\overline{Y}_{obs}$ (mean of observed data). Therefore, Gelman et al. (2005) proposed to compare the posterior replicates of the completed data, that is, $Y_{com}^{rep} = (Y_{obs}^{rep}, Y_{mis}^{rep})$ with the original completed data, that is, $Y_{com} = (Y_{obs}, Y_{mis})$. To further clarify these notations, Y_{mis} denotes the imputation of the missing data under the working model, and Y_{obs}^{rep} and Y_{mis}^{rep} denote the replicates of the whole data under the same working model. The latter two can also be treated as the re-imputation of the observed and missing values of the data.

An apparent advantage of this strategy (checking the completed data instead of only observed data) is that the reference distribution of the latter is not needed because R has no role here. However, the power (to detect the model deficiency) is expected to be reduced because Y_{mis}^{rep} is expected to be similar to Y_{mis} regardless of whether the model is adequate. If the rate of missingness is high, then such checking is expected to reveal less discrepancy.

Gelman et al. (2005) mainly used graphics to plot the difference between the completed data and their replicates. He and Zaslavsky (2012) further proposed to use some numerical summaries to quantify the checks. More specifically, the idea is to assess the posterior distribution of the completed-data discrepancy, the difference between function of completed data and their replicates under the model, that is, $Q(Y_{com}^{rep}) - Q(Y_{obs}, Y_{mis})$, where Y_{com}^{rep} and Y_{mis} are drawn from $f(Y_{com}^{rep}, Y_{mis}|Y_{obs}) = \int f(Y_{com}^{rep}|\theta)f(Y_{mis}, \theta|Y_{obs})f(\theta|Y_{obs})d\theta$. In addition, they proposed to calculate the posterior predictive p-value for the imputation model as:

$$\begin{aligned} p_{B,\text{com}} &= Pr(Q(Y_{com}^{rep}) \geq Q(Y_{obs}, Y_{mis})|Y_{obs}) \\ &= \int I(Q(Y_{com}^{rep}) \geq Q(Y_{obs}, Y_{mis}))f(Y_{com}^{rep}, Y_{mis}|Y_{obs})dY_{mis}dY_{com}^{rep} \\ &= \int I(Q(Y_{com}^{rep}) \geq Q(Y_{obs}, Y_{mis}))f(Y_{com}^{rep}, Y_{mis}, \theta|Y_{obs})d\theta dY_{mis}dY_{com}^{rep} \\ &= \int I(Q(Y_{com}^{rep}) \geq Q(Y_{obs}, Y_{mis}))f(Y_{com}^{rep}|\theta)f(Y_{mis}, \theta|Y_{obs})d\theta dY_{mis}dY_{com}^{rep}. \end{aligned} \quad (12.11)$$

In Eq. (12.11), $f(Y_{com}^{rep}|\theta)$ is the complete-data model, and $f(Y_{mis}, \theta|Y_{obs})$ is the joint posterior distribution of missing values and parameters given observed data. The estimate for $p_{B,com}$ can be obtained by simulation: for $l = 1, \ldots, L$, draw θ^l from $f(\theta|Y_{obs})$ and impute Y_{mis}^l from $f(Y_{mis}|Y_{obs}, \theta = \theta^l)$, as in a typical multiple imputation for Y_{mis}, and then simulate $Y_{com}^{rep,l}$ from $f(Y_{com}^{rep}|\theta = \theta^l)$; estimate $p_{B,com}$ as the proportion of the L draws for which $Q(Y_{com}^{rep,l}) \geq Q(Y_{obs}, Y_{mis}^l)$. In addition, the posterior distribution of $Q(Y_{com}^{rep}) - Q(Y_{obs}, Y_{mis})$ can be summarized by a credible interval estimated from the empirical distribution across L simulations. The mean, median, or other summary of such an interval (i.e., $\overline{Q}(Y_{com}^{rep}) - \overline{Q}(Y_{obs}, Y_{mis})$) quantifies the average discrepancy.

In a typical imputation project, analysts are ultimately concerned with substantive analysis results using imputations. Therefore, choosing discrepancy functions Q that are estimands of these analyses links the model diagnostics with the analytic objectives. For example, if a proposed analysis is a regression, analysis-specific discrepancy functions Q might include the regression coefficients, t-statistics (standardized regression coefficients), or significance levels, all of which are functions of the data. The magnitude or range of the completed-data discrepancy, $Q(Y_{com}^{rep}) - Q(Y_{obs}, Y_{mis})$, quantifies the bias of the imputation analysis if the entire dataset were generated from the assumed model. The associated $p_{B,com}$-value would help us decide if the discrepancy is caused by chance alone. The discrepancies can be considered in the context of the scientific objective of the analysis. For instance, a very extreme $p_{B,com}$-value may be of little concern if the average discrepancy is "practically" insignificant for the analysis of interest.

Example 12.5. *A simulation study for comparing completed data with their replicates*

We use a simple simulation study to illustrate the performance of the diagnostic strategy. We consider two variables, an incomplete Y and a fully observed X. The complete-data model is $Y = 1 + X + 0.5X^2 + \epsilon$, where $X \sim Uniform(-3, 3)$ and $\epsilon \sim N(0, 1)$. Missing data in Y are generated by MAR with $logit(f(R = 0)) = \beta_0 + 0.5X$, where β_0 is chosen so that approximately 20%, 40%, 60%, or 80% of cases are missing. We consider two imputation methods: (I) a normal linear model including X as the predictor; (II) a normal linear model including both X and X^2 as predictors. For imputation diagnostics, we consider a variety of completed-data statistics and analyses (choices of Q) including:

A Mean and variance of Y.

B Percentiles of Y (5%, 25%, 50%, 75%, and 95%).

C The coefficient estimates and t-statistics from each of the following completed-data analyses.

C.A A quadratic regression of Y on X.

C.B A linear regression of Y on X.

C.C A quadratic regression of X on Y.

C.D A linear regression of X on Y.

The completed-data sample size is 1000 and the simulation includes 300 replicates. We estimate all completed-data discrepancies by simulation with $L = 5000$.

When the imputation model includes both X and X^2 as predictors (model (II)) and therefore matches the data-generating model, the average discrepancies are small and the associated posterior predictive $p_{B,com}$-values are very close to 0.5 across different scenarios as expected. These results (not shown) do not indicate any misfit of the imputation model.

Table 12.3 (see Table IV from He and Zaslavsky 2012) shows the simulation results when the imputation model (I) only includes X as the predictor (a suboptimal model) with 20% and 80% missingness rates. We present the average of statistics from completed data and their replicates, namely, $\overline{Q}(Y_{obs}, Y_{mis})$ and $\overline{Q}(Y_{com}^{rep})$ as well as the average of the associated $p_{B,com}$-values across simulations in each scenario. For the marginal mean and variance of Y, the average statistics for the completed data and their replicates are indistinguishable and the associated $p_{B,com}$-values are close to 0.5, showing little evidence of model inadequacy. This is because both the marginal mean and variance are sufficient statistics of the first two moments of complete data, which tend to be replicated well even by the misspecified imputation model.

With $p_{mis} \approx 20\%$, average discrepancies for some of the percentiles (i.e., 5%, 50%, 75%, and 95%) are substantial, showing that the model fails to replicate well the marginal distribution of Y. This is because the complete data are asymmetrically distributed but the linear model imputes from a symmetric distribution. The associated $p_{B,com}$-values are also very extreme, suggesting the discrepancy is unlikely to be due to chance alone.

For both linear regression analyses, the regression coefficients are essentially the same in the completed data and replicates, showing no sign of model inadequacy. This is because the linear regression imputation preserves the linear correlation between the two variables. For the quadratic regression of Y on X, however, the coefficients for the intercept and quadratic term are very different between the completed data and their replicates, clearly identifying the nonlinear term/curvature (X^2) that is omitted from the imputation model. The corresponding $p_{B,com}$-values are rather extreme. The quadratic regression analysis of X on Y also suggests evidence of model inadequacy for the linear and quadratic terms. Diagnostic results for the t-statistics (not shown) are consistent with those from the regression coefficients. These diagnostics would suggest that a nonlinear term between Y and X might be necessary to improve the model adequacy.

With more (80% instead of 20%) missing data, the diagnostics are less likely to detect imputation model inadequacy, with smaller average discrepancies of all statistics and less extreme $p_{B,com}$-values. Nonetheless, some diag-

TABLE 12.3
Example 12.5. Simulation results with 20% and 80% missingness when the imputations are based on a suboptimal normal linear regression model for Y on X

Estimand	$p_{mis} \approx 20\%$			$p_{mis} \approx 80\%$		
	$\overline{Q}(Y_{obs}, Y_{mis})$	$\overline{Q}(Y_{com}^{rep})$	$p_{B,com}$	$\overline{Q}(Y_{obs}, Y_{mis})$	$\overline{Q}(Y_{com}^{rep})$	$p_{B,com}$
Mean of Y	2.41	2.41	0.500	2.20	2.20	0.500
Variance of Y	4.88	4.88	0.495	3.35	3.35	0.492
5%th-tile of Y	−0.52	−1.21	1	−0.62	−0.84	0.874
25%th-tile of Y	0.81	0.86	0.332	0.92	0.95	0.401
50%th-tile of Y	1.97	2.41	0	2.10	2.20	0.152
75%th-tile of Y	3.71	3.95	0.029	3.39	3.45	0.294
95%th-tile of Y	6.71	6.02	1	5.32	5.20	0.757
Linear regression Y on X						
Intercept coefficient	2.40	2.40	0.500	2.19	2.19	0.500
Linear coefficient	0.85	0.85	0.501	0.55	0.55	0.500
Linear regression X on Y						
Intercept coefficient	−1.26	−1.25	0.488	−1.06	−1.06	0.488
Linear coefficient	0.52	0.52	0.513	0.48	0.48	0.509
Quadratic regression Y on X						
Intercept coefficient	1.33	2.40	0	1.90	2.19	0.003
Linear coefficient	0.85	0.85	0.492	0.55	0.55	0.497
Quadratic coefficient	0.36	0	1	0.10	0	1
Quadratic regression X on Y						
Intercept coefficient	−1.18	−1.26	0.878	−1.03	−1.06	0.666
Linear coefficient	0.42	0.52	0.011	0.42	0.48	0.130
Quadratic coefficient	0.02	0	0.986	0.01	0	0.891

nostics still signal model inadequacy, such as those from the quadratic regression of Y on X. As expected, smaller total sample sizes reduce the power for detecting model inadequacy. If we reduce the complete-data sample size in the simulation (results not shown), the magnitudes of the average discrepancies are similar but the $p_{B,com}$-values of the former are less extreme because of the greater variation of the statistics.

The simulation results also illustrate the importance of the choice of Q. Letting Q be the quadratic regression coefficient of Y on X or vice versa detects the inadequacy of the linear regression imputation model because it identifies a key feature, the curvature of the relationship, omitted from the imputation model. On the other hand, an analysis more coherent with the same imputation model, such as the linear regression of Y on X or vice versa, does not help detect the discrepancy. Therefore, researchers and practitioners should try to identify statistics Q corresponding to features of analytic interest that are not reflected in imputation models. This is related to the notion that sufficient statistics are generally not the optimal discrepancy functions in PPC (Gelman et al. 2013).

More generally, the comparison strategy can be viewed as diagnosing models by re-imputing completed data using any desired method for multiple imputation. It then goes beyond the Bayesian diagnostics framework. In Eq. (12.11), Y_{com}^{rep} and Y_{mis} can be obtained as imputations in the concatenated (stacked) dataset $\begin{pmatrix} Y_{mis} & Y_{obs} \\ Y_{com}^{rep} & \end{pmatrix}$. To see this more specifically, suppose $Y = (Y_1, Y_2)$, where Y_1 includes incomplete variables whose imputation method (conditional on Y_2) is to be assessed while Y_2 is fully observed. We can first create a duplicated dataset in which Y_2 is retained but the incomplete Y_1 is made completely missing. Then we can concatenate the original and duplicate sets together as $\begin{pmatrix} (Y_{1,obs}, & Y_{1,mis}), & Y_2 \\ Y_{1,com}^{rep}, & & Y_2 \end{pmatrix}$, where $Y_{1,mis}$ and $Y_{1,com}^{rep}$ are missing. Imputations for $Y_{1,mis}$ and replicates $Y_{1,com}^{rep}$ can be obtained by applying the imputation method for the original dataset to the concatenated set, and the comparison follows naturally. The re-imputation can be conducted L times to calculate all the quantities needed for diagnostics.

Note that the duplication of Y_2 does not affect estimation of the conditional distribution $f(Y_1|Y_2)$ in the imputation model, but has the operational advantage of allowing the imputation program to automatically produce $Y_{1,com}^{rep}$. Even if all of the variables in Y are incomplete, that is, Y_2 is a null set and the bottom half of the concatenated dataset are all missing, the re-imputation strategy still works based on our experiences.

This re-imputation strategy facilitates imputation checking by practitioners who implement imputation through standard software, only requiring re-application of the imputation code after simple data manipulation but requiring no specific knowledge of the form of the model or the algorithms used in imputation, which sometimes can be rather sophisticated. In addition, the working imputation models checked do not have to be fully Bayesian as originally required for PPC. For example, imputations based on PMM methods can also be checked using the re-imputation strategy.

Example 12.6. *Comparison of models for multiple binary variables using the re-imputation strategy*

In Example 6.3 we considered several multiple imputation methods for four binary incomplete variables (CARERCVD_B, SELFHEALTH_B, COVERAGE, and DELAYM). The post-imputation analysis focused on the distribution of the CARERCVD_B for the whole sample and in different subgroups. We use the re-imputation strategy to calculate the posterior predictive statistics for these estimates under the considered imputation models. That is, we concatenate the original dataset and one holding its replicates together and apply the imputation methods to calculate the posterior predictive summaries and p-values, using R cat for MMI and SMI and using R jomo for LMI. The number of replicates L is set as 1000. Table 12.4 shows the results. For each method, the column $\overline{Q}(Y_{obs}, Y_{mis})$ includes the multiple imputation estimates under $M = 1000$ imputations, the column $\overline{Q}(Y_{com}^{rep})$ lists the corresponding average from the replicates, and $p_{B,com}$ lists the posterior predictive p-values. For the MMI method, which is based on an incorrect log-linear model assuming mutual independence among all four binary variables, many of the $p_{B,com}$ values are rather extreme (close to 0 or 1) and indicate a bad fit of the model. The corresponding differences between $\overline{Q}(Y_{obs}, Y_{mis})$ and $\overline{Q}(Y_{com}^{rep})$ are also large. Note that for difference mean estimates of CARERCVD_B, their corresponding averages of replicates ($\overline{Q}(Y_{com}^{rep})$) are all around 0.585. This is expected because under the assumption of mutual independence, the mean of CARERCVD_B is not related to the other three variables so it is constant in replicates. For the SMI method, which assumes the most general log-linear model, all of the $p_{B,com}$-values are close to 0.5, showing no evidence of the lack-of-fit of the model. For the LMI method, the two $p_{B,com}$-values are somewhat extreme, which correspond to the mean when $S = 1, C = 0, D = 0$ and when $S = 0, C = 0, D = 0$. Note that these two estimates are also different from those in the SMI method. This might indicate some lack-of-fit of the latent variable model in these two combinations of the four variables. Based on these $p_{B,com}$-values, we would prefer to use the SMI method and the associated results.

He and Zaslavsky (2012) also considered a few extensions of calculating and using $p_{B,com}$. For example, a double simulation strategy can be implemented to integrate out Y_{mis} and Y_{mis}^{rep}, which are generated under the same working imputation model (adequate or not), and keeping them in the comparison would add variance and reduce the power of assessment. Algebraic approximations were also proposed if a large number of simulations poses some computational burden. They applied the diagnosing strategy to several real datasets with multiple incomplete variables. This strategy can be effective if alternative imputation models are applied and compared. It might be reasonable to choose a model which yields fewer extreme posterior predictive p-values. Additional investigations of this strategy can be found in Cabras et al. (2011) and Nguyen et al. (2015). In addition, the idea of PPC can be applied to data with latent structure or other problems which can be viewed as missing data problems (e.g., Rizopoulos et al. 2010).

In summary, the main advantages of comparing completed data with their replicates in the context of practical imputation analyses include: (a) multiple imputation can be regarded as a form of replication; (b) the posterior predictive p-value is a nice numerical summary, which quantifies the evidence that the discrepancy is not due to chance alone; (c) through the re-imputation strategy, the calculation is easy to implement and suitable for the practitioners who may lack the statistical resources to apply alternative Bayesian model checking strategies.

12.5 Assessing the Fraction of Missing Information

In Chapter 3, we briefly covered the fraction of missing information (FMI) as one of the key statistics in multiple imputation analysis. To recap it, let B_M, U_M, and T_M denote the between-imputation, within-imputation, and total variance estimate of $\hat{Q} = \overline{Q}_M$ from M imputations, respectively. Several key

TABLE 12.4
Example 12.6. Analysis results

Mean of Y	MMI			SMI			PMI		
	$\overline{Q}(Y_{obs}, Y_{mis})$	$\overline{Q}(Y^{rep}_{com})$	$p_{B,com}$	$\overline{Q}(Y_{obs}, Y_{mis})$	$\overline{Q}(Y^{rep}_{com})$	$p_{B,com}$	$\overline{Q}(Y_{obs}, Y_{mis})$	$\overline{Q}(Y^{rep}_{com})$	$p_{B,com}$
Overall	0.585	0.585	0.492	0.555	0.555	0.485	0.555	0.557	0.525
$S = 1, C = 1, D = 1$	0.567	0.586	0.655	0.553	0.554	0.497	0.547	0.543	0.456
$S = 1, C = 1, D = 0$	0.759	0.585	0	0.784	0.780	0.491	0.776	0.743	0.241
$S = 1, C = 0, D = 1$	0.468	0.582	0.820	0.347	0.353	0.504	0.384	0.421	0.642
$S = 1, C = 0, D = 0$	0.409	0.583	0.912	0.334	0.346	0.514	0.430	0.632	0.939
$S = 0, C = 1, D = 1$	0.507	0.585	0.996	0.443	0.442	0.522	0.443	0.446	0.555
$S = 0, C = 1, D = 0$	0.640	0.584	0.026	0.647	0.647	0.526	0.650	0.663	0.670
$S = 0, C = 0, D = 1$	0.484	0.583	0.897	0.345	0.347	0.505	0.338	0.327	0.431
$S = 0, C = 0, D = 0$	0.675	0.583	0.129	0.759	0.757	0.505	0.677	0.537	0.084

Note: Y: CARERCVD_B; S: SELFHEALTH_B; C:COVERAGE; D:DELAYM.

statistics that measure the ratio of the between-imputation variance to total variance (Table 3.2) are listed here:

1. $\lambda_M = \frac{(1+1/M)B_M}{T_M}$ (the ratio of between-imputation variance to total variance).

2. $r_M = \frac{(1+1/M)B_M}{U_M} = \lambda_M/(1-\lambda_M)$ (the relative increase of the variance due to nonresponse).

3. $\nu_M = (M-1)/\lambda_M^2$ (degrees of freedom).

4. $\gamma_M = \frac{r_M+2/(\nu_M+3)}{1+r_M}$ (FMI).

Historically, FMI has often been used to indicate the severity of missing data problems. For example, Li and Raghunathan et al. (1991) classified FMI up to 0.2 as "modest", 0.3 as "moderately large", and 0.5 as "high". High values indicate a difficult problem in which the multiple imputation inferences are highly dependent on the imputation models. FMI can also be used to determine the number of imputations (M) in practice (Section 3.3).

In this section we focus on using FMI as a diagnostic measure of imputation models. This is because although FMI typically depends on the rate of missingness, it also depends on the analysis of interest and the extent to which the imputation model is predictive of the missing values. Intuitively, if the imputation model is more predictive of the missing values, the corresponding FMI for certain estimands tends to be smaller. Before showing the relation between FMI and imputation model predictability, we first present additional technical discussion about FMI.

Although FMI is generally understood as the ratio of the between-imputation variance to total variance, the exact definition is a subtle issue. This is because both the between-imputation and within-imputation variance are unknown quantities and need to be estimated with a finite number of imputations. One version that is simpler to understand is the so-called population of fraction of missing information (Rubin 1987):

$$\gamma_\infty = \frac{B_\infty}{B_\infty + U_\infty}, \tag{12.12}$$

where B_∞ and U_∞ are the between- and within-imputation variance (estimate) when the number of imputations $M \to \infty$.

For a finite M, Rubin (1987) proposed an estimator, γ_M, for γ_∞, and it can be expressed using only B_M and U_M as:

$$\gamma_M = \frac{(1+\frac{1}{M})\frac{B_M}{U_M}}{(1+\frac{1}{M})\frac{B_M}{U_M}+1} + \frac{\frac{2}{(M-1)(1+\frac{1}{(1+\frac{1}{M})\frac{B_M}{U_M}})^2+3}}{(1+\frac{1}{M})\frac{B_M}{U_M}+1}. \tag{12.13}$$

However, researchers and practitioners need to realize that γ_M is an estimate of γ_∞: if we change M or the random seed for the same M, then the output for γ_M can be different.

Obviously $\gamma_M \to \gamma_\infty$ as $M \to \infty$. With a finite M, it can be shown that

$$E(\gamma_M) \approx \frac{(M+1)\gamma_\infty}{M+\gamma_\infty} > \gamma_\infty, \tag{12.14}$$

$$SE(\gamma_M) \approx \gamma_\infty(1-\gamma_\infty)\frac{M(M+1)}{(M+\gamma_\infty)^2}\sqrt{\frac{2}{M-1}}, \tag{12.15}$$

$$CV(\gamma_M) \approx \sqrt{\frac{2}{M-1}}(1-\gamma_\infty)\frac{M}{M+\gamma_\infty} \approx \sqrt{\frac{2}{M-1}}(1-\gamma_\infty), \tag{12.16}$$

where CV stands for coefficient variation.

Eq. (12.14) shows that γ_M tends to overestimate γ_∞ when M is small or moderate. In addition, γ_M can be rather noisy. This is because B_M is more variable as a function of M, compared with $\overline{Q}_M$ and U_M (Example 3.2). Harel and Schafer (2003) showed that γ_M is considerably more noisy than $\overline{Q}_M$, and M often needs to be very large (say hundreds) to obtain a reliable estimate of γ_∞. Similar arguments were made by Raghunathan et al. (2018, Page 30). For instance, in order to have the $CV \leq 10\%$ with $\gamma_\infty = 20\%$, we need to set $M > 120$ for using γ_M as the estimate. Practitioners need to pay attention to this issue if γ_M is used as an imputation diagnostic measure.

There can be alternative estimators for γ_∞. For example, Harel (2007) proposed to use $\gamma_H = \frac{B_M}{B_M+U_M}$ to estimate γ_∞. Another possible estimator would be $\gamma_Y = \frac{(1+1/M)B_M}{(1+1/M)B_M+U_M}$. Additional research on comparing the performances of various estimators of γ_∞ would be interesting.

12.5.1 Relating the Fraction of Missing Information with Model Predictability

Various literature on multiple imputation analysis have argued for the association between FMI and the imputation model predictability. For example, Wagner (2010) suggested using FMI as a tool for monitoring the quality of survey data in terms of nonresponse. Andridge and Thompson (2015) used FMI to select auxiliary variables in imputation models and provided some theoretical justifications under simple setups. The general belief is that if the imputation model is more predictive of the missing values, the corresponding FMI for certain estimands (e.g., mean of the missing variable) tends to be smaller. However, the exact behavior can be more complicated because FMI for the mean estimate is also related to the relation between the covariates and the missingness mechanism (Andridge and Thompson 2015). And the behavior of FMI can be different for different estimands.

Here we attempt to provide some analytical reasoning. We first assume that the incomplete data are confined to a single variable Y, and the covariates X are fully observed. Let $\hat{Q}_{com}$ and $\hat{Q}_{MI}$ be the estimate based on complete data (before data are missing) and multiply imputed data, respectively. Examples of Q can include the marginal mean of Y and the regression coefficient of Y on X. In the following arguments, we assume that $M = \infty$ so that we focus on γ_∞. We also assume that the imputation model (i.e., the complete-data model) is correctly specified so that the variance estimate is unbiased. Let θ denote the model parameter. Under these assumptions, it can be seen that $Var(\hat{Q}_{MI}) = B_\infty + U_\infty$, and $Var(\hat{Q}_{com}) = U_\infty$. Therefore,

$$\gamma_\infty = \frac{Var(\hat{Q}_{MI}) - Var(\hat{Q}_{com})}{Var(\hat{Q}_{MI})}. \tag{12.17}$$

On the other hand, we have:

$$\begin{aligned} Var(\hat{Q}_{com}) &= Var(E(\hat{Q}_{com}|X,\theta)) + E(Var(\hat{Q}_{com}|X,\theta)), && (12.18) \\ Var(\hat{Q}_{MI}) &= Var(E(\hat{Q}_{MI}|X,\theta)) + E(Var(\hat{Q}_{MI}|X,\theta)). && (12.19) \end{aligned}$$

The purpose of Eqs (12.18) is to decompose the total variance associated with $\hat{Q}_{com}$ into two parts: one is the variance explained by the inclusion of covariate X and the specified model (parameterized by θ) as $Var(E(\hat{Q}_{com}|X,\theta))$; the other is the remainder as $E(Var(\hat{Q}_{com}|X,\theta))$, which cannot be explained by the covariate and model. The same technique is applied for $Var(\hat{Q}_{MI})$ in Eq. (12.19).

We surmise that in many cases,

$$Var(E(\hat{Q}_{com}|X,\theta)) \approx Var(E(\hat{Q}_{MI}|X,\theta)). \tag{12.20}$$

This is because under a correctly specified imputation model, the multiple imputation-based estimate would be consistent. That is, both $E(\hat{Q}_{com}|X,\theta)$ and $E(\hat{Q}_{MI}|X,\theta)$ would converge to some functions of θ and X (or statistics of X). For example, if Q is a direct function of θ, then both $Var(E(\hat{Q}_{com}|X,\theta))$ and $Var(E(\hat{Q}_{MI}|X,\theta))$ are 0, treating θ as a fixed quantity.

The relationship between $E(Var(\hat{Q}_{com}|X,\theta))$ and $E(Var(\hat{Q}_{MI}|X,\theta))$ is more complicated. Suppose that the missingness rate of Y is p_{mis}; then we might assume that

$$E(Var(\hat{Q}_{com}|X,\theta)) \approx E(Var(\hat{Q}_{MI}|X,\theta)) \times (1 - p_{mis}). \tag{12.21}$$

That is, after accounting for the predictiveness of the model, the "residual" variation of the estimate might be proportional to the inverse of the original sample size. In principle, this relation between the two terms is related to where Y is missing over the distribution of X (i.e., nonresponse mechanism). Therefore Eq. (12.21) perhaps holds better when the missingness is close to MCAR.

After plugging Eqs. (12.18)-(12.21) in Eq. (12.17), we obtain that

$$\gamma_\infty \approx \frac{p_{mis}\left(1 - \frac{Var(E(\hat{Q}_{com}|X,\theta))}{Var(\hat{Q}_{com})}\right)}{1 - p_{mis}\frac{Var(E(\hat{Q}_{com}|X,\theta))}{Var(\hat{Q}_{com})}}. \tag{12.22}$$

Eq. (12.22) aims to relate two factors with FMI: p_{mis} (the missingness rate); $\frac{Var(E(\hat{Q}_{com}|X,\theta))}{Var(\hat{Q}_{com})}$, which might be perceived as the variance fraction explained by the complete-data model for $\hat{Q}_{com}$. It suggests that FMI is in general less than or equal to p_{mis}. As $\frac{Var(E(\hat{Q}_{com}|X,\theta))}{Var(\hat{Q}_{com})}$ increases (i.e., the model predictability increases), FMI tends to decrease. We use several simple examples to illustrate this pattern.

Example 12.7. *A univariate missing data problem with MCAR*

As in Example 3.1, suppose $Y \sim N(\mu, \sigma^2)$ so $\theta = (\mu, \sigma^2)$. Let $Q = \mu$, the marginal mean of Y. Then $Var(E(\hat{Q}_{com}|X, \theta)) = Var(E(\hat{\mu}_{com}|\mu, \sigma^2)) = Var(\mu) = 0$. That is, the model does not provide any predictive information. This is obvious because there is no covariate X involved. Therefore $\gamma_\infty \approx \frac{p_{mis}(1-0)}{1-p_{mis}0} = p_{mis}$ (Rubin 1987). See also Example 3.5.

We conduct a simulation study to gauge the behavior of γ_M as an estimate of γ_∞ in this setting, using the simulation setup of Example 3.4. We generate $Y \sim N(1, 1)$ for 1000 observations and randomly set 50% of the observations as missing. We report γ_M varying the number of imputations $M = \{2, 5, 10, 20, 50, 100, 200, 100\}$. The experiment is replicated 1000 times. Fig. 12.4 includes Box plots of γ_M's from 1000 replicates across different number of imputations. When M is small or moderate (say <= 20), the median of γ_M's is slightly larger than $\gamma_\infty = 0.5$, and the estimates have a rather noisy distribution. As M increases, the estimates converge to 0.5 and are more stable.

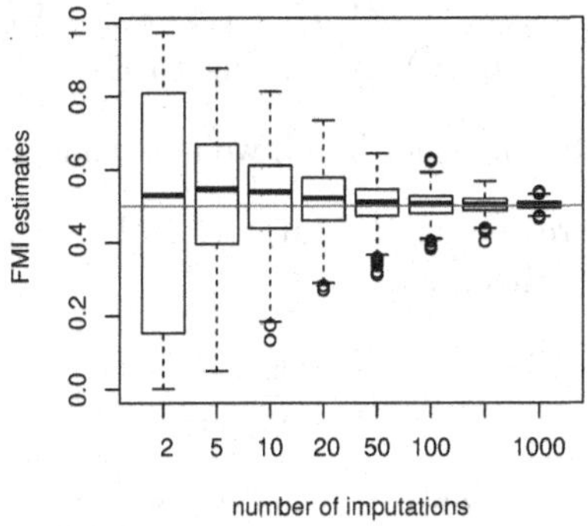

FIGURE 12.4
Example 12.6. Box plots of γ_M across different numbers of imputations.

Example 12.8. *Regression coefficient from an MAR missing data problem with one covariate X*

For simplicity, assume that X only has one predictor and the complete-data model is $Y = X\beta + \epsilon$ (an intercept-free regression model), where $\epsilon \sim N(0, \sigma^2)$. Here $\theta = (\beta, \sigma^2)$. Let $Q = \beta$, the regression coefficient; then $\hat{\beta}_{com} = (X^tX)^{-1}X^tY$. It can be shown that $Var(E(\hat{Q}_{com}|X, \theta)) = Var(E(\hat{\beta}_{com}|X, \beta, \sigma^2)) = Var(\beta) = 0$ because $\hat{\beta}_{com}$ is an unbiased estimate of β. According to Eq. (12.22), $\gamma_\infty \approx p_{mis}$, the same as in Example 12.7.

By going back to the details of the multiple imputation estimator, $\hat{\beta}_{MI}$, we know that asymptotically $\hat{\beta}_{MI} = \hat{\beta}_{CC} = (X_{obs}^t X_{obs})^{-1} X_{obs}^t Y_{obs}$ under MAR, where X_{obs} indicates the subset of X with observed Y-values. Therefore $Var(\hat{\beta}_{MI}) \approx Var(\hat{\beta}_{CC}) = (X_{obs}^t X_{obs})^{-1}\sigma^2$. In addition, $Var(\hat{\beta}_{com}) = (X^tX)^{-1}\sigma^2$. According to Eq. (12.17), $\gamma_\infty \approx \frac{(X_{obs}^t X_{obs})^{-1} - (X^tX)^{-1}}{(X_{obs}^t X_{obs})^{-1}}$, which certainly depends on how the missingness is related to X. However to assess its magnitude, suppose the complete-data sample size is n and observed-data sample size is n_{obs}, $\gamma_\infty = O(\frac{1/n_{obs} - 1/n}{1/n_{obs}}) = O(\frac{n - n_{obs}}{n}) = O(p_{mis})$, where $O(a)$ means is of the same order as a. In general, if MCAR holds well or MAR is weak, then $\gamma_\infty \approx p_{mis}$. Otherwise, γ_∞ might deviate from p_{mis} yet we expect that the difference is small or modest. In this example, however, the model predictiveness does not add much information to the multiple imputation estimate of β because γ_∞ is not reduced much. This is because all the X information needed for estimating β in the presence of missing data is already contained in X_{obs}.

Example 12.9. *Marginal mean from an MAR missing data problem with one covariate X*

Following from Example 12.8, now let $Q = \overline{Y}$, the marginal mean of Y. In this case, $Var(\hat{Q}_{com}) = \beta^2 Var(\overline{X}) + \sigma^2/n$, $Var(E(\hat{Q}_{com}|X, \theta)) = Var(\overline{Y}_{com}|X, \beta, \sigma^2) = \beta^2 Var(\overline{X})$. According to Eq. (12.22), we have $\gamma_\infty \approx \frac{p_{mis}(1 - \frac{\beta^2 Var(\overline{X})}{\beta^2 Var(\overline{X}) + \sigma^2/n})}{1 - p_{mis}\frac{\beta^2 Var(\overline{X})}{\beta^2 Var(\overline{X}) + \sigma^2/n}}$. On the other hand, we note that in here $\frac{\beta^2 Var(\overline{X})}{\beta^2 Var(\overline{X}) + \sigma^2/n}$ can be conveniently expressed as R^2-statistics in linear regression. Therefore Eq. (12.22) can also be written as

$$\gamma_\infty \approx \frac{p_{mis}(1 - R^2)}{1 - p_{mis}R^2}. \tag{12.23}$$

Since regression R^2 is a commonly used measure for the goodness-of-fit of the model, Eq. (12.23) establishes the connection between FMI and model predictability in typical regression models. A similar relationship can be derived under a bivariate normal model with one variable subject to nonresponse under MCAR (Little and Rubin 2020, Chapter 7, Eqs. (7.13) and (7.14)); for nonresponse under MAR, Andridge and Thompson (2015) provided a more general formula of FMI. In the latter situation, FMI is also related to the nonresponse mechanism: a covariate that is weakly predictive of the outcome but strongly associated with the nonresponse mechanism can actually increase FMI.

In spite of the additional impact from the missingness mechanism, Eq. (12.23) might be useful in general settings for illustrating the relationship between the missingness rate and model predictability with FMI for estimating the mean of the missing variable. As the missingness rate increases, FMI increases; as the R^2 (the model predictability) increases, FMI decreases.

Example 12.10. *A simulation study for assessing the pattern of FMI*

We conduct a simulation study to understand more about Eq. (12.23). We consider a complete-data model, $Y = -2 + X_1 + X_2 + X_1X_2 + \epsilon$, where $X_1 \sim N(1,1)$, $X_2 \sim N(1,1)$, and $\epsilon \sim N(0,1)$. Two types of missingness mechanisms are considered: one is MCAR; the other is MAR in which the nonresponse mechanism can be described as $logit(f(R=0)) = -1.4 + 0.7X_1 + 0.3X_2$. In both cases, the missingness rate is around 40%. We consider three imputation models: (a) a normal linear model only including X_1 as the predictor; (b) a normal linear model including both X_1 and X_2 as predictors; (c) a normal linear model including X_1, X_2, and their interactions X_1X_2 as predictors, which match with the data-generating model. We know that in advance the predictability of these imputation models gets improved gradually. Multiple imputation analyses are based on $M = 50$ sets of imputations. We consider the mean of Y in the imputation analysis. We calculate FMI using Eq. (12.13) (FMI_{RUBIN}) and its approximation based on Eq. (12.23) (FMI_{Approx}). The former can be directly obtained from imputation software (in this case R mice). For the latter, the R^2 statistics are obtained from running each regression model using completed Y's and then averaging over 50 imputed datasets. The complete data sample size $n = 1000$, and the simulation study is based on 1000 replications.

Table 12.5 shows the simulation results. Under MCAR, all three imputation methods yield little bias. Estimated FMI from all three models is below 40%, the missingness rate. This suggests that some predicative power is gained by including covariates in the imputation. Across models, FMI decreases from around 30% to 6% as the predictability of the imputation model increases; as R^2 increases, FMI decreases. Under MAR, the mean estimate from the imputation model only including X_1 is highly biased; the mean estimate from the imputation model including both X_1 and X_2 is marginally biased; and the mean estimate from the correct imputation model has little bias as expected. The patten of FMI is similar: FMI decreases as the predictability of model gets improved.

As reasoned before, the approximation of FMI by Eq. (12.23) works somewhat better under MCAR than MAR. In both scenarios, however, the magnitudes of two FMI estimates are similar. Note that the purpose of this simulation study is not to advocate the use of Eq. (12.23) for calculating FMI. The main purpose is to provide some quantitative insights on how FMI is affected by the model predictability. In addition, although the standard error estimate generally decreases as the model gets improved, the change of its magnitude is rather small, making it inconvenient to diagnosing models. The change of FMI appears to be a better indicator of model predictability. Finally, if an

FMI estimate is really small (say close to 0) for a moderate or large amount of missingness rate, then we should check whether the multiple imputation is improper (Section 3.4.2) and thus yields a smaller-than-expected between-imputation variance.

Example 12.11. *FMI estimates for different imputation models of missing gestational age*

As in Example 12.4, we apply the four imputation models to impute the missing DGESTAT and calculate its mean estimate, standard error, and associated FMI. The results are shown in Table 12.6 based on $M = 100$ imputations. The mean estimate of DGESTAT barely changes across all four models. The standard error decreases slightly from Model I to II or III. However, there is a steady trend that FMI estimate decreases from Models I to III, indicating that the model predictability increases as the imputation includes DBIRWT (Model II) and further includes its squared term (Model III). In Model IV, the standard error increases to the level of Model I, and the FMI estimate increases. This is largely due to the fact that imputed values are only taking the observed values by PMM and become less smooth (e.g., Fig. 5.9), which leads to a larger variance estimate and has less to do with the model predictability.

TABLE 12.5
Example 12.10. Simulation results

Model	RBIAS (%)	$\overline{\text{SE}}$	FMI_{RUBIN}	FMI_{Approx}	R^2
MCAR					
With X_1	−0.47	0.118	0.293	0.286	0.399
With X_1 and X_2	−0.23	0.107	0.123	0.118	0.800
With X_1X_2	−0.08	0.103	0.0667	0.0626	0.900
MAR					
With X_1	−55.9	0.108	0.326	0.313	0.352
With X_1 and X_2	−7.05	0.100	0.167	0.136	0.777
With X_1X_2	−0.03	0.104	0.0839	0.0659	0.900

TABLE 12.6
Example 12.11. Analysis results

Model	Mean	SE	FMI
I	38.66	0.01428	0.128
II	38.64	0.01403	0.0869
III	38.64	0.01408	0.07222
IV	38.64	0.0143	0.0980

As illustrated in Examples 12.10 and 12.11, it might be helpful to monitor FMI estimates when alternative imputation models are applied and compared. We may select the model which yields the lowest FMI for the estimands of interest. For example, Andridge and Thompson (2015) considered nonignorable missing data models to obtain a maximum likelihood estimate of FMI for separate sets of candidate imputation models and look for the point at which changes in FMI level off and adding further auxiliary variables do not improve the imputation model. Although we only consider a single incomplete variable when we derive Eqs. (12.22) and (12.23), we surmise that this idea also applies to the setting of multivariate incomplete variables. Of course, researchers and practitioners need to be reminded that a reliable estimate of FMI needs a large number of imputations. In addition, we have shown that the behavior of FMI can be different across different estimands of interest (e.g., regression coefficient vs. mean). A single FMI might not be sufficient to quantify the quality of the whole model. More research is needed to understand the behavior of FMI with different estimands of interest and quantify it whenever possible.

12.6 Prediction Accuracy

There often exists an (incorrect) notion that imputation is nothing different from prediction. Based on such a viewpoint, an intuitive assessment tool seems to be assessing the discrepancy between true data and the imputed values, for example, by the square root of the average of the squared difference between the two (i.e., root mean-squared-error). This is tempting because such a criterion is often used in assessing prediction models. In the context of imputation diagnostics, model predictability is important as it can be related to the fraction of missing information (Section 12.5). However, it is not the only factor because a key feature of multiple imputation analysis is to preserve the uncertainty due to missing data so that analysis results would not have inflated precision. Such uncertainty is preserved by a proper multiple imputation procedure drawing missing values from their posterior predictive distributions.

Example 12.12. *A simulation study assessing prediction accuracy for regression prediction and multiple imputation models*

We devise a simulation study to understand the property of prediction accuracy for multiple imputations. We consider two scenarios:

I. Complete-data model: $Y = -2+X+\epsilon$, where $X \sim N(1,1)$ and $\epsilon \sim N(0,1)$. We consider two imputation methods: (a) a regression prediction (RP) method (Example 2.5); (b) a multiple imputation method. Both methods assume the correct normal linear model.

II Complete-data model: $Y = -2 + X + X^2 + \epsilon$, where $X \sim N(1, 1)$ and $\epsilon \sim N(0, 1)$. We consider two multiple imputation methods: (a) a misspecified model with only the linear effect of X; (b) a correct model with both the linear and quadratic effects of X.

In both scenarios, MCAR is applied to generate around 40% of missing cases. The complete-data sample size is 1000, and the simulation includes 1000 replicates. For the multiple imputation methods applied, $M = 50$ is used. Since missing values before deletion are known in the simulation, we calculate the average of prediction error (AMPE) (i.e, the root MSE) over missing values as $AMPE = \sqrt{\frac{1}{n_{mis}} \sum_{i=1}^{n_{mis}} (y_{i,mis} - y_{i,impute})^2}$, where n_{mis} is the number of missing cases in the dataset, $y_{i,mis}$ is the value unobserved, and $y_{i,impute}$ is the imputed value. For RP, the AMPE are averaged over all 1000 replicates. For multiple imputation methods, they are also averaged over all 50 sets of imputations.

In Scenario I, the AMPE of RP is 1.003, and that of the multiple imputation is 2.018. Under this setup, the predictor error is minimized by the mean prediction of missing values based on the least-squared estimates of the model parameter. However as shown before (Chapters 2 and 3), RP is not a good imputation method in terms of statistical inference although it has a higher prediction accuracy.

In Scenario II, the AMPE based on the linear model is 6.04, and that based on the quadratic model is 2.013. The latter model is clearly a better fit by the simulation design. In this case, since both methods are multiple imputation-based, the difference on AMPE does reflect the difference of model quality.

We emphasize again that the focus of multiple imputation analysis is on the inference for the estimand of interest, not the prediction. In multiple imputation, noise (error) has to be added to model-based predictions to preserve the uncertainty of missing data. Use of the prediction accuracy may provide some insights about the model fit, yet strictly favoring the model that has the best ability of recovering the true data (i.e., the highest prediction accuracy) can be misleading (Van Buuren 2018, Section 2.6). In addition, prediction accuracy can only be used in a simulation setup while other imputation diagnostics discussed before can be readily applied to real data at hand.

12.7 Comparison among Different Missing Data Methods

In preceding sections, we discussed comparing some diagnostic statistics among multiple candidate imputation models. It is also important to compare among them in terms of the targeted analysis results (e.g., regression coefficients, standard errors, p-values, etc.). When necessary, subject-matter input can be solicited to assess which model yields results that make better scientific sense.

As is often done in practice, multiple imputation analysis results can also be compared with some ad hoc methods such as the complete-case (CC) analysis. Although the truth behind missing values is never known, under some reasonable assumptions such as MAR, it is often expected that using multiple imputation would improve the precision of the analysis results in general (i.e., smaller standard error estimates) compared with CC that discards information from partially observed subjects. Results that display opposite patterns might need some attention. In some scenarios, depending on the missingness mechanism, regression coefficient estimates of CC can be unbiased (Carpenter and Kenward 2013, Table 1.9), which provide a possible reference check for multiple imputation analysis.

It would also be beneficial to compare the multiple imputation analysis with other principled missing data methods such as the likelihood-based methods (Section 2.4). For example, when the JM strategy is used for imputation, analysis results can be compared with the direct Bayesian estimates obtained from some software packages such as WinBUGS. When the imputation is implemented for survey data, sometimes a comparison can be made with the traditional nonresponse weighting method (e.g., He et al. 2010).

Finally and again, "All models are wrong, but some are useful" (George E. P. Box). The goal of assessing and diagnosing imputation models is not to identify the correct or best imputation model per se. It is rather to aid the multiple imputation analysis using some carefully chosen and diagnosed models that are useful, making both statistical and scientific sense with some reasonable assumptions.

12.8 Summary

In this chapter we have discussed several topics related to assessing and diagnosing imputation models. Imputation diagnostics is a comprehensive process, and there might exist no single, mechanical procedure (algorithm) to follow. A few recommendations are made, spanning from the stage of formulating models to that of running analysis and checking results using completed data.

In the major methods discussed that include comparison of conditional distributions using propensity score, comparing completed data with their replicates, and monitoring the fraction of missing information, we aim to extract some numerical summaries of the quality of imputation models. These methods can be used together and used with other diagnosing techniques such as plots which were heavily used in examples from previous chapters. In addition, more research is needed to understand the property of these diagnostic methods, especially to extend them to handle multivariate missing data problems with complex data features and structure.

13

Multiple Imputation Analysis for Nonignorable Missing Data

13.1 Introduction

For most parts of the book we assume that the missingness is ignorable or MAR. However, researchers and practitioners might wonder whether this assumption always holds. If there is a strong belief against MAR in some cases, what would be the implication, and how would we use an appropriate imputation analysis strategy? Chapter 13 briefly touches on this topic. Section 13.2 provides some conceptual discussions of missingness not at random (MNAR) and its implications for statistical inference. Section 13.3 suggests using the inclusive imputation strategy as a remedy under MNAR. Section 13.4 presents several major modeling frameworks under MNAR. Section 13.5 discusses imputation analysis strategies based on these MNAR models including using sensitivity analysis. Section 13.6 provides a summary.

13.2 The Implication of Missing Not at Random

In Chapters 1 and 2 we introduced the statistical framework for describing missingness mechanisms. Here is a little recap. To put these issues in the algebraic form, note that the observed information from data Y includes both Y_{obs} (observed data elements) and R (response indicators). The joint distribution function of Y_{obs} and R together can be expressed as $f(Y_{obs}, R|\theta, \phi)$. This model is governed by parameter θ (for describing the distribution of complete data Y) and parameter ϕ (for describing the distribution of R that characterizes why certain parts of Y are missing). In general, θ is of main scientific interest and ϕ is of less interest.

The concept of ignorable missingness plays an important role in statistical inference with missing data. For multiple imputation, we would like to draw imputations from the posterior predictive distribution of missing values, given the observed data and the process that generated the missing data. That is, we draw Y_{mis} from $f(Y_{mis}|Y_{obs}, R)$. If the nonresponse is ignorable, then this

DOI: 10.1201/9780429156397-13

distribution does not depend on R, that is, $f(Y_{mis}|Y_{obs}, R) = f(Y_{mis}|Y_{obs})$. This indicates that R can be "ignored" from the right side of condition. The general implication is that

$$f(Y|Y_{obs}, R = 1) = f(Y|Y_{obs}, R = 0),$$

so the distribution of the complete data is the same in the response and nonresponse groups (e.g., Example 1.9). Thus we can set up the complete-data model from the observed data, and use this model to create imputations for the missing values. In previous chapters of the book, we simply use $f(Y_{mis}|Y_{obs})$ to characterize the posterior predictive distribution of missing values under the ignorability assumption.

In practical terms, the assumption of ignorability is essentially the belief from users that the available (observed) data Y_{obs} are sufficient to correct for the effects of missing data. This assumption cannot be tested on the data itself, but it might be pondered using suitable external validation data or subject-matter knowledge. On the other hand, if we strongly believe that the ignorability does not hold, that is, $f(Y|Y_{obs}, R = 1) \neq f(Y|Y_{obs}, R = 0)$, then we need to draw imputations from $f(Y_{mis}|Y_{obs}, R)$ instead of $f(Y_{mis}|Y_{obs})$. This is typically much harder because we need to consider and model R in addition to Y_{obs}.

Example 13.1. *A normal linear regression with either the outcome or predictor missing*

To give an example of how R might impact missing data estimation, we consider a normal linear regression for response Y with one predictor X:

$$Y = \beta_0 + \beta_1 X + \epsilon, \tag{13.1}$$

where $\epsilon \sim N(0, \sigma^2)$. First suppose some of the Y's are missing and X is fully observed. In Example 2.4, which describes the problem by assuming X and Y follow a bivariate normal distribution, we show that if the missingness of Y is MAR, then using complete cases can obtain unbiased regression coefficients of β_0 and β_1. Let R be the response indicator of Y; this fact can also be generally demonstrated by noting that $f(Y|X, R = 1) = \frac{f(Y,X,R=1)}{f(X,R=1)} = \frac{f(R=1|Y,X)f(Y,X)}{f(R=1|X)f(X)} = \frac{f(R=1|Y,X)}{f(R=1|X)} f(Y|X)$. When $f(R = 1|Y, X) = f(R = 1|X)$, that is, the missingness of Y is MAR (including MCAR as a special case), then $f(Y|X, R = 1) = f(Y|X)$, implying that using complete cases can obtain the same conditional distribution (i.e, the regression coefficients) as the complete data. When $f(R = 1|Y, X) \neq f(R = 1|X)$, that is, the missingness of Y is MNAR, then $f(Y|X, R = 1) \neq f(Y|X)$ in general, implying that using complete cases would obtain biased regression coefficients.

On the other hand, suppose Y is fully observed and X has some missing values. Let R be the response indicator of X, and this becomes a missing covariate problem in a regression. By the same argument, when $f(R = 1|Y, X) = f(R = 1|X)$, then using complete cases can obtain unbiased regression coefficients.

However, here this condition means that the missingness of X is MNAR because X is not fully observed. Yet when $f(R = 1|Y, X) \neq f(R = 1|X)$, the coefficient estimates from complete cases would be biased. This condition includes both (a) MNAR where R is related to both Y and X and (b) MAR (excluding MCAR) where R is only related to Y (e.g., Examples 2.5 and 2.6). In summary, as long as the missingness of X is not related to Y, then the complete-case analysis can yield unbiased regression coefficients. See similar arguments in Section 4.5.

Little and Rubin (2020, Example 3.3) provided additional comments on the behavior of complete-case analysis with missing covariates in regression. Carpenter and Kenward (2013, Table 1.9) documented the pattern of biases of complete-case analysis in both linear and logistic regressions with two predictors, where missingness can happen either in the outcome or regressors under different mechanisms. These research showed that in general, the impact of MNAR on missing data analysis is complex and it is difficult to identify a systematic pattern if there exists any.

13.3 Using Inclusive Imputation Strategy to Rescue

If the nonresponse mechanism is believed to be nonignorable, then the first strategy is to expand the imputation model by including possibly more variables so that the ignorability assumption might be more plausible. This follows the principle of the inclusive imputation strategy. In Example 1.10, we showed that by including more variables, the nonresponse mechanism for the income variable in NHIS 2016 seems to be less related to its own values. When the ignorability assumption holds well, then estimates from a well specified imputation model would have nice properties as we have shown throughout the book.

On the other hand, even if variables are not related to the missingness mechanism and yet related to the missing variable, including them in the imputation model would still improve the estimates. This is shown in Example 5.8 under an MAR mechanism. Similar patterns would hold for MNAR scenarios, as will be shown in the following example.

Example 13.2. *A simulation study to assess the performance of the inclusive imputation strategy under MNAR*

We use the same complete-data model as in Example 5.8. Suppose the complete-data analysis includes both Y and X, where $Y = -2 + X + \epsilon$, where $X \sim N(1, 1)$ and $\epsilon \sim N(0, 1)$. We also consider an auxiliary variable Z, which is not relevant to the analysis yet is correlated with Y: $Z = 2 + Y + \epsilon$, where $\epsilon \sim N(0, 1)$. We consider two MNAR mechanisms: (a): $logit(f(R = 0)) = -1.4 - Y$; (b) $logit(f(R = 0)) = -1.4 - 0.75Y + 0.25X$. Under both mechanisms, smaller Y-values (or those with larger X-values) are more likely to be missing, and the missingness rate is around 40%. We intentionally do not let R be related to Z.

As in Example 5.8, we consider two imputation methods based on $M = 50$ imputations. One is the inclusive approach, imputing Y based on a linear model including both X and Z as predictors. The other is the exclusive (or restrictive) approach, imputing Y based on a linear model only including X as the predictor. The estimands of interest include the mean of Y, the slope coefficient for regressing Y on X, and the slope coefficient for regressing X on Y. The complete-data sample size is 1000, and the simulation includes 1000 replicates.

Table 13.1 shows the simulation results. When the missingness mechanism is only related to Y, complete-case (CC) analysis has a large bias for estimating the mean of Y. Using the exclusive imputation reduces the relative bias from around −62% to −36%. By including Z, the inclusive imputation further reduces the relative bias to around −19%. However, the bias cannot be completely removed even by using the inclusive imputation method. For estimating the slope coefficient of regressing Y on X, there is little difference between CC and exclusive imputation method, both of which have biased estimates. The similarity between the two methods is due to the fact that the imputation is based on the model fitting for complete cases. Again, including Z in the imputation reduces the bias and improves the performance, reducing the MSE and increasing the coverage rate. For estimating the regression coefficient of X on Y, CC yields little bias because the missingness of Y is only depending on Y (here Y is a predictor in the regression). Estimates from both imputation methods have moderate biases and are worse than CC, although the inclusive imputation is still somewhat better than the exclusive imputation.

When the missingness is related to both Y and X, the comparative pattern is generally similar for all methods. Since the missingness mechanism is partially related to X, both imputation methods perform better than the scenario where the missingness mechanism is only related to Y. The imputation model including both X and Z still maintains some advantages over that only including X. Note that here since the missingness is not solely related to Y, CC has some bias (RBIAS=−3%) when regressing X on Y.

13.4 Missing Not at Random Models

Even after trying the inclusive imputation strategy, there might still exist certain cases that the ignorability assumption is doubtful. Some typical scenarios are summarized in Van Buuren (2018), and we state them as follows.

(A) Important variables that govern the missing data mechanisms are not

TABLE 13.1
Example 13.2. Simulation results

MNAR depends on Y only						
Method	RBIAS (%)	SD	$\overline{\text{SE}}$	MSE	CI	COV (%)
Mean of Y						
BD	0	0.043	0.045	0.00188	0.175	96.1
CC	−62.3	0.053	0.051	0.3898	0.200	0
Exclusive MI	−35.7	0.049	0.048	0.1295	0.202	0
Inclusive MI	−19.1	0.047	0.047	0.0385	0.186	1.6
Slope coefficient of regressing Y on X						
BD	0	0.032	0.032	0.000995	0.124	95.0
CC	−14.4	0.042	0.041	0.02239	0.162	6.2
Exclusive MI	−14.1	0.042	0.041	0.02172	0.164	7.4
Inclusive MI	−7.7	0.037	0.037	0.00737	0.146	44.4
Slope coefficient of regressing X on Y						
BD	0	0.016	0.016	0.000251	0.062	95.0
CC	0	0.024	0.024	0.000597	0.095	95.2
Exclusive MI	7.3	0.021	0.021	0.00178	0.083	58.5
Inclusive MI	6.1	0.020	0.020	0.00131	0.078	67.5
MNAR depends on both Y and X						
Mean of Y						
CC	−44.2	0.055	0.054	0.2132	0.213	0
Exclusive MI	−27.9	0.049	0.049	0.080	0.194	0
Inclusive MI	−14.5	0.046	0.048	0.023	0.187	14.1
Slope coefficient of regressing Y on X						
CC	−6.6	0.041	0.041	0.00607	0.157	62.1
Exclusive MI	−6.4	0.041	0.040	0.00571	0.160	65.4
Inclusive MI	−3.5	0.037	0.036	0.00254	0.143	84.3
Slope coefficient of regressing X on Y						
CC	3.0	0.023	0.022	0.000745	0.087	89.1
Exclusive MI	5.0	0.020	0.019	0.00103	0.076	72.8
Inclusive MI	4.1	0.019	0.019	0.000767	0.074	82.0

available. For example, in clinical trial settings, patients can drop out either because they feel their symptoms have improved, remain the same, or become even worse. However, the symptoms (or related trial outcomes) for the drop-outs are not observed. Another example is the income nonresponse problem in surveys: subjects with a certain level of income might be less likely to respond to income questions.

(B) There is reason to believe that respondents differ from nonrespondents, even after accounting for the observed information. In clinical trials, some factors can be included to mitigate the nonignorability, such as age and patients' comorbidity at baseline. For the income nonresponse problem, for example, we might include survey participants' known demographic variables to reduce the extent of the nonignorability. Yet how much these variables can help is typically unknown.

(C) The data are truncated such as variables measured under the detection limits. Some of the related discussion can be found in Section 11.3.1.5.

In these scenarios, since using the inclusive imputation strategy based on an MAR model might not fully mitigate the problem of MNAR (Example 13.2), it might be helpful to explore and analyze the data using some models under the nonignorable missingness assumption. For illustration, we focus on one incomplete variable Y and use R to denote its response indicator. The joint distribution of Y and R is easy to process if we decompose it into separate components. There exist two main strategies to decompose $f(Y, R)$: one is based on the selection model, and the other is based on the pattern mixture model (Rubin 1987). In addition, we can postulate some latent variables that are shared between Y and R to induce a correlation between the two, leading to shared parameter models. This section briefly describes these modeling strategies.

13.4.1 Selection Models

The selection model decomposes the joint distribution $f(Y, R)$ as

$$f(Y, R) = f(Y)f(R|Y).$$

In brief, the selection model weights the marginal distribution $f(Y)$ in the population with the response weights (propensity), $f(R|Y)$. Both $f(Y)$ and $f(R|Y)$ need to be specified. A well-known selection model where $f(Y)$ is normal and $f(R|Y)$ is a probit regression model is the Heckman selection model, first proposed by Heckman (1976) in the econometrics literature. To describe it specifically, we need to introduce a standard normal latent variable R^*: $R = 1$ if $R^* > 0$ and $R = 0$ if $R^* <= 0$. In addition, suppose we have fully observed covariates X and W in addition to the missing variable; then the Heckman selection model can be expressed as

$$\begin{aligned} y_i &= x_i\beta + \sigma\epsilon_{1i}, \\ r_i^* &= w_i\gamma + \epsilon_{2i}, \end{aligned} \tag{13.2}$$

where x_i and w_i are the i-th row of the X- and W-covariate matrix, respectively, for $i = 1, \ldots n$, and $\begin{pmatrix} \epsilon_{1i} \\ \epsilon_{2i} \end{pmatrix} \sim N\left\{\begin{pmatrix} 0 \\ 0 \end{pmatrix}, \begin{pmatrix} 1 & \rho \\ \rho & 1 \end{pmatrix}\right\}$.

In Eq. (13.2), r_i^*, a standard normal latent variable, is fully unobserved; yet $r_i = I(R_i^* > 0)$ is the observed response indicator for subject i. That is, we only observe y_i for $r_i = 1$ (or $r_i^* > 0$). We assume fully observed covariate X

and W for the outcome and response models, respectively. Eq. (13.2) implies nonignorable missingness when $\rho \neq 0$ and ignorable missingness when $\rho = 0$.

Example 13.3. *Distributions of observed data under a univariate Heckman selection model*

To illustrate the idea of the Heckman selection model, we consider a simplified case of Model (13.2). We assume no covariate is available and set $x_i\beta$ as μ_β and $w_i\gamma$ as μ_γ, which is the mean of Y and R^*, respectively. We simulate some data from this univariate selection model by setting $\mu_\beta = 0$, $\mu_\gamma = 0$ (around 50% of missing cases), and $\sigma = 1$. We assign a range of values for ρ ($\rho = 0, 0.2, 0.4, 0.6, 0.8$) and generate Y's with sample size 10000. Fig. 13.1 shows histograms of Y_{obs} and their QQ plots. When $\rho = 0$, Y_{obs} is essentially a simple random sample of the original Y. The mean of Y_{obs} (highlighted by the vertical bar in the histogram) is essentially 0. As ρ increases from 0.2 to 0.8, the mean of Y_{obs} moves away from 0 towards the positive direction. When $\rho = 0.8$, $\overline{Y}_{obs}$ from the simulated sample is 0.647, considerably larger than 0. This pattern is expected because a positive ρ-value determines that larger values of Y are more likely to be observed. In addition, the distribution of Y_{obs} is also gradually deviating from the normality by reviewing the histograms and QQ plots. The nonnormality of Y_{obs} is the most apparent when $\rho = 0.8$.

From a historical perspective, Heckman (1976) first proposed the selection modeling framework to model the selection of women into the labor force. Then this approach and various extensions have been widely used in econometrics and other fields. Note that for a univariate missing data problem in Example 13.3, the three parameters μ_β, μ_γ, and ρ are not identifiable based on Y_{obs} alone. Therefore covariates X and W have to be included for identifying the model parameters including β, γ, and ρ in Eq. (13.2). Several estimation methods have been proposed to solve the Eq. (13.2), such as the Heckman's two-step approach and maximum likelihood approach. In practice, a frequently used constraint for avoiding collinearity issues and stabilizing the estimation is the inclusion of different sets of covariates for the outcome and response equations, intentionally denoted by X and W in Eq. (13.2). This is the so-called exclusion-restriction rule (Puhani 2000).

There exist other ways for specifying selection models. For example, the data-generating model for Y and R in Example 13.2 can be generalized as

$$\begin{aligned} Y &= X\beta + \epsilon, \\ logit(f(R=0)) &= \gamma_0 + \gamma_Y Y + \gamma_X X, \end{aligned} \qquad (13.3)$$

where $\gamma_Y \neq 0$ indicates an MNAR mechanism via a logistic link function. Note that in Eq. (13.2), the equation for R is defined through a probit link function. Suppose X includes the fully observed baseline variables and Y is the follow-up measurement; Model (13.3) had been used for modeling nonrandom dropouts in longitudinal data (e.g., Diggle and Kenward 1994).

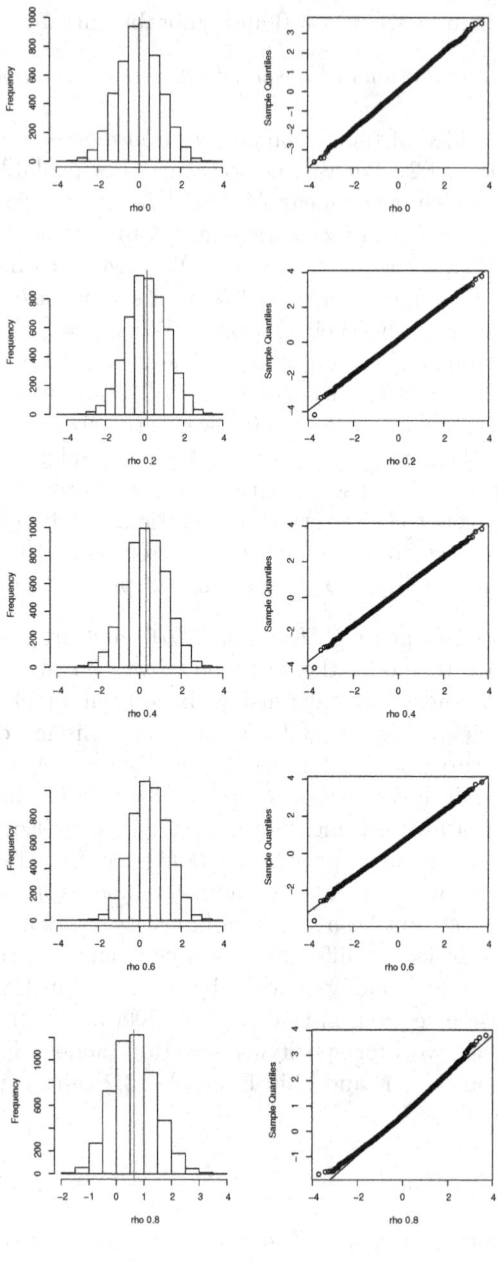

FIGURE 13.1
Example 13.3. Left: the histogram of Y_{obs}; right: the QQ plot of Y_{obs}. Data are generated by setting $\rho = (0, 0.2, 0.4., 0.6, 0.8)$ from a univariate Heckman selection model. Vertical bar: the mean of Y_{obs}.

13.4.2 Pattern Mixture Models

An alternative factorization of $f(Y, R)$ leads to pattern mixture models as follows:

$$f(Y, R) = f(Y|R)f(R) = f(Y|R = 1)f(R = 1) + f(Y|R = 0)f(R = 0).$$

The pattern mixture model emphasizes that the combined distribution of Y and R is a mixture of the distribution of Y in the respondents and nonrespondents. Assuming $f(R = 1)$ is known (more generally depending on covariates such as $f(R = 1|X)$), the pattern mixture model needs a specification of the distribution $f(Y|R = 1)$ for the respondents, which can be conveniently modeled using observed data. It also needs a specification of the distribution $f(Y|R = 0)$ for the nonrespondents, for which we do not have data at all and thus can only speculate.

Under the pattern mixture model, although there is essentially no observed data for us to specify the model $f(Y|R = 0)$, it is often helpful to employ a model specification that can be related to the model for observed data, $f(Y|R = 1)$. This relation can help us identify the joint model or run some sensitivity analysis (Section 13.5.2).

Example 13.4. *A simple pattern mixture model for univariate missing data problems*

In Example 3.1, we considered a univariate missing data problem assuming MCAR. Specifically, we assume that $y_{obs,i} \sim N(\mu_{obs}, \sigma^2_{obs})$ for the observed cases $i = 1, \ldots, n_{obs}$. In the pattern mixture modeling framework, this is the model $f(Y|R = 1)$. To complete the full model, we need to specify $f(Y|R = 0)$ and $f(R = 1)$. The latter is straightforward as we can assume $f(R = 1) \sim Bernoulli(p)$. For $f(Y|R = 0)$, we can assume that $y_{mis,i} \sim N(\mu_{mis}, \sigma^2_{mis})$ so that the two sub-models both take the form of normal distributions. However, based on Y_{obs} alone, there is no data for us to infer about μ_{mis} and σ^2_{mis} unless we relate μ_{mis} and σ^2_{mis} with μ_{obs} and σ^2_{obs} in some ways. In addition, we are not limited by specifying $f(Y|R = 0)$ as a normal distribution because in theory $f(Y|R = 0)$ can take any type of distributions such as a uniform distribution or t-distribution. Therefore, specifying $f(Y|R = 0)$ in this case is purely speculation. This simple example also implies that models for nonignorable missing data in general can suffer from identifiability issues due to the lack of information from $f(Y|R = 0)$.

The selection model and pattern mixture model are based on two approaches to factorizing $f(Y, R)$. By the Bayes rule, they can be technically converted to each other (Van Buuren 2018, Section 3.8.4). However, sometimes the conversion does not have a neat form, which is demonstrated in the following example.

Example 13.5. *Conversion between selection and pattern mixture models*

We first consider a simple selection model used in Example 13.3. In that model, $f(Y) \sim N(\mu_\beta, \sigma^2)$, which is the marginal distribution of the complete data. By the Bayes rule, we have the distribution of missing values, $f(Y|R = 0) = f(Y|R^* < 0) = f(Y)\frac{f(R^*<0|Y)}{f(R^*<0)}$. After some algebra, we obtain

$$f(Y|R=0) = f(Y)\frac{\Phi(\frac{-\mu_\gamma - \rho\frac{Y-\mu_\beta}{\sigma}}{\sqrt{1-\rho^2}})}{1-\Phi(\mu_\gamma)},$$

which is not a normal distribution when $\rho \neq 0$ because the factor $\Phi(\frac{-\mu_\gamma - \rho\frac{Y-\mu_\beta}{\sigma}}{\sqrt{1-\rho^2}})$ involves Y. Similarly, the distribution of observed values, $f(Y|R = 1)$, is not a normal distribution either when $\rho \neq 0$. When $\rho = 0$ (MCAR), it is obvious that both $f(Y|R = 0)$ and $f(Y|R = 1)$ have the same normal distribution as $f(Y)$.

If we have a normal pattern mixture model as is used in Example 13.4, then it can be shown that under this model in general $f(R = 1|Y)$ is a logistic function with a linear term of Y and a quadratic term of Y^2. This is reduced to a logistic function with only a linear term on Y if $\sigma^2_{obs} = \sigma^2_{mis}$. On the other hand, under the simplified selection model in Example 13.3, we have $f(R = 1|Y)$ as a probit function with a linear term of Y (i.e., a probit regression model). Therefore, the selection model in Example 13.3 and the pattern mixture model in Example 13.4 are not generally equivalent.

In more general cases, Kaciroti and Raghunathan (2014) provided the selection modeling parameterizations for a variety of the pattern mixture models for different outcomes in the exponential family of distributions.

Pattern mixture models are generally not identifiable. So why can selection models (e.g., Eq. (13.2)) be estimated? This is because they have some hidden restrictions for the parameters. To see that, we note that Model (13.2) implies

$$E(y_i|r_i = 1) = x_i\beta + \frac{\phi(w_i\gamma)}{\Phi(w_i\gamma)}\rho\sigma, \quad (13.4)$$

$$E(y_i|r_i = 0) = x_i\beta + \frac{-\phi(w_i\gamma)}{1-\Phi(w_i\gamma)}\rho\sigma, \quad (13.5)$$

where $\phi(\cdot)$ and $\Phi(\cdot)$ are the probability density function and cumulative distribution function of the standard normal distribution, respectively. At first sight, there is an algebraic connection between $E(Y|R = 1)$ and $E(Y|R = 0)$ based on Eqs. (13.4) and (13.5). However, as we emphasized in Section 13.4.2, under nonignorable missingness, in general there shall exist no information from $f(Y|R = 1)$ helping us identify $f(Y|R = 0)$ in the framework of pattern mixture modeling. Here the connection between the two components, that is, the hidden restriction which makes the selection model estimable, is primarily due to the normality assumption we impose for $f(Y|X)$ and $f(R^*|W)$. As a result, the estimates can be rather sensitive to the distributional assumptions, such as the normality and symmetry of the error terms. Such limitations have been well documented, for example, in Little and Rubin (2020, Chapter 15) and Rubin (1987, Example 6.2). More generally, identifiable parametric selection models are due to some hidden restrictions of parameters (Molenberghs and Kenward 2007, Section 16.3).

13.4.3 Shared Parameter Models

The third framework uses unobservable latent variables to link missingness with potentially missing data. In a shared parameter model, the joint distribution of the data is expressed in terms of latent variables, latent classes, or random effects (denoted by b). Therefore, the application of shared parameter models is mostly seen in longitudinal data, clustered data, or survival analysis. In the presence of such unobserved b in both the models for Y and R, the resulting missingness mechanism is in general MNAR, yet the form of this mechanism is not as obvious as in typical selection models or pattern mixture models.

For example, we can write the joint distribution of Y, R, and b as

$$f(Y, R, b|X, Z, \theta, \phi, \delta), \tag{13.6}$$

where Z denotes the covariates corresponding to random effects (or latent variables); b and δ denotes the associated parameter.

Eq. (13.6) has the selection model factorization:

$$f(Y, R, b|X, Z, \theta, \phi, \delta) = f(Y|X, b, \theta)f(R|Y, b, X, \phi)f(b|Z, \delta),$$

and the pattern mixture model factorization:

$$f(Y, R, b|X, Z, \theta, \phi, \delta) = f(Y|R, b, X, \theta)f(R|b, X, \phi)f(b|Z, \delta).$$

In these factorizations, $f(b|Z, \delta)$ models the distributions of random effects b given covariates Z. An important simplification arises when Y and R are assumed to be independent conditioning on the random effects b. We then obtain the shared parameter decomposition as

$$f(Y, R, b|X, Z, \theta, \phi, \delta) = f(Y|X, b, \theta)f(R|b, X, \phi)f(b|Z, \delta).$$

Example 13.6. *A simple shared parameter model for clustered data*

Consider clustered data y_{ij} for $i = 1, \ldots, m$ and $j = 1, \ldots n_i$, where i indexes groups/clusters and j indexes the repeated observation within each cluster. In Example 9.1 we considered a one-way random effects model for y_{ij}'s under a MCAR mechanism. We can also consider a shared parameter model as

$$y_{ij} = \mu + b_i + \epsilon_{ij}, \tag{13.7}$$

$$logit(f(r_{ij} = 0)) = \beta_0 + \beta_1 b_i, \tag{13.8}$$

where $b_i \sim N(0, \tau^2)$ and $\epsilon_{ij} \sim N(0, \sigma^2)$. Eq. (13.7) describes the complete data using a random effects model where b_i's are the random effects. Eq. (13.8) describes the nonresponse mechanism through a logistic mixed model, which shares the same random effects, b_i's, with the outcome Y's. Since random effects are unobserved and also follow a distribution, missing data in Y's are MNAR. See also Little and Rubin (2020, Example 6.25).

Although shared parameter models are not discussed much in this book, they are widely used in real applications. For example, an application of shared parameter models for small area estimation from survey data can be found in Malec et al. (1999).

13.5 Analysis Strategies

For nonignorable missing data, intuitively we might be able to incorporate R into the model to create imputations and conduct imputation analysis. That is, we directly draw Y_{mis} from $f(Y_{mis}|Y_{obs}, R)$ based on some aforementioned MNAR models. However, nonignorable missing data models can generally suffer from model identifiability problems. This direct imputation strategy might render the imputation inferences sensitive to the modeling assumptions. An alternative strategy is to adapt the imputed data (based on an ignorability assumption) to make them more realistic, that is, more general than the unverifiable MAR assumption. Since such adaptations are based on unverifiable assumptions as well, it is recommended to carefully assess the impact of different possibilities on the final inferences by means of sensitivity analysis (Rubin 1987).

13.5.1 Direct Imputation

A straightforward strategy is to impute missing data using some seemingly identifiable selection models such as Eqs. (13.2) and (13.3). In essence this idea is no different from the multiple imputation methods based on ignorable missing data models. The key is to derive or approximate the posterior predictive distribution of missing values given observed information, $f(Y_{mis}|Y_{obs}, R)$.

Unlike the multiple imputation analysis under MAR, there is a lack of software packages that are designed for direct imputation using MNAR models. In many cases, practitioners need to posit their own speculations about the departure from MAR models and write the code. Since WinBUGS has been used to model nonignorable missing data in the literature (e.g., Mason et al. 2012a; 2012b), we use WinBUGS to demonstrate the direct imputation strategy.

Example 13.7. *A simulation study for the direct imputation based on a selection model*

We use the setup in Example 13.2, which follows the selection model in Eq. (13.3), to generate missing data under MNAR. The specified imputation model matches with this selection model. In the scenario where the missingness of Y is only related to Y, we only keep the regressor Y in the response equation for R in the imputation model. The imputation is implemented using WinBUGS. However, we find out that in the scenario where the missingness of Y is related to both Y and X, the posterior samples of model parameters have some convergence problems, which result in suboptimal multiple imputation estimates. We suspect this might be due to the fact that selection models can typically have identifiability issues (Section 13.4.2). However, there is no apparent convergence problem when the missingness is only related to Y.

Fig. 13.2 shows the trace/history plots of the parameter draws of the selection model from the imputation algorithm using one simulation replicate. The chain was running for 20000 iterations including a burn-in period of 10000 iterations and a thinning interval of 1 iteration. When the missingness is only related to Y (the top row), posterior draws of parameters appear to converge well. However, when the missingness is related to both Y and X (the bottom row), the convergence behavior of parameters is questionable. This is especially true for the γ's: the posterior means are not close to the true values used in the simulation. And parameter draws are more volatile than those from the model in which the missingness is only related to Y.

Table 13.2 shows the multiple imputation analysis results, which are based on $M = 50$ imputations obtained by selecting every 200 iterations after a burn-in period of 10000 iterations. They can be compared with the results shown in Table 13.1. When the missingness of Y is only related to Y, the imputation analysis results are good, yielding little bias and coverage rates around the nominal level for all estimands of interest. Note that for estimating the slope coefficient of regressing X on Y, the results are even more efficient than those from CC in Table 13.1. However, when the missingness of Y is related to both X and Y, the multiple imputation analysis results are not good based on the simulation, showing considerable biases and rather large standard deviations for all the estimands of interest. The coverage rates are lower than the nominal level. Therefore, the problem of nonconvergence of the imputation algorithm results in suboptimal multiple imputation inferences.

Sample code of the WinBUGS program is included. When the missingness of Y depends on both Y and X, we have tried a few strategies including standardizing X and/or Y, as well as specifying more informative prior distributions. However, these efforts do not appear to improve the convergence significantly. In the selection models, since Y is totally missing when $R = 0$, it is expected that the estimation of $logit(f(R = 0)) = \gamma_0+\gamma_1 Y+\gamma_2 X$ can be difficult. In addition, it might be difficult to separate the effect of X from Y in the missingness model. To see that, if we plug in $Y = \beta_0+\beta_1 X+\epsilon$ into the missingness model, then it becomes $logit^{-1}(f(R = 0)) = \gamma_0 + \beta_0\gamma_1 + \beta_1\gamma_1 X + \gamma_2 X + \gamma_1\epsilon$.

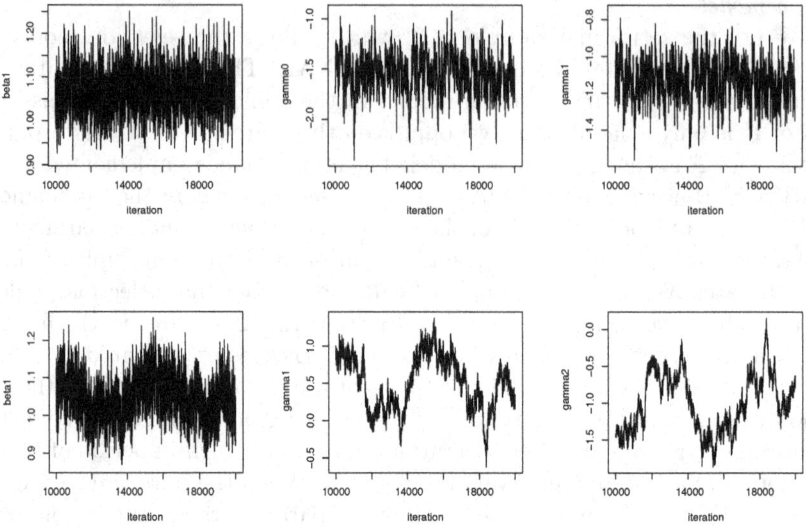

FIGURE 13.2
Example 13.7. History plots of some selected selection model parameters against iterations in the imputation algorithm. Top row: β_1, γ_0, and γ_1 from the model where the missingness is only related to Y; bottom row: β_1, γ_1, and γ_2 from the model where the missingness is related to both Y and X. Labels of parameters follow those specified in the WinBUGS program. Posterior estimates are chosen from one simulation replicate.

Here the parameters, β's and γ's, are convoluted together in this missingness model. As a result, estimates for β's might be quite sensitive to those of γ's, yet the former plays a leading role for imputing missing Y-values. In this simulation example, the missingness rate is around 40%, which is not low and might also contribute to the difficulty of estimation.

```
for( i in 1 : M ){
# outcome model
y_miss[i] ~ dnorm(mu[i], tau)
mu[i] <- beta0+beta1*x[i]
# missingness model
m[i] ~ dbern(p[i])
# missingness depends only on y_miss;
logit(p[i]) <-gamma0+gamma1*y_miss[i]
# missingness depends on both y_miss and x
logit(p[i]) <-gamma0+gamma1*y_miss[i]+gamma2*x[i]}
```

There might exist delicate statistical techniques for stabilizing and improving the estimation. We do not explore it further. However, the main point here is that this seemingly simple selection model can encounter practical difficulties using a conventional Bayesian imputation strategy.

Turning back to the Heckman selection model in Eq. (13.2), Galimard et al. (2016) proposed an imputation algorithm which is not fully Bayesian. This algorithm is based on the Heckman's two-step estimator and its variance correction. In their simulation study, the proposed imputation method showed advantages over complete-case analysis and the imputation method simply assuming ignorable missing data. Galimard (2018) extended the imputation method to binary missing data assuming nonignorable missingness modeled by a selection modeling framework. Ogundimu and Collins (2019) explored the imputation method based on the maximum likelihood estimator of the selection model, which is supposedly more efficient than the two-step estimator used in Galimard et al. (2016). In addition, they considered an extension of Model (13.2) in which ϵ_{1i}'s and ϵ_{2i}'s follow a more general t-distribution. Such an extension was demonstrated to be more robust against the absence of the exclusion restriction in their simulation study.

Example 13.8. *Bayesian multiple imputation using Heckman selection model for missing HbA1C data*

The fully Bayesian imputation algorithm for Model (13.2) is complicated to derive. This is because R^* is a latent variable and has to be simulated

TABLE 13.2
Example 13.7. Simulation results

	MNAR depends on Y only					
Method	RBIAS (%)	SD	$\overline{\text{SE}}$	MSE	CI	COV (%)
	Mean of Y					
Selection model	0.9	0.066	0.065	0.0044	0.257	94.6
	Slope coefficient of regressing Y on X					
Selection model	0.2	0.044	0.043	0.00196	0.171	94.4
	Slope coefficient of regressing X on Y					
Selection model	−0.4	0.022	0.022	0.00047	0.085	95.0
	MNAR depends on both Y and X					
	Mean of Y					
Selection model	−23.4	0.31	0.30	0.1532	1.276	81.3
	Slope coefficient of regressing Y on X					
Selection model	−5.6	0.083	0.083	0.01013	0.334	85.9
	Slope coefficient of regressing X on Y					
Selection model	−6.7	0.032	0.042	0.00213	0.170	92.0

from its posterior distribution in the imputation algorithm. In addition, in Examples 13.3 and 13.5 we already show that $f(Y|R^* < 0)$ is not a normal distribution so that specific simulation techniques (e.g., rejection sampling) might need to be employed. With the help of WinBUGS, a Bayesian multiple imputation under Model (13.2) can be conducted.

We illustrate the method and analysis using a clinical dataset measuring patients' HbA1C (glycated haemoglobin) level (Hsu et al. 2020). This illustrative dataset consists of patients who had high-grade carotid artery stenosis and were scheduled to undergo carotid artery interventions at the Veterans Affairs Palo Alto Health Care System. Both clinical literature and subject-matter opinions suggest that the pre-operative HbA1c level can be highly associated with post-operative complications, so it may be important to have information on a pre-operative HbA1c level (e.g., their mean estimate) in order to assess necessary post-operative interventional care. Additional information about the patient selection and clinical background can be found in Zhou et al. (2012).

The dataset consists of 180 patients, 50 (27.8%) of whom had missing pre-operative HbA1c values. Fully observed covariates include age (mean=69.9 yrs; sd=7.6), diabetic status (yes/no; 40.6% yes), body mass index (BMI, mean=28.6; sd=5.3), and white (yes/no; 83.3% yes). We conduct some exploratory analyses. The distribution of HbA1c is skewed to the right, and we apply a log transformation to make the distribution close to normality (top row of Fig. 13.3). Based on the observed cases ($n = 130$), we apply a normal linear regression to identify the variables predictive of log(HbA1c) values. Using all the cases, we apply a probit regression on the response/selection indicator to identify the variables associated with the selection probability. Here the use of probit-link, instead of logit-link, is to facilitate an easy comparison with the fit of the Heckman selection model.

Table 13.3 (under the column "Observed data") shows the exploratory analysis results. Based on the linear regression, age is negatively associated with log(HbA1c) with a marginally significant effect (p-values not shown), and diabetic status is positively associated with log(HbA1c) with a significant effect. This is consistent with some simple statistics (e.g., the mean of HbA1c is 7.38 and 5.74 for diabetic and nondiabetic patients, respectively). Based on the probit regression of the response indicator, only diabetic status is significantly predictive of the selection probability. This is also consistent with some simple statistics (e.g., the proportion of diabetic patients is 46% and 26% among respondents and nonrespondents, respectively). This simple analysis suggests that nondiabetic patients with a normal (i.e., lower) HbA1c level were more likely to have missing HbA1c values than diabetic patients with an abnormal (i.e., higher) HbA1c level.

We then fit a Heckman selection model to the data including the covariates using R sampleSelection. Table 13.3 (under the column "Selection model") shows the model estimates from the maximum likelihood estimation. For the outcome model (i.e., predicting log(HbA1c)): the age effect is still marginally significant (p-values not shown), the diabetic status is still a significant factor, and BMI becomes marginally significant with a positive association. For the response model: the diabetic status is still a significant factor. More importantly, the fitted selection model also produces a highly significant ρ estimate of 0.958, showing some evidence that the missingness is MNAR after controlling for these covariates. That is, patients with lower HbA1c levels are more likely to be missing after controlling for the covariates. In this example, the same set of covariates is used in both the outcome and response models, which apparently does not cause any estimation problem.

TABLE 13.3
Example 13.8. Outcome and response model estimates

	Observed data		Selection model		Bayesian estimation	
Parameter	Outcome	Response	Outcome	Response	Outcome	Response
Intercept	1.852	0.199	1.742	0.216	1.738	0.220
	(0.131)	(1.157)	(0.150)	(1.106)	(0.158)	(1.132)
Age	−0.00278	−0.00472	−0.00297	−0.00506	−0.00300	−0.00422
	(0.00148)	(0.0133)	(0.00171)	(0.0133)	(0.00174)	(0.0133)
BMI	0.00283	0.0245	0.00482	0.0161	0.00490	0.0158
	(0.00226)	(0.0208)	(0.00257)	(0.0190)	(0.00267)	(0.0193)
Diabetes	0.231	0.461	0.253	0.741	0.255	0.720
	(0.0234)	(0.217)	(0.0269)	(0.202)	(0.0274)	(0.194)
White	0.00792	−0.171	−0.00513	−0.0638	−0.00476	−0.109
	(0.0304)	(0.288)	(0.0351)	(0.263)	(0.0360)	(0.257)
σ			0.163		0.0510	
			(0.0119)		(0.012)	
ρ				0.958		0.947
				(0.0218)		(0.0301)

Note: Regression coefficient estimates (standard errors in parentheses) are shown. For Bayesian estimation, the posterior means (posterior standard deviations in parentheses) are shown. The posterior estimates are based on 1000 iterations of the MCMC chain, selected from every 100 iterations after a burn-in period of 5000 iterations.

We implement a Bayesian imputation algorithm under the Heckman selection model, which can simultaneously impute missing log(HbA1c) and provide posterior draws of the model parameters. Diffuse prior distributions are used for model parameters. Table 13.3 (under the column "Bayesian estimation") shows the model estimates, which are based upon 1000 posterior draws sampled per 100 iterations after a burn-in period of 5000 iterations. Estimates for covariate coefficients as well as for ρ are similar to the maximum likelihood estimates. The only difference appears to be in σ, the standard deviation of the residuals from the outcome model. The Bayesian estimate is considerably smaller than that from the maximum likelihood estimate and seems to be rather consistent after trying several different prior distributions for σ.

The sample code for the WinBUGS program is included. We standardize covariates to stabilize the estimation. Unlike in Example 13.7, posterior draws of model parameters converge well here. A key strategy in coding is to decompose $f(Y, R^*)$ as $f(R^*)f(Y|R^*)$. Both distributions are normal and imply linear regression models, which can be easily specified in WinBUGS. Note that the regression coefficient for the linear model from $f(Y|R^*)$ is $\sigma\rho$, and the variance is $\sigma^2(1-\rho^2)$. On the other hand, decomposing $f(Y, R^*)$ as $f(Y)f(R^*|Y)$ is not simple to specify in WinBUGS.

```
model
{
for(i in 1:n)
{
k[i] <-1-2*ma1c[i]
c0[i] <-0
c1[i] <-10000
lower[i]<-min(k[i]*c0[i], k[i]*c1[i])
upper[i] <- max(k[i]*c0[i], k[i]*c1[i])
# nonresponse model
ys[i]~ dnorm(mus[i], 1)I(lower[i],upper[i])
mus[i] <-
b[1]+b[2]*(Age[i]-mean(Age[]))/sd(Age[])+
b[3]*(Diabetes[i]-mean(Diabetes[]))/sd(Diabetes[])
+b[4]*(BMI[i]-mean(BMI[]))/sd(BMI[])+
b[5]*(white[i]-mean(white[]))/sd(white[])
mu[i] <- a[1]+a[2]*(Age[i] -mean(Age[]))/sd(Age[])+
a[3]*(Diabetes[i] -mean(Diabetes[]))/sd(Diabetes[])+
a[4]*(BMI[i]-mean(BMI[]))/sd(BMI[])
+a[5]*(white[i]-mean(white[]))/sd(white[])
z_resid[i] <- ys[i]-mus[i];
ln_hga1c[i] ~ dnorm(mu_resid[i], tau)
mu_resid[i] <- mu[i]+theta*z_resid[i];
}
b[1] ~ dnorm( 0, 1.0E-6)
b[2] ~ dnorm( 0, 1.0E-6)
b[3] ~ dnorm( 0, 1.0E-6)
b[4] ~ dnorm( 0, 1.0E-6)
b[5] ~ dnorm( 0, 1.0E-6)
a[1] ~ dnorm( 0, 1.0E-6)
a[2] ~ dnorm( 0, 1.0E-6)
a[3] ~ dnorm( 0, 1.0E-6)
a[4] ~ dnorm( 0, 1.0E-6)
a[5] ~ dnorm( 0, 1.0E-6)
theta ~ dnorm(0, 1.0E-6)
tau <- 1/(sigma*sigma)
sigma ~ dunif(0, 100)
```

```
rho <- sqrt(theta*theta/(theta*theta+1/tau))
b.cons  <- b[1]-b[2]*mean(Age[])/sd(Age[]) -
b[3]*mean(Diabetes[])/sd(Diabetes[])
-b[4]*mean(BMI[])/sd(BMI[]) - b[5]*mean(white[])/sd(white[])
b.Age <- b[2]/sd(Age[])
b.Diabetes  <- b[3]/sd(Diabetes[])
b.BMI  <- b[4]/sd(BMI[])
b.white <- b[5]/sd(white[])
a.cons  <- a[1]-a[2]*mean(Age[])/sd(Age[]) -
a[3]*mean(Diabetes[])/sd(Diabetes[])
-a[4]*mean(BMI[])/sd(BMI[]) - a[5]*mean(white[])/sd(white[])
a.Age <- a[2]/sd(Age[])
a.Diabetes <- a[3]/sd(Diabetes[])
a.BMI <- a[4]/sd(BMI[])
a.white <- a[5]/sd(white[])
}
```

We compare the multiple imputation analysis using the selection model with a typical imputation assuming MAR. For the former, we select $M = 50$ sets of imputations by running the MCMC chain for a burn-in period of 10000 iterations and then take each set of imputations every 200 iterations. For the latter, the model is a linear regression model for log(HbA1c) conditional on covariates, and we implement it using R mice. Table 13.4 presents the estimates for the overall mean of HbA1c from both imputation methods as well as from CC. The estimate from the selection model-based imputation is the lowest among three methods. This is consistent with the assumed MNAR mechanism: patients with lower HbA1c values are less likely to have their HbA1c values measured, even after adjusting for the factor of diabetic status. The fraction of missing information estimate is 0.10 for the selection model-based imputation and 0.21 for the MAR linear model-based imputation. Both values are less than the rate of missingness (27.8%). In addition, Fig. 13.3 (bottom row) plots the histograms of one randomly chosen completed log(HbA1c) values using the two imputation models. The imputed values under the selection model seem to make the distribution more symmetric (and normal). This is not a coincidence, as will be explained next.

TABLE 13.4
Example 13.8. Estimates for the mean of HbA1c

CC		Selection model		MAR model	
Mean	95% CI	Mean	95% CI	Mean	95% CI
6.50	(6.28, 6.72)	6.00	(5.79, 6.22)	6.41	(6.20, 6.61)

Note: The results are from $M = 50$ sets of imputations. MAR model: a linear regression imputation assuming MAR.

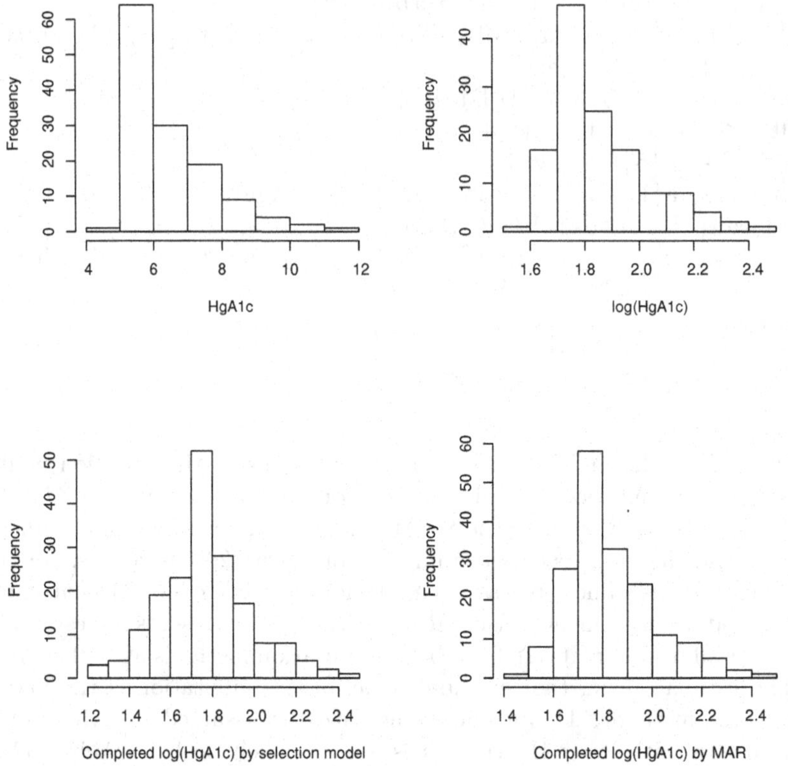

FIGURE 13.3
Example 13.8. Top left: the histogram of observed HbA1c; top right: the histogram of observed log(HbA1c); bottom left: the histogram of completed log(HbA1c) under the selection model; bottom right: the histogram of completed log(HbA1c) under an MAR imputation model. Completed data are based on one set of imputations.

Overall, imputation directly using selection models seems straightforward. Yet it can suffer from estimation problems (Example 13.7). Concerns about this strategy exist in the literature (Little and Rubin 2020, Example 15.7). One major issue is that the identifiability of the selection model is heavily relying on the distributional assumption of complete data, which are not fully observed and thus difficult to verify. As we pointed out before, in the Heckman selection model (13.2), an important assumption is that the original Y is normally distributed, which is unverifiable based on only observed Y. To see the point further, we merge the histograms of Y and Y_{obs} generated in Example 13.3 (by setting $\rho = 0.8$) in Fig. 13.4. The distribution of Y_{obs} (with tilted bars) is skewed to the right as determined from the selection model. The imputation attempts to fill in the missing values (with transparent bars) based on the assumption that Y is normal. However the normality assumption of Y is difficult to verify under the nonignorable missing data mechanism. If there exists no prior knowledge about the distribution of Y, an equally plausible assumption might be that nonrespondents have similar distribution as that of the respondents, which is essentially based on an MCAR assumption. This can also be seen by comparing the completed data under the selection model with those under an MAR model in the bottom row of Fig. 13.3. The former is closer to a normal distribution and the latter is more similar to the original distribution of observed log(HbA1c).

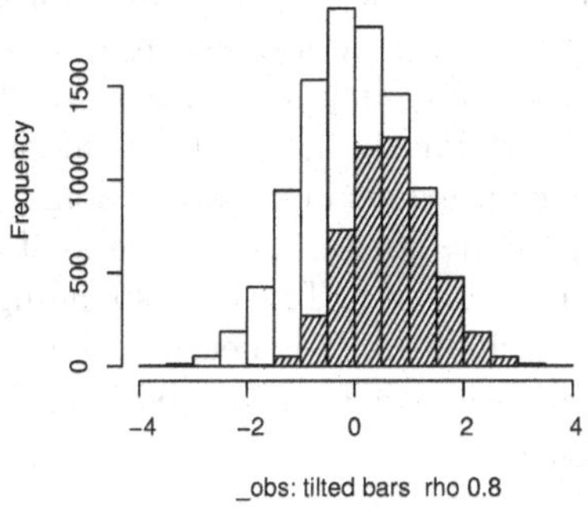

FIGURE 13.4
Example 13.3. Overlaid histograms of simulated Y and Y_{obs} when $\rho = 0.8$.

This simple illustration again emphasizes the fact that the ignorable (or nonignorable) missingness assumption is not testable based on observed data alone. Although we can sometimes obtain estimates that relate the missingness with the missing variable (e.g., ρ in Model (13.2) and γ_Y in Model (13.3)), these estimates can be rather sensitive by changing a few extreme values that lead to some changes of the distribution (Diggle and Kenward 1994; Kenward 1998). More importantly, a seemingly identifiable nonignorable missing data model (i.e., selection model) can lead to erroneous inferences if the assumption is wrong. Therefore, a particular selection model should not be used to provide definitive conclusions from a study; instead it can only be used to provide conclusions that would apply if certain specific (untestable) assumptions about the missingness mechanism were true. More discussion about the identifiability of selection models can be found, for example, in Molenberghs and Kenward (2007) and Daniels and Hogan (2008).

In addition to the selection model (13.2) discussed so far, pattern mixture models with some restrictions for parameters can also be identified (e.g., Little 1993; Little and Rubin 2020). Therefore it is possible to apply direct imputation methods using these pattern mixture models. However in general, we surmise that conducting direct imputation using identifiable nonignorable models (selection models, restricted pattern mixture models, or shared parameter models) can produce very unreliable results when models are misspecified. It might perform better than the imputation assuming MAR if the data are indeed MNAR given the observed information. However, if the true missingness mechanism does not follow the assumed model, the performance pattern of the direct imputation results is largely unclear. The success of the direct imputation hinges on the hidden restriction and strong assumption of the distribution of complete data. If these assumptions are severely distorted in some cases, bad imputation analysis results might be generated.

In summary, we believe that the use of nonignorable models for direct imputation is still acceptable in practice. Yet practitioners need to realize that this is based on a strong modeling assumption which is not verifiable by observed data alone. Support for the use of these models might come from subject-matter knowledge or previous literature of similar scientific problems. In addition, this strategy might not be used alone as if the results from the direct imputation is the final answer. The corresponding results only reflect one possible missingness mechanism. It is better to be included as one of the alternative imputation analyses, that is, a sensitivity analysis that will be discussed in the next section.

13.5.2 Sensitivity Analysis

Rubin (1977) pioneered the idea of sensitivity analysis for nonignorable missing data problems. The sensitivity analysis targets to present the sensitivity of analysis results to different specifications (ranging from plausible to perhaps more extreme) for the missingness mechanisms. Within the context of multiple imputation analysis, the display of sensitivity is accomplished by conducting the completed-data analysis based on each of several models for nonresponse and comparing the corresponding conclusions. In principle, if nonignorable missingness is still suspected even after attempts to make it MAR have been tried, the appropriate multiple imputation analysis should consist of a primary analysis assuming MAR and sensitivity analysis exploring the robustness of the imputation inference to departures from MAR (Carpenter and Kenward 2013). In practice, some adjustments of imputed values under MAR or sensitivity analyses (Example 13.9-13.11) might be done using existing software packages designed for MAR imputation.

Example 13.9. *Directly modifying imputed values under MAR*

Rubin (1987, Chapter 6) proposed several simple approaches to sensitivity analysis. In the simplest case, suppose Y_{imp}^{MAR} are imputations based on an ignorable missing data model; then we might assume that the corresponding imputations under a nonignorable model Y_{imp}^{MNAR} can be characterized as $Y_{imp}^{MNAR} = a \times Y_{imp}^{MAR} + b$, where a and b indicate the scale and location changes due to the departure from MAR. The choices of a and b are arbitrary yet might also depend on some subject-matter input and common sense. The advantage of these simple adjustments is that the results are easier to communicate, and they also avoid the use of sophisticated models by directly operating on the imputed values.

In another example, Rubin and Schenker (1991) discussed how an approximate Bayesian bootstrap (ABB) can be modified to handle nonignorable missing data. Instead of drawing n_{obs} cases of Y_{obs} randomly with replacement (i.e., with equal probability), they suggested drawing n_{obs} cases of Y_{obs} with probability proportional to Y_{obs}^c so that the probability of selection for $y_j \in Y_{obs}$ is $\frac{y_j^c}{\sum_{j=1}^{n_{obs}} y_j^c}$. Here c is a power parameter that is assumed to be fixed. This idea skews the nonrespondents to have typically larger (when $c > 0$ and $y_j > 0$) values of Y than respondents. Siddique and Belin (2008b) incorporated this idea into predictive mean matching to handle the nonignorable missing variable with covariates.

More formal modeling strategies can be applied in sensitivity analysis. Since the pattern mixture modeling framework decomposes the complete-data distribution as a mixture of distribution of responses and nonresponses, most of the relevant ideas specify the departure from MAR in sensitivity analysis using pattern mixture models. To sketch the idea, note that the pattern mixture model can be expressed as

$$\begin{aligned} f(Y, R|\theta, \phi) &= f(Y|R, \theta) f(R|\phi) \\ &= f(Y|R=1, \theta_{R=1}) f(R=1|\phi) + f(Y|R=0, \theta_{R=0}) f(R=0|\phi) \end{aligned} \tag{13.9}$$

where $\theta = (\theta_{R=1}, \theta_{R=0})$ governs the model describing the distribution of Y, and ϕ governs the model describing the distribution of R. For the multiple imputation analysis of Y, parameter ϕ in general has no role. An MAR imputation model essentially equates $f(Y|R = 1, \theta_{R=1})$ (for respondents) with $f(Y|R = 0, \theta_{R=0})$ (for nonrespondents). This includes setting both the distributional forms the same as well as setting $\theta_{R=1} = \theta_{R=0}$. On the other hand, to describe a possible departure from MAR, we can still let $f(Y|R = 1, \theta_{R=1})$ and $f(Y|R = 0, \theta_{R=0})$ have the same distributional form, yet they take different values for $\theta_{R=1}$ and $\theta_{R=0}$. In a simple case, this can be formulated as letting $\theta_{R=0} = K(\delta, \theta_{R=1})$, where $K(\cdot)$ is a known function and δ connects $\theta_{R=0}$ with $\theta_{R=1}$. Here δ can be referred to as the sensitivity parameter: when $\delta = 0$, $\theta_{R=0} = \theta_{R=1}$, reducing the situation to MAR. In a sensitivity analysis for nonignorable missing data, we can set δ to different values. Or we can generate δ from some prior distributions with known parameters to better reflect its uncertainty. Since $\theta_{R=1}$ can be estimated using observed data, the corresponding $\theta_{R=0}$ is also known by fixing δ. Multiple imputations for missing Y can then be generated using $f(Y|R = 0, \theta_{R=0})$.

Note that by using the pattern mixture model, the estimate for $\theta_{R=1}$ in sensitivity analysis is not changed since it is purely obtained from the model fit using observed data.

Example 13.10. *A sensitivity analysis based on a univariate normal mixture model*

In Example 13.4, we consider a normal mixture model for incomplete data. This can be expressed as: $f(Y|R = 1) \sim N(\mu_1, \sigma_1^2)$ and $f(Y|R = 0) \sim N(\mu_0, \sigma_0^2)$, $R \sim Bernoulli(p)$. When data are not MCAR, there is no information about μ_0 and σ_0^2 based on only observed Y-values. One way of building up this information is to add some connections between the two sets of parameters. We can set $\mu_0 = \mu_1(1 + c_\mu)$ and $\sigma_0 = c_\sigma \sigma_1$, where $c_\sigma > 0$. Here c_μ is the relative change of the nonrespondent mean compared with the respondent mean, and c_σ is the proportional change of the corresponding standard deviation. Both c_μ and c_σ can be termed as "connection" (or sensitivity) parameters, which describe the extent of nonignorability. The multiple imputation sensitivity analysis could proceed as follows:

1. Choose a set of plausible values of c_μ and c_σ.

2. Obtain posterior draws of μ_1 and σ_1 based on observed data, plug in selected c_μ and c_σ to obtain corresponding μ_0 and σ_0, and then impute Y_{mis}.

3. Use combining rules to obtain multiple imputation inferences of estimands of interest for selected c_μ and c_σ. Present the results across a range of c_μ and c_σ to provide the sensitivity of results to the chosen c_μ and c_σ.

We illustrate the idea using a simulated example. We simulate 1000 cases using the model: $f(Y|R = 1) \sim N(1, 1)$ and $f(Y|R = 0) \sim N(1.5, 1.5)$ with $R \sim Bernoulli(0.5)$. The nonrespondents have a larger mean and variance than respondents. We carry out a sensitivity analysis by specifying a range of connection parameters, $c_\mu = \{-0.5, -0.25, 0, 0.25, 0.5\}$ and $c_\sigma = \{0.25, 0.5, 1, 2, 4\}$. The estimates of the overall mean and its 95% confidence intervals (CI) are presented in Table 13.5 from $M = 50$ imputations. Before the data are deleted, the mean of Y is 1.249 and its 95% CI is (1.18, 1.319). The mean estimate from the CC analysis is 1.032 and its 95% CI is (0.947, 1.118). Clearly CC is biased and its confidence interval is rather wide. For the sensitivity analysis: when c_μ is fixed and c_σ increases, the mean estimate changes little yet its confidence interval is increasingly wider as expected; when c_σ is fixed and c_μ increases, the mean estimate increases as well. Note that the model becomes MCAR as we set $c_\mu = 0$ and $c_\sigma = 1$, where the imputation results are rather close to those of CC. Of course the main goal of this sensitivity analysis is not to guess which number should be the correct estimate; instead it is to provide a range of possible estimates with varying sensitivity parameters.

TABLE 13.5
Example 13.10. Multiple imputation sensitivity analysis results from a simulation

	c_μ				
c_σ	−0.5	−0.25	0	0.25	0.5
0.25	0.764	0.897	1.031	1.164	1.297
	(0.713, 0.815)	(0.843, 0.952)	(0.969, 1.092)	(1.094, 1.234)	(1.217, 1.377)
0.5	0.764	0.898	1.031	1.164	1.297
	(0.708, 0.821)	(0.838, 0.958)	(0.965, 1.097)	(1.090, 1.238)	(1.214, 1.381)
1	0.765	0.898	1.031	1.165	1.298
	(0.689, 0.841)	(0.820, 0.976)	(0.949, 1.114)	(1.076, 1.254)	(1.201, 1.395)
2	0.766	0.899	1.032	1.166	1.299
	(0.639, 0.893)	(0.771, 1.027)	(0.902, 1.163)	(1.031, 1.300)	(1.160, 1.438)
4	0.768	0.901	1.034	1.168	1.301
	(0.528, 1.008)	(0.660, 1.142)	(0.793, 1.276)	(0.924, 1.411)	(1.055, 1.547)

Note: Mean estimates and 95% CIs (in parentheses) are shown with varying c_μ and c_σ.

Example 13.11. *A sensitivity analysis based on a mixture of regression models*

The framework in Example 13.10 can be extended to cases where covariates are included. For example, suppose that we have a fully observed covariate X. We further assume that for respondents, $f(Y|X, R = 1)$ is a normal linear regression model as $f(Y|X, R = 1) \sim N(\alpha_0 + \alpha_1 X, \sigma^2)$. Then it might be natural to also assume a linear regression model for nonrespondents as $f(Y|X, R = 0) \sim N(\beta_0 + \beta_1 X, \tau^2)$. Of course there exists no observed data to estimate β_0, β_1, and τ^2. Following the idea in Example 13.10, we can posit some connections between β_0, β_1, and τ^2 with their counterparts (α_0, α_1, and σ^2) from respondents. A straightforward parametrization would be to set $\beta_0 = \alpha_0(1 + c_{\beta_0})$, $\beta_1 = \alpha_1(1 + c_{\beta_1})$, and $\tau = c_\tau \sigma$. Then the complete-data model (for both Y and R) can be specified as follows:

$$\begin{aligned} f(Y|X, R = 1) &\sim N(\alpha_0 + \alpha_1 X, \sigma^2), \\ f(Y|X, R = 0) &\sim N((1 + c_{\beta_0})\alpha_0 + (1 + c_{\beta_1})\alpha_1 X, c_\tau^2 \sigma^2). \end{aligned} \quad (13.10)$$

In Model (13.10), we can refer to α_0, α_1, and σ^2 as "baseline MAR" parameters, and refer to c_{β_0}, c_{β_1}, and c_τ as "departure" (sensitivity) parameters. The latter parameters characterize changes of intercept, slope, and error variance of the linear model for the departure of MAR. Another parameterization is to set $a = (1 + c_{\beta_1})$ and $b = (c_{\beta_0} - c_{\beta_1})\alpha_0$ so that a and b define the shift and tilt of the regression line for nonrespondents (Raghunathan et al. 2018, Section 10.2).

For a sensitivity analysis, we can specify some plausible values for the departure parameters, ranging from "weak MNAR" to "strong MNAR" and then conduct multiple imputation analysis based on these values. Note that the posterior distributions of α_0, α_1, and σ^2 can be obtained using the DA algorithm for normal linear regression models. Given chosen values for the departure parameters, the posterior distributions of β_0, β_1, and τ^2 can then be easily obtained to draw imputations. A sketch of the process is as follows:

1. Choose plausible values for the departure parameters as $c_{\beta_0}^{(t)}$, $c_{\beta_1}^{(t)}$, and $c_{\tau}^{(t)}$.

2. Draw α_0, α_1 and σ^2 from their posterior distributions, denoted as $\alpha_0^{*(m)}$, $\alpha_1^{*(m)}$, and $\sigma^{*2(m)}$. This can be done from running a Bayesian linear regression of Y on X using observed cases.

3. For chosen values of departure parameters, obtain $\beta_0^{*(m)} = \alpha_0^{*(m)}(1+c_{\beta_0}^{(t)})$, $\beta_1^{*(m)} = \alpha_1^{*(m)}(1+c_{\beta_1}^{(t)})$, and $\tau^{*2(m)} = c_{\tau}^{2(t)}\sigma^{*2(m)}$.

4. Impute the missing values as $y_i^{*(m)} \sim N(\beta_0^{*(m)} + \beta_1^{*(m)} X_{i,mis}^T, \tau^{*2(m)})$ independently for missing y_i's

5. Repeat Steps 2 to 4 independently M times, conduct multiple imputation analysis for estimands of interest, and report the results for $c_{\beta_0}^{(t)}$, $c_{\beta_1}^{(t)}$, and $c_{\tau}^{(t)}$.

6. Repeat Steps 1 to 5 T times for $t = 1, \ldots T$ for different values of the departure parameters.

If fixed values of the "departure" parameters seem to be somewhat arbitrary, we might specify some prior distributions to relate them to the baseline parameters. For example, we might specify priors such as $c_{\beta_0} \sim N(0, \delta_0)$, $c_{\beta_1} \sim N(0, \delta_1)$, and $log(c_{\tau}) \sim N(0, \delta_2)$. Note that δ_0, δ_1, and δ_2 have to be specified so that we can draw the "departure" parameters multiple times and each time nail down a specification of an MNAR model. The sensitivity analysis then consists of a collection of multiple imputation analysis results with different drawn departure parameters. By using prior distributions for the departure parameters, the associated uncertainty of the missingness mechanism might be more naturally accounted for in the sensitivity analysis. It is generally recommended to set up the departure parameters and their prior distributions in an interpretable way. In this simple setup, δ_0 and δ_1 can be viewed as a shift of the intercept and slope when referring to the baseline MAR model. More discussions can be found in Rubin (1987, Chapter 6). Carpenter and Kenward (2013) provided several examples in more realistic settings such as longitudinal studies and clinical trials.

Similar ideas of sensitivity analysis can be applied to a mixture of generalized linear models if the missing variable is not continuous. The modification can be applied to parameters of the regression model from responses. Or imputed values under the MAR model can be changed directly to reflect the MNAR mechanism (e.g., Raghunathan et al. 2018, Section 10.2).

We use the missing HbA1c dataset (Example 13.8) to illustrate the strategy. Although there exist multiple covariates including age, diabetic status, BMI, and race in the outcome model for predicting HbA1c, we focus only on the coefficient with diabetic status for the purpose of illustration. Note that the diabetic status is the only significant predictor in both the outcome and response models assuming MAR (Table 13.3). Let α_1 and β_1 denote the regression coefficient of the diabetic status in the outcome model from respondents and nonrespondents, respectively. We set $\beta_1 = (1 + c_{\beta_1})\alpha_1$ and vary c_{β_1} from -1 to 1 by changing it for every 20%.

Table 13.6 shows the results for estimating the overall mean of the sample based on $M = 50$ imputations. Note that when $c_{\beta_1} = 0$, the results correspond to those of the MAR imputation in Example 13.8. The sensitivity of the results to the change of the diabetic effect is clear: when such effect varies 20% in nonrespondents, the change for the overall mean is between 0.02 and 0.03. Setting $c_{\beta_1} = -100\%$ assumes that there exists no diabetic effect on HbA1c for nonrespondents in this sample. Even for this seemingly extreme assumption, the mean estimate is 6.30 and still higher than that obtained by the selection model-based imputation (6.0) in Example 13.8. This suggests that the selection model has to assume that additional factors are different for nonrespondents. By examining Table 13.3, these factors are the intercept and BMI. On the other hand, setting c_{β_1} as positive renders the situation that patients with higher HbA1c values are more likely to be missing in this dataset. Although the multiple imputation sensitivity analysis can provide the corresponding estimates, this assumption is unlikely to hold in the context of actual studies.

In selection models, there exist parameters governing the missingness mechanism (e.g., ρ in the Heckman selection model). Multiple imputation analyses by changing such parameters to different values can also be done. However, unlike the sensitivity analysis based on the pattern mixture modeling framework, changing the missingness mechanism parameters in selection models would generally change the model fit and estimates for both respondents and nonrespondents. This can be seen, for example, by examining Eqs (13.4) and (13.5), where the mean of both observed and missing cases would change once ρ is changed; whereas in Example 13.10, changing the connection parameter would not change the mean estimate from observed cases. Additional discussion can be found in Daniels and Hogan (2008, Chapter 8), where it presented a formal definition of sensitivity parameter (analysis) to deal with nonignorable drop-outs in longitudinal data and preferred to use the pattern mixture model for sensitivity analysis.

However, we believe that it might still be helpful to conduct multiple imputation analysis using selection models by changing the missingness mechanism parameter. One motivation is driven by the fact that sensitivity parameters are usually not standardized for pattern mixture models. For instance, in Example 13.10, c_μ can theoretically range from $-\infty$ and ∞, which might not be very informative. On the other hand, the Heckman selection model (13.2) uses the correlation coefficient ρ between the variable subject to missingness and an unobservable latent variable associated with the missingness probability to specify the missingness mechanism. The correlation coefficient ρ is always between -1 and $+1$ and standardized. Note that in past literature, the Heckman selection model has been used to develop a sensitivity analysis approach for nonrandom samples through a profile likelihood method (Copas and Li 1997), in which the sensitivity analysis parameter is a function of the correlation coefficient and thus standardized.

Example 13.12. *Multiple imputation sensitivity analysis using the Heckman selection model*

We use the missing HbA1c dataset (Example 13.8) to illustrate the idea. In Example 13.8 we performed a direct imputation analysis, which also provided Bayesian estimates for all the model parameters including ρ. Here we fix ρ and vary its value from -0.9 to 0.9. Note that at different ρ-value, estimates for other model parameters also change in the selection model. Table 13.7 shows the corresponding results for estimating the overall mean of HbA1c values based on $M = 50$ imputations. When ρ is negative, it implies that patients with higher HbA1c values are more likely to be missing, which is somewhat counterintuitive. On the other hand, when ρ is positive, it implies that patients with lower HbA1c values are more likely to be missing. Results under different

TABLE 13.6
Example 13.11. Multiple imputation sensitivity analysis results for the mean of HbA1c using a pattern mixture model

c_{β_1} (%)	Mean	95% CI
−100	6.30	(6.10, 6.49)
−80	6.32	(6.12, 6.51)
−60	6.34	(6.14, 6.53)
−40	6.36	(6.16, 6.55)
−20	6.38	(6.18, 6.58)
0	6.40	(6.20, 6.60)
20	6.43	(6.22, 6.63)
40	6.46	(6.25, 6.66)
60	6.48	(6.27, 6.70)
80	6.51	(6.29, 6.73)
100	6.54	(6.31, 6.77)

ρ's are consistent with this pattern. When $\rho = 0.9$, the mean estimate (6.04) is close to the direct imputation estimate (6.00) from Example 13.8.

TABLE 13.7
Example 13.12. Multiple imputation sensitivity analysis results for the mean of HbA1c using a Heckman selection model

ρ	Mean	95% CI
−0.9	6.95	(6.69, 7.20)
−0.7	6.74	(6.52, 6.96)
−0.5	6.61	(6.40, 6.82)
−0.3	6.53	(6.33, 6.74)
−0.1	6.44	(6.24, 6.64)
0.1	6.36	(6.16, 6.56)
0.3	6.29	(6.09, 6.49)
0.5	6.21	(6.01, 6.41)
0.7	6.14	(5.94, 6.34)
0.9	6.04	(5.84, 6.25)

In addition, Fig. 13.5 plots the histograms of one randomly chosen completed log(HbA1c) values from imputation models with some chosen ρ-values. When $\rho = -0.9$, the distribution is even more right-skewed than that of the observed cases (Fig. 13.3). This also explains that when ρ is taking a highly negative value, the 95% CI of the mean estimate is wider than that of other ρ-values: imputed values are more likely to come from the right-tail of the log-transformed distribution. As ρ increases, the distribution is becoming more symmetric and close to normality, which is also consistent with the direct imputation results from Example 13.8.

13.6 Summary

Missingness mechanisms can be speculated about and characterized by certain models if missing data are believed to be nonignorable. This is plausible in many real situations. For example, subjects in clinical trials may drop out from studies because side effects associated with the treatment prohibited them from continued study participation, or they feel that their status has been sufficiently improved and it is not worth continuing the trial.

However, handling missing data under the nonignorable missingness assumption is much harder. Nonignorable missing data models are in general subject to identifiability issues, and the corresponding inferences can be rather

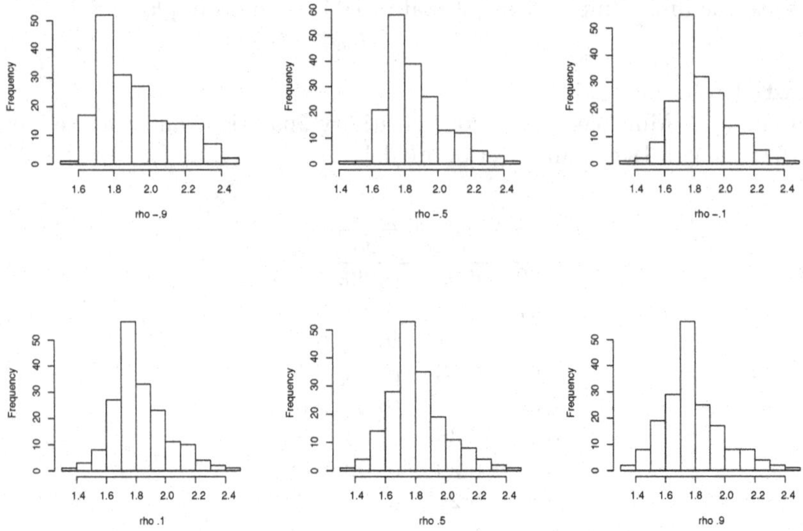

FIGURE 13.5
Example 13.12. Histograms of completed log(HbA1c) using a Heckman selection model by setting ρ at different values. Completed data are based on one set of imputations.

sensitive to alternative model specifications. In this chapter we briefly go over three main multiple imputation strategies: (a) include more variables so that the MAR assumption becomes more plausible; (b) conduct a direct imputation analysis using a well specified noningorable model; (c) perform a sensitivity analysis by varying parameters in the nonignorable model, which can include the MAR model used in (a) as a special case. Among them, the strategy of conducting a sensitivity analysis is commonly accepted in the research community and literature, as it is recommended to be included in reporting of findings from clinical trials (e.g., National Research Council 2010).

Note that the sensitivity analysis strategy is very general and can be applied for non-imputation missing data methods. For example, Janicki and Malec (2013) proposed a Bayesian model averaging approach to analyzing categorical data with nonignorable nonresponse, incorporating a range of ignorable/nonignorable missingness assumptions into the overall inference.

Since drop-outs from longitudinal studies are likely to be nonignorable, most related literature has focused on this area. For example, references can be found in Molenberghs and Kenward (2007), Daniels and Hogan (2008), Molenberghs et al. (2015), and Little and Rubin (2020). Note that many missing data methods introduced in these references are not imputation-based. Using multiple imputation, Demirtas and Schafer (2003) investigated the utility of random coefficient pattern mixture models as sensitivity analysis tools for longitudinal data when drop-outs are thought to be nonignorable. The basis of every one of such models is extrapolation. Their research demonstrated that alternative models that fit equally well may lead to very different estimates for parameters of interest. Their research also showed that minor model misspecifications can introduce substantive biases. Therefore, researchers and practitioners should understand the limitations of nonignorable models and interpret the results from fitting these models to longitudinal data with caution.

In addition, the relevant literature for nonignorable missing data methods is mostly focused on one variable collected in either a cross-sectional or longitudinal setting. In the presence of multiple incomplete variables, it can be rather difficult to specify models for multiple Y's and R's under the nonignorablity assumption. One promising research idea is to embed nonignorable models in the flexible FCS framework for multivariate missing data problems. Jolani (2012) proposed a random indicator method for directly imputing the missing variable suspect of nonignorableness, which is implemented in R mice. Galimard et al. (2016) proposed to include the missingness indicator of the variable subject to nonignorableness in the imputation model for other missing variables in the FCS framework. Giusti and Little (2011) presented a sensitivity analysis within the framework of FCS.

Finally, it has often been said that the best way to handle missing values is to not to have them. This saying, although true, offers somewhat limited help to those engaged in real studies, where even the best study designs and protocols may not reduce missingness to inconsequential levels. In the context of handling possibly nonignorable missing data, a more useful maxim may be, "The best way to handle nonignorable is to make it more ignorable." (Demirtas and Schafer 2003) This might be achieved in two stages. In the data collection stage, additional variables that may help to predict nonresponse of subjects can be collected. For instance, study participants in a trial may simply be asked at each occasion "How likely is it that are you willing to remain in this study through the next follow-up period?" with the answer given on an ordinal scale ranging from "very likely" to "very unlikely". Introducing this additional variable might convert a nonignorable situation to one that is close to being ignorable. In the data analysis stage, a multiple imputation analysis that is based on a carefully constructed ignorable model, which includes good predictors of missingness (e.g., the added variable in data collection) and all other available information, is perhaps more preferable than a direct imputation using a sophisticated nonignorable model. The former can also be embedded in a reasonable sensitivity analysis.

14

Some Advanced Topics

14.1 Overview

In past chapters we have presented basic concepts of multiple imputation analysis, discussed some commonly used modeling strategies, and illustrated these ideas using either simulated or real data, as well as examples from past literature. The field of multiple imputation is evolving rapidly. In the era of big data, we expect that new and complex missing data problems will arise and they will motivate more ideas and strategies to be developed for multiple imputation. Regretfully it is impossible for us to cover all the related topics. In the ending chapter we choose and discuss a few involved topics that might be important bases for further expansion of multiple imputation research and application. Section 14.2 discusses the issue of imputation uncongeniality, which we have briefly touched on in previous chapters. Section 14.3 discusses possible expansions of multiple imputation combining rules to deal with many types of completed-data analysis. Section 14.4 discusses the use of multiple imputation in the presence of high-dimensional data. Section 14.5 provides a summary and concludes this book.

14.2 Uncongeniality in Multiple Imputation Analysis

Chapter 3 presented some theoretical justifications for multiple imputation analysis. Using Rubin's combining rules, completed-data analysis results can be combined easily to yield a single set of inference. However in theory, the validity of these rules hinges on the coherence between the models for imputation and the procedures used for completed-data analysis. A technical term for this coherence is "imputation congeniality" (Meng 1994). Imputation congeniality is not always guaranteed. On the contrary, imputation uncongeniality (i.e., incoherence) often happens when the person creating the imputations (i.e., imputer) and the person analyzing the imputed data (i.e, analyst) are different. For example, this situation will apply when multiply imputed databases from organizations are analyzed by external users (Chapter 10). Even if both tasks are conducted by the same person in many projects, he/she could

DOI: 10.1201/9780429156397-14

analyze completed data using procedures with different assumptions and setups from the models used for creating imputations. Imputation uncongeniality can have important impacts on the property of multiple imputation estimates based on Rubin's combining rules. Fig. 14.1 shows a schematic plot of imputation uncogeniality.

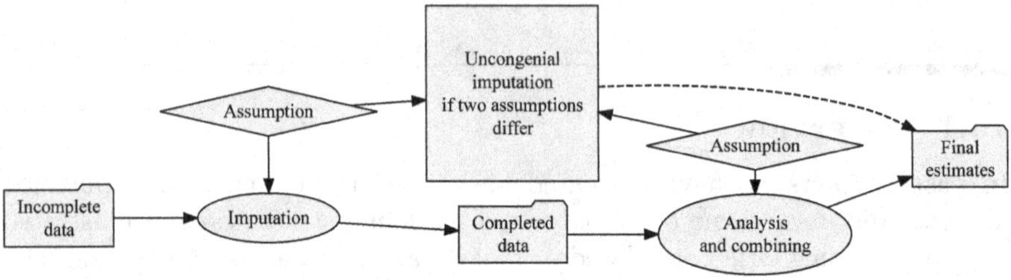

FIGURE 14.1
A schematic plot of imputation uncongeniality

Uncongeniality of imputation models has been mentioned multiple times in past chapters. For example, GLM-type models can be used to impute incomplete covariates for a survival analysis using Cox models (Chapter 8), where the imputation model and analysis procedure are clearly incoherent in terms of model forms. In imputation for multilevel data, for instance, treating random effects as fixed effects (an uncongenial model) would lead to overestimation of the variance (Example 9.1). In imputation analysis for survey data, imputation models excluding survey weights or design-related information are uncongenial because this information is included in the design-based inference of completed data (Chapter 10). When the imputation includes information from an external data source and yet the analysis excludes these data (Example 11.3), imputation uncongeniality happens and some adjustments might need to be made (Reiter 2008; Little and Rubin 2020, Page 281).

Formulating uncongeniality using mathematics can be complicated. An in-depth discussion of the theoretical underpinning is beyond the scope of this book. In practical terms, imputation uncongeniality can happen if the imputation model is either more restrictive or more general than the analysis procedures. The notion of "restrictive" and "general" can be in terms of variables included in the model, distributional assumptions of variables and modeling structure, as well as estimation efficiency. For example, if important variables are omitted in the imputation model (e.g., failing to include the response variable when imputing a missing covariate in a targeted analysis), then the restrictive imputation model can lead to estimates with biases and less efficiency (Section 4.4). On the other hand, if the imputation model includes a surrogate variable that is not directly involved in the analysis yet associated with the missing variable, the imputation inference can be more efficient or less biased via the inclusive imputation strategy (Section 5.6).

Example 14.1. *Imputation model includes more or fewer variables than the analysis procedure*

Example 5.8 showed some patterns when variables included in the imputation model and analysis procedure are different. To recap it, suppose the imputer used a regression model to impute the missing values in Y with two other fully observed variables X and Z. First, suppose the analyst is interested in the relationship only between Y and X so that Z is a surrogate. In Example 5.8, we compared the performance between the inclusive imputations strategy (with both X and Z in the model) and the exclusive strategy (with only X in the model). When Z is unrelated to Y (Scenarios I and III in Example 5.8), the inclusive strategy has used a "noise" variable in the imputation process. The estimate is still consistent because the imputation model is more general than the analysis procedure. However, it could be less efficient when Z is related to the missingness mechanism of Y (Scenario III in Example 5.8). When Z is related to Y and/or the missingness mechanism (Scenarios II and IV in Example 5.8), the inclusive imputation strategy can lead to better inferences (i.e., correct the bias and/or improve the efficiency) than the exclusive strategy.

One lingering question is, "What would happen if the imputation model is misspecified in the inclusive imputation strategy?" In addition, suppose the analyst is interested in the relationship among all three variables. If the imputer uses only Y and X, will the exclusive strategy lead to biased estimates because the imputer has assumed no relationship between (Y, X) and Z? On the basis of Example 5.8, we design a simulation study to gain some insights into the behavior of the exclusive and inclusive strategies in these two scenarios. In both cases, we generate Y according to $Y = -2 + X + \epsilon$, where $X \sim N(1, 1)$ and $\epsilon \sim N(0, 1)$. The missingness of Y follows MCAR with around 40% incomplete cases:

I. Z is correlated with Y: $Z = -2 + Y + Y^{1/2} + \epsilon$, where $\epsilon \sim N(0, 1)$. Compared with the Scenario II in Example 5.8, the relationship between Z and Y is now nonlinear.

II Z is correlated with Y: $Z = 2 + Y + \epsilon$, where $\epsilon \sim N(0, 1)$. The data generating model is identical to that in Scenario II of Example 5.8. However, the completed-data regression analysis now includes all three variables.

For simplicity, Z is not related to the missingness mechanism of Y in both scenarios. The inclusive imputation strategy assumes a normal linear model for Y on X and Z. Note that this model is misspecified in Scenario I here. The exclusive imputation strategy assumes a normal linear model for Y on X, which is correct in both scenarios. The simulation consists of 1000 replicates, each of which has complete-data sample size $n = 1000$. Multiple imputation is conducted using $M = 50$ sets of imputations.

Table 14.1 presents the simulation results. In Scenario I, both imputation methods yield little bias and nominal coverage rates for the estimands in the assessment. The gain of efficiency from the inclusive imputation over exclusive imputation is minimal, compared with the results from Scenario II in Example 5.8. However, this shows that a misspecified inclusive imputation model does not appear to bring obvious damage to the inference. In Scenario II, when the regression analysis model now includes both X and Z and the interest is on the slope coefficient on Z, the exclusive imputation yields biased estimates, and yet the inclusive imputation performs well as expected.

TABLE 14.1
Example 14.1. Simulation results

Scenario I: Z related to Y in a nonlinear manner						
Method	Bias	SD	$\overline{\text{SE}}$	MSE	CI	COV (%)
Mean of Y						
BD	0	0.043	0.045	0.00188	0.175	96.1
CC	−0.003	0.058	0.058	0.00335	0.226	94.6
Exclusive MI	−0.004	0.051	0.052	0.00262	0.203	96.3
Inclusive MI	0.002	0.051	0.051	0.00256	0.199	95.5
Slope coefficient of regressing Y on X						
BD	0	0.032	0.032	0.000995	0.124	95.0
CC	−0.002	0.041	0.041	0.00165	0.160	95.0
Exclusive MI	−0.002	0.041	0.041	0.00167	0.163	94.9
Inclusive MI	0	0.041	0.040	0.00170	0.158	94.0
Scenario II: Z related to Y in a linear manner						
Coefficient of Z when regressing Y on X and Z						
BD	0	0.016	0.016	0.000251	0.062	95.4
CC	0	0.021	0.020	0.000427	0.080	94.8
Exclusive MI	−0.2	0.018	0.025	0.04035	0.098	0
Inclusive MI	−0.001	0.021	0.020	0.000429	0.080	94.7

Results from both Example 5.8 and Example 14.1 show the advantage of applying the inclusive imputation strategy when the uncongeniality happens in terms of variables included. Past research (e.g., Collins et al. 2001, Raghunathan and Siscovick 1996, and Schafer 2003) has demonstrated the benefits for multiple imputation inference in terms of preserving (approximate) unbiasedness of the estimates and reasonable coverage rates even when the "additional" information is not 100% correct. Results shown in Scenario I are consistent with these conclusions. However, if the additional variables are also incomplete, the exact pattern of imputation inferences from including them is not clear.

A more subtle situation of imputation uncongeniality is related to the efficiency of different estimators used in imputation models and analysis procedures. We use the notation system in Chapter 3 here. To understand the basic issues, recall that the combined multiple imputation variance estimator T_M is decomposed into three parts: $\overline{U}_M$, B_M, and $\frac{1}{M}B_M$ (Section 3.2.2). According to Kim and Shao (2014), first note that $\hat{Q}_M = \hat{Q}_{com} + \hat{Q}_\infty - \hat{Q}_{com} + \hat{Q}_M - \hat{Q}_\infty$, where $\hat{Q}_{com}$ is the estimator for Q with complete-data, and $\hat{Q}_\infty$ is the multiple imputation estimator with an infinite number of imputations. In an ideal situation, in which the analysis model/procedure and imputation model are congenial, this variance decomposition can be written as

$$Var(\hat{Q}_M) = Var(\hat{Q}_{com}) + Var(\hat{Q}_\infty - \hat{Q}_{com}) + Var(\hat{Q}_M - \hat{Q}_\infty), \quad (14.1)$$

where the first term is estimated by $\overline{U}_M$, the second term is estimated by B_M, and the third term is estimated by $\frac{1}{M}B_M$. Thus Rubin's combining rule for the variance estimation is (asymptotically) unbiased. However, this is based on a critical assumption:

$$Var(\hat{Q}_\infty) = Var(\hat{Q}_{com}) + Var(\hat{Q}_\infty - \hat{Q}_{com}). \quad (14.2)$$

When imputation uncongeniality happens, Eq. (14.2) does not hold and using the Rubin's variance formula might yield biased results. These situations can occur due to the fact that $\hat{Q}_\infty$ (as the asymptotic basis for the analysis model/procedure) is not as efficient as $\hat{Q}_{com}$ (as the basis for the imputation model), and thus there is a residual covariance between $\hat{Q}_{com}$ and $\hat{Q}_\infty - \hat{Q}_{com}$ (Nielsen 2003).

Example 14.2. *The proportion of a missing continuous variable less than some threshold: estimators with different efficiency*

Following the ideas from Kim and Shao (2014, Section 4.5) and Raghunathan (2016, Section 9.1), we consider an example of estimating the proportion of a normally distributed variable less than some threshold. Suppose that Y is an incomplete variable and X is a fully observed covariate. Also assume that X and Y follow a bivariate normal distribution as $X \sim N(3, 1)$ and $Y = -2 + X + \epsilon$, where $\epsilon \sim N(0, 1)$. For simplicity, we assume that the missingness of Y is MCAR with the nonresponse rate around 40%. The estimand of interest is $\theta = Pr(Y < 1)$. Note that X is a surrogate variable because it is not involved in the completed-data analysis.

With complete data of sample size n, then a straightforward (method-of-moment) estimator is $\hat{\theta}_1 = \frac{\sum_{i=1}^n y_i}{n}$, and its variance estimator is $\hat{V}_1 = \frac{\hat{\theta}_1(1-\hat{\theta}_1)}{n}$. Additionally, since marginally Y follows a normal distribution, we can construct another estimator $\hat{\theta}_2 = \Phi(\frac{1-\overline{Y}}{s_Y})$, where $\overline{Y}$ and s_Y are the mean and standard deviation of Y, respectively. Its variance estimator can be shown to be $\hat{V}_2 = \phi^2(\frac{1-\overline{Y}}{s_Y})(\frac{1}{n} + \frac{(1-\overline{Y})^2}{2s_Y^2(n-1)})$ using the delta method. In theory $\hat{\theta}_2$ is more efficient than $\hat{\theta}_1$ because the former uses the information from the correct distributional assumption yet the latter does not use such information.

To show the performance of these two estimators under multiple imputation in conjunction with the issue of imputation uncongeniality, we conduct a simulation study with $n = 1000$ based on 1000 simulations. We consider two multiple imputation models: one is a univariate normal model for the incomplete Y; the other is to include X as the regressor and uses the normal linear regression model for imputation. For both models we conduct $M = 50$ imputations.

Table 14.2 presents the simulation results: $\hat{\theta}_1$ and $\hat{\theta}_2$ applied before data are deleted (BD), after missing cases are removed (CC), and after missing data are imputed (UNI-MI for the univariate imputation model and BIV-MI for the normal regression imputation model.) All estimators yield little bias for the true proportion (0.5). We focus on the performance of variance estimation.

TABLE 14.2
Example 14.2. Simulation results

Estimator	SD	$\overline{\text{SE}}$	CI	COV (%)
BD-$\hat{\theta}_1$	0.016	0.016	0.062	94.7
BD-$\hat{\theta}_2$	0.012	0.013	0.049	96.1
CC-$\hat{\theta}_1$	0.020	0.020	0.080	96.3
CC-$\hat{\theta}_2$	0.016	0.016	0.064	94.5
UNI-MI-$\hat{\theta}_1$	0.018	0.020	0.078	97.7
UNI-MI-$\hat{\theta}_2$	0.016	0.016	0.064	94.8
BIV-MI-$\hat{\theta}_1$	0.016	0.018	0.072	97.7
BIV-MI-$\hat{\theta}_2$	0.014	0.015	0.057	95.2

The results show the following patterns:

1. Based on both BD and CC results, $\hat{\theta}_2$ is more efficient than $\hat{\theta}_1$ as the former has shorter CIs and smaller variances. This is consistent with the theory.

2. In both UNI-MI-$\hat{\theta}_1$ and BVI-MI-$\hat{\theta}_1$, their averages of standard errors are larger than the empirical standard deviations (i.e., $\overline{\text{SE}}$ >SD), indicating some overestimation of the true variances. Correspondingly, the associated coverage rates are somewhat higher than the nominal level (close to 98%). This is caused by imputation uncongeniality due to the fact that the normality assumption is used in both imputation models yet $\hat{\theta}_1$ itself does not assume any normality for Y. This phenomenon was also encountered in the proportion estimates in Example 5.1, where the coverage rate from the imputation estimator is higher than the nominal level.

On the other hand, the variance estimator for the efficient estimator $\hat{\theta}_2$ has little bias in both imputation models (i.e., $\overline{\text{SE}} \approx$ SD for UNI-MI-$\hat{\theta}_2$ and BVI-MI-$\hat{\theta}_2$), and the coverage rate is very close to the nominal level. Here the imputation model and $\hat{\theta}_2$ used in the analysis are congenial because the latter also uses the correct normality assumption posited in the imputation.

3. Comparing UNI-MI-$\hat{\theta}_1$ with CC-$\hat{\theta}_1$, both of them use the method-of-moment estimator. The variance (standard deviation) of the MI estimate is less than that of the CC estimate ($0.018 < 0.020$), and the length of CI is shorter ($0.078 < 0.080$). This is because the normality assumption is used in imputing the missing Y hence somewhat improving the efficiency of the method-of-moment estimator. This phenomenon is more obvious if we compare BVI-MI-$\hat{\theta}_1$ with CC-$\hat{\theta}_1$ because the former method also includes extra information from the surrogate variable X. This phenomenon is again caused by imputation uncongeniality and termed "superefficiency" (Rubin 1996). See also Example 8.2 where marginal survival estimates based on imputed event times are more efficient than simply using the Kaplan-Meier method for handling censored observations.

   In general, when the imputation method uses more information than the completed-data analysis and this information is correct, the imputation analyses tend to be more efficient than anticipated: confidence intervals will be shorter yet have greater than the nominal coverage.

4. Comparing UNI-MI-$\hat{\theta}_2$ with CC-$\hat{\theta}_2$, their results are very close. This is because both assume a univariate normal model for Y and then we would expect that both estimators are asymptotically equivalent under MCAR. However, comparing BVI-MI-$\hat{\theta}_2$ with CC-$\hat{\theta}_2$, the former is more efficient because the imputation model uses additional information from the surrogate variable X. This is similar to what we have shown in Example 5.8 in terms of the advantages of the inclusive imputation strategy.

In this rather delicate example, if $\hat{\theta}_1$ is used for the post-imputation analysis, then both imputation models are uncongenial with the analysis procedure. If $\hat{\theta}_2$ is used for the post-imputation analysis, then the univariate imputation model is congenial with the analysis procedure yet the bivariate regression imputation model is uncongenial. These results suggest that the essence of imputation uncongeniality is related to the asymmetry of the modeling assumptions used in the imputation and analysis phases as well as the efficiency of the estimators.

From a historical perspective, the issue of imputation uncongeniality originally appeared as an early crisis for the multiple imputation approach, yet gradually has turned out to be a major force behind a better understanding of this missing data strategy. This started with an early criticism of multiple imputation by Fay (1992), where he pointed out that the validity of the multiple imputation analysis can depend on the form of post-imputation analyses.

Meng (1994) defended the use of multiple imputation and pointed out that in these counterexamples, the imputation models omitted key relations that were necessary to the analysis model. Meng (1994) also coined the term "multiple imputation uncongeniality", which implies some incoherence between models for imputation and procedures for completed-data analysis. Additional discussion can be found in Fay (1993), Rao (1996), and Rubin (1996). The theoretical property of Rubin's multiple imputation estimator had been further investigated and discussed in the literature, such as Wang and Robins (1998), Robins and Wang (2000), Nielsen (2003), Kim et al. (2006), Tsiatis (2006), Kim and Shao (2014), Xie and Meng (2017), and Bartlett and Hughes (2020).

Along with these theoretical investigations, some alternative methods to the multiple imputation variance estimators have been proposed. However, the alternative methods are typically more difficult to compute and require more information than the simple sufficient statistics (i.e., the completed-data estimates and their variances), which are readily available from general statistical software packages for multiple imputation analysis (Carpenter and Kenward 2013, Section 2.8).

What would be the implications for researchers and practitioners? In general, uncongeniality is of less concern as long as the imputations are obtained using good fitting predictive models that contain at least the information/feature reflected in the analysis model/procedure, and consistent estimators are used in completed-data analysis. In these scenarios, the use of uncongenial imputation models may lead to unbiased or consistent point estimates with overestimated variances, which might not be a serious issue provided that mildly conservative inference is usually acceptable. On the other hand, if the imputer ignores some important features that are exhibited in the analysis model/procedure while generating imputations, then multiple imputation inference can be biased or the variance is underestimated, both of which are usually unacceptable. These conclusions are based on the assumption that the repeated sampling calculations are performed under the analysis model.

14.3 Combining Analysis Results from Multiply Imputed Datasets: Further Considerations

In Chapter 3, we reviewed Rubin's combining rules for multiple imputation estimates. To refresh it, the idea is to approximate the posterior distribution $f(Q|Y_{obs})$ (e.g., for a scalar estimand Q assuming MAR) using a t-distribution based on the M sets of statistics, which are completed-data estimates and their variances that can be expressed as $\{\hat{Q}_m, \hat{U}_m\}_{m=1}^M$. One main attractiveness of multiple imputation analysis rests on the fact that it only requires M sets of point estimates and their variances from the complete-data procedure, and the combining steps are simple algebraic operations. Are these rules general enough for any type of estimand Q? In past examples we have used straightforward estimands such as means and regression coefficients. What about more complex estimands and completed-data analyses? Are there any other alternative rules for combining results from multiply imputed datasets? This section provides some discussions.

14.3.1 Normality Assumption in Question

As discussed in Chapter 3, the combining rules are derived based on the asymptotic theory. They might work well if the posterior distribution of the quantity (estimand) of interest, $f(Q|Y_{obs})$, is close to a normal distribution. If the normality assumption is doubtful, one apparent strategy is to apply normalizing (or variance-stabilizing) transformations to the estimand and use the delta method to obtain the variance formula on the transformed scale. We apply the combining rules for estimates on the transformed scale and calculate the variance, and we then back-transform the inference (i.e., the point estimate and confidence interval) to the original scale.

Table 14.3 (largely based on Table 5.2 from Van Buuren 2018) listed several commonly used estimands which might need a transformation before applying the combining rules. However, the use of transformation mainly follows some standard statistical practices. The exact performance from applying these transformations to nonnormal estimands for combining estimates remains largely unclear and warrants more research.

TABLE 14.3
Suggested transformations toward normality for some statistics used in the combining rules

Statistics	Transformation	Supporting literature
Correlation	Fisher Z	Schafer (1997)
Odds Ratio	Log	Agresti (1990)
Relative Risk	Log	Agresti (1990)
Hazard ratio	Log	Marshall et al. (2009)
Explained variance R^2	Fisher Z on R	Harel (2009)
Survival probabilities	Complementary log-log	Marshall et al. (2009)
Survival distributions	Log	Marshall et al. (2009)
Proportion	Logit	Rubin and Schenker (1987)
	Arc-sine-square-root	Raghunathan (2016, Section 2)
Standard deviation	Log	Raghunathan (2016, Section 2)

Example 14.3. *Combining regression R^2 from multiply imputed data*
In fitting linear regression models, the variance fraction explained by predictors, which can be quantified by regression R^2, is often used as a metric to measure the model fit. Harel (2009) demonstrated a multiple imputation combining technique for estimated R^2 of linear regression models from incomplete data. As included in Table 14.3, Harel (2009) proposed to apply a Fisher Z transformation on the square root of R^2. That is, let $\hat{Q} = \frac{1}{2}log(\frac{1+\hat{R}}{1-\hat{R}})$ and then with large samples, $Var(\hat{Q}) \approx 1/(n-3)$, where n is the sample size. Rubin's combining rules can be applied to $\hat{Q}$ first, and then the point estimate and its confidence interval can be transformed back.

We compare this approach (Tran) with two other alternatives. One is to apply Rubin's rules to $\hat{R}^2$ on the original scale and obtain its approximated variance using the delta method (Delta). It can be shown that $Var(\hat{R}^2) \approx \frac{4\hat{R}^2(1-\hat{R}^2)^2}{n-3}$. The other approach is also to operate on the original scale, yet obtain the variance of $\hat{R}^2$ using the nonparametric bootstrap approach (Boot). We conduct a simple simulation study to gauge the performance of the three methods.

We consider a complete-data model, $Y = -2 + \beta X + \epsilon$, where $X \sim N(1,1)$ and $\epsilon \sim N(0,1)$. We change β so that the $R^2 (= \frac{\beta^2}{\beta^2+1})$ of this regression model varies. The complete-data sample size $n = 100$. We apply a missingness rate around 40% with MCAR. Multiple imputation is conducted using $M = 50$ sets of imputations. The bootstrap variance is calculated based on 200 bootstrap samples. The simulation study is based on 1000 replications.

Table 14.4 shows the simulation results. The true value of R^2 in the assessment is calculated as the average of before-deletion estimates. When R^2 is small or moderate (i.e., ranging from 0.05 to 0.20), both methods that combine $\hat{R}^2$ on the original scale (i.e., Delta and Boot) appear to have more biases, lower coverage rates, and wider confidence intervals than the method combining on the transformed scale (i.e., Tran). When R^2 increases, the difference among the three methods is little. However, it is fair to say that all three methods yield reasonable performances given that R^2 is usually used for exploratory data analysis and not for exact inference.

Additional procedures have been developed for combining more complex estimands, such as sums of squares in ANOVA analysis (Raghunathan 2016, Section 5), the profile likelihood information from logistic regression with sparse data (Heinze et al. 2013), and Wilson confidence intervals for binomial proportions (Lott and Reiter 2020). Here we provide an example of combining design effects in survey data.

Example 14.4. *Combining design effects from multiply imputed survey data*
We use the simulation setup in Example 10.2 to illustrate the idea. In that setup, the population mean of Y is -1, and we create a binary indicator $Y_D = I(Y <= 1)$, where $I(\cdot)$ is the identity function. We assess the design effects of the proportion estimates for Y_D from several methods including (1)

TABLE 14.4
Example 14.3. Simulation results

	$R^2 = 0.059$		
Method	Bias	CI	COV (%)
Delta	0.011	0.224	90.5
Boot	0.011	0.225	92.9
Tran	0.007	0.203	97.7
	$R^2 = 0.108$		
Delta	0.008	0.283	89.1
Boot	0.008	0.280	89.5
Tran	0.004	0.263	93.5
	$R^2 = 0.206$		
Delta	0.003	0.350	90.5
Boot	0.003	0.344	89.6
Tran	0.001	0.333	93.5
	$R^2 = 0.501$		
Delta	−0.005	0.351	92.9
Boot	−0.005	0.346	92.4
Tran	−0.004	0.343	94.4
	$R^2 = 0.800$		
Delta	−0.003	0.178	93.8
Boot	−0.003	0.177	93.7
Tran	−0.002	0.178	94.7
	$R^2 = 0.900$		
Delta	−0.002	0.094	94.3
Boot	−0.002	0.094	94.0
Tran	−0.001	0.095	95.3

complete-case analysis (CC), denoted by $Deff(\hat{p}_{CC})$; (2) single imputation (SI), which calculates the design effect for each of the completed datasets and then averages these M design effect estimates, denoted by $Deff(\hat{p}_{SI})$; (3) single imputation and adjusted by the fraction of missing information (FMI), which is $Deff(\hat{p}_{FMI}) = Deff(\hat{p}_{SI})/(1 - FMI)$; and (4) multiple imputation (MI) based on the formula $Deff(\hat{p}_{MI}) = \frac{V(\hat{p}_{MI})}{\frac{\hat{p}_{MI}(1-\hat{p}_{MI})}{n}}$. The corresponding effective sample size is also calculated as $n_{eff} = n/Deff$: for CC, the observed-data sample size is used in the numerator; for other methods, n is set as the complete-data sample size 1000. Design effect estimates from CC and SI can be readily obtained using R survey. The simulation is conducted using 1000 replicates.

For simplicity, we only show the estimates based on the MI-Z imputation (using Z as the predictor) and include the simulation results in

Table 14.5. Due to the unequal sample selection in generating the data, the design effect is greater than one even before some of the data are set as missing ($Deff_{BD} = 1.463$). Since missing data are MCAR, the design effect is retained using complete cases (i.e., $Deff_{CC} \approx Deff_{BD}$). After imputation, the design effect estimated from completed data also recovers the variance inflation due to the sampling design (i.e., $Deff_{SI} \approx Deff_{BD}$). However, for multiple imputation estimates, we cannot simply average the design effect estimates from the completed datasets. This can be seen from the fact that $Deff_{MI} = 1.788 > Deff_{SI} = 1.463$. This is because for estimating $Deff_{SI}$, we only use the within-imputation variance and ignore the between-imputation variance. However, if we inflate $Deff_{SI}$ by the factor $1/(1-FMI)$, we effectively account for the between-imputation variance (i.e., $Deff_{FMI} \approx Deff_{MI}$). In general, using $Deff_{FMI}$ might have some advantages in practice because for estimators other than proportions, the design effect calculation is rather complicated so the formula for $Deff_{MI}$ might not be simple. Yet we can simply adjust the average design effect from completed datasets ($Deff_{SI}$ that can be easily calculated from survey statistics software) by using FMI.

TABLE 14.5
Example 14.4. Simulation results

Method	Deff	n_{eff}
BD	1.463	686
CC	1.458	483
SI	1.463	686
FMI	1.814	554
MI	1.788	562

Finally, by assessing the effective sample size, the advantage of using multiple imputation is clear. In CC, although the design effect is retained, the effective sample size is much reduced due to nonresponses. However, by using multiple imputation, the effective sample size is increased but still less than that before missing data occur.

14.3.2 Beyond Sufficient Statistics

Rubin's combining rules operate under the frequentist framework for completed-data analysis. However, a Bayesian version of the combining rules can be implemented. Since multiple imputation can be viewed as an approximation to Bayesian estimation (Section 3.2.1), we can obtain posterior draws of Q using the approximation steps and obtain the inference based on these draws (Si and Reiter 2011). More specifically, we first draw B_∞ based on $\frac{(M-1)B_M}{B_\infty} \sim \chi^2_{M-1}$, then draw Q_{com} based on $Q_{com} \sim N(\overline{Q}_M, (1+1/M)B_\infty)$, and finally draw Q based on $Q \sim t_\nu(Q_{com}, \overline{U}_M)$. Si and Reiter (2011) assessed the performance of this Bayesian method using simulations and concluded that there is no apparent advantage compared with the traditional, frequentist combining rules.

One might also wonder whether it sufficient to use $\{\hat{Q}_m, \hat{U}_m\}_{m=1}^M$ as the basis of combining results from multiply imputed datasets. Recall that in Chapter 3, Eqs. (3.1)-(3.4) imply that the posterior distribution, $f(Q|Y_{obs})$, can also be approximated by a mixture distribution $\sum_{m=1}^M \frac{1}{M} f(Q|Y_{obs}, Y_{mis}^m)$, although Rubin's combining procedure approximates $f(Q|Y_{obs})$ by a t-distribution. The mixing strategy is more straightforward if the completed-data analyses are Bayesian instead of frequentist. In this case, instead of operating on $\{\hat{Q}_m, \hat{U}_m\}_{m=1}^M$, the analyst would obtain M-sets of posterior draws of Q. The next step for combining the results could be to mix these M sets of posterior samples together as the approximate posterior distribution of Q given observed data, $f(Q|Y_{obs})$. This strategy was suggested by Gelman et al. (2013, Chapter 21) and researched by Zhou and Reiter (2010).

Example 14.5. *Combining regression coefficients from multiply imputed datasets by mixing posterior draws and bootstrap samples*

We use a simple simulation study to demonstrate the idea of mixing posterior draws from multiple sets of completed data (Bayes). We also consider a bootstrap method (Boot). That is, for each imputed dataset, we generate the bootstrap sample of the estimate and then mix M-sets of bootstrap samples. In these two methods, the point estimate of Q is taken as the mean of the posterior or bootstrap samples after mixing, and the interval estimate of Q is taken as either the credible interval from the mixed posterior samples or the confidence interval of the mixed bootstrap samples. Strictly speaking, the bootstrap method is not a Bayesian approach. However, as we pointed out in Section 2.2.4, nonparametric bootstrap samples can be viewed as an approximation of the noninformative, nonparametric posterior samples. We compare these two methods with the standard combining procedure.

Similar to Example 2.5, we consider a complete-data model, $Y = -2 + X + \epsilon$, where $X \sim N(0,1)$ and $\epsilon \sim N(0,1)$. The complete-data sample size $n = 100$. We apply a missingness rate around 40% with MCAR. Multiple imputation is conducted using the normal linear regression imputation setting $M = \{5, 10, 20, 50, 100\}$ sets of imputations. The estimand of interest is the slope regression coefficient. For Bayesian estimation, we use noninformative prior distributions for the parameters. For bootstrap estimation, the estimates are from the least-squares fitting. The numbers of posterior draws and bootstrap sample include both 200 and 1000. The simulation study includes 1000 replications.

Table 14.6 shows the simulation results based on 200 draws from posterior samples and 200 bootstrap samples. The results based on 1000 draws and 1000 bootstrap samples are similar and therefore omitted here. For the point estimates, there is little bias for all three methods in all scenarios (results not shown). Using Rubin's rules for combining multiply imputed estimates (the Rubin method), the coverage rate is consistently around nominal level from $M = 5$ to $M = 50$. As M increases, the width of the CI decreases, implying that the imputation estimates become more efficient. However, for smaller M ($M = 5$ or 10), the coverage rates for both the Bayes and Bootstrap methods are notably lower than the nominal level. The widths of their CIs are also notably shorter than those from the Rubin method. Note that their variance estimates are somewhat smaller than the variance of the estimates (i.e., $\overline{\text{SE}} <$ SD). When M is large (say 50 or 100), the coverage rates of both the Bayes and Boostrap methods are close to the nominal level. The widths of their CIs are still somewhat shorter than those from using Rubin's combining rules.

TABLE 14.6
Example 14.5. Simulation results

Method	SD	$\overline{\text{SE}}$	CI	COV (%)
		$M = 5$		
Rubin	0.144	0.137	0.596	94.0
Bayes	0.144	0.128	0.494	91.0
Boot	0.144	0.126	0.490	90.4
		$M = 10$		
Rubin	0.142	0.136	0.561	94.3
Bayes	0.142	0.131	0.512	92.5
Boot	0.142	0.130	0.510	92.6
		$M = 20$		
Rubin	0.141	0.135	0.549	94.6
Bayes	0.141	0.133	0.522	93.3
Boot	0.141	0.132	0.520	92.7
		$M = 50$		
Rubin	0.139	0.134	0.540	94.9
Bayes	0.139	0.134	0.525	94.2
Boot	0.139	0.133	0.524	93.9
		$M = 100$		
Rubin	0.138	0.134	0.538	95.0
Bayes	0.138	0.134	0.527	94.5
Boot	0.138	0.133	0.525	94.0

Note: "Rubin" denotes combining the estimates using Rubin's rules. "CI" denotes confidence interval for the Rubin and Bootstrap methods, and it denotes credible interval for the Bayes method.

This simulation study suggests that the tested approaches of mixing distributions (either by Bayesian draws or bootstrap sample) would produce narrower intervals with somewhat lower coverage rates with small M. This phenomenon was noticed in Zhou and Reiter (2010) for the Bayesian method and in Schomaker and Heumann (2018) for the bootstrap method. This is partly because both approaches use the mixture distribution to combine the results. With a finite M, the mean of such distribution is $\overline{Q}$, and the variance can be estimated by $T_{\infty} = \overline{U}_{\infty} + \frac{M-1}{M}B_{\infty}$. When M increases, the total variance of the mixture distribution would converge to $\overline{U}_{\infty} + B_{\infty}$, the same as the Rubin's variance formula. With a small M, however, the total variance (from the mixture distribution) would suffer from the omission of $\frac{1}{M}B_{\infty}$ (Steele et al. 2010; Rashid et al. 2015), resulting in the underestimation of the variance with shorter intervals and lower coverage rates.

In summary, there appear to exist several options for combining estimates that might deviate from the normal distribution or do not have simple variance functions:

1. Derive the variance function using the delta method and then combine the estimates using Rubin's rules at the original scale.
2. Transform the estimate in a way (making it close to normality or stabilizing the variance) and use Rubin's combining rules at the transformed scale, and then transform both the point and interval estimates back to the original scale.
3. Use Rubin's rules to combine the estimates at the original scale, for which the variance can be estimated using the bootstrap samples if the variance function is not simple.
4. Instead of directly using Rubin's rules, combine the estimates at the distributional level using the mixture of bootstrap samples obtained from multiply imputed datasets.
5. For Bayesian analysis of completed data, obtain the posterior estimates and variances from each of the imputed datasets and then combine them using Rubin's rules.
6. For Bayesian analysis of completed data, instead of directly using Rubin's rules, combine the estimates at the distributional level using the mixture of posterior draws from multiply imputed datasets.

The idea of mixing distributions of estimates is rather simple to implement (Methods 4 and 6) because it does not involve the direct calculation of the variance of the estimate and respects the original distribution of the estimate. However, in order for the mixing approach to work well (i.e., have a nominal coverage), a large M seems to be necessary.

In addition, there exist multiple alternative approaches to combining bootstrap with multiple imputation methods if the analytical complete-variance estimator is not available. For example, as opposed to in Example 14.5 where the bootstrap is applied after multiple imputation, the bootstrap procedure can be applied to incomplete data first and then be followed by multiple imputation. The latter strategy can be more preferable if imputation uncongeniality happens. More detailed discussions can be found, for example, in Schomaker and Heumann (2018), Brand et al. (2019), and Bartlett and Hughes (2020). Additional research is warranted to assess and compare the performances of alternative combining methods.

14.3.3 Complicated Completed-Data Analyses: Variable Selection

Some complete-data analyses might not have explicit Q so that Rubin's rules cannot be easily applied. One of such examples is the variable (model) selection in the presence of missing data, which is often used for the purpose of creating parsimonious prediction models. Apparently with completed data, a specific variable selection procedure can be applied in a straightforward manner. Therefore, an attractive strategy would be to run the variable selection procedures on multiply imputed datasets separately and then combine the results. However in many cases, the variable selection results could vary across imputed datasets. That is, different variables can be selected across multiple completed datasets, although it is desirable to have a single set of variables as the final result. This might be due to the uncertainty from the missing data, from the imputation model, as well as from the variable selection procedures applied. In this context, the combining strategy is less obvious.

This problem has stimulated many research investigations. With stepwise backward elimination, the principled approach appears to be fitting the model under consideration to each dataset and combining results via Rubin's rules to obtain Wald tests for all variables. The least significant variable is removed from the model before moving on to the next step. It is the only method that preserves the type-I error. However, it may require intensive computation (Wood et al. 2008).

One practical option is to fit the final model based on the most frequently selected variables (Wood et al., 2008). The second approach is to combine variable selection with bootstrapping (Heymans et al. 2007; Long and Johnson 2014). The variable inclusion frequency is calculated over all bootstrap and multiply imputed datasets. This method takes into account the uncertainty caused both by missing data and by sampling variability. The third approach is to stack all imputed datasets and perform variable selection on that single dataset (Wood et al., 2008). Unlike the previous two approaches, the stacked approach does not lead to different sets of selected variables. When computing p-values in the backward elimination procedure, Wood et al. (2008) additionally proposed a weighting scheme to account for the fraction of missingness per variable and the repeated occurrence of individuals.

In addition to the aforementioned traditional variable selection procedures, numerous approaches have been proposed to take advantage of shrinkage selection methods such as LASSO (i.e., the least absolute shrinkage and selection operator proposed by Tibshirani 1996) and its variants. In a case study, Lachenbruch (2011) applied the least angle regression (Efron et al. 2004) to multiply imputed data and selected variables that had non-zero coefficients in at least 10 imputed datasets. Chen and Wang (2013) applied LASSO to the stacked dataset as suggested by Wood et al. (2008). Yet to ensure that the same variables are selected across the multiple datasets, they treated the estimated regression coefficients associated with the same variable across different imputed datasets as a group and selected or removed the whole group together, following the group LASSO penalty (Yuan and Lin 2006). Similarly, the multiple imputation strategy has been combined with random LASSO (Liu et al. 2016) and penalized generalized estimation equations (Geronimi and Saporta 2017) for selecting models in the presence of missing data. Using elastic net as a penalized likelihood variable selection method, Wan et al. (2015) adopted the same idea of concatenating the multiple datasets but their proposed weighting scheme is per observation rather than per variable.

Different from the aforementioned frequentist variable selection procedures, Yang et al. (2005) proposed to combine multiple imputation with a Bayesian stochastic search variable selection procedure (George and McCulloch 1993). They considered the following model:

$$Y = \beta_0 + \sum_{j=1}^{p} \gamma_j X_j \beta_j + \epsilon, \tag{14.3}$$

where j index the X-covariate and $\epsilon \sim N(0, \sigma^2)$. In Model (14.3), the indicator $\gamma_j = 1$ or $\gamma_j = 0$ corresponds to the inclusion or exclusion of X_j. By doing so, the variable selection process is transformed into a statistical inferential process under the Bayesian paradigm. One approach is to "simultaneously impute and select", noting that γ_j's are now latent variables (model parameters). This strategy is essentially an extension of variable selection via Gibbs sampling (George and McCulloch 1993) to the cases of incomplete data. Another strategy, which is more related to the typical multiple imputation process, is to "impute then select". This strategy runs the Gibbs sampling procedures on all multiply imputed datasets and then averages the posterior selection probability $f(\gamma_j|Y)$ across the imputed datasets for selection. The authors concluded that the performances of both strategies are largely similar, and the former strategy slightly outperforms the latter. In theory, this approach is consistent with the Bayesian motivation of multiple imputation analysis. However, the related imputation and variable selection procedures are not straightforward but they can be conveniently implemented using Bayesian software programs such as WinBUGS (e.g., Ntzoufras 2009).

Reviews and comparisons of various methods can be found in Zhao and Long (2017), Thao and Geskus (2019), and Van Buuren (2018, Section 5.4). In summary, how to combine and synthesize model/variable selection processes/results from multiply imputed data is a fast developing area, and many methodological questions remain. In practice we believe that it is important to conduct comparison analysis using various developed strategies, involve subject-matter opinions, and not rely on one particular method.

Here a brief review of the variable selection problem shows the challenges for multiple imputation analysis if the completed-data analysis goes beyond regular model estimates or statistics (e.g., means, regression coefficients, p-values, etc.). More complex examples might include, for example, ranking data, clustering analysis (e.g., Basagaña et al. 2013; Bruckers et al. 2017), and data visualization. The final endpoints of these analyses are usually beyond some well defined point and variance estimates that are the typical components in Rubin's rules. For these analyses, a single decision is preferably made from multiple sets of results with possibly different patterns. How to combine the results in an appropriate way warrants additional research.

14.4 High-Dimensional Data

There might exist many ways of understanding and interpreting the term "high-dimensional data". In our context, we are conceiving broadly of scenarios where traditional imputation models allowing each variable to be related with other variables would have too many parameters to be stably estimated. When the number of variables is large relative to the number of cases, missing data problems can be further exacerbated, and the traditional models can become overparameterized. For example, with 20 variables on 100 cases, if 10% of the values on each variable are randomly missing, we would expect only about $100 \times 0.9^{20} \approx 12$ cases with complete records. In the meanwhile, suppose we model the 20 variables using a multivariate normal model with a general covariance matrix (Section 6.3.1); the number of correlation parameters is $20 \times 19/2 = 190$ and already exceeds the number of cases. Overparameterized models can induce problems in both the estimation and imputation.

Many research investigations have been conducted to tackle this problem. For multivariate continuous variables, Schafer (1997) introduced a method to handle possible overparameterization using a ridge prior distribution under a multivariate normal model. The ridge prior is a limiting case of the normal inverted Wishart prior. When a data matrix Y follows a $N(\mu, \Sigma)$ distribution, a normal inverted Wishart prior for the mean vector μ and variance-covariance matrix Σ is implied by the specification that $\mu \sim N(\mu_0, \Sigma/\tau)$ and $\Sigma \sim Inverse - Wishart(m, \Lambda)$ for given prior mean vector μ_0, prior scalar precision parameter τ, and prior degrees of freedom m for prior covariance matrix Λ. When $\tau \rightarrow 0$, the resulting prior distribution is termed the ridge prior owing to an analogy with ridge regression (Hoerl and Kennard 1970). Posterior distributions are then given by $f(\mu|\Sigma, Y) \sim N(\overline{Y}, \Sigma/n)$ and $f(\Sigma|Y) \sim Inverse - Wishart(n + m, (\Lambda^{-1} + nS)^{-1})$ where n is the number of cases, $\overline{Y}$ is the sample mean vector of Y, and S is the sample variance-covariance matrix of Y. After standardization, a common choice of Λ^{-1} is $\Lambda^{-1} = aI$ for $a > 0$ and identity matrix I. The corresponding prior smoothes the sample correlation matrix toward the identity matrix, reflecting an assumption of prior independence across variables and thus reducing the number of parameters. In practice, the ridge parameter a can be tuned to stabilize the covariance matrix by downweighting the off-diagonal elements. For example, Schafer (1997, Section 6.3.4) provided an example with around 300 observations and 12 variables, choosing $a = 3$ worked well to stabilize the parameter estimates. The use of ridge priors in multivariate normal models is implemented in some software packages such as SAS PROC MI and R mice.

Song and Belin (2004) considered a class of common factor analysis models to reduce the number of parameters. For data matrix Y with n rows and p columns, where n represents the number of observations and p represents the number of variables, let Y_i (a $1 \times p$ vector), $i = 1, \ldots, n$, denote the ith observation of Y representing an independent random draw from an underlying sampling distribution. The factor model with k underlying factors ($k < p$) can be described as

$$Y_i = \alpha + Z_i\beta + \epsilon_i, \tag{14.4}$$

where α is a $1 \times p$ mean vector, Z_i is a $1 \times k$ factor score vector with $Z_i \sim N(0, I_k)$, β is a $k \times p$ factor loading matrix, $\epsilon_i \sim N(0, \Sigma)$, where $\Sigma = diag(\sigma_1^2, \ldots, \sigma_p^2)$, and Z_i and ϵ_i are independent. Common prior distributions can be assigned for α, β, and Σ. Note that Z_i's can also be viewed as random effects with known covariance matrix I_k across the subjects. With a preset k, DA algorithm can be developed for this structured factor model and for conducting missing data imputation. Regarding the number of factors chosen, Song and Belin (2004) showed that the model performed well when the assumed number of factors equals or exceeds the true underlying number of factors. This implies that it is helpful for picking a sufficiently enough number factors in practical analysis.

In general, categorical variables are more difficult to model than continuous variables. In Section 6.4.1 we presented log-linear models for imputing multivariate categorical data. However, as noted there, log-linear models can become very difficult to fit in high-dimensional data due to sparse cells in the cross-classifications. Vermunt et al. (2008) considered the latent-class models to reduce the number of parameters that are required in the contingency tables from the log-linear models. Let Y_{ij} be the value of categorical variable j for individual i, where $i = 1, \ldots, n$ and $j = 1, \ldots p$. Let $Y_i = (Y_{i1}, \ldots, Y_{ip})$. Suppose the possible values of Y_{ij} are in $\{1, \ldots, d_j\}$, where $d_j > 2$ is the total number of categories for variable j. The latent-class model can be considered as a finite mixture of multinomial models. Suppose that each individual i belongs to exactly one of K latent classes. For $i = 1, \ldots n$, let $Z_i \in \{1, \ldots, K\}$ indicate the class of individual i, and let $\pi_h = Pr(Z_i = h)$ and assume that $\pi = (\pi_1, \ldots, \pi_K)$ is the same for all individuals. Within any class, suppose that each of the p variables independently follows a class-specific multinomial distribution. This implies that individuals in the same latent class have the same cell probabilities. For any value y, let $\phi_{hjy} = p(Y_{ij} = y|Z_i = h)$ be the probability of $Y_{ij} = y$, given that individual i is in class h. Let $\phi = \{(\phi_{hjy} : y = 1, \ldots, d_j, j = 1, \ldots, p, h = 1, \ldots, K)\}$ be the collection of all ϕ_{hjy}. Mathematically, the finite mixture model can be expressed as

$$Y_{ij}|Z_i, \phi \sim Multinomial(\phi_{Z_{i}j1}, \ldots, \phi_{Z_{i}jd_j}) \text{ for all } i, j, \quad (14.5)$$
$$Z_i|\pi \sim Multinomial(\pi_1, \ldots, \pi_K), \quad (14.6)$$

where each multinomial distribution has sample size equal to one and the number of levels is implied by the dimension of the corresponding probability vector.

Vermunt et al. (2008) considered Models (14.5) and (14.6) for missing data imputation for a preset K. To make the imputation model more flexible and also incorporate the uncertainty of K, Si and Reiter (2013) considered a non-parametric Bayesian extension of the latent class models, which is essentially an infinite mixture of products of multinomial distributions (i.e, let $K \to \infty$). The prior distribution for the mixture probabilities $\pi = (\pi_1, \ldots, \pi_\infty)$ is modeled using the stick-breaking representation of the Dirichlet process (e.g., Dunson and Xing 2009). This model is termed the Dirichlet process mixture of products of multinomial distributions (DPMPM). For simplicity, more technical details are omitted. See also Manrique-Vallier and Reiter (2014) for an extension to handle structural zeros and missing data due to the high number of cross-classifications from categorical variables. The DPMPM imputation routines are implemented in R NPBayesImpute.

For mixed continuous and categorical variables, He and Belin (2014) proposed a multivariate normal model encompassing both continuous variables and latent variables that are used to model binary variables (i.e., the multivariate probit models). They introduced prior distribution families for unstructured covariance matrices in the multivariate probit models to reduce the dimension of the parameter space. Murray and Reiter (2016) proposed a class of Bayesian mixture models with local dependence. The model fuses Dirichlet process mixtures of multinomial distributions for categorical variables (Si and Reiter 2013) with Dirichlet process mixtures of multivariate normal distributions for continuous variables (Kim et al. 2014). The imputation routines are implemented in R MixedDataImpute.

The ideas reviewed so far all follow the JM framework, using flexible priors (parametric or nonparametric) or reduced model specifications to mitigate the problem of model overparametrization in the presence of high-dimensional data. Another direction is to use the FCS framework. For example, Burgette and Reiter (2010) used classification and regression trees (CART) as the conditional models for imputation, which is implemented in R mice. The tree-based imputation demonstrated improved performance over default, main-effects-only applications of FCS in simulation studies with complex dependencies. See also Akande et al. (2017). It is also possible that more advanced regression or machine learning techniques (e.g., LASSO) can be incorporated in the conditional regression model fitting and variable selection in the FCS imputation process to handle the challenge of high-dimensional data.

14.5 Final Thoughts

Multiple imputation analysis is a flexible and practical statistical tool for missing data problems. In multiple imputation, appropriate assumptions have to be made on the missingness mechanism and complete-data model. In addition, since multiple imputation consists of the imputation and analysis phases, attention has to be paid to the connection as well as possible differences between the imputation model and analysis procedure. However, the separation of the two phases allows for the possibility of using a richer set of variables or other information in the imputation phase than those used in the analysis phase to improve the imputation estimates. This idea follows the inclusive imputation strategy. The goal of multiple imputation analysis is to obtain optimal inferences about the estimand of interest while accounting for the uncertainty of missing data and underlying models.

Exploratory analysis, model building and checking, and diagnostics are all important in multiple imputation analysis. Following standard statistical conventions and seeking subject-matter input are also important. For data with complex structure and patterns, the imputation model may often need to be sophisticated to describe the relation among variables and their distributions. Some of these models have been implemented in imputation software programs to make the process automatic and simple to execute. Imputation software packages greatly enhance the use of multiple imputation analysis in practice. We encourage using these software packages and comparing among different options of models/methods whenever possible. On the other hand, it is very important to understand the ideas behind imputation models and analysis procedures. Mechanical uses of imputation programs can be problematic. Software packages are just tools and require a careful and deliberate user to make the best use of them.

Built on the recommendations from He (2010), we now summarize some of the key steps involved in a typical multiple imputation project for researchers and practitioners.

1. Understand the study design and analytic objective. Note that many statistical analysis problems that do not appear to have missing data can be approached from the missing data perspective.

2. Identify the data structure including which variables have missing data and the corresponding missing data pattern.

3. Make appropriate assumptions for the missing data mechanism.

4. Identify variables to be included in the imputation. The general principle is to include at least all variables involved in the planned analysis. For example, when imputing missing predictors, the outcome variables should be included in imputation to retain the association between the outcome and predictors. In addition, variables not used in the analysis yet having strong correlation with missing variables and/or the missingness need to be included.

5. Construct the imputation model for included variables. For example, do we use JM or FCS? It is important to seek a balance between sophistication and feasibility of models. For most empirical analyses, we recommend using existing models in the literature or those provided by available software. However it is important to understand the main idea behind these models.

6. Use the appropriate imputation software packages for implementation whenever possible.

7. Carry out imputation diagnostics and refine the imputation. If MNAR models are used, conduct sensitivity analysis.

8. Combine completed-data analysis results from multiply imputed data and report the final results.

9. Document the multiple imputation project. Additional work is needed if the goal is to release multiple completed datasets for public/shared use.

As briefly mentioned in this chapter, new research and expansion of multiple imputation analysis are never ending. These areas span from the basic combining procedures to imputing missing data with high-dimensional features. Due to limited space, this book does not cover many other interesting and important topics that use the idea of multiple imputation analysis. These areas include but are not limited to data synthesis for statistical disclosure control (e.g., Drechsler 2011), causal effect analysis (e.g., Rubin 2004), and efficient survey designs with planned missing data (e.g., Raghunathan and Grizzle 1995). Similar to all other relevant literature, we hope our book can be a catalyst for better understanding the idea of multiple imputation analysis and for increasing its use for various missing data problems!

Bibliography

[1] Abayomi, K., Gelman, A.E., and Levy, M. (2008), "Diagnostics for multivariate imputations", *Journal of the Royal Statistical Society: Series C (Applied Statistics)*, **57**, 273–291.

[2] Abmann, C., Wurbach, A., Gobmann, S., Geissler, F., and Bela, A. (2017), "Nonparameteric multiple imputation for questionnaires with individual skip patterns and constraints: The case of income imputation in the National Educational Panel Study", *Sociological Methods and Research*, **46**, 864–897.

[3] Aerts, M., Claeskens, G., Hens, N., and Molenberghs, G. (2002) "Local multiple imputation," *Biometrika*, **89**, 375-388.

[4] Agresti, A. (1990), *Categorical Data Analysis*, 1st Edition, New York: Wiley.

[5] Agresti, A. (2002), *Categorical Data Analysis*, 2nd Edition, New York: Wiley.

[6] Agresti, A. (2010), *Analysis of Ordinal Categorical Data*, 2nd Edition, New York: Wiley.

[7] Akande, O., Li, F., and Reiter, J.P. (2017), "An empirical comparison of multiple imputation methods for categorical data", *The American Statistician*, **71**, 162–170.

[8] Albert J. and Chib, S. (1993), "Bayesian analysis of binary and polychotomous response data", *Journal of the American Statistical Association*, **88**, 669–679.

[9] Alexander, G.R. and Allen, M.C. (1996), "Conceptualization, measurement, and use of gestational age. I. Clinical and public health practice", *Journal of Perinatology*, **16**, 53–59.

[10] Allison, P.D. (2001), *Missing Data*, Sage Publications.

[11] Allison, P.D. (2005), "Imputation of categorical variables with PROC MI", *Proceedings of the SAS Users Group International (SUGI)*, **30**, 113–30.

[12] An, H. and Little, R.J.A. (2008), "Robust model-based inference for incomplete data via penalized spline propensity prediction", *Communications in Statistics-Simulation and Computation*, **37**, 1718–1731.

[13] Anderson, P.K. and Gill, R.D. (1982), "Cox's regression model for counting process: A large sample study", *The Annals of Statistics*, **10**, 1100–1120.

[14] Andridge, R.R. (2011), "Quantifying the impact of fixed effects modeling of clusters in multiple imputation for cluster randomized trials", *Biometrical Journal*, **53**, 57–74.

[15] Andridge, R.R. and Little, R.J.A. (2010), "A review of hot deck imputation for survey nonresponse", *International Statistical Review*, **78**, 40–64.

[16] Andridge, R.R. and Thompson, K.J. (2015), "Using the fraction of missing information to identify auxiliary variables for imputation procedures via proxy pattern-mixture models", *International Statistical Review*, **83**, 472–492.

[17] Arnold, B.C., Castillo, E., and Sarabia, J.M. (1999), *Conditional Specification of Statistical Models*, Springer, New York.

[18] Arunajadai, S.G. and Rauh, V.A. (2012), "Handling covariates subject to limits of detection in regression", *Environmental and Ecological Statistics*, **19**, 369–391.

[19] Asparouhov, T. and Muthen, B. (2010), "Multiple imputation with Mplus",(http://statmodel2.com/download/Imputations7.pdf.)

[20] Atkinson, A.D. (2019), "Reference based sensitivity analysis for time-to-event data", Ph.D. Thesis, London School of Hygiene and Tropical Medicine, U.K.

[21] Austin, P. (2009), "Balance diagnostics for comparing the distribution of baseline covariates between treatment groups in propensity-score matched samples", *Statistics in Medicine*, **28**, 3083–3107.

[22] Ayanian J.Z., Chrischilles, E.A., Fletcher, R.H., Fouad, M.N., Harrington, D.P., Kahn, K.L., et al. (2004) "Understanding cancer treatment and outcomes: The cancer care outcomes research and surveillance consortium", *Journal of Clinical Oncology*, **22**, 2992–2996.

[23] Baccini, M., Cook, S., Frangakis, C.E., Li, F., Mealli, F., Rubin, D.B., and Zell, E.R. (2010), "Multiple imputation in the anthrax vaccine research program", *Chance*, **23**, 16–23.

[24] Barnard, J. and Meng, X.L. (1999), "Applications of multiple imputation in medical studies: From AIDS to NHANES", *Statistical Methods in Medical Research*, **8**, 17–36.

[25] Barnard, J. and Rubin, D.B. (1999), "Small sample degrees of freedom with multiple imputation", *Biometrika*, **86**, 948–955.

[26] Bartlett, J.W. and Hughes, R.A. (2020), "Bootstrap inference for multiple imputation under uncongeniality and misspecification", *Statistical Methods in Medical Research*, **29**, 3533–3546.

[27] Bartlett, J.W., Seaman, S.R., White, I.R., and Carpenter, J.R. (2015), "Multiple imputation of covariates by fully conditional specification: Accommodating the substantive model", *Statistical Methods in Medical Research*, **24**, 462–487.

[28] Basagaña, X., Barrera-Gómez, J., Benet, M., Antó, J.M., and Garcia-Aymerich, J. (2013), "A framework for multiple imputation in cluster analysis", *American Journal of Epidemiology*, **177**, 718–725.

[29] Belin, T.R., Diffendal, G.J., Mack, S., Rubin, D.B., Schafer, J.L., and Zaslavsky, A.M. (1993), "Hierarchical logistic regression models for imputation of unresolved enumeration status in undercount estimation (with discussion)", *Journal of the American Statistical Association*, **88**, 1149–1159.

[30] Belin, T.R., Hu, M.Y., Young, A.S., and Grusky, O. (1999), "Performance of a general location model with an ignorable missing-data assumption in a multivariate mental health services study", *Statistics in Medicine*, **18**, 3123–3135.

[31] Berglund, P. and Heeringa, S. (2014), *Multiple Imputation of Missing Data Using SAS*, SAS Institute, Inc.

[32] Berkson, J. (1950), "Are there two regressions?", *Journal of the American Statistical Association*, **45**, 164–180.

[33] Bernaards, C.A., Belin, T.R., and Schafer, J.L. (2007), "Robustness of a multivariate normal approximation for imputation of incomplete binary data", *Statistics in Medicine*, **26**, 1368–1382.

[34] Bodner, T. E. (2008), "What improves with increased missing data imputations?", *Structural Equation Modeling*, **15**, 651–675.

[35] Böhning, D., Seidel, W., Alfó, M., Garel, B., Patilea, V., Walther, G., DiZio, M., Guarnera, U., and Luzi, O. (2007), "Imputation through finite gaussian mixture models", *Computational Statistics and Data Analysis*, **51**, 5305–5316.

[36] Bondarenko, I. and Raghunathan, T.E. (2016), "Diagnostics for multiple imputations", *Statistics in Medicine*, **26**, 3007–3020.

[37] Box, G.E.P. and Cox, D.R. (1964), "An analysis of transformations", *Journal of the Royal Statistical Society: Series B (Statistical Methodology)*, **26**, 211-252.

[38] Brand, J.P.L. (1999), "Development, implementation, and evaluation of multiple imputation strategies for the statistical analysis of incomplete data sets", Ph.D. Thesis, Erasmuns University, Rotterdam, Netherland.

[39] Brand, J.P.L., Van Buuren, S., Le Cessie, S., and Van den Hout, W. (2019), "Combining multiple imputation and bootstrap in the analysis of cost-effectiveness trial data", *Statistics in Medicine*, **38**, 210–220.

[40] Brazi, F. and Woodward, M. (2004), "Imputations of missing values in practice: Results from imputations of serum cholesterol in 28 cohort studies", *American Journal of Epidemiology*, **160**, 34–45.

[41] Brick, J.M. and Kalton, G. (1996), "Handling missing data in survey research", *Statistical Methods in Medical Research*, **5**, 215–238.

[42] Bruckers, L., Molenberghs, G., and Dendale, P. (2017), "Clustering multiply imputed multivariate high-dimensional longitudinal profiles: Clustering multiply imputed data", *Biometrical Journal*, **59**, 998–1015.

[43] Buckley, J. and James, I. (1991) "Linear regression with censored data", *Biometrika*, **66**, 429–436.

[44] Buonaccorsi, J.P. (2010), *Measurement Error: Models, Methods and Applications.* Boca Raton: Chapman and Hall.

[45] Burgette, L.F. and Reiter, J.P. (2010), "Multiple imputation for missing data via sequential regression trees", *American Journal of Epidemiology*, **172**, 1070–1076.

[46] Burton, A., Altman, D.G., Royston, P., and Holder, R.L. (2006), "The design of simulation studies in medical statistics", *Statistics in Medicine*, **25**, 4279–4292.

[47] Cabras, S., Maria, C., and Alicia, Q. (2011), "Goodness-of-fit of conditional regression models for multiple imputation", *Bayesian Analysis*, **6**, 429–456.

[48] Carpenter, J.R., Goldstein, H., and Kenward, M.G. (2011), "REALCOM-IMPUTE software for multilevel multiple imputation with mixed response types", *Journal of Statistical Software*, **45**, 1–14.

[49] Carpenter, J.R. and Kenward, M.G. (2013), *Multiple Imputation and Its Application.* Chichester, UK: Wiley.

[50] Carpenter, J.R., Kenward, M., and Vansteelandt, S. (2006), "A comparison of multiple imputation and doubly robust estimation for analysis with missing data", *Journal of the Royal Statistical Society: Series A (Statistics in Society)*, **169**, 571–584.

[51] Carroll, R.J., Ruppert, D., Stefanski, L.A., and Crainiceanu, C.M. (2006), *Measurement Error in Nonlinear Models: A Modern Perspective*, 2nd Edition. London: Chapman and Hall.

[52] Casella, G. and Berger, G.L. (1992), *Statistical Inference.* Pacific Grove, California: Wadsworth and Brooks/Cole Advanced Books and Software.

[53] Cassel, C.M., Sarndal, C.E., and Wretman, J.H. (1976), "Some results on generalized difference estimation and generalized regression estimation for finite populations", *Biometrika*, **63**, 615–620.

[54] Chen, H., Quandt, S.A., Grzywacz, J.G., and Arcury, T.A. (2011), "A distribution-based multiple imputation method for handling bivariate pesticide data with values below the limit of detection", *Environmental Health Perspectives*, **119**, 351–356.

[55] Chen, H., Quandt, S.A., Grzywacz, J.G., and Arcury, T.A. (2013), "A Bayesian multiple imputation method for handling longitudinal pesticide data with values below the limit of detection", *Environmentrics*, **24**, 132–142.

[56] Chen, Q. and Wang, S. (2013), "Variable selection for multiply-imputed data with application to dioxin exposure study", *Statistics in Medicine*, **32**, 3646–3659.

[57] Chib, S. and Greenberg, E. (1995), "Understanding the Metropolis-Hastings algorithm", *The American Statistician*, **49**, 327–335.

[58] Cho, M. and Schenker, N. (1999), "Fitting the log-F accelerated failure time model with incomplete covariate data", *Biometrics*, **55**, 826–833.

[59] Citro, C.F. (2014), "From multiple modes for surveys to multiple data sources for estimates", *Survey Methodology*, **40**, 137–161.

[60] Clark, T.G. and Altman, D.G. (2003), "Developing a prognostic model in the presence of missing data: An ovarian cancer case study", *Journal of Clinical Epidemiology*, **56**, 28–37.

[61] Clayton, D. (1994), "Bayesian analysis of frailty models," Technical report, Medical Research Council Biostatistics Unit, Cambridge.

[62] Clogg, C.C., Rubin, D.B., Schenker, N., Schultz, B., and Weidman, L. (1991), "Multiple imputation of industry and occupation codes in census public-use samples using Bayesian logistic regression", *Journal of the American Statistical Association*, **86**, 68–78.

[63] Cochran, W.G. (1977), *Sampling Techniques*, 3rd Edition. New York: Wiley.

[64] Cole, S.R., Chu, H., and Greenland, S. (2006), "Multiple imputation for measurement error correction", *International Journal of Epidemiology*, **25**, 1074–1081.

[65] Collins, L.M., Schafer, J.L., and Kam, C.M. (2001), "A comparison of inclusive and restrictive strategies in modern missing data procedures", *Psychological Methods*, **6**, 330–351.

[66] Copas, J.B. and Li, H.G. (1997), "Inference for non-random samples", *Journal of the Royal Statistical Society: Series B (Statistical Methodology)*, **59**, 55–95.

[67] Cox, D.R. (1972), "Regression models and life-tables (with discussion)", *Journal of the Royal Statistical Society: Series B (Statistical Methodology)*, **34**, 187–220.

[68] Crainiceanu, C.M., Ruppert, D., and Wand, M.P. (2005), "Bayesian analysis for penalized spline regression using WinBUGS", *Journal of Statistical Software*, **14**, 1–24.

[69] Daniels, M.J. and Hogan, J.W. (2008), *Missing Data in Longitudinal Studies. Strategies for Bayesian Modeling and Sensitivity Analysis.* London: Chapman and Hall.

[70] Davern, M., Klerman, J.A., Baugh, D.K., Call, K.T., and Greenberg, G.D. (2009), "An examination of the Medicaid undercount in the current population surveys: Preliminary results from record linkage," *Health Services Research*, **44**, 965–987.

[71] David, M., Little, R.J.A., Samuhel, M.E., and Triest, R.K. (1986), "Alternative methods for CPS income imputation", *Journal of the American Statistical Association*, **81**, 29–41.

[72] Davis, W.W., Parsons, V.L., Xie, D. et al. (2010) "State-based estimates of mammography screening rates based on information from two health surveys", *Public Health Reports*, **125**, 567–78.

[73] Davison, A.C. and Hinkley, D.V. (1997), *Bootstrap Methods and Their Application.* Cambridge University Press.

[74] Dawid, A.P. (2006), "Conditional independence", *Wiley Online Encyclopedia of Statistical Sciences.*

[75] Dellaportas, P. and Smith, A.F.M. (1993), "Bayesian inference for generalized linear and proportional hazards models via Gibbs sampling", *Journal of the Royal Statistical Society: Series C (Applied Statistics)*, **42**, 443–459.

[76] Demirtas, H. (2009), "Rounding strategies for multiply imputed binary data", *Biometrical Journal*, **51**, 677–688.

[77] Demirtas, H. (2010), "A distance-based rounding strategy for post-imputation ordinal data", *Journal of Applied Statistics*, **37**, 489–500.

[78] Demirtas, H., Freels, S.A., and Yucel, R.M. (2008) "Plausibility of multivariate normality assumption when multiply imputing non-Gaussian continuous outcomes: A simulation assessment", *Journal of Statistical Computation and Simulation*, **78**, 69–84.

[79] Demirtas, H. and Hedeker, D. (2007), "Gaussianization-based quasi-imputation and expansion strategies for incomplete correlated binary responses", *Statistics in Medicine*, **26**, 782–799.

[80] Demirtas, H. and Hedeker, D. (2008), "An imputation strategy for incomplete longitudinal ordinal data", *Statistics in Medicine*, **27**, 4086–4093.

[81] Demirtas, H. and Schafer, J.L. (2003), "On the performance of random-coefficient pattern-mixture models for non-ignorable drop-out", *Statistics in Medicine*, **22**, 2553–2575.

[82] Dempster, A.P., Laird, N.M., and Rubin, D.B. (1977), "Maximum likelihood estimation from incomplete data via the EM algorithm (with discussion)", *Journal of the Royal Statistical Society: Series B (Statistical Methodology)*, **39**, 1–38.

[83] Diggle, P.J. and Kenward, M.G. (1994), "Informative drop-out in longitudinal data analysis (with discussion)", *Journal of the Royal Statistical Society: Series C (Applied Statistics)*, **43**, 49–93.

[84] Ding, P. (2016), "On the conditional distribution of the multivariate t distribution", *The American Statistician*, **70**, 293–295.

[85] Drechsler, J. (2011), *Synthetic Datasets for Statistical Disclosure Control.* New York: Springer.

[86] Duan, N., Manning, W.G., Morris, C.N., and Newhouse, J.P. (1983), "A comparison of alternative models for the demand of medical care", *Journal of Business and Economic Statistics*, **1**, 115–126.

[87] Dunson, D. B., and Xing, C. (2009), "Nonparametric Bayes modeling of multivariate categorical data", *Journal of the American Statistical Association*, **104**, 1042–1051.

[88] Eekhout I, De Vet H.C.W., De Boer, M.R., Twisk J.W.R., and Heymans, M.W. (2018), "Passive imputation and parcel summaries are both valid to handle missing items in studies with many multi-item scales", *Statistical Methods in Medical Research*, **27**, 1128–1140.

[89] Eekhout I, De Vet H.C.W., Twisk J.W.R., Brand, J.P.L., De Boer, M.R., and Heymans, M.W. (2014), "Missing data in a multi-item instrument were best handled by multiple imputation at the item score level", *Journal of Clinical Epidemiology*, **67**, 335–342.

[90] Efron, B. (1975), "The efficiency of logistic regression compared to normal discriminant analysis", *Journal of the American Statistical Association*, **70**, 892–898.

[91] Efron, B., Hastie, T., Johnstone, I., and Tibshirani, T. (2004), "Least angle regression", *The Annals of Statistics*, **32**, 407–499.

[92] Efron, B. and Tibshirani, R. (1993), *An Introduction to Bootstrap.* London: Chapman and Hall.

[93] Eilers, P.H.C. and Marx, B. D. (1996), "Flexible smoothing with B-splines and penalties", *Statistical Science*, **11**, 89–121.

[94] Elliott, M.R. and Stettler, N. (2007), "Using a mixture model for multiple imputation in the presence of outliers: The"Healthy for Life"project", *Journal of the Royal Statistical Society: Series C, (Applied Statistics)*, **56**, 63–78.

[95] Elliott, M.R., Raghunathan, T.E., and Schenker, N. (2018), "Combining estimates from multiple surveys", *Wiley StatsRef: Statistics Reference Online.*

[96] Erler, N.S., Rizopoulos, D., Rosmalen, J.V., Jaddoe, V.W., Franco, O.H., and Lesaffre, E.M. (2016), "Dealing with missing covariates in epidemiologic studies: A comparison between multiple imputation and a full Bayesian approach", *Statistics in Medicine*, **35**, 2955–2974.

[97] Ezzati-Rice, T., Massey, J., Waksberg, J., Chu, A., and Maurer, K. (1992), "Sample design: Third National Health and Nutrition Examination Survey", National Center for Health Statistics, *Vital Health Statistics, Series 2*, **Sep;(113)**, 1–35.

[98] Ezzati-Rice, T., Johnson, W., Khare, M., Little, R.J.A., Rubin, D., and Schafer, J. (1995), "A simulation study to evaluate the performance of model-based multiple imputation in NCHS health examination surveys", *Proceedings of 1995 Annual Research Conference, U.S. Bureau of Census*, 257–266.

[99] Faucett, C.L., Schenker, N., and Taylor, J.M.G. (2002), "Survival analysis using auxiliary variables via multiple imputation with application to AIDS clinical trial data", *Biometrics*, **58**, 37–47.

[100] Fay, R.E. (1992), "When are inferences from multiple imputation valid?", *Proceedings for the Survey Research Methods Section, American Statistical Association*, 227–232.

[101] Fay, R.E. (1993), "Valid inference from imputed survey data", *Proceedings for Survey Research Methods Section, American Statistical Association*, 41–48.

[102] Fay, R.E. and Herriot, R. (1979), "Estimates of income for small places: An application of James-Stein procedures to census data", *Journal of the American Statistical Association*, **74**, 269–277.

[103] Fellegi, I.P. and Holt, D. (1976), "A systematic approach to automatic edit and imputation", *Journal of the American Statistical Association*, **71**, 17–35.

[104] Fisher, R.A. (1938), "The statistical utilization of multiple measurements," *Annals of Eugenics*, **8**, 376–386.

[105] Fitzmaurice, G.M., Davidian, M., Verbeke, G., and Molenberghs, G. (2009), *Longitudinal Data Analysis*. Boca Raton: Chapman and Hall.

[106] Fraley, C. and Raftery, A.E. (2002), "Model-based clustering, discriminant analysis, and density estimation", *Journal of the American Statistical Association*, **97**, 611–631.

[107] Freedman, L.S., Midthune, D., Carroll, R.J., and Kipnis, V. (2008), "A comparison of regression calibration, moment reconstruction, and imputation for adjusting for covariate measurement error in regression", *Statistics in Medicine*, **27**, 5195–5216.

[108] Fuller, W.A. (2006), *Measurement Error Models*, 2nd Edition. New York: Wiley.

[109] Galimard, J.E., Chevret, S., Protopopescu, C., and Resche-Rigon, M. (2016), "A multiple imputation approach for MNAR mechanisms compatible with Heckmans' model", *Statistics in Medicine*, **35**, 2907–2920.

[110] Galimard, J.E., Chevret, S., Protopopescu, C., and Resche-Rigon, M. (2018), "Heckman imputation models for binary or continuous MNAR outcomes and MAR predictors", *BMC Medical Research and Methodology*, **35**, 2907–2920.

[111] Gelfand, A.E., Hills, S.E., Racine-Poon, A., and Smith, A.F.M. (1990), "Illustration of Bayesian inference in normal data models using Gibbs sampling", *Journal of the American Statistical Association*, **85**, 972–985.

[112] Gelman, A.E. (2003), "A Bayesian formulation of exploratory analysis and goodness-of-fit testing", *International Statistical Review*, **71**, 369–382.

[113] Gelman, A.E. (2007), Comment on "Bayesian checking of the second levels of hierarchical model" by Bayarri M.J. and Castellanos, M.E., *Statistical Science*, **22**, 349–352.

[114] Gelman, A.E., Carlin, J.B., Stern, H.S., and Rubin, D.B. (2013), *Bayesian Data Analysis*, 3rd Edition. London: Chapman and Hall.

[115] Gelman, A.E. and Hill J. (2007), *Data Analysis Using Regression and Multilevel/Hierarchical Models.* Cambridge University Press.

[116] Gelman, A.E., Jakulin, A., Grazia Pittau, M., and Su, Y.S. (2008), "A weakly informative default prior distribution for logistic and other regression models", *Annals of Applied Statistics*, **2**, 1360–1383.

[117] Gelman, A.E., King, G., and Liu, C. (1998), "Not asked and not answered: Multiple imputation for multiple surveys", *Journal of the American Statistical Association*, **93**, 846–857.

[118] Gelman, A.E., Mechelen, IV., Verbeke, G., Heitjan, D.F. and Meulders, M. (2005), "Multiple imputation for model checking: Completed-data plots with missing and latent data", *Biometrics*, **61**, 74–85.

[119] Gelman, A.E. and Meng, X.L., eds (2004), *Applied Bayesian Modeling and Causal Inference from Incomplete-Data Perspectives.* Chichester, UK: Wiley.

[120] Gelman, A.E., Meng, X.L., and Stern, H.S. (1996) "Posterior predictive assessment of model fitness via realized discrepancies (with discussion)", *Statistica Sinica*, **6**, 733–807.

[121] George, E. I. and McCulloch, R. E. (1993), "Variable selection via Gibbs sampling", *Journal of the American Statistical Association*, **88**, 881–889.

[122] Geronimi, J. and Saporta, G. (2017), "Variable selection for multiply-imputed data with penalized generalized estimating equations", *Computational Statistics and Data Analysis*, **110**, 103–114.

[123] Ghosh-Dastidar, B. and Schafer, J.L. (2003), "Multiple edit/multiple imputation for multivariate continuous data", *Journal of the American Statistical Association*, **98**, 807–817.

[124] Gilks, W., Richardson, S., and Spiegelhalter, D. eds (1996), *Markov Chain Monte Carlo in Practice*, Suffolk, UK, Chapman and Hall.

[125] Giorgi, R., Belot, A., Gaudart, J, and Launoy, G. (2008), "The performance of multiple imputation for missing covariate data within the context of regression relative survival analysis", *Statistics in Medicine*, **27**, 6310–6331.

[126] Giusti, C. and Little, R.J.A. (2011), "An analysis of nonignorable nonresponse to income in a survey with a rotating panel design", *Journal of Official Statistics*, **27**, 211–229.

[127] Goldstein, H. (2011), *Multilevel Statistical Models*, 4th Edition. Chichester, UK: Wiley.

[128] Goldstein, H., Carpenter, J.R., and Browne, W.J. (2014), "Fitting multilevel multivariate models with missing data in responses and covariates that may include interactions and non-linear terms", *Journal of the Royal Statistical Society: Series A (Statistics in Society)*, **177**, 553–564.

[129] Goldstein, H., Carpenter, J., Kenward, M.G., and Levin, K.A. (2009), "Multilevel models with multivariate mixed response types", *Statistical Modelling*, **9**, 173–197.

[130] Gorfine, M., Lipshtat, N., Freedman, L.S., and Prentice, R.L. (2007), "Linear measurement error models with restricted sampling", *Biometrics*, **63**, 137–142.

[131] Gottschall, A.C., West, S.G. and Enders, C.K. (2012), "A comparison of item-level and scale-level multiple imputation for questionnaire batteries", *Multivariate Behavioral Research*, **47**, 1–25.

[132] Graham, J.W. (2012), *Missing Data: Analysis and Design.* New York: Springer.

[133] Graham, J.W., Olchowski, A.E., and Gilreath, T.D. (2007), "How many imputations are really needed? Some practical clarifications of multiple imputation theory", *Prevention Sciences*, **8**, 206–213.

[134] Green, P.J. (1987), "Penalized likelihood for general semi-parametric regression models", *International Statistical Review*, **55**, 245–260.

[135] Greenwood, M. (1926), "The natural duration of cancer", *Reports on Public Health and Medical Subjects*, **33**, London: Her Majesty's Stationary Office, 1–26.

[136] Groves, R.M. (1989), *Survey Errors and Survey Costs.* New York: Wiley.

[137] Groves, R.M., Fowler, F.J., Couper, M.P., Lepkowski, J.M., Singer, E. and Tourangeau, R. (2009), *Survey Methodology*, 2nd Edition. New Jersey: Wiley.

[138] Guo, W. (2002), "Functional mixed effect models", *Biometrics*, **58**, 121–128.

[139] Guo, Y. and Little, R.J.A. (2011), "Regression analysis with covariates that have heteroscedastic measurement error", *Statistics in Medicine*, **30**, 2278–2294.

[140] Gurrin, L.C., Moss, T.J., Sloboda, D.M., Hazelton, M.L., Challis, J.R.G., and Newnham, J.P. (2003) "Using WinBUGS to fit nonlinear mixed models with an application to pharmacokinetic modelling of insulin response to glucose challenge in sheep exposed antenatally to glucocorticoids", *Journal of Biopharmaceutical Statistics*, **13**, 117–139.

[141] Gustafson, P. (2004), *Measurement Error and Misclassification in Statistics and Epidemiology.* Boca Raton: Chapman and Hall.

[142] Harel, O. (2007), "Inferences on missing information under multiple imputation and two-stage multiple imputation", *Statistical Methodology*, **4**, 75–89.

[143] Harel, O. (2009), "The estimation of r^2 and adjusted r^2 in incomplete data sets using multiple imputation", *Journal of Applied Statistics*, **36**, 1109–1118.

[144] Harel, O. and Schafer, J.L. (2003), "Multiple imputation in two stages", *Federal Committee on Statistical Methodology Conference.*

[145] Hastie, T. and Tibshirani, R. (1990), *Generalized Additive Models.* London: Chapman and Hall.

[146] Hastie, T., Tibshirani, R., and Friedman, J. (2008), *The Elements of Statistical Learning.* 2nd Edition. New York: Springer.

[147] Hastings, W.K. (1970), "Monte Carlo sampling methods using Markov chains and their applications", *Biometrika*, **57**, 97–109.

[148] He, R. and Belin, T.R. (2014), "Multiple imputation for high-dimensional mixed incomplete continuous and binary data", *Statistics in Medicine*, **33**, 2251–2262.

[149] He, Y. (2010) "Missing data analysis using multiple imputation: Getting to the heart of the matter", *Circulation: Cardiovascular Quality and Outcomes*, **3**, 98–105.

[150] He, Y., Cai, B., Shin, H-C., Beresovsky, V., Parsons, V., Irimata, K., Scanlon, P., Parker, J. (2020) "The National Center for Health Statistics' 2015 and 2016 Research and Development Surveys", National Center for Health Statistics, *Vital Health Statistics, Series 1* **Oct;(59)**, 1–60.

[151] He, Y., Landrum, M.B., and Zaslavsky, A.M. (2014), "Combining information from two data sources with misreporting and incompleteness to assess hospice-use among cancer patients: A multiple imputation approach", *Statistics in Medicine*, **33**, 3710-3724.

[152] He, Y. and Raghunathan, T. E. (2006), "Tukey's *gh* distribution for multiple imputation", *The American Statistician*, **60**, 251–256.

[153] He, Y. and Raghunathan, T.E. (2009), "On the performance of sequential regression multiple imputation methods with nonnormal error distributions", *Communications in Statistics: Simulation and Computation*, **38**, 856–883.

[154] He, Y. and Raghunathan, T.E. (2012), "Multiple imputation using multivariate *gh* transformations", *Journal of Applied Statistics*, **39**, 2177–2198.

[155] He, Y., Shimizu, I., Schappert, S., Xu, J., Valverde, R., Beresovsky, V., Khan, D., and Schenker N. (2016), "A note on the effect of data clustering on the multiple imputation variance estimator: An addendum to "The relative impacts of design effects and multiple imputation on variance estimates: A case study with the 2008 National Ambulatory Medical Care Survey"", *Journal of Official Statistics*, **32**, 147–164.

[156] He, Y., Yucel, R., and Raghunathan, T.E. (2011), "A functional multiple imputation approach to incomplete longitudinal data", *Statistics in Medicine*, **30**, 1137–1156.

[157] He, Y., Yucel, R.M., and Zaslavsky, A.M. (2008), "Misreporting, missing data, and multiple imputation: Improving accuracy of cancer registry databases", *Chance*, **21**, 55–58.

[158] He, Y. and Zaslavsky, A.M. (2009), "Combining information from cancer registry and medical records data to improve analyses of adjuvant cancer therapies", *Biometrics*, **65**, 946–952.

[159] He, Y. and Zaslavsky (2012), "Diagnosing imputation models by applying target analyses to posterior replicates of completed data", *Statistics in Medicine*, **31**, 1–18.

[160] He, Y., Zaslavsky, A.M., Harrington, D.P., Catalano, P., and Landrum, M.B. (2010), "Multiple imputation in a large-scale complex survey: A practical guide", *Statistical Methods in Medical Research*, **19**, 653–670.

[161] Heckman, J.J. (1976), "The common structure of statistical models of truncation, sample selection, and limited dependent variables and a simple estimator for such models", *Annals of Economic and Social Measurement*, **5**, 475–592.

[162] Hedeker, D., and Gibbons, R. D. (2006). *Longitudinal Data Analysis.* New York: Wiley.

[163] Heinze, G., Ploner, M., and Bayea, J. (2013), "Confidence intervals after multiple imputation: Combining profile likelihood information from logistic regressions", *Statistics in Medicine*, **32**, 5062–5076.

[164] Heitjan, D.F. (1993), "Ignorability and coarse data: Some biomedical examples", *Biometrics*, **49**, 1099–1109.

[165] Heitjan, D.F. (1994), "Ignorability in general incomplete-data models", *Biometrika*, **81**, 701–708.

[166] Heitjan, D.F. and Little, R.J.A. (1991), "Multiple imputation for the fatal accident reporting system", *Journal of the Royal Statistical Society: Series C (Applied Statistics)*, **40**, 13–29.

[167] Heitjan, D.F. and Rubin, D.B. (1991), "Ignorability and coarse data", *The Annals of Statistics*, **19**, 2244–2253.

[168] Hemming, K. and Hutton, J.L. (2012), "Bayesian sensitivity models for missing covariates in the analysis of survival data", *Journal of Evaluation in Clinical Practice*, **18**, 238–246.

[169] Hershberger, S.L. and Fisher, D.G. (2003), "A note on determining the number of imputations for missing data", *Structural Equation Modeling*, **10**, 648–650.

[170] Heymans, M. W., van Buuren, S., Knol, D. L., van Mechelen, W., and de Vet, H. C. (2007), "Variable selection under multiple imputation using the bootstrap in a prognostic study", *BMC Medical Research Methodology*, **7**, 33.

[171] Hoerl, A.E. and Kennard, R. (1970), "Ridge regression: Biased estimation for nonorthogonal problems", *Technometrics*, **12**, 55–67.

[172] Horton, N.J., Lipsitz, S.R., and Parzen, M. (2003), "A potential for bias when rounding in multiple imputation", *The American Statistician*, **57**, 229–232.

[173] Horvitz, D.G. and Thompson, D.J. (1952), "A generalization of sampling without replacement from a finite population." *Journal of the American Statistical Association*, **47**, 663–685.

[174] Hsu, C.H., He, Y., Hu, C., Zhou, W. (2020), "A multiple imputation-based sensitivity analysis approach for data subject to missing not at random", *Statistics in Medicine*, **39**, 3756–3771.

[175] Hsu, C.H., He, Y., Li, Y., Long, Q., and Frise, R. (2016), "Doubly robust multiple imputation using kernel-based techniques", *Biometrical Journal*, **58**, 588–606.

[176] Hsu, C.H., Long, Q., Li, Y., and Jacobs, E. (2014), "A nonparametric multiple imputation approach for data with missing covariate values with applications to colorectal adenoma data", *Journal of Biopharamaceutical Statistics*, **24**, 634–648.

[177] Hsu, C.H. and Taylor, J.M.G. (2009), "Nonparametric comparison of two survival functions with dependent censoring via nonparametric multiple imputation", *Statistics in Medicine*, **28**, 462–475.

[178] Hsu, C.H. and Taylor J.M.G. (2010) "A robust weighted Kaplan-Meyer approach for data with dependent censoring using linear combinations of prognostic covariates", *Statistics in Medicine*, **29**, 2215–2223.

[179] Hsu, C.H., Taylor, J.M.G., and Hu, C. (2015), "Analysis of accelerated failure time data with dependent censoring using auxiliary variables via nonparametric multiple imputation", *Statistics in Medicine*, **34**, 2768–2780.

[180] Hsu, C.H., Taylor, J.M.G., Murray, S., and Commenges, D. (2006), "Survival analysis using auxiliary variables via non-parametric multiple imputation", *Statistics in Medicine*, **25**, 3503–3517.

[181] Hsu, C.H., Taylor, J.M.G., Murray, S., and Commenges, D. (2007), "Multiple imputation for interval censored data with auxiliary variables", *Statistics in Medicine*, **26**, 769–781.

[182] Hsu, C.H. and Yu, M. (2018) "Cox regression analysis with missing covariates via non-parametric multiple imputation", *Statistical Methods in Medical Research*, **28**, 1676–1688.

[183] Hu, J., Mitra, R., and Reiter, J. (2013), "Are independent parameter draws necessary for mulitple imputation?", *The American Statistician*, **67**, 143–149.

[184] Hughes, R.A., White, I.R., Seaman, S.R., Carpenter, J.R., Tilling, K. and Sterne, J.A.C. (2014), "Joint modelling rationale for chained equations", *BMC Medical Research Methodology*, **14**, 28.

[185] Hutton, J.L. and Solomon, P.J. (1997), "Parameter orthogonality in mixed regression models for survival data", *Journal of the Royal Statistical Society: Series B (Statistical Methodology)*, **59**, 125–136.

[186] Ibrahim, J. G., Chen, M. H. and Lipsitz, S. R. (2001), "Missing responses in generalised linear mixed models when the missing data mechanism is non-ignorable", *Biometrika*, **88**, 551–564.

[187] Jackson, D., White, I.R., Searman, S., Evans, H., Baisley, K., and Carpenter, J.R. (2014), "Relaxing the independent censoring assumption in the Cox proportional hazards model using multiple imputation", *Statistics in Medicine*, **33**, 4681–4694.

[188] Janicki, R. and Malec, D. (2013), "A Bayesian model averaging approach to analyzing categorical data with nonignorable nonresponse", *Computational Statistics and Data Analysis*, **57**, 600–614.

[189] Johnson, R.A. and Wichern, D.W. (2007), *Applied Multivariate Statistical Analysis*, 6th Edition. New Jersey: Pearson Prentice Hall.

[190] Jolani, S. (2012), "Dual imputation strategies for analyzing incomplete data", Ph.D. Thesis. University of Utrecht, Utrecht, Netherland.

[191] Jolani, S., Van Buuren, S., and Frank, L.E. (2013), "Combining the complete-data and nonresponse models for drawing imputations under MAR", *Journal of Statistical Computation and Simulation*, **83**, 868–879.

[192] Kaciroti, N.A., and Raghunathan, T.E. (2014), "Bayesian sensitivity analysis of incomplete data: Bridging pattern-mixture and selection models", *Statistics in Medicine*, **33**, 4841–4857.

[193] Kalbfleisch, J.D. and Prentice, R.L. (2002), *The Statistical Analysis of Failure Time Data*, 2nd Edition. Hoboken, NJ: Wiley.

[194] Kalton, G. and Kasprzyk, D. (1986), "The treatment of missing survey data", *Survey Methodology*, **12**, 1–16.

[195] Kang, J.D.Y. and Schafer, J.L. (2007), "Demystifying double robustness: A comparison of alternative strategies for estimating population means from incomplete data", *Statistical Science*, **26**, 523-539.

[196] Kaplan, E.L. and Meier, P. (1958), "Nonparametric estimation from incomplete observations", *Journal of the American Statistical Association*, **53**, 457–481.

[197] Karvanen, J., Saarela, O. and Kuulasmaa, K. (2010) "Nonparametric multiple imputation of left censored event times in analysis of follow-up data", *Journal of Data Science*, **8**, 151–172.

[198] Kenward, M.G. (1998), "Selection models for repeated measurements with nonrandom dropout: An illustration of sensitivity", *Statistics in Medicine*, **17**, 2723–2732.

[199] Keogh, R.H. and White, I.R. (2014), "A toolkit for measurement error correction, with a focus on nutritional epidemiology", *Statistics in Medicine*, **33**, 2137–2155.

[200] Kim, H.J., Reiter, J.P., Wang, Q., Cox, L.H., and Karr, A.F. (2014), "Multiple imputation of missing or faulty values under linear constraints", *Journal of Business and Economic Statistics*, **32**, 375–386.

[201] Kim, J.K. (2004), "Finite sample properties of multiple imputation estimators", *The Annals of Statistics*, **32**, 766–783.

[202] Kim, J.K., Brick, J.M., Fuller, W.A., and Kalton, G. (2006), "On the bias of the multiple imputation estimator in survey sampling", *Journal of the Royal Statistical Society: Series B (Statistical Methodology)*, **68**, 509–521.

[203] Kim, J.K. and Shao, J. (2014), *Statistical Methods for Handling Incomplete Data.* Boca Raton, FL: Chapman and Hall.

[204] Kim, S., Sugar, C.A., and Belin, T.R. (2015), "Evaluating model-based imputation methods for missing covariates in regression models with interactions", *Statistics in Medicine*, **34**, 1876–1888.

[205] Kim, S., Belin, T.R., and Sugar, C.A. (2018), "Multiple imputation with non-additively related variables: Joint-modeling and approximations", *Statistical Methods in Medical Research*, **27**, 1683–1694.

[206] Kish, L. (1965), *Survey Sampling.* New York: Wiley.

[207] Kleinke, K., Reinecke, J., Salfran, D., and Spiess, M. (2020), *Applied Multiple Imputation, Advantages, Pitfalls: New Developments and Applications in R.* Switzerland: Springer.

[208] Korn, E.L. and Graubard, B.I. (1999), *Analysis of Health Surveys.* New York: Wiley.

[209] Kropko, J., Goodrich, B., Gelman, A., and Hill, J. (2014),"Multiple imputation for continuous and categorical data: Comparing joint multivariate normal and conditional approaches," *Political Analysis*, **22**, 497–519.

[210] Lachenbruch, P.A. (2011), "Variable selection when missing values are present: A case study", *Statistical Methods in Medical Research*, **20**, 429–444.

[211] Laird, N.M. and Ware, J.H. (1982), "Random-effects models for longitudinal data", *Biometrics*, **38**, 963–974.

[212] Lavori, P.W., Dawson, R., and Shera, D. (1995), "Multiple imputation strategy for clinical trials with truncation of patient data", *Statistics in Medicine*, **14**, 1913–1925.

[213] Lee, K. J. and Carlin, J. B. (2010), "Multiple imputation for missing data: Fully conditional specification versus multivariate normal imputation,", *American Journal of Epidemiology*, **171**, 624–632.

[214] Lewis, T., Goldberg, E., Schenker, N., Beresovsky, V., Schappert, S., Decker, S., Sonnenfeld, N., and Shimizu, I. (2014), "The relative impacts of design effects and multiple imputation on variance estimates: A case study with the 2008 National Ambulatory Medical Care Survey", *Journal of Official Statistics*, **30**, 147–161.

[215] Li, F., Baccini, M., Mealli, F., Zell, E. Z., Frangakis, C. E., and Rubin, D. B. (2014), "Multiple imputation by ordered monotone blocks with application to the anthrax vaccine research program", *Journal of Computational and Graphical Statistics*, **23**, 877–892.

[216] Li, K.H., Meng, X.L., Raghunathan, T.E., and Rubin, D.B. (1991), "Significance levels from repeated p-values with multiply-imputed data", *Statistica Sinica*, **1**, 65–92.

[217] Li, K.H., Raghunathan, T.E., and Rubin, D.B. (1991), "Large-sample significance levels from multiply-imputed data using moment-based statistics and an F reference distribution", *Journal of the American Statistical Association*, **86**, 1065–1073.

[218] Licht, C. (2010), "New methods for generating significance levels from multiply-imputed data", Ph.D. Thesis. University of Bamberg, Germany.

[219] Lipsitz, S.R., Parzen, M, and Zhao, L.P. (2002), "A degees-of-freedom approximation in multiple imputation", *Journal of Statistical Computation and Simulation*, **72**, 309–318.

[220] Little, R.J.A. (1988), "Missing-data adjustments in large surveys", *Journal of Business and Economics Statistics*, **6**, 287–296.

[221] Little, R.J.A. (1992), "Regression with missing X's: a review", *Journal of the American Statistical Association*, **87**, 1227–1237.

[222] Little, R.J.A. (1993), "Pattern-mixture models for multivariate incomplete data", *Journal of the American Statistical Association*, **84**, 125–134.

[223] Little, R.J.A. (2004), "To model or not to model? Competing modes of inference for finite population sampling." *Journal of the American Statistical Association*, **99**, 546–556.

[224] Little, R.J.A. and An, H. (2004), "Robust likelihood-based analysis of multivariate data with missing values", *Statistica Sinica*, **14**, 949–969.

[225] Little, R.J.A. and Rubin, D.B. (1987), *Statistical Analysis with Missing Data*, 1st Edition. New York: Wiley.

[226] Little, R.J.A. and Rubin, D.B. (2002), *Statistical Analysis with Missing Data*, 2nd Edition. New York: Wiley.

[227] Little, R.J.A. and Rubin, D.B. (2020), *Statistical Analysis with Missing Data*, 3rd Edition. New York: Wiley.

[228] Little, R.J.A. and Smith, P.J. (1987), "Editing and imputation for quantitative survey data", *Journal of the American Statistical Association*, **82**, 58–68.

[229] Little, R.J.A. and Yau, L. (1996), "Intent-to-treat analysis in longitudinal studies with drop-outs", *Biometrics*, **54**, 1324–1333.

[230] Liu, B., Parsons, V.L., Feuer, E.J., et al. (2019) "Small area estimation of cancer risk factors and screening behaviors in US counties by combining two large national health surveys", *Preventing Chronic Disease*, **16**, E119.

[231] Liu, C. (1995), "Missing data imputation using the multivariate t distribution", *Journal of Multivariate Analysis*, **53**, 139–158.

[232] Liu, J., Gelman, A., Hill, J., Su, Y. and Kropko J. (2014), "On the stationary distribution of iterative imputations", *Biometrika*, **101**, 155–173.

[233] Liu, M., Taylor, J.M.G., and Belin, T.R. (2000), "Multiple imputation and posterior simulation for multivariate missing data in longitudinal studies", *Biometrics*, **56**, 1157–1163.

[234] Liu, Y., Wang, Y., Feng, Y., and Wall, M.M. (2016), "Variable selection and prediction with incomplete high-dimensional data", *Biostatistics*, **10**, 418–450.

[235] Lohr, S. and Raghunathan, T.E. (2017), "Combining survey data with other data sources", *Statistical Science*, **32**, 293–312.

[236] Long, Q., Hsu, C.H., and Li, Y. (2012), "Doubly robust nonparametric multiple imputation for ignorable missing data", *Statistica Sinica*, **22**, 149–172.

[237] Long, Q. and Johnson, B. A. (2014), "Variable selection in the presence of missing data: Resampling and imputation", *Biostatistics*, **16**, 596–610.

[238] Lott, A. and Reiter, J.P. (2020), "Wilson confidence intervals for binomial proportions with multiple imputation for missing data", *The American Statistician*, **74**, 109–115,

[239] Lunn, D.J., Jackson, C., Best, N., Thomas, A., and Spiegelhalter, D. (2013), *The BUGS Book: A Practical Introduction to Bayesian Analysis*. Boca Raton, FL: Chapman and Hall.

[240] Lunn, D.J., Spiegelhalter, D., Thomas, A., and Best, N. (2009), "The BUGS project: Evolution, critique, and future directions", *Statistics in Medicine*, **28**, 3049–3067.

[241] Lunn, D.J., Thomas, A., Best, N., and Spiegelhalter, D. (2000), "WinBUGS-A Bayesian modelling framework: concepts, structure, and extensibility", *Statistics and Computing*, **10**, 325–337.

[242] Malec, D., Davis, W.W., and Cao, X. (1999), "Model-based small area estimates of overweight prevalence using sample selection adjustment", *Statistics in Medicine*, **18**, 3189–3200.

[243] Manrique-Vallier, D. and Reiter, J. (2014), "Bayesian multiple imputation for large-scale categorical data with structural zeros", *Survey Methodology*, **40**, 125–134.

[244] Marron, J.S. and Wand, M.P. (1992), "Exact mean integrated squared error", *The Annals of Statistics*, **20**, 712–736.

[245] Marshall, A., Altman, D.G., Holder, R.L., and Royston, P. (2009), "Combining estimates of interest in prognostic modelling studies after multiple imputation: Current practice and guidelines", *BMC Medical Research and Methodology*, **9**, 57.

[246] Martin, J.A., Hamilton, B.E., Sutton, P.D., Ventura, S.J., Menacker, F., and Munson, M.L. (2003). "Births: Final data for 2002", National Center for Health Statistics, *National Vital Statistics Reports*, **52(10)**, 1–113.

[247] Mason, A., Richardson, S. and Best, N. (2012a), "Two-pronged strategy for using DIC to compare selection models with non-ignorable missing responses", *Bayesian Analysis*, **7**, 109–146.

[248] Mason, A., Richardson, S., Plewis, I., and Best, N. (2012b), "Strategy for modelling non-random missing data mechanisms in observational studies using Bayesian methods", *Journal of Official Statistics*, **28**, 279–302.

[249] McCullagh, P. and Nelder, J.A. (1989), *Generalized Linear Models*, 2nd Edition. London: Chapman and Hall.

[250] McCulloch, C.E., Searle, S.R., and Neuhaus, J.M. (2008), *Generalized, Linear, and Mixed Models*, 2nd Edition. Hoboken, New Jersey: Wiley.

[251] McLachlan, G.J. and Peel, D. (2000) *Finite Mixture Models.* New York: Wiley.

[252] Meng, X.L. (1994), "Multiple imputation with uncongenial sources of input (with discussion)", *Statistical Science*, **10**, 538-573.

[253] Meng, X.L. and Rubin, D.B. (1992), "Performing likelihood ratio tests with multiply imputed data sets", *Biometrika*, **79**, 103–111.

[254] Messer, K. and Natarajan, L. (2008), "Maximum likelihood, multiple imputation and regression calibration for measurement error adjustment", *Statistics in Medicine*, **27**, 6332–6350.

[255] Molenberghs, G., Fitzmaurice, C., Kenward, M.G., Tsiatis, A., and Verbeke, G., eds. (2015), *Handbook of Missing Data Methodology.* Boca Raton: Chapman and Hall.

[256] Molenberghs, G. and Kenward, M.G. (2007), *Missing Data in Clinical Studies.* Chichester, UK: Wiley.

[257] Moons, K.G.M., Donders, A.R.T., Stijnen, T., and Harrell, F.E. (2006), "Using the outcome for imputation of missing predictor values was preferred", *Journal of Clinical Epidemiology*, **59**, 1092–1101.

[258] Moriarity, C. and F, Scheuren. (2001), "Statistical matching: A paradigm for assessing the uncertainty in the procedure", *Journal of Official Statistics*, **17**, 407–422.

[259] Morris, T.P., White, I.R., and Crowther, M.J. (2019), "Using simulation studies to evaluate statistical methods", *Statistics in Medicine*, **38**, 2074–2102.

[260] Morris, T.P., White, I.R., and Royston, P. (2014), "Tuning multiple imputation by predictive mean matching and local residual draws", *BMC Medical Research Methodology*, **14**, 75.

[261] Murray, J.S. and Reiter, J.P. (2016), "Multiple imputation of missing categorical and continuous values via Bayesian mixture models with local dependence", *Journal of the American Statistical Association*, **111**, 1466–1479.

[262] Nadaraya, E.A. (1964), "On estimating regression", *Theory of Probability and Its Applications*, **9**, 141–142.

[263] National Center for Health Statistics (2009), "2008 NAMCS Micro-Data File Documentation," Division of Health Care Surveys, National Center for Health Statistics, Centers for Disease Control and Prevention, U.S. Department of Health and Human Services, Hyattsville, MD.

[264] National Center for Health Statistics (2017), "2016 National Health Interview Survey (NHIS) Public Use Data Release: NHIS Survey Description," Division of Health Interview Statistics, National Center for Health Statistics, Centers for Disease Control and Prevention, U.S. Department of Health and Human Services, Hyattsville, MD.

[265] National Research Council (2010), *The Prevention and Treatment of Missing Data in Clinical Trials.* Washington, D.C.: National Academy Press.

[266] Neuhaus, J.M. (1999), "Bias and efficiency loss due to misclassified responses in binary regression", *Biometrika*, **86,** 843–855.

[267] Nevalainen, J., Kenward, M.G., and Virtanen, S.M. (2009), "Missing values in longitudinal dietary data: A multiple imputation approach based on a fully conditional specification", *Statistics in Medicine*, **28**, 3657–3669.

[268] Nielsen, S.F. (2003), "Proper and improper multiple imputation", *International Statistical Review*, **71**, 593–627.

[269] Nguyen, C.D., Carlin, J.B., and Lee, K.J. (2013), "Diagnosing problems with imputation models using the Kolmogorov-Smirnov test: A simulation study", *BMC Medical Research Methodology*, **13**, 144.

[270] Nguyen, C.D., Lee, K.J., and Carlin, J.B. (2015), "Posterior predictive checking for multiple imputation models", *Biometrical Journal*, **57**, 676–694.

[271] Ntzoufras, I. (2009), "Bayesian Modeling Using WinBUGS". Hoboken, NJ: Wiley.

[272] Ogundimu, E.O. and Collins, G.S. (2019), "A robust imputation method for missing responses and covariates in sample selection models", *Statistical Methods in Medical Research*, **28**, 102–116.

[273] Olkin, I. and Tate, R.F. (1961), "Multivariate correlation models with mixed discrete and continuous variables", *Annals of Mathematical Statistics*, **32**, 448–465.

[274] Ono, M. and Miller, H.P. (1969), "Income nonresponses in the current population survey", *Proceedings in Sociological Statistics Section, American Statistical Association*, 277–288.

[275] Orchard, T. and Woodbury, M.A. (1972), "A missing information principle: Theory and applications", *Proceedings of the 6th Berkeley Symposium on Mathematical Statistics and Probability*, **1**, 697–715.

[276] Pan, Q., Wei, R., Shimizu, I., and Jamoom, E. (2014), "Determining sufficient number of imputations using variance of imputation variances: Data from 2012 NAMCS physician work flow mail survey", *Applied Mathematics*, **2014**, 3421–3430.

[277] Pan, W. (2000), "A multiple imputation approach to Cox regression with interval-censored data", *Biometrics*, **56**, 199–203.

[278] Pan, Y., He, Y., Song, R., Wang, G., and An, Q. (2020), "A passive and inclusive strategy to impute missing values of a composite categorical variable with an application to determine HIV transmission categories", *Annals of Epidemiology*, **51**, 41–47.

[279] Parker, J.D. and Schenker, N. (2007), "Multiple imputation for national public-use datasets and its possible application for gestational age in United States natality files", *Paediatric and Perinatal Epidemiology*, **21**, 97–105.

[280] Parsons, V.L., Moriarity, C., Jonas K, et al. (2014), "Design and estimation for the National Health Interview Survey, 2006–2015", National Center for Health Statistics, *Vital Health Statistics, Series 2* **Apr;(165)**, 1–53.

[281] Paulin, G. D. and Sweet, E. M. (1996), "Modeling income in the U.S. Consumer Expenditure Survey", *Journal of Official Statistics*, **12**, 403–419.

[282] Plummer, M. (2015), "Cuts in Bayesian graphic models", *Statistical Computing*, **25**, 37–43.

[283] Prentice, R.L. and Pyke, R. (1979) "Logistic disease incidence model and case-control studies", *Biometrika*, **66**, 403–411.

[284] Priebe, C.E. (1994), "Adaptive mixtures", *Journal of the American Statistical Association*, **89**, 796–806.

[285] Puhani, P. (2000), "The Heckman correction for sample selection and its critique", *Journal of Economics Surveys*, **14**, 53–68.

[286] Qi, L., Wang, Y-F., Chen, R., Siddique, J., Robbins, J., and He, Y. (2018), "Strategies for imputing missing covariates in accelerated failure time models", *Statistics in Medicine*, **37**, 3417–3436.

[287] Qi, L., Wang, Y-F., and He, Y. (2010), "A comparison of multiple imputation and fully augmented weighted estimators for Cox regression with missing covariates", *Statistics in Medicine*, **29**, 2592–2604.

[288] Quartagno, M., Carpenter, J.R., and Goldstein, H. (2020) "Multiple imputation with survey weights: A multilevel approach," *Journal of Survey Statistics and Methodology*, **8**, 965–989.

[289] Raghunathan, T.E. (2006), "Combining information from multiple surveys for assessing health disparities", *Allgemeines Statistiches Archiv*, **90**, 515–526.

[290] Raghunathan, T.E. (2016), *Missing Data Analysis in Practice.* Boca Raton, FL: Chapman and Hall.

[291] Raghunathan, T.E., Berglund, P.A., Solenberger, P.W. (2018), *Multiple Imputation in Practice With Examples Using IVEware.* Boca Raton, FL: Chapman and Hall.

[292] Raghunathan, T.E. and Grizzle, J.E. (1995), "A split questionnaire survey design", *Journal of the American Statistical Association*, **90**, 54–63.

[293] Raghunathan, T.E., Lepkowski, J.M., Van Hoewyk, J., and Solenberger, P.W. (2001) "A multivariate technique for multiply imputing missing values using a sequence of regression models", *Survey Methodology*, **27**, 85–95.

[294] Raghunathan, T.E. and Siscovick, D.S. (1996), "A multiple-imputation analysis of a case-control study of the risk of primary cardiac arrest among pharmacologically treated hypertensives", *Journal of the Royal Statistical Society: Series C (Applied Statistics)*, **45**, 335–352.

[295] Raghunathan, T.E. and Siscovick, D.S. (1998), "Combining exposure information from various sources in an analysis of a case-control study", *Journal of the Royal Statistical Society: Series D (The Statistician)*, **47**, 333-347.

[296] Raghunathan, T.E, Xie, D., Schenker, N., Parsons, V.L., Davis, W.W., Feuer, E.J., and Dodd, K.W. (2007), "Combining information from two surveys to estimate county-level prevalence rates of cancer risk factors and screening", *Journal of the American Statistical Association*, **102**, 474–486.

[297] Rammon, J., He, Y., and Parker, J.D. (2019), "Accounting for study participants who are ineligible for linkage: A multiple imputation approach to analyzing the linked National Health and Nutrition Examination Survey and Centers for Medicare and Medicaid Services' Medicaid Data", *Health Services and Outcome Research Methodology*, **19**, 87–105.

[298] Rao, J.N.K. (1996), "On variance estimation with imputed survey data", *Journal of the American Statistical Association*, **91**, 499–506.

[299] Rao, J.N.K. (2003), *Small Area Estimation*, 1st Edition. Hoboken, NJ: Wiley.

[300] Rashid, S., Mitra, R., and Steele, R.J. (2015), "Using mixtures of t densities to make inferences in the presence of missing data with a small number of multiply imputed data sets", *Computational Statistics and Data Analysis*, **92**, 84–96.

[301] Rathouz, P. (2007) "Identifiability assumptions for missing covariate data in failure time regression models", *Biostatistics*, **8**, 345–356.

[302] Reiter, J.P. (2007), "Small-sample degrees of freedom for multi-component significance tests with multiple imputation for missing data", *Biometrika*, **94**, 502–508.

[303] Reiter, J.P. (2008), "Multiple imputation when records used for imputation are not used or disseminated for analysis", *Biometrika*, **95**, 933–946.

[304] Reiter, J.P. and Raghunathan, T.E. (2007), "The multiple adaptations of multiple imputation", *Journal of the American Statistical Association*, **102**, 1462–1471.

[305] Reiter, J.P., Raghunathan, T.E., and Kinnery, S.K. (2006), "The importance of modeling the sampling design in multiple imputation for missing data", *Survey Methodology*, **32**, 143–149.

[306] Rezvan, H.P., Lee, K.J., and Simpson, J.A. (2015), "The rise of multiple imputation: A review of the reporting and implementation of the method in medical research", *BMC Medical Research and Methodology*, **15**, 30.

[307] Rizopoulos, D., Verbeke, G., and Molenberghs, G. (2010), "Multiple imputation based residuals and diagnostic plots for joint models of longitudinal and survival outcomes", *Biometrics*, **66**, 20–29.

[308] Robins, J.M., Rotnitzky, A., and Zhao, L.P. (1994), "Estimation of regression coefficients when some regressors are not always observed", *Journal of the American Statistical Association*, **89**, 846–886.

[309] Robins, J.M., Rotnitzky, A., and Zhao, L.P. (1995), "Analysis of semiparametric regression models for repeated outcomes in the presence of missing data", *Journal of the American Statistical Association*, **90**, 106–121.

[310] Robins, J.M. and Wang, N. (2000), "Inference for imputation estimators", *Biometrika*, **87**, 113–124.

[311] Rosenbaum, P.R. and Rubin, D.B. (1983), "The central role of the propensity score in observational studies for causal effects", *Biometrika*, **70**, 41–55.

[312] Royston, P. (2004), "Multiple imputation of missing values", *Stata Journal*, **4**, 227–241.

[313] Royston, P. (2009), "Multiple imputation of missing values: further update of ice, with an emphasis on categorical variables", *Stata Journal*, **9**, 466–477.

[314] Rubin, D.B. (1976), "Inference and missing data", *Biometrika*, **63**, 581–592.

[315] Rubin, D.B. (1977), "Formalizing subjective notions about the effect of nonrespondents in sample surveys", *Journal of the American Statistical Association*, **72**, 538–543.

[316] Rubin, D.B. (1978), "Multiple imputation in sample surveys–A phenomenological Bayesian approach to nonresponse", *Proceedings of the Survey Research Methods Section, American Statistical Association*, 20–34.

[317] Rubin, D.B. (1984), "Bayesianly justifiable and relevant frequency calculations for the applied statistician", *The Annals of Statistics*, **12**, 1151–1172.

[318] Rubin, D.B. (1986), "Statistical matching using file concatenation with adjusted weights and multiple imputations", *Journal of Business and Economics Statistics*, **4** 87–94.

[319] Rubin, D.B. (1987), *Multiple Imputation for Nonresponse in Surveys*, 1st Edition. New York: Wiley.

[320] Rubin, D.B. (1996), "Multiple imputation after 18+ years", *Journal of the American Statistical Association*, **81**, 366–374.

[321] Rubin, D.B. (2003), "Nested multiple imputation of NMES via partially incompatible MCMC," *Statistica Neerlandica*,**57**, 3–18.

[322] Rubin, D.B. (2004), "Direct and indirect causal effects via potential outcomes", *Scandinavian Journal of Statistics*, **31**, 161–170.

[323] Rubin, D.B. and Schenker, N. (1986), "Multiple imputation for interval estimation from simple random samples with ignorable non-response", *Journal of the American Statistical Association*, **81**, 366–374.

[324] Rubin, D.B. and Schenker, N. (1987), "Interval estimation from multiply-imputed data: A case study using census agriculture industry codes", *Journal of Official Statistics*, **3**, 375–387.

[325] Rubin, D.B. and Schenker, N. (1991), "Multiple imputation in health-care databases: An overview and some applications", *Statistics in Medicine*, **10**, 585–598.

[326] Rubin, D.B., Stern, H.S., and Vehovar, V. (1995), "Handling "Don't Know" survey responses: The case of the Slovenian Plebiscite", *Journal of the American Statistical Association*, **90**, 822–828.

[327] Ruan, P.K. and Gray, R.J. (2008), "Analyses of cumulative incidence functions via non-parametric multiple imputation", *Statistics in Medicine*, **27**, 5709–5724.

[328] Ruppert, D., Wand, M.P., and Carroll, R.J. (2003), *Semiparametric Regression.* Cambridge University Press.

[329] Sande, I.G. (1982), "Imputation in surveys: Coping with reality,", *The American Statistician*, **36**, 145–152.

[330] Sarndal, C.E. and Lundstrom (2005), *Estimation in Surveys with Nonresponse.* Chichester, UK: Wiley.

[331] Sarndal, C.E., Swensson, B., and Wretman, J. (1992), *Model Assisted Survey Sampling.* New York: Springer.

[332] Schafer, J.L. (1997), *Analysis of Incomplete Multivariate Data.* London: Chapman and Hall.

[333] Schafer, J.L. (2003), "Multiple imputation in multivariate problems when the imputation and analysis models differ," *Statistica Neerlandica*, **57**, 19–35.

[334] Schafer, J.L. (2001), "Multiple imputation models and procedures for NHANES III", Division of Health and Nutrition Examination Statistics, National Center for Health Statistics, Centers for Disease Control and Prevention, U.S. Department of Health and Human Services, Hyattsville, MD.

[335] Schafer, J.L. and Graham, J.W. (2002), "Missing data: Our view of the state of the art", *Psychological Methods*, **7**, 545–571.

[336] Schafer, J.L. and Olsen, M.K. (1998), "Multiple imputation for multivariate missing-data problems: A data analyst's perspective", *Multivariate Behavioral Research*, **33**, 545–571.

[337] Schafer, J.L. and Yucel, R.M. (2002), "Computational strategies for multivariate linear mixed-effects models with missing values", *Journal of Computational and Graphic Statistics*, **11**, 421–442.

[338] Schenker, N., Borrud, L.G., Burt, V.L., Curtin, L.R., Flegal, K.M., Hughes, J., Johnson, C.L., Looker, A.C., and Mirel, L. (2011), "Multiple imputation of missing dual-energy X-ray absorptiometry data in the National Health and Nutrition Examination Survey", *Statistics in Medicine*, **30**, 260–276.

[339] Schenker, N. and Raghunathan, T.E. (2007), "Combining information from multiple surveys to enhance estimation of measures of health", *Statistics in Medicine*, **26**, 1802–1811.

[340] Schenker, N., Raghunathan, T.E., and Bondarenko, I. (2010), "Improving on analyses of self-reported data in a large-scale health survey by using information from an examination-based survey", *Statistics in Medicine*, **29**, 533–545.

[341] Schenker, N., Raghunathan, T.E., Chiu, P.L., Makuc, D.M., Zhang, G., and Cohen, A.J. (2006), "Multiple imputation of missing income data in the National Health Interview Survey", *Journal of the American Statistical Association*, **101**, 924–933.

[342] Schenker, N. and Taylor, J.M.G. (1996), "Partially parametric techniques for multiple imputation", *Computational Statistics and Data Analysis*, **22**, 425–446.

[343] Schenker, N., Treiman, D.J., and Weidman, L. (1993), "Analyses of public use decennial census data with multiply imputed industry and occupation codes", *Journal of the Royal Statistical Society: Series C (Applied Statistics)*, **42**, 545–556.

[344] Schenker, N. and Welsh, A.H. (1988), "Asymptotic results for multiple imputation", *The Annals of Statistics*, **16**, 1550–1566.

[345] Schomaker, M. and Heumann, C. (2018), "Bootstrap inference when using multiple imputation", *Statistics in Medicine*, **37**, 2252–2266.

[346] Schwartz, S.L., Gelfand, A.E., and Miranda, M.L. (2010), "Joint Bayesian analysis of birth weight and censored gestational age using finite mixture models", *Statistics in Medicine*, **29**, 1710–1723.

[347] Seaman, S.R., Bartlett, J.W., and White, I.R. (2012), "Multiple imputation of missing covariates with non-linear effects and interactions: An evaluation of statistical methods", *BMC Medical Research Methodology*, **12**, 46.

[348] Seaman, S.R. and Hughes, R.A. (2018), "Relative efficiency of joint-model and full-conditional-specification multiple imputation when conditional models are compatible: The general location model", *Statistical Methods in Medical Research*, **27**, 1603–1614.

[349] Seaman, S.R., White, I.R., Copas, A.J., and Li, L. (2012), "Combining multiple imputation and inverse-probability weighting", *Biometrics*, **68**, 129–137.

[350] Sekar, C. and Deming, E.W. (1949), "On a method of estimating birth and death rates and the extent of registration", *Journal of the American Statistical Association*, **44**, 101–115.

[351] Si, Y. and Reiter, J.P. (2011), "A comparison of posterior simulation and inference by combining rules for multiple imputation", *Journal of Statistical Theory and Practice*, **5**, 335–347.

[352] Si, Y. and Reiter, J. P. (2013), "Nonparametric Bayesian multiple imputation for incomplete categorical variables in large-scale assessment surveys", *Journal of Educational and Behavioral Statistics*, **38**, 499–521.

[353] Siddique, J. and Belin, T.R. (2008a), "Multiple imputation using an iterative hot-deck with distance-based donor selection", *Statistics in Medicine*, **27**, 83–102.

[354] Siddique, J. and Belin, T.R. (2008b), "Using an approximate Bayesian bootstrap to multiply impute nonignorable missing data", *Computational Statistics and Data Analysis*, **53**, 405–415.

[355] Song, J. and Belin, T.R. (2004), "Imputation for incomplete high-dimensional multivariate normal data using a common factor model", *Statistics in Medicine*, **23**, 2827–2843.

[356] Spiegelhalter, D., Best, N., Carlin, B., and Van der Linde, A. (2002), "Bayesian measures of model complexity and fit (with discussion)", *Journal of the Royal Statistical Society: Series B (Statistical Methodology)*, **64**, 583–639.

[357] Spiegelhalter, D., Best, N., Carlin, B., and Van der Linde, A. (2014), "The deviance information criterion (12 years on)", *Journal of the Royal Statistical Society: Series B (Statistical Methodology)*, **76**, 485–493.

[358] Steele, R.J., Wang, N., and Raftery, A.E. (2010), "Inference from multiple imputation for missing data using mixtures of normals", *Statistical Methodology*, **7**, 351–364.

[359] Sternberg, M. (2017), "Multiple imputation to evaluate the impact of an assay change in national surveys", *Statistics in Medicine*, **36**, 2697–2719.

[360] Struthers C.A. and Kalbfleisch, J.D. (1986), "Misspecified proportional hazards models," *Biometrika*, **73**, 363–369.

[361] Stuart, E.A. (2010), "Matching methods for causal inference: A review and a look forward", *Statistical Science*, **25**, 1–21.

[362] Sullivan, T.R., Salter, A.B., Ryan, P., and Lee, K.J. (2015), "Bias and precision of the "multiple imputation, then deletion" method for dealing with missing outcome data", *American Journal of Epidemiology*, **182**, 528–534.

[363] Taffel, S., Johnson, D., and Heuse, R. (1982), "A method of imputing length of gestation on birth certificates", National Center for Health Statistics, *Vital and Health Statistics, Series 2*, **May;(93)**, 1–11.

[364] Tang, L., Ununtzer, J., Song, J., and Belin, T.R. (2005), "A comparison of imputation methods in a longitudinal randomized clinical trial", *Statistics in Medicine*, **24**, 2111–2128.

[365] Tanner, M.A., and Wong, W.H. (1987), "The calculation of posterior distribution by data augmentation (with discussion)", *Journal of the American Statistical Association*, **82**, 528–550.

[366] Taylor, J.M.G., Munoz, A., Bass, S.M., Chimiel, J.S., Kingsley, L.A., and Saah, A.J. (1990), "Estimating the distribution of times for HIV seroconversion to AIDS using multiple imputation", *Statistics in Medicine*, **9**, 505–514.

[367] Taylor, J.M.G., Murray, S., and Hsu, C.H. (2002), "Survival estimation and testing via multiple imputation", *Statistics and Probability Letters*, **58**, 221–232.

[368] Thao, L. and Geskus, R. (2019), "A Comparison of model selection methods for prediction in the presence of multiply impute data", *Biometrical Journal*, **61**, 343–356.

[369] Tibshirani, R. (1996), "Regression shrinkage and selection via the Lasso", *Journal of the Royal Statistical Society: Series B (Statistical Methodology)*, **58**, 267–288.

[370] Tsiatis, A.A. (2006), *Semiparametric Theory and Missing Data.* New York: Springer.

[371] Tukey, J. W. (1977), "Modern techniques in data analysis", NSF-sponsored regional research conference at Southeastern Massachusetts University, North Dartmouth, MA.

[372] Valliant, R., Dever, J.A., and Kreuter, F. (2013), *Practical Tools for Designing and Weighting Survey Samples.* New York: Springer.

[373] Valliant, R., Dorfman, A.H., and Royall, R.M. (2000), *Finite Population Sampling and Inference: A Prediction Approach.* New York: Wiley.

[374] Van Buuren, S. (2007), "Multiple imputation of discrete and continuous data by fully conditional specification", *Statistical Methods in Medical Research,* **16**, 219–242.

[375] Van Buuren, S. (2011), "Multiple imputation of multilevel data", in Hox, J. and Roberts, J., Eds, *The Handbook of Advanced Multilevel Analysis,* Chap, 10, 173–196. Milton Park, UK: Routledge.

[376] Van Buuren, S. (2012), *Flexible Imputation for Missing Data,* 1st Edition, Boca Raton, FL: Chapman and Hall.

[377] Van Buuren, S. (2018), *Flexible Imputation for Missing Data,* 2nd Edition. Boca Raton, FL: Chapman and Hall.

[378] Van Buuren, S., Boshuizen, H.C., and Knook, D.L. (1999), "Multiple imputation of missing blood pressure covariates in survival analysis", *Statistics in Medicine,* **18**, 681–694.

[379] Van Buuren, S., Brand, J.P.L., Groothuis-Oudshoorn, C.G.M., and Rubin, D.B. (2006), "Fully conditional specification in multivariate imputation", *Journal of Statistical Computation and Simulation,* **76**, 1049–1064.

[380] Van Der Palm, D., Van Der Ark, L.A., and Vermunt, J.K. (2016), "A comparison of incomplete-data methods for categorical data", *Statistical Methods in Medical Research,* **25**, 754–774.

[381] Vermunt, J.K., VanGinkel, J.R., VanDerArk, L.A., and Sijtsma, K. (2008), "Multiple imputation of incomplete categorial data using latent class analysis", *Sociological Methodology,* **38**, 369–397.

[382] Vink, G., Frank, E.L., Pannekoek, J., and Van Buuren, S. (2014), "Predictive mean matching imputation of semicontinuous variables", *Statistica Neerlandica,* **68**, 61–90.

[383] Vink, G. and Van Buuren, S. (2013), "Multiple imputation of squared terms", *Sociological Methods and Research,* **42**, 598–607.

[384] Von Hippel, P.T. (2007), "Regression with missing Ys: An improved strategy for analyzing multiply imputed data", *Sociological Methodology*, **37**, 83–117.

[385] Von Hippel, P.T. (2009), "How to impute interactions, squares, and other transformed variables", *Sociological Methodology*, **39**, 265–291.

[386] Von Hippel, P.T. (2013), "Should a normal imputation model be modified to impute skew variables?", *Sociological Methods and Research*, **42**, 105–138.

[387] Wagner, J. (2010), "The fraction of missing information as a tool for monitoring the quality of survey data", *Public Opinion Quarterly*, **74**, 223–243.

[388] Wagstaff, D.A. and Harel, O. (2011), "A closer examination of three small-sample approximations to the multiple-imputation degrees of freedom", *The Stata Journal*, **11**, 403–419.

[389] Wahba, G. (1978), "Improper priors, spline smoothing and the problem of guarding against model errors in regression", *Journal of the Royal Statistical Society: Series B (Statistical Methodology)*, **40**, 364–372.

[390] Wan, Y., Datta, S., Conklin, D. J., and Kong, M. (2015), "Variable selection models based on multiple imputation with an application for predicting median effective dose and maximum effect", *Journal of Statistical Computation and Simulation*, **85**, 1902–1916.

[391] Wang, N. and Robins, J.M. (1998), "Large-sample theory for parametric multiple imputation procedures", *Biometrika*, **85**, 935–948.

[392] Watson, G.S. (1964), "Smooth regression analysis", *Sankhya*, **A26**, 359–372.

[393] Wei, G.C.G., and Tanner, M.A. (1991), "Applications of multiple imputation to the analysis of censored regression data", *Biometrics*, **47**, 1297–1309.

[394] Wei, L.J. (1992) "The accelerated failure time model: A useful alternative to the Cox regression model in survival analysis", *Statistics in Medicine*, **11**,1871–1879.

[395] White, I.R., Daniel, R., and Royston, P. (2010), "Avoiding bias due to perfect prediction in multiple imputation of incomplete categorical variables", *Computational Statistics and Data Analysis*, **54**, 2267–2275.

[396] White, I.R. and Royston, P. (2009), "Imputing missing covariate values for the Cox model", *Statistics in Medicine*, **28**, 1982–1998.

[397] White, I.R., Royston, P., and Wood, A.M. (2011), "Multiple imputation using chained equations: Issues and guidance for practice", *Statistics in Medicine*, **28**, 1982–1998.

[398] Wood, A.M., White, I.R., and Royston, P. (2008), "How should variable selection be performed with multiply imputed data"? *Statistics in Medicine*, **27**, 3227–3246.

[399] Wu, H. and Wu, L. (2001), "A multiple imputation method for missing covariates in non-linear mixed-effects models with application to HIV dynamics", *Statistics in Medicine*, **20**, 1755–1769.

[400] Wu, L. (2010), *Mixed Effects Models for Complex Data.* Boca Raton, FL: Chapman and Hall.

[401] Wu, W., Jia, F., and Enders, C. (2015), "A comparison of imputation strategies for ordinal missing data on Likert scale variables", *Multivariate Behavioral Research*, **50**, 484–503.

[402] Xia, Y. and Yang, Y. (2016), "Bias introduced by rounding in multiple imputation for ordered categorical variables", *The American Statistician*, **70**, 358–364.

[403] Xie, X. and Meng, X.L. (2017), "Dissecting multiple imputation from a multiphase inference perspective: What happens when God's, imputer's, and analyst's models are uncongenial? (with discussion)", *Statistica Sinica*, **27**, 1485–1594.

[404] Yang, X., Belin, T.R., and Boscardin, W.J. (2005), "Imputation and variable selection in linear regression models with missing covariates", *Biometrics*, **61**, 498–506.

[405] Yu, L.M., Burton, A., and Rivero-Arias, O. (2007), "Evaluation of software for multiple imputation of semi-continuous data", *Statistical Methods in Medical Research*, **16**, 243–258.

[406] Yuan, M. and Lin, Y. (2006), "Model selection and estimation in regression with grouped variables", *Journal of the Royal Statistical Society: Series B (Statistical Methodology)*, **68**, 49–67.

[407] Yucel, R.M. (2008), "Multiple imputation inference for multivariate multilevel continuous data with ignorable non-response", *Philosophical Transaction of the Royal Society A*, **336**, 2389–2403.

[408] Yucel, R.M. (2011), "Random covariance and mixed-effects models for imputing multivariate multilevel continuous data", *Statistical Modelling*, **11**, 351–370.

[409] Yucel, R.M., He, Y., and Zaslavsky, A.M. (2008), "Using calibration to improve rounding in imputation", *The American Statistician*, **62**, 125–129.

[410] Yucel, R.M., He, Y., and Zaslavsky, A.M. (2011), "Gaussian-based routines to impute categorical variables in health surveys", *Statistics in Medicine*, **30**, 3447–3460.

[411] Yucel, R.M. and Zaslavsky, A.M. (2005), "Imputation of binary treatment variables with measurement error in administrative data", *Journal of the American Statistical Association*, **100**, 1123–1132.

[412] Yucel, R.M., Zhao, E., Schenker, N., and Raghunathan, T.E. (2018), "Sequential hierarchical regression imputation", *Journal of Survey Statistics and Methodology*, **6**, 1–22.

[413] Zhang, D., Lin, X., Raz, J., and Sowers, M. (1998), "Semiparametric stochastic mixed models for longitudinal data", *Journal of the American Statistical Association*, **93**, 710–719.

[414] Zhang, G. and Little, R.J.A. (2009), "Extensions of the penalized spline of propensity prediction method of imputation", *Biometrics*; **65**, 911–918.

[415] Zhang, G. and Little, R.J.A. (2011), "A comparative study of doubly robust estimation of the mean with missing data", *Journal of Statistical Computation and Simulation*, **81**, 2039–2058.

[416] Zhang, G., Parker, J.D., and Schenker, N. (2016), "Multiple imputation for missingness due to nonlinkage and program characteristics: A case study of the National Health Interview Survey linked to Medicare claims", *Journal of Survey Statistics and Methodology*, **4**, 319–338.

[417] Zhang, G., Schenker, N., Parker, J.D., and Liao, D. (2014), "Identifying implausible gestational ages in preterm babies with Bayesian mixture models", *Statistics in Medicine*, **33**, 3710–3724.

[418] Zhang, X., Boscardin, J., Belin, T., Wan, X., He, Y., and Zhang, K. (2015), "A Bayesian method for analyzing combinations of continuous, ordinal, and nominal categorical data with missing values", *Journal of Multivariate Analysis*, **135**, 43–58.

[419] Zhao, Y. and Long, Q. (2017) "Variable selection in the presence of missing data: Imputation-based methods", *Wiley Interdisciplinary Reviews: Computational Statistics*, **9**, e1402.

[420] Zhou, H., Elliott, M.R., and Raghunathan, T.E. (2016), "A two-step semiparametric method to accommodate sampling weights in multiple imputation", *Biometrics*, **72**, 242–252.

[421] Zhou, H., Elliott, M.R., and Raghunathan, T.E. (2016), "Synthetic multiple-imputation procedure for multistage complex samples", *Journal of Official Statistics*, **32**, 231–256.

[422] Zhou, H., Elliott, M.R., and Raghunathan, T.E. (2016), "Multiple imputation in two-stage cluster samples using the weighted finite population Bayesian bootstrap", *Journal of Survey Statistics and Methodology*, **4**, 139–170.

[423] Zhou, M., He, Y., Yu, M., and Hsu, C.H. (2017), "A nonparametric multiple imputation approach for missing categorical data," *BMC Medical Research and Methodology*, **17**, 87.

[424] Zhou, W., Hitchner, E., Gillis, K., Sun, L., Floyd, R., Lane, B., and Rosen, A. (2012), "Prospective neurocognitive evaluation of patients undergoing carotid interventions", *Journal of Vascular Surgery*, **56**, 1571–1578.

[425] Zhou, X. and Reiter, J.P. (2010), "A note on Bayesian inference after multiple imputation", *The American Statistician*, **64**, 159–163.

[426] Zhu, J. and Raghunathan, T.E. (2015), "Convergence properties of a sequential regression multiple imputation algorithm", *Journal of the American Statistical Association*, **110**, 1112–1124.

Authors Index

Subect Index

Printed in the United States
by Baker & Taylor Publisher Services